THE ESSENTIALS OF STATISTICS

A Tool for Social Research

THE ESSENTIALS OF STATISTICS

A Tool for Social Research

Third Edition

Joseph F. Healey

Christopher Newport University

Australia • Brazil • Japan • Korea • Mexico • Singapore • Spain • United Kingdom • United States

WADSWORTH
CENGAGE Learning

The Essentials of Statistics: A Tool for Social Research, Third Edition
Joseph F. Healey

Senior Publisher: Linda Schreiber-Ganster

Acquiring Sponsoring Editor: Erin Mitchell

Developmental Editor: Liana Sarkisian

Assistant Editor: Mallory Ortberg

Associate Media Editor: John Chell

Marketing Program Manager: Janay Pryor

Manufacturing Planner: Judy Inouye

Rights Acquisitions Specialist: Thomas McDonough

Text Researcher: Pablo D'Stair

Copy Editor: Gary Von Euer

Cover Designer: Caryl Gorska

Cover Image: Carl Pendle / Photographer's Choice

Art Direction, Production Management, and Composition: PreMediaGlobal

Library of Congress Control Number: 2011945427

ISBN-13: 978-1-111-82956-8

ISBN-10: 1-111-82956-X

Wadsworth
20 Davis Drive
Belmont, CA 94002-3098
USA

Cengage Learning is a leading provider of customized learning solutions with office locations around the globe, including Singapore, the United Kingdom, Australia, Mexico, Brazil and Japan. Locate your local office at **www.cengage.com/global.**

Cengage Learning products are represented in Canada by Nelson Education, Ltd.

For your course and learning solutions, visit **www.cengage.com.**

Purchase any of our products at your local college store or at our preferred online store **www.cengagebrain.com.**

Printed in the United States of America
1 2 3 4 5 6 7 16 15 14 13 12

Brief Contents

Detailed Contents

Chapter 7 Hypothesis Testing I 162

Chapter 8 Hypothesis Testing II 193

Preface

Statistics are part of the everyday language of sociology and other social sciences (including political science, social work, public administration, criminal justice, urban studies, and gerontology). These disciplines are research based and routinely use statistics to express knowledge and to discuss theory and research. To join the conversations being conducted in these disciplines, you must be literate in the vocabulary of research, data analysis, and scientific thinking. Knowledge of statistics will enable you to understand the professional research literature, conduct quantitative research yourself, contribute to the growing body of social science knowledge, and reach your full potential as a social scientist.

Although essential, learning (and teaching) statistics can be a challenge. Students in social science statistics courses typically have a wide range of mathematical backgrounds and an equally diverse set of career goals. They are often puzzled about the relevance of statistics for them, and not infrequently, there is some math anxiety to deal with. This text introduces statistical analysis for the social sciences while addressing these realities.

This text makes minimal assumptions about mathematical background (the ability to read a simple formula is sufficient preparation for virtually all the material in this text), and a variety of special features help students analyze data successfully. This text has been written especially for sociology and social work programs but is sufficiently flexible to be used in any program with a social science base.

This text is written at a level intermediate between a strictly mathematical approach and a mere "cookbook." I emphasize interpreting and understanding statistics in the context of social science research, but I have not sacrificed comprehensive coverage or statistical correctness. Mathematical explanations are kept at an elementary level, as is appropriate in a first exposure to social statistics. For example, in this text, I do not treat formal probability theory per se.[1] Rather, the background necessary for an understanding of inferential statistics is introduced, informally and intuitively, in Chapters 5 and 6 while considering the concepts of the normal curve and the sampling distribution.

This text does not claim that statistics are "fun" or that the material can be mastered without considerable effort. At the same time, students are not overwhelmed with abstract proofs, formula derivations, or mathematical theory, which can needlessly frustrate the learning experience at this level.

[1] A presentation of probability is available on the website for this text for those who are interested.

Goal of This Text and Changes in This Edition

The goal of this text is to develop basic statistical literacy. The statistically literate person understands and appreciates the role of statistics in the research process, is competent to perform basic calculations, and can read and appreciate the professional research literature in his or her field as well as any research reports he or she may encounter in everyday life.

These three aspects of statistical literacy provide a framework for discussing the features of this text:

1. **An Appreciation of Statistics.** A statistically literate person understands the relevance of statistics for social research, can analyze and interpret the meaning of a statistical test, and can select an appropriate statistic for a given purpose and a given set of data. This textbook develops these qualities, within the constraints imposed by the introductory nature of the course, in these ways.

- *The relevance of statistics.* Chapter 1 includes a discussion of the role of statistics in social research and stresses their usefulness as ways of analyzing and manipulating data and answering research questions. Each example problem is framed in the context of a research situation. A question is posed and then, with the aid of a statistic, answered. The relevance of statistics for answering questions is thus stressed throughout this text. This central theme of usefulness is further reinforced by a series of "Application" boxes, each of which illustrates some specific way statistics can be used to answer questions, and by the "Using Statistics" boxes, which open each chapter.

  Almost all end-of-chapter problems are labeled by the social science discipline or subdiscipline from which they are drawn: **SOC** for sociology, **SW** for social work, **PS** for political science, **CJ** for criminal justice, **PA** for public administration, and **GER** for gerontology. By identifying problems with specific disciplines, students can more easily see the relevance of statistics to their own academic interests. (Not incidentally, they will also see that the disciplines have a large subject matter in common.)
- *Interpreting statistics.* For most students, interpretation—saying what statistics mean—is a big challenge. The ability to interpret statistics can be developed only by exposure and experience. To provide exposure, I have been careful in the example problems to express the meaning of the statistic in terms of the original research question. To provide experience, the end of-chapter problems almost always call for an interpretation of the statistic calculated. To provide examples, many of the answers to odd-numbered computational problems in the back of this text are expressed in words as well as numbers. The "Becoming a Critical Consumer" boxed inserts provide additional examples of how to express the meaning of statistics.
- *Using statistics: Ideas for research projects.* Appendix E offers ideas for independent data-analysis projects for students. The projects require students to use SPSS to analyze a data set and can be assigned at intervals throughout the semester or at the end of the course. Each project provides an opportunity for students to practice and apply their statistical skills and, above all, to exercise their ability to understand and interpret the meaning of the statistics they produce.

2. **Computational Competence.** Students should emerge from their first course in statistics with the ability to perform elementary forms of data analysis—to execute a series of calculations and arrive at the correct answer. While computers and calculators have made mere arithmetic less of an issue today, computation is inseparable from statistics, and I have included a number of features to help students cope with these mathematical challenges.

- *"One Step at a Time" boxes* for each statistic break down computation into individual steps for maximum clarity and ease.
- *Extensive problem sets* are provided at the end of each chapter. For the most part, these problems use fictitious data and are designed for ease of computation.
- *Solutions* to odd-numbered computational problems are provided so students can check their answers.
- *SPSS for Windows* continues as a feature (as in previous editions) to give students access to the computational power of the computer. This is explained in more detail later.

3. **The Ability to Read the Professional Social Science Literature.** The statistically literate person can comprehend and critically appreciate research reports written by others. The development of this skill is a particular problem at the introductory level because (1) the vocabulary of professional researchers is so much more concise than the language of this textbook and (2) the statistics featured in the literature are more advanced than those covered at the introductory level. This text helps to bridge this gap by:

- Always expressing the meaning of each statistic in terms of answering a social science research question.
- Providing a series of boxed inserts called "Becoming a Critical Consumer" that help students decipher the statistics they are likely to encounter in everyday life as well as in the professional literature. Many of these inserts include excerpts from the popular media, the research literature, or both.

Additional Features. A number of other features make this text more meaningful for students and more useful for instructors.

- *Readability and clarity.* The writing style is informal and accessible to students without ignoring the traditional vocabulary of statistics. Problems and examples have been written to maximize student interest and to focus on issues of concern and significance. For the more difficult material (such as hypothesis testing), students are first walked through an example problem before being confronted by formal terminology and concepts. Each chapter ends with a summary of major points and formulas and a glossary of important concepts. A list of frequently used formulas inside the front cover and a glossary of symbols inside the back cover can be used for quick reference.
- *Organization and coverage.* This text is divided into four parts, with most of the coverage devoted to univariate descriptive statistics, inferential statistics, and bivariate measures of association. The distinction between description and inference is introduced in the first chapter and maintained throughout this text. In selecting statistics for inclusion, I have tried to strike a balance between the essential concepts with which students must be familiar and the amount of

material students can reasonably be expected to learn in their first (and perhaps only) statistics course—all the while bearing in mind that different instructors will naturally wish to stress different aspects of the subject. Thus, this text covers a full gamut of the usual statistics, with each chapter broken into subsections so instructors can choose the particular statistics they wish to include. Some less frequently used techniques and statistical procedures can be found on the companion website.

- *Learning objectives.* Learning objectives are stated at the beginning of each chapter. These are intended to serve as "study guides" and to help students identify and focus on the most important material.
- *Using Statistics.* At the beginning of each chapter, some applications of the statistics to be introduced are presented. These give students some context for appreciating the material and some further examples of the usefulness of statistics for answering research questions.
- *Review of mathematical skills.* A comprehensive review of all the mathematical skills that will be used in this text is included as a prologue. Students who are inexperienced or out of practice with mathematics can study this review early in the course and/or refer to it as needed. A self-test is included so students can check their level of preparation for this course.
- *Statistical techniques and end-of-chapter problems are explicitly linked.* After a technique is introduced, students are directed to specific problems for practice and review. The "how to do it" aspects of calculation are reinforced immediately and clearly.
- *End-of-chapter problems are organized progressively.* Simpler problems with small data sets are presented first. Often, explicit instructions or hints accompany the first several problems in a set. The problems gradually become more challenging and require more decision making by the student (e.g., choosing the most appropriate statistic for a certain situation). Thus, each problem set gradually and progressively develops problem-solving abilities.
- *Computer applications.* This text integrates SPSS for Windows the leading social science statistics package, to help students take advantage of the power of the computer. Appendix F provides an introduction to SPSS for Windows. The demonstrations at the end of each chapter explain how to use the statistical package to produce the statistics presented in that chapter. Student exercises analyzing data with SPSS for Windows are also included. The student version of SPSS for Windows is available as a supplement to this text.
- *Realistic, up-to-date data.* The database for computer applications in this text is a shortened version of the 2010 General Social Survey. This database will give students the opportunity to practice their statistical skills on "real life" data. The database is described in Appendix G.
- *Companion website.* The website for this text includes a more extensive treatment of probability and some less frequently used techniques, a table of random numbers, hypothesis tests for ordinal-level variables, "find the test" flowcharts, and a number of other features.
- *Instructor's Manual Testbank.* The *Instructor's Manual* includes chapter summaries, a test item file of multiple-choice questions, answers to even-numbered

computational problems, and step-by-step solutions to selected problems. In addition, the *Instructor's Manual* includes cumulative exercises (with answers) that can be used for testing purposes.

- *Study Guide.* The *Study Guide*, written by Professor Roy Barnes, contains additional examples to illuminate basic principles, review problems with detailed answers, SPSS projects, and multiple-choice questions and answers that complement but do not duplicate the test item file.

Summary of Key Changes in this Edition. The following are the most important changes in this edition:

- The text has been redesigned to enhance readability.
- All data have been updated.
- The former Chapter 3 has been merged with Chapter 2.
- A new feature called "Statistics in Everyday Life" has been added. Several are included in every chapter, and they provide examples of the relevance and usefulness of statistics in a wide variety of everyday situations, ranging from telephone solicitations to the effect of the Internet on the crime of rape. These inserts demonstrate to students that statistics are not just abstract mathematical constructs but have practical value in the real world of government, education, business, media, sports, politics, and so on.
- The sections of some chapters have been divided into subsections for ease of comprehension.
- The data sets used in this text have been updated to the 2010 General Social Survey.
- There is more emphasis on interpretation and less on "mere computation" throughout this text.
- The steps in the "One Step at a Time" feature have been renumbered and reorganized to improve clarity and ease of use. The numbering of chapter sections has been deleted to streamline the layout.
- Titles have been added to the "Applying Statistics" features to highlight the topic.

This text has been thoroughly reviewed for clarity and readability. As with previous editions, my goal is to offer a comprehensive, flexible, and student-oriented book that will provide a challenging first exposure to social statistics.

Acknowledgments

This text has been in development, in one form or another, for almost 30 years. An enormous number of people have made contributions, both great and small, to this project, and at the risk of inadvertently omitting someone, I am bound to at least attempt to acknowledge my many debts.

This edition reflects the thoughtful guidance of Erin Mitchell of Cengage, and I thank her for her contributions. Much of whatever integrity and quality this book has is a direct result of the very thorough (and often highly critical) reviews that have been conducted over the years. I am consistently impressed by the sensitivity of my colleagues to the needs of the students, and for their assistance in preparing this edition, I would like to thank these reviewers: Susanne Bohmer, Edmonds Community College; Bennett Judkins, Lee University; and Junmin Wang, University of Memphis. Whatever failings are contained in this text are, of course, my responsibility and are probably the result of my occasional decisions not to follow the advice of my colleagues.

I would like to thank the instructors who made statistics understandable to me (professors Satoshi Ito, Noelie Herzog, and Ed Erikson) and all my colleagues at Christopher Newport University for their support and encouragement (especially professors F. Samuel Bauer, Stephanie Byrd, Cheryl Chambers, Robert Durel, Marcus Griffin, Mai Lan Gustafsson, Kai Heidemann, Ruth Kernodle, Michael Lewis, Marion Manton, Eileen O'Brien, Lea Pellet, Eduardo Perez, Andria Timmer, and Linda Waldron). Also, I thank all of my students for their patience and thoughtful feedback, and I am grateful to the literary executor of the late Sir Ronald A. Fisher, F.R.S., to Dr. Frank Yates, F.R.S., and to Longman Group Ltd., London, for permission to reprint Appendixes B, C, and D from their book *Statistical Tables for Biological, Agricultural and Medical Research* (6th edition, 1974).

Finally, I want to acknowledge the support of my family and rededicate this work to them. I have the extreme good fortune to be a member of an extended family that is remarkable in many ways and that continues to increase in size. Although I cannot list everyone, I would especially like to thank the older generation (my mother, Alice T. Healey, may she rest in peace), the next generation (my sons, Kevin and Christopher, and my daughters-in-law, Jennifer and Jessica), the new members (my wife Patricia Healey, Christopher Schroen, Jennifer Schroen, and Kate and Matt Cowell), and the youngest generation (Benjamin, Caroline, and Isabelle Healey and Abigail Cowell).

Prologue: Basic Mathematics Review

You will probably be relieved to hear that this text—your first exposure to statistics for social science research—does not stress computation per se. While you will encounter many numbers to work with and numerous formulas to solve, the major emphasis will be on understanding the role of statistics in research and the logic by which we attempt to answer research questions empirically. You will also find that in this text, the example problems and many of the homework problems have been intentionally simplified so the computations will not unduly impede the task of understanding the statistics themselves.

On the other hand, you may regret to learn that there is, inevitably, some arithmetic that you simply cannot avoid if you want to master this material. It is likely that some of you have not taken any math in a long time, others have convinced themselves that they just cannot do math under any circumstances, and still others are simply rusty and out of practice. All of you will find that mathematical operations that might seem complex and intimidating at first glance can be broken down into simple steps. If you have forgotten how to cope with some of these steps or are unfamiliar with these operations, this prologue is designed to ease you into the skills you will need to do all the computations in this textbook. Also, you can use this section for review whenever you feel uncomfortable with the mathematics in the chapters to come.

Calculators and Computers

A calculator is a virtual necessity for this text. Even the simplest, least expensive model will save you time and effort and is definitely worth the investment. However, I recommend that you consider investing in a more sophisticated calculator with memories and preprogrammed functions, especially the statistical models that can automatically compute means and standard deviations. Calculators with these capabilities are available for around $20.00 to $30.00 and will almost certainly be worth the small effort it will take to learn to use them.

In the same vein, there are several computerized statistical packages (or **statpaks**) commonly available on college campuses that can further enhance your statistical and research capabilities. The most widely used of these is a statistical package called **SPSS**. This program comes in a student version that is available bundled with this text (for a small fee). Statistical packages like SPSS are many times more powerful than even the most sophisticated handheld calculators, and it will be well worth your time to learn how to use them because they will eventually save you time and effort. SPSS is introduced in Appendix F of this text, and exercises at the end of almost every chapter will show you how to use the program to generate and interpret the statistics covered in that chapter.

There are many other programs that can help you generate accurate statistical results with a minimum of effort and time. Even spreadsheet programs such as Microsoft Excel, which is widely available, have some statistical capabilities. You should be aware that all these programs (other than the simplest calculators) will require some effort to learn, but the rewards will be worth the effort.

In summary, you should find a way at the beginning of this course—with a calculator, a statpak, or both—to minimize the tedium and hassle of mere computing. This will permit you to devote maximum effort to the truly important goal of increasing your understanding of the meaning of statistics in particular and social research in general.

Variables and Symbols

Statistics are a set of techniques by which we can describe, analyze, and manipulate variables. A **variable** is a trait that can change value from case to case or from time to time. Examples of variables include height, weight, level of prejudice, and political party preference. The possible values or scores associated with a given variable might be numerous (for example, income) or relatively few (for example, gender). I will often use symbols, usually the letter X, to refer to variables in general or to a specific variable.

Sometimes we will need to refer to a specific value or set of values of a variable. This is usually done with the aid of subscripts. Thus, the symbol X_1 (read "X-sub-one") would refer to the first score in a set of scores, X_2 ("X-sub-two") to the second score, and so forth. Also, we will use the subscript i to refer to all the scores in a set. Thus, the symbol X_i ("X-sub-eye") refers to all the scores associated with a given variable (for example, the test grades of a particular class).

Operations

You are all familiar with the four basic mathematical operations of addition, subtraction, multiplication, and division and the standard symbols ($+$, $-$, $\times$, $\div$) used to denote them. I should remind you that multiplication and division can be symbolized in a variety of ways. For example, the operation of multiplying some number a by some number b can be symbolized in (at least) six different ways:

$a \times b$

$a \cdot b$

$a * b$

ab

$a(b)$

$(a)(b)$

In this text, we will commonly use the "adjacent symbols" format (that is, *ab*), the conventional times sign (×), or adjacent parentheses to indicate multiplication. On most calculators and computers, the asterisk (*) is the symbol for multiplication.

The operation of division can also be expressed in several different ways. In this text, we will use either of these two methods:

$$a/b \text{ or } \frac{a}{b}$$

Several of the formulas with which we will be working require us to find the square of a number. To do this, multiply the number by itself. This operation is symbolized as X^2 (read "X squared"), which is the same thing as $(X)(X)$. If X has a value of 4, then

$$X^2 = (X)(X) = (4)(4) = 16$$

or we could say that "4 squared is 16."

The square root of a number is the value that when multiplied by itself results in the original number. Thus, the square root of 16 is 4 because (4)(4) is 16. The operation of finding the square root of a number is symbolized as:

$$\sqrt{X}$$

A final operation with which you should be familiar is summation, or the addition of the scores associated with a particular variable. When a formula requires the addition of a series of scores, this operation is usually symbolized as ΣX_i . Σ is the uppercase Greek letter sigma and stands for "the summation of." Thus, the combination of symbols ΣX_i means "the summation of all the scores" and directs us to add the value of all the scores for that variable. If four people had family sizes of 2, 4, 5, and 7, then the summation of these four scores for this variable could be symbolized as:

$$\Sigma X_i = 2 + 4 + 5 + 7 = 18$$

The symbol Σ is an operator, just like the + and × signs. It directs us to add all the scores of the variable indicated by the X symbol.

There are two other common uses of the summation sign, but unfortunately, the symbols denoting these uses are not at first glance sharply different from each other or from the symbol used earlier. A little practice and some careful attention to these various meanings should minimize the confusion. The first set of symbols is ΣX_i^2, which means "the sum of the squared scores." This quantity is found by *first* squaring each of the scores and *then* adding the squared scores together. A second common set of symbols will be $(\Sigma X_i)^2$, which means "the sum of the scores squared." This quantity is found by *first* summing the scores and *then* squaring the total.

These distinctions might be confusing at first, so let us see if an example helps to clarify the situation. Suppose we had a set of three scores: 10, 12, and 13. Thus:

$$X_i = 10, 12, 13$$

The sum of these scores would be indicated as:

$$\Sigma X_i = 10 + 12 + 13 = 35$$

The sum of the squared scores would be:

$$(\Sigma X_i)^2 = (10)^2 + (12)^2 + (13)^2 = 100 + 144 + 169 = 413$$

Take careful note of the order of operations here. The scores are first squared one at a time and then the squared scores are added. This is a completely different operation from squaring the sum of the scores:

$$(\Sigma X_i)^2 = (10 + 12 + 13)^2 = (35)^2 = 1225$$

To find this quantity, first the scores are summed and then the total of all the scores is squared. The squared sum of the scores (1,225) is *not* the same as the sum of the squared scores (413). In summary, the operations associated with each set of symbols can be stated as follows:

Symbols	Operations
ΣX_i	Add the scores
ΣX_i^2	First square the scores and then add the squared scores
$(\Sigma X_i)^2$	First add the scores and then square the total

Operations With Negative Numbers

A number can be either positive (if it is preceded by a + sign or by no sign at all) or negative (if it is preceded by a – sign). Positive numbers are greater than zero, and negative numbers are less than zero. It is very important to keep track of signs because they will affect the outcome of virtually every mathematical operation. This section briefly summarizes the relevant rules for dealing with negative numbers.

First, adding a negative number is the same as subtraction. For example:

$$3 + (-1) = 3 - 1 = 2$$

Second, subtraction changes the sign of a negative number:

$$3 - (-1) = 3 + 1 = 4$$

Note the importance of keeping track of signs here. If you neglected to change the sign of the negative number in the second expression, you would get the wrong answer.

For multiplication and division, you need to be aware of various combinations of negative and positive numbers. Ignoring the case of all positive numbers, this leaves several possible combinations. A negative number times a positive number results in a negative value:

$$(-3)(4) = -12$$

or

$$(3)(-4) = 12$$

A negative number multiplied by a negative number is always positive:

$$(-3)(-4) = 12$$

Division follows the same patterns. If there is one negative number in the calculations, the answer will be negative. If both numbers are negative, the answer will be positive. Thus,

$$\frac{-4}{2} = -2$$

and

$$\frac{4}{-2} = -2$$

but

$$\frac{-4}{-2} = 2$$

Negative numbers do not have square roots because multiplying a number by itself cannot result in a negative value. Squaring a negative number always results in a positive value (see the multiplication rules earlier).

Accuracy and Rounding Off

A possible source of confusion in computation involves accuracy and rounding off. People work at different levels of accuracy and precision and for this reason alone may arrive at different answers to problems. This is important because our answers can be least slightly different if you work at one level of precision and I (or your instructor or your study partner) work at another. You may sometimes think you have gotten the wrong answer when all you have really done is round off at a different place in the calculations or in a different way.

There are two issues here: when to round off and how to round off. In this text, I have followed the convention of working with as much accuracy as my calculator or statistics package will allow and then rounding off to two places of accuracy (two places beyond, or to the right of, the decimal point) only at the very end. If a set of calculations is lengthy and requires the reporting of intermediate sums or subtotals, I will round the subtotals off to two places as I go.

In terms of how to round off, begin by looking at the digit immediately to the right of the last digit you want to retain. If you want to round off to 100ths (two places beyond the decimal point), look at the digit in the 1,000ths place (three places beyond the decimal point). If that digit is 5 or more, round up. For example, 23.346 would round off to 23.35. If the digit to the right is less than 5, round down. So, 23.343 would become 23.34.

Let us look at some more examples of how to follow these rules of rounding. If you are calculating the mean value of a set of test scores and your calculator shows a final value of 83.459067 and you want to round off to two places, look at the digit three places beyond the decimal point. In this case, the value is 9 (greater than 5), so we would round the second digit beyond the decimal point up and report the mean as 83.46. If the value had been 83.453067, we would have reported our final answer as 83.45.

Formulas, Complex Operations, and the Order of Operations

A mathematical formula is a set of directions stated in general symbols for calculating a particular statistic. To "solve a formula," you replace the symbols with the proper values and then perform a series of calculations. Even the most complex formula can be simplified by breaking the operations down into smaller steps. Working through these steps requires some knowledge of general procedure and the rules of precedence of mathematical operations. This is because the order in which you perform calculations may affect your final answer. Consider this expression:

$$2 + 3(4)$$

If you add first, you will evaluate the expression as

$$5(4) = 20$$

but if you multiply first, the expression becomes

$$2 + 12 = 14$$

Obviously, it is crucial to complete the steps of a calculation in the correct order.

The basic rules of precedence are to find all squares and square roots first, then do all multiplication and division, and finally complete all addition and subtraction. Thus, the expression

$$8 + 2 \times 2^2/2$$

would be evaluated as

$$8 + 2 \times \frac{4}{2} = 8 + \frac{8}{2} = 8 + 4 = 12$$

The rules of precedence may be overridden by parentheses. Solve all expressions within parentheses before applying the rules stated earlier. For most of the complex formulas in this text, the order of calculations will be controlled by the parentheses. Consider this expression:

$$(8 + 2) - 4(3)^2/(8 - 6)$$

Resolving the parenthetical expressions first, we would have:

$$(10) - 4 \times 9/2 = 10 - 36/2 = 10 - 18 = -8$$

Without the parentheses, the same expression would be evaluated as:

$$\begin{aligned} 8 + 2 - 4 \times 3^2/8 - 6 &= 8 + 2 - 4 \times 9/8 - 6 \\ &= 8 + 2 - 36/8 - 6 \\ &= 8 + 2 - 4.5 - 6 \\ &= 10 - 10.5 \\ &= -0.5 \end{aligned}$$

A final operation you will encounter in some formulas in this text involves denominators of fractions that themselves contain fractions. In this situation, solve the fraction in the denominator first and then complete the division. For example,

$$\frac{15 - 9}{6/2}$$

would become

$$\frac{15 - 9}{6/2} = \frac{6}{3} = 2$$

When you are confronted with complex expressions such as these, do not be intimidated. If you are patient with yourself and work through them step-by-step, beginning with the parenthetical expression, even the most imposing formulas can be managed.

Exercises

You can use the following problems as a self-test on the material presented in this review. If you can handle these problems, you are more than ready to do all the arithmetic in this text. If you have difficulty with any of these problems, please review the appropriate section of this prologue. You might also want to use this section as an opportunity to become more familiar with your calculator. The answers are given immediately following these exercises, along with commentary and some reminders.

1. Complete each of these:

a. $17 \times 3 =$

b. $17(3) =$

c. $(17)(3) =$

d. $17/3 =$

e. $(42)^2 =$

f. $\sqrt{113}$

2. For the set of scores (X_i) of 50, 55, 60, 65, and 70, evaluate each of these expressions:

$\Sigma X_i =$

$\Sigma X_i^2 =$

$(\Sigma X_i)^2 =$

3. Complete each of these:

a. $17 + (-3) + (4) + (-2) =$

b. $15 - 3 - (-5) + 2 =$

c. $(-27)(54) =$

d. $(113)(-2) =$

e. $(-14)(-100) =$

f. $-34/-2 =$

g. $322/-11$

h. $\sqrt{-2} =$

i. $(-17)^2 =$

4. Round off each of the following to two places beyond the decimal point:

a. 17.17532

b. 43.119

c. 1,076.77337

d. 32.4641152301

e. 32.4751152301

5. Evaluate each of these:

a. $(3 + 7)/10 =$

b. $3 + 7/10 =$

c. $\dfrac{(4 - 3) + (7 + 2)(3)}{(4 + 5)(10)} =$

d. $\dfrac{22 + 44}{15/3} =$

Answers to Exercises

1. **a.** 51 **b.** 51 **c.** 51
(The obvious purpose of these first three problems is to remind you that there are several different ways of expressing multiplication.)
d. 5.67 (Note the rounding off.) **e.** 1,764
f. 10.63

2. The first expression is "the sum of the scores," so this operation would be:

$$\Sigma X_i = 50 + 55 + 60 + 65 + 70 = 300$$

The second expression is the "sum of the squared scores." Thus:

$$\Sigma X_i^2 = (50)^2 + (55)^2 + (60)^2 + (65)^2 + (70)^2$$
$$\Sigma X_i^2 = 2500 + 3025 + 3600 + 4225 + 4900$$
$$\Sigma X_i^2 = 18{,}250$$

The third expression is "the sum of the scores squared":

$$(\Sigma X_i)^2 = (50 + 55 + 60 + 65 + 70)^2$$
$$(\Sigma X_i)^2 = (300)^2$$
$$(\Sigma X_i)^2 = 90{,}000$$

Remember that ΣX_i^2 and $(\Sigma X_i)^2$ are two completely different expressions with very different values.

3. **a.** 16 **b.** 19 (Remember to change the sign of –5.)
c. –1458 **d.** –226 **e.** 1400 **f.** 17
g. –29.27
h. Your calculator probably gave you some sort of error message for this problem because negative numbers do not have square roots.
i. 289

4. **a.** 17.18 **b.** 43.12 **c.** 1,076.77
d. 32.46 **e.** 32.48

5. **a.** 1 **b.** 3.7 (Note again the importance of parentheses.)
c. 0.31 **d.** 13.2

1 Introduction

LEARNING OBJECTIVES

By the end of this chapter, you will be able to:

1. Describe the limited but crucial role of statistics in social research.
2. Distinguish between three applications of statistics and identify situations in which each is appropriate.
3. Identify and describe three levels of measurement and cite examples of variables from each.

USING STATISTICS

One of the most important themes of this text is that statistics are useful. They can be misused, misinterpreted, and misunderstood, but they provide extremely useful tools for analyzing and understanding the world and everyday life as well as for communicating the results of social science research. To stress this idea, each chapter will open with a brief list of situations in which statistics can be (and should be) applied to good advantage. In this introductory chapter, we will focus on the most general examples, but in the rest of this book, this opening section will highlight the usefulness of the specific statistics presented in that chapter.

Statistics can be used to:

- Demonstrate the connection between smoking and cancer.
- Measure the political preferences of the general public, including the popularity of specific candidates for office.
- Analyze the connection between the availability of guns and the homicide rate.
- Track attitudes about gay marriage and other controversial issues over time.
- Compare the cost of living (housing prices and rents, the cost of groceries and gas, health care, and so forth) between different localities (cities, states, and even nations).

Why Study Statistics?

Students sometimes wonder about the value of studying statistics. What, after all, do numbers and statistics have to do with understanding people and society? The value of statistics will become clear as we move from chapter to chapter, but for now, we can demonstrate their importance by considering **research**. Scientists conduct research to answer questions, examine ideas, and test theories. Research is a disciplined inquiry that can take numerous forms. Statistics are relevant for **quantitative** research projects or projects that collect information in the form of numbers or **data**. Statistics, then, are a set of mathematical techniques used by social scientists to organize and manipulate data for the purpose of answering questions and testing theories.

What is so important about learning how to manipulate data? On one hand, some of the most important and enlightening works in the social sciences do not utilize statistics at all. There is nothing magical about data and statistics. The mere presence of numbers guarantees nothing about the quality of a scientific inquiry. On the other hand, data can be the most trustworthy information available to the researcher and, consequently, deserve special attention. Data that have been carefully collected and thoughtfully analyzed can be the strongest, most objective foundation for building theory and enhancing our understanding of the social world.

Let us be very clear about one point: It is never enough merely to gather data (or, for that matter, any kind of information). Even the most carefully collected information does not and cannot speak for itself. Researchers must be able to effectively use statistics to organize, evaluate, and analyze the data. Without a good understanding of the principles of statistical analysis, the researcher will be unable to make sense of the data. Without the appropriate application of statistics, data will be useless.

Statistics are an indispensable tool for the social sciences. They provide the scientist with some of the most useful techniques for evaluating ideas and testing theory. The next section describes the relationships between theory, research, and statistics in more detail.

The Role of Statistics in Scientific Inquiry

Figure 1.1 graphically represents the role of statistics in the research process. The diagram is based on the thinking of Walter Wallace and illustrates a useful conception of how the knowledge base of any scientific enterprise grows and develops. One point the diagram makes is that scientific theory and research continually shape each other. Statistics are one of the most important means by which research and theory interact. Let us take a closer look at the wheel.

A Journey Through the Research Process

Theory. Because Figure 1.1 is circular, it has no beginning or end, and we could start our discussion at any point. For the sake of convenience, let us begin at the top and follow the arrows around the circle. A **theory** is an explanation of the relationships between phenomena. People naturally (and endlessly) wonder about problems in society (like prejudice, poverty, child abuse, and serial murders), and they develop explanations ("low levels of education cause prejudice") in their attempts to understand. Unlike our everyday, informal explanations, scientific theory is subject to a rigorous testing process. Let us take racial prejudice as an example to illustrate how the research process works.

FIGURE 1.1 The Wheel of Science

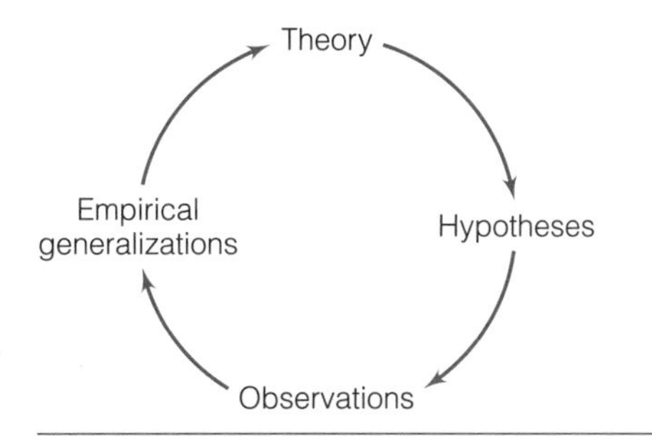

Source: Adapted from Walter Wallace, *The Logic of Science in Sociology* (Chicago: Aldine-Atherton, 1971).

What causes racial prejudice? One possible answer to this question is provided by the *contact hypothesis*. This theory was stated over 50 years ago by social psychologist Gordon Allport, and it has been tested on a number of occasions since that time.[1] The theory states that contact in which members of different groups have equal status and are engaged in cooperative behavior will result in a reduction of prejudice on all sides. The more equal and cooperative the contact, the more likely people will see each other as individuals and not as representatives of a particular group. For example, we might predict that members of a racially mixed athletic team that cooperate with each other to achieve victory would experience a decline in prejudice. On the other hand, when different groups compete for jobs, housing, or other resources, prejudice will tend to increase. The contact hypothesis is not a complete explanation of prejudice, of course, but it will serve to illustrate a sociological theory.

Variables and Causation. Note that the hypothesis is stated in terms of causal relationships between two variables. A **variable** is any trait that can change values from case to case; examples include gender, age, income, and political party affiliation. A theory may identify some variables as causes and others as effects or results. In the language of science, the causes are **independent variables** and the effects or result variables are **dependent variables**. In our theory, equal-status contact is the independent variable (or cause) and prejudice is the dependent variable (the result or effect). In other words, the theory argues that prejudice *depends on* (or is caused by) the extent to which a person participates in equal-status, cooperative contacts with members of other groups.

Diagrams can be a useful way of representing the relationships between variables:

Equal-Status Contact → Prejudice
Independent Variable → Dependent Variable

$$X \rightarrow Y$$

---

[1]Allport, Gordon, 1954. *The Nature of Prejudice*. Reading, Massachusetts: Addison-Wesley. For recent attempts to test this theory, see: McLaren, Lauren. 2003. "Anti-Immigrant Prejudice in Europe: Contact, Threat Perception, and Preferences for the Exclusion of Migrants." *Social Forces*. 81: 909–937; Pettigrew, Thomas. 1997. "Generalized Intergroup Contact Effects on Prejudice." *Personality and Social Psychology Bulletin*. 23: 173–185; and Sigelman, Lee, and Susan Welch. 1993. "The Contact Hypothesis Revisited: Black–White Interaction and Positive Racial Attitudes." *Social Forces*. 71: 781–795.

The arrow represents the direction of the causal relationship and X and Y are general symbols for the independent and dependent variables, respectively.

Hypotheses. So far, we have a theory of prejudice and independent and dependent variables. Is the theory true or false? To find out, we need to compare our theory with the facts; we need to do some research. Our next steps would be to define our terms and ideas. One problem we often face in research is that theories are complex and abstract, and we need to be very specific to conduct a valid test. Often, we do this by deriving a **hypothesis** from the theory: a statement about the relationship between our variables that is specific and exact.

For example, to test the contact hypothesis, we would have to say exactly what we mean by prejudice and we would need to describe "equal-status, cooperative contacts" in detail. We would also review the research literature to help develop and clarify our definitions and our understanding of these concepts.

As the hypothesis takes shape, we begin the next step of the research process, during which we decide exactly how we will gather data. We must decide how cases will be selected and tested, how variables will be measured, and a host of related matters. Ultimately, these plans will lead to the observation phase (the bottom of the wheel of science), where we actually measure social reality. Before we can do this, we must have a very clear idea of what we are looking for and a well-defined strategy for conducting the search.

Making Observations and Using Statistics. To test the contact hypothesis, we would begin with people from different racial or ethnic groups. We might place some subjects in cooperative situations and others in competitive situations and measure levels of prejudice before and after each type of contact. We might administer a survey that asked subjects to agree or disagree with statements such as "Greater efforts must be made to racially integrate public schools." Our goal would be to see if the people exposed to the cooperative contact situations actually become less prejudiced.

Now, finally, we come to statistics. During the observation phase, we will collect a great deal of numerical information or data. If we had a sample of 100 people, we would have 200 completed surveys measuring prejudice: 100 completed before the contact situation and 100 filled out afterward. Try to imagine dealing with 200 completed surveys. If we had asked each respondent just five questions to measure their prejudice, we would have a total of 1,000 separate pieces of information to deal with. What do we do? We have to organize and analyze this information, and statistics will become very valuable at this point. They will supply us with many ideas about "what to do" with the data, and we will begin to look at some of the options in the next chapter. For now, let me stress two points about statistics.

First, statistics are crucial. Simply put, without statistics, quantitative research is impossible. We need statistics to analyze data and to shape and refine our theories of the social world. Second—and somewhat paradoxically—the role of statistics is limited. As Figure 1.1 makes clear, scientific research proceeds through several mutually interdependent stages. Statistics become directly relevant only at the end of the observation stage. However, before any statistical analysis can be legitimately applied, the preceding phases of the process must have been successfully completed. If the researcher has asked poorly conceived questions or has made serious errors of design or method, then even the most sophisticated statistical analysis is useless. Statistics cannot substitute for rigorous conceptualization, detailed and careful planning, or creative use of theories. Statistics cannot salvage a poorly designed research project. They cannot make sense out of garbage.

On the other hand, inappropriate statistical applications can limit the usefulness of an otherwise carefully done project. Only by successfully completing *all* phases of the process can a research project hope to contribute to understanding. A reasonable knowledge of the uses and limitations of statistics is as essential to the education of the social scientist as is training in theory and methodology.

Empirical Generalizations. Our statistical analysis would focus on assessing our theory, but we would also look for other trends in the data. For example, if we found that equal-status, cooperative contact reduces prejudice in general, we might go on to ask if the pattern applies to males as well as to females and to the well educated as well as to the poorly educated. As we probed the data, we might begin to develop some generalizations based on the empirical patterns we observe. For example, what if we found that contact reduced prejudice for younger respondents but not for older respondents? Could it be that younger people are less "set in their ways" and have attitudes and feelings that are more open to change? As we developed tentative explanations, we would begin to revise or elaborate our theory.

A New Theory? If we changed our theory based on our empirical generalizations, a new research project would probably be needed to test the revised theory, and the wheel of science would begin to turn again. We (or perhaps some other researchers) would go through the entire process once again with this new—and, hopefully, improved—theory. This second project might result in further revisions that would require still more research, and the wheel of science would continue to turn as long as scientists were able to suggest revisions or develop new insights. Every time the wheel turned, our understandings of the phenomena under consideration would (hopefully) improve.

Summary: The Dialogue Between Theory and Research. This description of the research process does not include white-coated, clipboard-carrying scientists who, in a blinding flash of inspiration, discover some fundamental truth about reality and shout "Eureka!" The truth is that in the normal course of science, we are rarely in a position to declare a theory true or false. Rather, evidence will gradually accumulate over time, and ultimate judgments of truth will be the result of many years of hard work, research, and debate.

Let us briefly review our imaginary research project. We began with a theory and examined the various stages of the research project that would be needed to test that theory. We wound up back at the top of the wheel—ready to begin a new project guided by a revised theory. We saw how theory can motivate research and how our observations might cause us to revise theory and, thus, motivate a new research project: Theory stimulates research and research shapes theory. This constant interaction is the lifeblood of science and the key to enhancing our understandings of the social world.

The dialogue between theory and research occurs at many levels and in multiple forms, and statistics are one of the most important links between the two realms. Statistics permit us to analyze data, to identify and probe trends and relationships, to develop generalizations, and to revise and improve our theories. As you will see throughout this text, statistics are limited in many ways. They are also indispensable—an essential tool for conducting research and shaping theory. (*For practice in describing the relationship between theory and research and the role of statistics in research, see problems 1.1 and 1.2.*)

The Goals of This Text

Clearly, statistics are a crucial part of social science research, and every social scientist needs some training in statistical analysis. In this section, we address how much training is necessary and the purpose of that training.

First, this textbook takes the view that statistics are tools—useful but not ends in themselves. Thus, we will not take a "mathematical" approach to the subject, although we will cover enough material so you can develop a basic understanding of why statistics "do what they do." Instead, statistics will be presented as tools that can be used to answer important questions, and our focus will be on how these techniques are applied in the social sciences.

Second, you will soon become involved in advanced coursework in your major fields of study, and you will find that much of the professional literature assumes at least basic statistical literacy. Furthermore, after graduation, many of you will find yourselves in positions—either in a career or in graduate school—where some understanding of statistics will be very helpful or perhaps even required. Very few of you will become statisticians per se, but you must have a grasp of statistics in order to read and critically appreciate the research literature of your discipline. As a student in the social sciences and in many careers related to the social sciences, you simply cannot realize your full potential without a background in statistics.

Within these constraints, this textbook is an introduction to statistics as they are utilized in the social sciences. The general goal of this text is to develop an appreciation—a healthy respect—for statistics and their place in the research process. You should emerge from this experience with the ability to use statistics intelligently and to know when other people have done so. You should be familiar with the advantages and limitations of the more commonly used statistical techniques, and you should know which techniques are appropriate for a given purpose. Lastly, you should develop sufficient statistical and computational skills and enough experience in the interpretation of statistics to be able to carry out some elementary forms of data analysis by yourself.

Descriptive and Inferential Statistics

As noted earlier, statistics are tools used to manipulate data and answer research questions. Two general types of statistical techniques, introduced in this section, are available to accomplish this task.

Descriptive Statistics

These techniques have several different uses:

- We can use *univariate* ("one variable") descriptive statistics to summarize or describe the distribution of a single variable.
- We can use *bivariate* ("two variable") statistics to describe the relationship between two variables and *multivariate* (more than two variable) statistics to describe the relationship between more than two variables.

STATISTICS
IN EVERYDAY LIFE

American society is increasingly connected to the Internet. In 2010, about 79% of all American adults used the Internet at least occasionally—an increase from 53% in 2000. However, connectedness is very dependent on social class: 95% of the most affluent Americans used the Internet vs. only 63% of the least affluent. What are the implications of this pattern? Are the less affluent being left behind? Will their lower access to this increasingly essential resource affect their education and job prospects and thus perpetuate their lower income? What additional information would you need to address these issues?

Source: U.S. Bureau of the Census. 2011. *Statistical Abstract of the United States, 2011.* Accessed on June 7, 2011, from http://www.census.gov/compendia/statab/2011edition.html.

Univariate Descriptive Statistics. Many of these statistics are probably familiar to you. For example, percentages, averages, and graphs can all be used to describe single variables.

To illustrate the usefulness of univariate descriptive statistics, consider the following situation: Suppose you wanted to summarize the income distribution of a community of 10,000 families. How would you do it? Obviously, you could not simply list all the incomes and let it go at that. You would want to use a summary measure of the overall distribution—perhaps a graph, an average, or the percentage of incomes that are low, moderate, or high. Whatever method you choose, its function is to reduce these thousands of items of information into a few clear, concise, and easily understood numbers. The process of using a few numbers to summarize many—called **data reduction**—is the basic goal of univariate descriptive statistics. Part I of this text is devoted to these statistics.

Bivariate and Multivariate Descriptive Statistics. The second type of descriptive statistics is designed to help us understand the relationship between two or more variables. These statistics—called **measures of association**—quantify the strength and direction of a relationship. We can use these statistics to investigate two matters of central theoretical and practical importance to any science: causation and prediction. Measures of association help us disentangle the connections between variables and trace the ways in which some variables might affect others. We can also use them to predict scores on one variable from scores on another.

For example, suppose you were interested in the relationship between the amount of study time and grades and had gathered data from a group of college students. By calculating the appropriate measure of association, you could determine the strength of the relationship and its direction. Suppose you found a strong positive relationship. This would indicate that "study time" and "grade" were closely related (strength of the relationship) and that as one increased, the other also increased (direction of the relationship). You could make predictions from one variable to the other ("the longer the study time, the higher the grade").

Measures of association can give us valuable information about relationships between variables and help us understand how one variable causes another. However, one important point to keep in mind about these statistics is that they cannot, by themselves, *prove* that two variables are causally related. Even if a measure of association shows a very strong relationship between study time and grades, we cannot conclude that one variable causes the other. Correlation is not the same thing as causation, and the mere existence of a correlation cannot prove that a causal relationship exists. We will consider bivariate associations or correlations in Part III of this text, and multivariate analysis will be covered in Part IV.

Inferential Statistics

This second class of statistical techniques becomes relevant when we wish to generalize to a **population**: the total collection of all cases in which the researcher is interested and that he or she wishes to understand better. Examples of possible populations would be all voters in the United States, all parliamentary democracies, all unemployed people in Atlanta, or all sophomore college football players in the Midwest.

Populations can theoretically range from inconceivable in size ("all humanity") to quite small (all 35-year-olds living in downtown Cleveland) but are usually fairly large, and social scientists almost never have the resources or time to test every case in a population. Hence, the need for **inferential statistics**, which involve using information from **samples**—carefully chosen subsets—to make inferences about a population. Because they have fewer cases, samples are cheaper to assemble and—if the proper techniques are followed—generalizations based on these samples can be very accurate representations of the population.

The concepts and procedures used in inferential statistics are probably unfamiliar to most people, but we are experienced consumers of inferential statistics—most familiar, perhaps, in the form of public-opinion polls and election projections. When a public-opinion poll reports that 42% of the American electorate plans to vote for a certain presidential candidate, it is reporting a generalization to a population ("the American electorate"—which numbers over 130 million people) from a carefully drawn sample (usually about 1,500 respondents). Inferential statistics will occupy our attention in Part II of this book. (*For practice in describing different statistical applications, see problems 1.3 and 1.7.*)

Level of Measurement

We have many statistical techniques from which to choose, as you will begin to see in the next chapter. How do we select the best statistic for a given situation? The most basic and important guideline for choosing statistics is **level of measurement** of a variable or its mathematical characteristics. Some variables, such as age or income, have numerical scores (years or dollars) and can be analyzed in a variety of ways with many different statistics. For example, we could summarize these variables with a mean or average and make such statements as

STATISTICS IN EVERYDAY LIFE

In 2009, a sample of Americans was asked if they thought that marriages between same-sex couples should be recognized by law. About 40% of the respondents said yes and 57% said no. In 1996, in response to the same question, only 27% said yes and 68% said no. Some say that American society is becoming increasingly polarized on the issue of gay marriage. Do these statistics support that view? How?

Source: Jones, Jeffrey. 2009. "Majority of Americans Continue to Oppose Gay Marriage" Gallup. Accessed on June 7, 2011 from http://www.gallup.com/poll/118378/Majority-Americans-Continue-Oppose-Gay-Marriage.aspx

BECOMING A CRITICAL CONSUMER: Introduction

The most important goal of this text is to develop your ability to understand, analyze, and appreciate statistical information. To assist in reaching this goal, I have included a series of boxed inserts called "Becoming a Critical Consumer" to help you exercise your statistical expertise. In this feature, we will examine the everyday statistics you might encounter in the media and in casual conversations with friends as well as the professional social science research literature. In this first installment, I briefly outline the activities that will be included in this feature. We will start with social science research and then examine statistics in everyday life.

As you probably already know, articles published in social science journals are often mathematically sophisticated and use statistics, symbols, and formulas that may be at this point in your education completely indecipherable. Compared to my approach in this text, the language of the professional researcher is more compact and dense. This is partly because space in research journals and other media is expensive and partly because the typical research project requires the analysis of many variables. Thus, a large volume of information must be summarized in very few words. Researchers may express in just a word or two a result or an interpretation that will take us a paragraph or more to state in this text. Also, professional researchers assume a certain level of statistical knowledge in their audience: They write for colleagues, not for undergraduate students.

How can you bridge the gap that separates you from this literature? It is essential to your education that you develop an appreciation for this knowledge base, but how can you understand the articles, which seem so challenging? The (unfortunate but unavoidable) truth is that a single course in statistics will not close the gap entirely. However, the information and skills developed in this text will enable you to read much of the social science research literature and give you the ability to critically analyze statistical information. I will help you decode research articles by explaining their typical reporting style and illustrating my points with actual examples from a variety of social science disciplines.

As you develop your ability to read professional research reports, you will simultaneously develop your ability to critically analyze the statistics you encounter in your daily routine. *In this age of information, statistical literacy is not just for academics or researchers.* A critical perspective on statistics in everyday life—as well as in the social science research literature—can help you think more critically and carefully; assess the torrent of information, opinion, facts, and factoids that washes over us every day; and make better decisions on a broad range of issues. Therefore, these boxed inserts will also examine how to analyze the statistics you are likely to encounter in your everyday non-academic life. What (if anything) do statements like the following really mean?

- The Democratic candidate for governor is projected to win 55% of the vote in the election.
- The average life expectancy has reached 77 years.
- The number of cohabiting couples in this town has increased by 300% since 1980.
- There is a strong correlation between church attendance and divorce: the more frequent the church attendance, the lower the divorce rate.

Which of these statements sounds credible? How would you evaluate the statistical claims in each? The truth is elusive and multifaceted—how can we know it when we see it? The same skills that help you read the professional research literature can also be used to sort out statistical information you encounter in the media and in casual conversation. These boxed inserts will help you develop a more critical and informed approach to all statistics.

Statistical literacy will not always lead you to the truth, of course, but it will enhance your ability to analyze and evaluate information, to sort through claims and counterclaims, and to appraise them sensibly.

"The average income of this city is $43,000" or "The average age of students on this campus is 19.7."

Other variables, such as sex or zip codes, have "scores" that are really just labels, not numbers at all, and we have fewer options for analyzing them. The average would be meaningless as a way of describing these variables. Your personal zip code might *look* like a number, but it is merely an arbitrary label that happens to be expressed in digits. These "numbers" cannot be added or divided, and statistics such as the average cannot be applied to this variable. The average zip code of a group of people is meaningless.

Thus, different variables have different mathematical characteristics, and our statistics must match these qualities. Determining the level of measurement of a variable is one of the very first steps in any statistical analysis, and we will consider this matter at some length. We will consider the level of measurement of variables throughout this text and each time a statistical technique is introduced.

There are three levels of measurement. In order of increasing sophistication, they are nominal, ordinal, and interval-ratio. Each is discussed separately.

The Nominal Level of Measurement

Variables measured at the nominal level have "scores" or categories that are not numerical. Examples of variables at this level include gender, zip code, race, religious affiliation, and place of birth. At this lowest level of measurement, the only mathematical operation permitted is comparing the relative sizes of the categories (e.g., "There are more females than males in this dorm"). The categories or scores of nominal-level variables cannot be ranked with respect to each other and cannot be added, divided, or otherwise manipulated mathematically. Even when the scores or categories are expressed in digits (such as zip codes and street addresses), all we can do is compare relative sizes of categories (e.g., "The most common zip code of students on this campus is 22033"). The categories themselves do not form a mathematical scale; they are different from each other, but we cannot say that one category is more or less or higher or lower than another. Males and females differ in terms of gender, but neither has more or less gender than the other. In the same way, a zip code of 54398 is different from but not higher than a zip code of 13427.

Nominal variables are rudimentary, but there are criteria and procedures that we need to observe in order to measure them adequately. In fact, these criteria apply to variables measured at all levels, not just the nominal. First, the categories of nominal-level variables must be mutually exclusive of each other so there is no ambiguity about the category or score of any given case. Each case must have one and only one score. Second, the categories must be exhaustive. In other words, there must be a category—at least an "other" or miscellaneous category—for every possible score that might be found.

Third, the categories of nominal variables should be relatively homogeneous. That is, our categories should include cases that are truly comparable. Or, to put it another way, we need to avoid lumping apples with oranges. There are no hard-and-fast guidelines for judging if a set of categories is appropriately homogeneous. The researcher must make that decision in terms of the specific

TABLE 1.1 Four Scales for Measuring Religious Preference

Scale A (not mutually exclusive)	Scale B (not exhaustive)	Scale C (not homogeneous)	Scale D (an adequate scale)
Protestant	Protestant	Protestant	Protestant
Episcopalian	Catholic	Non-Protestant	Catholic
Catholic	Jew		Jew
Jew			None
None			Other
Other			

purpose of the research, and categories that are too broad for some purposes may be perfectly adequate for others.

Table 1.1 demonstrates some errors in measuring religious preference. Scale A in the table violates the criterion of mutual exclusivity because of overlap between the categories Protestant and Episcopalian. Scale B is not exhaustive because it does not provide a category for people with no religious preference (None) or for people who belong to religions other than the three listed. Scale C uses a category (Non-Protestant, which would include Catholics, Jews, Buddhists, and so forth) that seems too broad for most research projects. Scale D represents the way religious preference is often measured in North America, but these categories may be too general for some research projects and not comprehensive enough for others. For example, an investigation of issues that have strong moral and religious content (assisted suicide, abortion, and capital punishment, for example) might need to distinguish between the various Protestant denominations, and an effort to document religious diversity would need to add categories for Buddhists, Muslims, and numerous other religious faiths.

As is the case with zip codes, numerical labels are sometimes used to identify the categories of a variable measured at the nominal level. This practice is especially common when the data are being prepared for computer analysis. For example, the various religions might be labeled with a 1 indicating Protestant, a 2 signifying Catholic, and so on. Remember that these scores are merely labels or names and have no numerical quality to them. They cannot be added, subtracted, multiplied, or divided. The only mathematical operation permissible with nominal variables is counting and comparing the number of cases in each category of the variable.

The Ordinal Level of Measurement

Variables measured at the ordinal level are more sophisticated than nominal-level variables. They have scores or categories that can be ranked from high to low, so in addition to classifying cases into categories, we can describe the categories in terms of "more or less" with respect to each other. Thus, with ordinal-level variables, not only can we say that one case is different from another, but we can also say that one case is higher or lower, more or less, than another.

For example, the variable socioeconomic status (SES) is usually measured at the ordinal level. The categories of the variable are often ordered according to the following scheme:

4. Upper class

3. Middle class

2. Working class

1. Lower class

Individual cases can be compared in terms of the categories into which they are classified. Thus, an individual classified as a 4 (upper class) would be ranked higher than an individual classified as a 2 (working class), and a lower-class person (1) would rank lower than a middle-class person (3). Other variables that are usually measured at the ordinal level include attitude and opinion scales, such as those that measure prejudice, alienation, or political conservatism.

The major limitation of the ordinal level of measurement is that a particular score represents only a position with respect to some other score. We can distinguish between high and low scores, but the distance between the scores cannot be described in precise terms. Although we know that a score of 4 is more than a score of 2, we do not know if it is twice as much as 2.

Because we do not know what the exact distances are from score to score on an ordinal scale, our options for statistical analysis are limited. For example, addition (and most other mathematical operations) assumes that the intervals between scores are exactly equal. If the distances from score to score are equal, $2 + 2$ will always be 4. If they are not equal, $2 + 2$ might equal 3 or 5 or even 15. Thus, strictly speaking, such statistics as the average or mean (which requires that the scores be added together and then divided by the number of scores) are not permitted with ordinal-level variables. The most sophisticated mathematical operation fully justified with an ordinal variable is the ranking of categories and cases (although, as we will see, it is not unusual for social scientists to take liberties with this criterion).

The Interval-Ratio Level of Measurement[2]

The scores of variables measured at the interval-ratio level are actual numbers. First of all, this means there are equal intervals from score to score. For example, asking people how old they are will produce an interval-ratio-level variable (age) because the unit of measurement (years) has equal intervals (the distance from year to year is 365 days). Similarly, if we asked people how many siblings they have, we would produce a variable with equal intervals: Two siblings are one more than one and 13 is one more than 12.

Second, interval-ratio variables have true zero points. That is, the score of zero for these variables is not arbitrary; it indicates the absence or complete lack of whatever is

---

[2]Many statisticians distinguish between the interval level (equal intervals) and the ratio level (true zero point). I find the distinction unnecessarily cumbersome in an introductory text and will treat these two levels as one.

being measured. For example, the variable "number of siblings" has a true zero point because it is possible to have no siblings at all. Similarly, it is possible to have zero years of education, no income at all, a score of zero on a multiple-choice test, and be zero years old (although not for very long). Other examples of interval-ratio variables include the number of children in a family, life expectancy, and years married. All mathematical operations are permitted for variables measured at the interval-ratio level.

Summary and Final Points

Table 1.2 summarizes this discussion by presenting the basic characteristics of the three levels of measurement. Note that the number of permitted mathematical operations increases as we move from nominal to ordinal to interval-ratio levels of measurement. Ordinal-level variables are more sophisticated and flexible than nominal-level variables, and interval-ratio-level variables permit the broadest range of mathematical operations.

Let us end this section by making several points. The first stresses the importance of level of measurement, and the others discuss some common points of confusion in applying this concept.

First, knowing the level of measurement of a variable is crucial because it tells us which statistics are appropriate and useful. Not all statistics can be used with all variables. As displayed in Table 1.2, different statistics require different mathematical operations. For example, computing an average requires addition and division, and finding a median (or middle score) requires that the scores be ranked from high to low. Addition and division are appropriate only for interval-ratio-level variables, and ranking is possible only for variables that are at least ordinal in level of measurement. Your first step in dealing with a variable and selecting appropriate statistics is *always* to determine its level of measurement.

Second, note that some variables can be measured at more than one level. In particular, interval-ratio level variables are sometimes measured at the ordinal level,

TABLE 1.2 Basic Characteristics of the Three Levels of Measurement

Levels	Examples	Measurement Procedures	Mathematical Operations Permitted
Nominal	Sex, race, religion, marital status	Classification into categories	Counting number in each category, comparing sizes of categories
Ordinal	Social class, attitude, and opinion scales	Classification into categories plus ranking of categories with respect to each other	All of the foregoing plus judgments of "greater than" and "less than"
Interval-ratio	Age, number of children, income	All of the foregoing plus description of distances between scores in terms of equal units	All of the foregoing plus all other mathematical operations (addition, subtraction, multiplication, division, square roots, etc.)

so when you think about level of measurement, always examine the way in which the scores of the variable are *actually* stated. To illustrate, consider income. Table 1.3 displays two ways of asking people how much money they earn. The question in the top row asks for exact income. This method will produce a true interval-ratio-level variable, with scores that have a true zero point (that is, a score for someone who has no income at all) and equal intervals (dollars). If this is the way income is *actually* measured, the variable will be interval-ratio in level of measurement.

A second method for measuring income is illustrated in the second row of Table 1.3. Here, income is measured with five categories that are unequal in size: The first is $10,000 wide, the second is $14,999 wide, and so forth. Also, the categories do not have a true zero point. With this form of the question, we would not be able to distinguish between someone with an income as high as $9,999 and someone with no income at all.

Researchers will measure a variable such as income in different ways depending on the purpose of the research. If researchers want exact, precise information about income, they might use the question in the top row of Table 1.3. If they are more concerned about providing a convenient, simple format for respondents, they might use the question in the bottom row of Table 1.3. Either way of measuring income could be valid, but you should realize that the choices will produce very different variables and, thus, will affect the choices for statistical analysis. The point is that level of measurement can vary depending on the choices made by the researcher, and you always need to be careful to look at the way in which the variable is *actually* measured.

Third, there is a mismatch between the variables that are usually of most interest to social scientists (race, sex, marital status, attitudes, and opinions) and the most powerful and interesting statistics (such as the mean). The former are typically nominal or, at best, ordinal in level of measurement, but the more sophisticated statistics require measurement at the interval-ratio level. This mismatch creates some very real difficulties for social science researchers. On one hand, researchers will want to measure variables at the highest, most precise level of measurement. For example, if income is measured in exact dollars, researchers can make very precise descriptive statements about the differences between people. For example: "Ms. Smith earns $12,547 more than Mr. Jones." If the same variable is measured in broad, unequal categories, such as those in the bottom row of Table 1.3, comparisons between individuals would be less precise and provide less information: "Ms. Smith earns more than Mr. Jones."

TABLE 1.3 Measuring Income at Two Different Levels of Measurement

Level	Form of the Question:
Interval-Ratio	What was your total family income last year? _____
Ordinal	Please choose the category that best describes your total family income last year: a. Less than $10,000 ___ b. $10,001 to $25,000 ___ c. $25,001 to $50,000 ___ d. $50,001 to $100,000 ___ e. $100,001 or more ___

On the other hand, given the nature of the disparity, researchers are more likely to treat variables as if they were higher in level of measurement than they actually are. In particular, variables measured at the ordinal level, especially when they have many possible categories or scores, are often treated as if they were interval-ratio and analyzed with the more powerful, flexible, and interesting statistics available at the higher level. This practice is common, but researchers should be cautious in assessing statistical results and developing interpretations when the level of measurement criterion has been violated.

In conclusion, level of measurement is a very basic characteristic of a variable, and we will always consider it when presenting statistical procedures. Level of measurement is also a major organizing principle for the material that follows, and you should make sure you are familiar with these guidelines. *(For practice in determining the level of measurement of a variable, see problems 1.4 through 1.9.)*

ONE STEP AT A TIME **Determining the Level of Measurement of a Variable**

1. Inspect the scores of the variable as they are *actually* stated, keeping in mind the definition of the three levels of measurement (see Table 1.2).
2. Change the order of the scores. Do they still make sense? If the answer is *yes*, the variable is **nominal**. If the answer is *no*, proceed to step 3.

   Illustration: Gender is a nominal variable and its scores can be stated in any order:

   1. male
   2. female

   or

   1. female
   2. male

   Each statement of the scores is just as sensible as the other. For a nominal-level variable, no score is higher or lower than any other score and the order in which they are stated is arbitrary.
3. Is the distance between the scores unequal or undefined? If the answer is *yes*, the variable is **ordinal**. If the answer is *no*, proceed to Step 4.

   Illustration: Consider the following scale, which measures support for capital punishment:

   1. Strongly Support
   2. Somewhat Support
   3. Neither Support or Oppose
   4. Somewhat Oppose
   5. Strongly Oppose

   People who "strongly support" the death penalty are more in favor than people who "slightly support" it, but the distance from one level of support to the next (from a score of 1 to a score of 2) is undefined. We do not have enough information to ascertain how much more or less one score is than another.
4. If you answered *no* in steps 2 and 3, the variable is **interval-ratio**. Variables at this level have scores that are actual numbers: They have an order with respect to each other and are a defined, equal distance apart. They also have a true zero point. Examples of interval-ratio variables include age, income, and the number of siblings.

Source: This system for determining level of measurement was suggested by Michael G. Bisciglia, Louisiana State University.

SUMMARY

1. The purpose of statistics is to organize, manipulate, and analyze data so researchers can test theories and answer questions. Along with theory and methodology, statistics are a basic tool used by social scientists to enhance their understanding of the social world.
2. There are two general classes of statistics. Descriptive statistics are used to summarize the distribution of a single variable and the relationships between two or more variables. We use inferential statistics to generalize to populations from random samples.
3. Variables may be measured at any of three different levels. At the nominal level, we can compare category sizes. At the ordinal level, scores can be ranked from high to low. At the interval-ratio level, all mathematical operations are permitted.

GLOSSARY

Data. Information expressed as numbers.

Data reduction. Summarizing many scores with a few statistics.

Dependent variable. A variable that is identified as an effect or outcome. The dependent variable is thought to be caused by the independent variable.

Descriptive statistics. The branch of statistics concerned with (1) summarizing the distribution of a single variable or (2) measuring the relationship between two or more variables.

Hypothesis. A specific statement, derived from a theory, about the relationship between variables.

Independent variable. A variable that is identified as a cause. The independent variable is thought to cause the dependent variable.

Inferential statistics. The branch of statistics concerned with making generalizations from samples to populations.

Level of measurement. The mathematical characteristic of a variable and the major criterion for selecting statistical techniques. Variables can be measured at any of three levels, each permitting certain mathematical operations and statistical techniques. The characteristics of the three levels are summarized in Table 1.2.

Measures of association. Statistics that summarize the strength and direction of the relationship between variables.

Population. The total collection of all cases in which the researcher is interested.

Quantitative research. Research project that collect data or information in the form of numbers.

Research. Any process of gathering information systematically and carefully to answer questions or test theories. Statistics are useful for research projects that collect numerical information or data.

Sample. A carefully chosen subset of a population. In inferential statistics, information is gathered from a sample and then generalized to a population.

Statistics. A set of mathematical techniques for organizing and analyzing data.

Theory. A generalized explanation of the relationship between two or more variables.

Variable. Any trait that can change values from case to case.

PROBLEMS

1.1 In your own words, describe the role of statistics in the research process. Using the wheel of science as a framework, explain how statistics link theory with research.

1.2 Find a research article in any social science journal. Choose an article on a subject of interest to you, and do not worry about being able to understand all the statistics that are reported.

a. How much of the article is devoted to statistics?

b. Is the research based on a sample from some population? How large is the sample? How were subjects or cases selected? Can the findings be generalized to some population?

c. What variables are used? Which are independent and which are dependent? For each variable, determine the level of measurement.

d. What statistical techniques are used? Try to follow the statistical analysis and see how much you can understand. Save the article and read it again after you finish this course and see if you understand it any better.

1.3 Distinguish between descriptive and inferential statistics. Describe a research situation that would use both types.

1.4 Following are some items from a public-opinion survey. For each item, indicate the level of measurement.

a. What is your occupation? __________

b. How many years of school have you completed? _________

c. If you were asked to use one of these four names for your social class, which would you say you belonged to?

___________ Upper ___________ Middle

___________ Working ___________ Lower

d. What is your age? _____

e. In what country were you born? _____

f. What is your grade-point average? _____

g. What is your major? _____

h. The only way to deal with the drug problem is to legalize all drugs.

________ Strongly agree

________ Agree

________ Undecided

________ Disagree

________ Strongly disagree

i. What is your astrological sign? ___________

j. How many brothers and sisters do you have? ___________

1.5 Following are brief descriptions of how researchers measured a variable. For each situation, determine the level of measurement of the variable.

a. Race or ethnicity. Respondents were asked to check all that apply:

________ Black

________ White

________ Hispanic

________ Asian or Pacific Islander

________ Native American

________ Other (Please specify: ______________)

b. Honesty. Subjects were observed as they passed by a spot where an apparently lost wallet was lying. The wallet contained money and complete identification. Subjects were classified into one of the following categories:

_____ Returned the wallet with money

_____ Returned the wallet but kept the money

_____ Did not return wallet

c. Social class. Subjects were asked about their family situation when they were 16 years old. Was their family:

_____ Very well off compared to other families?

_____ About average?

_____ Not so well off?

d. Education. Subjects were asked how many years of schooling they had completed.

e. Racial integration on campus. Students were observed during lunchtime at the cafeteria for a month. The number of students sitting with students of other races was counted for each meal period.

f. Number of children. Subjects were asked: "How many children have you ever had? Please include any that may have passed away."

g. Student seating patterns in classrooms. On the first day of class, instructors noted where each student sat. Seating patterns were remeasured every two weeks until the end of the semester. Each student was classified as

____ same seat as last measurement;

____ adjacent seat;

____ different seat, not adjacent;

____ absent.

h. Physicians per capita. The number of physicians was counted in each of 50 cities. The researchers used population data to compute the number of physicians per capita.

i. Physical attractiveness. A panel of 10 judges rated each of 50 photos of a mixed-race sample of males and females for physical attractiveness on a scale from 0 to 20, with 20 being the highest score.

j. Number of accidents. The number of traffic accidents for each of 20 intersections was recorded. Also, each accident was rated as

_____ minor damage, no injuries;

_____ moderate damage, personal injury requiring hospitalization;

_____ severe damage and injury.

1.6 Classify each of the first 20 items in the General Social Survey (see Appendix G) in terms of level of measurement.

1.7 For each of the following research situations, identify the level of measurement of all variables. Also, decide which statistical applications are used: descriptive statistics (single variable), descriptive statistics (two or more variables), or inferential statistics. Remember that it is quite common for a given situation to require more than one type of application.

a. The administration of your university is proposing a change in parking policy. You select a random sample of students and ask each one if he or she favors or opposes the change.

b. You ask everyone in your social research class for their highest grade in any math course and the grade on a recent statistics test. You compare the two sets of scores to see if there is any relationship.

c. Your aunt is running for mayor and hires you to question a sample of voters about their concerns. Specifically, she wants a profile of voters that will tell her what percentage belong to each political party, what percentage are male or female, and what percentage favor or oppose the widening of the main street in town.

d. Several years ago, a state reinstituted the death penalty for first-degree homicide. Did this reduce the homicide rate? A researcher has gathered information on the number of homicides in the state for the two-year periods before and after the change.

e. A local automobile dealer is concerned about customer satisfaction. He wants to mail a survey form to all customers for the past year and ask them if they are satisfied, very satisfied, or not satisfied with their purchases.

1.8 Identify all variables in these research projects and classify them by their level of measurement. Which variables are independent and which are dependent?

a. A graduate student asks 500 female students if they have experienced any sexual harassment on campus. Each student is asked to estimate the frequency of these incidents as either "often," "sometimes," "rarely," or "never." The researcher also gathers data on age and major to see if there is any connection between these variables and the frequency of sexual harassment.

b. A supervisor in the Solid Waste Management Division of city government is assessing two different methods of trash collection. One area of the city is served by trucks with two-man crews who do "backyard" pickups, and the rest of the city is served by "high-tech" single-person trucks with curbside pickup. The assessment measures include the number of complaints received from the two different areas over a six-month period, the amount of time per day required to service each area, and the cost per ton of trash collected.

c. The adult bookstore near campus has been raided and closed by the police. Your social research class has decided to poll a sample of students to find out if he or she supports or opposes the closing of the store, how many times each has visited the store, and if he or she agrees or disagrees that "pornography causes sexual assaults on women." The class also collects information on the sex, age, religion, political philosophy, and major to see if opinions are related to these characteristics.

d. For a research project in a political science course, a student has collected information about the quality of life and the degree of political democracy in 50 nations. Specifically, she used infant mortality rates to measure quality of life and the percentage of all adults who are permitted to vote in national elections as a measure of democratization. Her hypothesis is that quality of life is higher in more democratic nations.

e. A highway engineer wonders if a planned increase in the speed limit on a heavily traveled local avenue will result in any change in the number of accidents. He plans to collect information on traffic volume, the number of accidents, and the number of fatalities for the six-month periods before and after the change.

f. Students are planning a program to promote safe sex and awareness of other health concerns for college students. To measure the effectiveness of the program, they plan to give a survey measuring knowledge to a random sample of students before and after the program.

g. States have drastically cut their budgets for mental health care. Will this increase the number of homeless people? A researcher contacts a number of agencies serving the homeless in each state and develops an estimate of the size of the population before and after the cuts.

h. Does tolerance for diversity vary by race or ethnicity? Samples of white, black, Asian, Hispanic, and Native Americans have been given a survey that measures their interest in and appreciation of cultures and groups other than their own.

1.9 Identify all variables in these research situations and classify them by their level of measurement. Which variables are independent and which are dependent?

a. A researcher is wondering about racial preferences in dating among college students and asks a large sample of undergraduates about their own racial self-identification, gender, and age, and to rank some racial-ethnic categories (stated as: white, black, Latino, Asian) in terms of desirability as potential dates.

b. For adolescents, is one's level of sexual activity related to academic success in high school? A sample of teenagers was interviewed about the number of different romantic relationships they have had, the number of times they have had sexual intercourse, and their high school GPA.

c. Several hundred voting precincts across the nation have been classified in terms of the percentage of minority voters and voting turnout as well as percentage of local elected officials who are members of minority groups. Do precincts with higher percentages of minority voters have lower turnout? Do precincts with higher percentages of minority representation among local elected officials have higher turnout?

d. As nations become more affluent (as measured by per capita income), does the percentage of children enrolled in school increase? Is this relationship different for boys and girls?

e. Does the level of support for gun control vary by level of education? Does this relationship vary by gender, region of the country, or political party preference? Support for gun control was measured by a five-point scale that ranged from "strongly in favor" to "strongly opposed."

YOU ARE THE RESEARCHER

Introduction

The best way—maybe the only way—to learn statistics and to appreciate their importance is to apply and use them. This includes selecting the correct statistic for a given purpose, doing the calculations, and interpreting the result. I have included extensive end-of-chapter problems to give you multiple opportunities to select and calculate statistics and say what they mean. Most of these problems have been written so they can be solved with just a simple hand calculator. I have purposely kept the number of cases unrealistically low so the tedium of mere calculation would not interfere with the learning process. These problems present an important and useful opportunity to develop your statistical skills.

As important as they are, these end-of-chapter problems are simplified and several steps removed from the complex realities of social science research. To provide a more realistic statistical experience, I have included a feature called "You Are the Researcher," in which you will walk through many of the steps of a research project, making decisions about how to apply your growing knowledge of research and statistics and interpreting the statistical output you generate.

To conduct these research projects, you will analyze a shortened version of the 2010 General Social Survey (GSS). This data set can be downloaded from our website ([www.cengagebrain.com]). The GSS is a public-opinion poll that has been conducted on nationally representative samples of citizens of the United States since 1972. The full survey includes hundreds of questions covering a broad range of social and political issues. The version supplied with this text has a limited number of variables and cases but still consists of actual real-life data, so you have the opportunity to practice your statistical skills in a more realistic context.

Even though the version of the GSS we use for this text is shortened, it is still a large data set with almost 1,500 respondents and 50 different variables—too large for even the most advanced hand calculator. To analyze the GSS, you will learn how to use a computerized statistical package called SPSS (Statistical Package for the Social Sciences). A statistical package is a set of computer programs designed to analyze data. The advantage of these packages is that because the programs are already written, you can capitalize on the power of the computer with minimal computer literacy and virtually no programming experience. Be sure to read Appendix F before attempting any data analysis.

In most of the research exercises, which begin in Chapter 2, you will make the same kinds of decisions as professional researchers and move through some of the steps of a research project. You will select variables and statistics, generate and analyze output, and express your conclusions. When you finish these exercises, you will be well prepared to conduct your own research project (within limits, of course) and perhaps make a contribution to the ever-growing social science research literature.

Part I Descriptive Statistics

Part I consists of four chapters, each devoted to a different application of univariate descriptive statistics. Chapter 2 covers "basic" descriptive statistics, including percentages, ratios, rates, frequency distributions, and graphs. This material is relatively elementary and at least vaguely familiar to most people. Although these statistics are "basic," they are not necessarily simple or obvious, and the explanations and examples should be considered carefully before attempting the end-of-chapter problems or using them in actual research.

Chapters 3 and 4 cover measures of central tendency and dispersion, respectively. Measures of central tendency describe the typical case or average score (e.g., the mean), while measures of dispersion describe the amount of variety or diversity among the scores (e.g., the range, or the distance from the high score to the low score). These two types of statistics are presented in separate chapters to stress the point that centrality and dispersion are independent, separate characteristics of a variable. You should realize, however, that both measures are necessary and commonly reported together (along with some of the statistics presented in Chapter 2). To reinforce this idea, many of the problems at the end of Chapter 4 require the computation of a measure of central tendency from Chapter 3.

Chapter 5 is a pivotal chapter in the flow of the text. It takes some of the statistics from Chapters 2 through 4 and applies them to the normal curve, a concept of great importance in statistics. The normal curve is a line chart (see Chapter 2), which can be used to describe the position of scores using means (Chapter 3) and standard deviations (Chapter 4). Chapter 5 also uses proportions and percentages (Chapter 2).

In addition to its role in descriptive statistics, the normal curve is a central concept in inferential statistics, the topic of Part II of this text. Thus, Chapter 5 ends the presentation of univariate descriptive statistics and lays essential groundwork for the material to come.

2 Descriptive Statistics Percentages, Ratios and Rates, Tables, Charts, and Graphs

LEARNING OBJECTIVES

By the end of this chapter, you will be able to:

1. Explain the purpose of descriptive statistics in making data comprehensible.
2. Compute and interpret percentages, proportions, ratios, rates, and percentage change.
3. Construct and analyze frequency distributions for variables at each of the three levels of measurement.
4. Construct and analyze bar and pie charts, histograms, and line graphs.

USING STATISTICS

The statistical techniques presented in this chapter are used to summarize the scores on a single variable. They can be used to:

- Express the percentage of people in a community that belong to various religions, including people with no religious preference and atheists.
- Compare the popularity of cohabitation among young adults today with previous generations.
- Report changes in the crime rate over time.
- Organize information into easy-to-read tables, charts, and graphs.
- Create a line graph to display the falling use of cigarettes in the United States—down from 40% in the 1960s to 20% today.

Research results do not speak for themselves. They must be organized and manipulated so that whatever meaning they have can be quickly and easily understood by the researcher and by his or her readers. Researchers use statistics to clarify their results and communicate effectively. In this chapter, we consider some commonly used techniques for presenting research results, including percentages, rates, tables, and graphs. These univariate descriptive statistics are not mathematically complex (although they are not as simple as they might seem at first glance), but they are extremely useful for presenting research results clearly and concisely.

Percentages and Proportions

Consider this statement: "Of the 269 cases handled by the court, 167 resulted in prison sentences of five years or more." While there is nothing wrong with this statement, the same fact could have been more clearly conveyed if it had been reported as a percentage: "About 62% of all cases resulted in prison sentences of five or more years."

Percentages and proportions supply a frame of reference for reporting research results in the sense that they standardize the raw data: percentages to the base 100 and proportions to the base 1.00. The mathematical definition of **percentages** is:

FORMULA 2.1

$$\text{Percentage: } \% = \left(\frac{f}{N}\right) \times 100$$

Where f = frequency, or the number of cases in any category
N = the number of cases in all categories

To illustrate the computation of percentages, consider the data presented in Table 2.1. How can we find the percentage of cases that resulted in sentences of five years or more? Note that there are 167 cases in this category ($f = 167$) and a total of 269 cases in all ($N = 269$). Thus:

$$\text{Percentage } (\%) = \left(\frac{f}{N}\right) \times 100 = \left(\frac{167}{269}\right) \times 100 = (0.6208) \times 100 = 62.08\%$$

Using the same procedures, we can also find the percentage of cases that resulted in sentences of less than five years:

$$\text{Percentage } (\%) = \left(\frac{f}{N}\right) \times 100 = \left(\frac{72}{269}\right) \times 100 = (0.2677) \times 100 = 26.77\%$$

Percentages are easier to read and comprehend than frequencies. This advantage is particularly obvious when comparing groups of different sizes. For example, can you tell from Table 2.2 which college has the higher relative number of social science majors? Because the total enrollments are so different, comparisons are

TABLE 2.1 Disposition of 269 Criminal Cases (fictitious data)

Sentence	Frequency (f)	Percentage (%)
Five years or more	167	62.08
Less than five years	72	26.77
Suspended	20	7.44
Acquitted	10	3.72
	N = 269	100.01%*

*The slight discrepancy in the total of this column is due to a rounding error. See the Preface (Basic Mathematical Review) for more on rounding.

TABLE 2.2 Declared Major Fields on Two College Campuses (fictitious data)

Major	College A	College B
Business	103	3,120
Natural sciences	82	2,799
Social sciences	137	1,884
Humanities	93	2,176
	$N = 415$	$N = 9{,}979$

TABLE 2.3 Declared Major Fields on Two College Campuses (fictitious data)

Major	College A	College B
Business	24.82%	31.27%
Natural sciences	19.76%	28.05%
Social sciences	33.01%	18.88%
Humanities	22.41%	21.81%
	100.00% (415)	100.01% (9,979)

difficult to make from the raw frequencies. Percentages eliminate the difference in size of the two campuses by standardizing both distributions to the base of 100. The same data are presented in percentages in Table 2.3.

Table 2.3 makes it easier to identify differences as well as similarities between the two colleges. College A has a much higher percentage of social science majors (even though the absolute number of social science majors is less than at College B) and about the same percentage of humanities majors. How would you describe the differences in the remaining two major fields?

Social scientists use **proportions** as well as percentages. Proportions vary from 0.00 to 1.00: They standardize results to a base of 1.00 instead of the base of 100 for percentages. When we find a percentage, we divide the number of cases in a category (f) by the total number of all cases (N) and multiply the result by 100. To find a proportion, we do exactly the same division but do *not* multiply by 100:

FORMULA 2.2

$$\text{Proportion} = \frac{f}{N}$$

Percentages can be converted to proportions by dividing by 100; conversely, proportions can be converted to percentages by multiplying by 100. The two statistics give us the same information and are interchangeable. For example, we could state the relative number of sentences of more than 5 years in Table 2.1 as a proportion:

$$\text{Proportion} = \frac{f}{N} = \frac{167}{269} = 0.62$$

How can we choose between these statistics? Percentages are easier for most people (including statisticians) to comprehend and generally would be preferred, especially when the goal is to express results clearly and quickly. Proportions are used less frequently—generally when we are dealing with probabilities (see Chapter 5). The preference for percentages is based solely on ease of communication; these statistics are equally valid ways of expressing results and you should be proficient in both. *(For practice in computing and interpreting percentages and proportions, see problems 2.1 and 2.2.)*

Here are some further guidelines on the use of percentages and proportions:

1. When the number of cases is small (say, fewer than 20), it is usually preferable to report the actual frequencies rather than percentages or proportions. With a small number of cases, percentages are unstable and can change drastically with relatively minor changes in the size of the data set. For example, if you begin with a group of 10 males and 10 females (that is, 50% of each gender) and then add another female, the percentage distributions will change to 52.38% female and 47.62% male. Of course, as the number of observations increases, each additional case will have a smaller impact. If we started with 500 males and 500 females and then added one more female, the percentage of females would change very slightly to 50.05%.

2. Always report the number of observations along with proportions and percentages. This permits the reader to judge the adequacy of the sample size and, conversely, helps to prevent the researcher from lying with statistics. Statements such as "Two out of three people questioned prefer courses in statistics to any other course" might impress you, but the claim would lose its gloss if you learned that only three people were tested. *You should be extremely suspicious of reports that fail to report the number of cases that were tested.*

Applying Statistics 2.1 Communicating with Statistics

Not long ago, in a large social service agency, the following conversation took place between the executive director of the agency and a supervisor of one of the divisions:

> *Executive director:* Well, I don't want to seem abrupt, but I've only got a few minutes. Tell me, as briefly as you can, about this staffing problem you claim to be having.
>
> *Supervisor:* Ma'am, we just don't have enough people to handle our workload. Of the 177 full-time employees of the agency, only 50 are in my division. But 6,231 of the 16,722 cases handled by the agency last year were handled by my division.
>
> *Executive director (smothering a yawn):* Very interesting. I'll certainly get back to you on this matter.

How could the supervisor have presented his case more effectively? Because he wants to compare two sets of numbers (his staff versus the total staff and the workload of his division versus the total workload of the agency), proportions or percentages would be a more forceful way of presenting results. What if the supervisor had said, "Only 28.25% of the staff is assigned to my division, but we handle 37.26% of the total workload of the agency"? Is this a clearer message?

The first percentage is found by:

$$\% = \left(\frac{f}{N}\right) \times 100 = \frac{50}{177} \times 100$$
$$= (.2825) \times 100 = 28.25\%$$

The second percentage is found by:

$$\% = \left(\frac{f}{N}\right) \times 100 = \left(\frac{6231}{16{,}722}\right) \times 100$$
$$= (.3726) \times 100 = 37.26$$

3. Since these statistics require division, it might seem that we are violating level-of-measurement guidelines (see Table 1.2) when we calculate percentages and proportions for ordinal and—especially—nominal level variables. However, this is not the case because these statistics are only expressions of the relative size of a category of a variable. For example, when we say that "43% of the sample is female," we are making a statement about the size of one category (female) of the variable (gender) relative to the sample as a whole. Because they do not require the division of the *scores* of the variable (as would be the case in computing the average score on a test, for example), percentages and proportions can be computed for variables at any level of measurement.

Ratios, Rates, and Percentage Change

Ratios, rates, and percentage change provide some additional ways of summarizing results simply and clearly. Although they are similar to each other, each statistic has a specific application and purpose.

Ratios

Ratios are especially useful for comparing the relative size of different categories of a variable and are determined by dividing the frequency of one category by the frequency of another. The formula for a ratio is:

FORMULA 2.3

$$\text{Ratio} = \frac{f_1}{f_2}$$

Where f_1 = the number of cases in the first category
f_2 = the number of cases in the second category

To illustrate the use of ratios, suppose you were interested in the relative sizes of the various religious denominations and found that a particular community

ONE STEP AT A TIME **Finding Percentages and Proportions**

Step	*Operation*
1.	Determine the values for f (number of cases in a category) and N (number of cases in all categories). Remember that f will be the number of cases in a *specific category* (e.g., males on your campus) and N will be the number of cases in *all* categories (e.g., all students—males and females—on your campus) and that f will be smaller than N—except when the category and the entire group are the same (e.g., when all students are male). Proportions cannot exceed 1.00, and percentages cannot exceed 100.00%.
2.	For a proportion, divide f by N.
3.	For a percentage, multiply the value you calculated in step 2 by 100.

Applying Statistics 2.2 Ratios

In Table 2.2, how many natural science majors are there compared to social science majors at College B? This question could be answered with frequencies, but a more easily understood way of expressing the answer would be with a ratio. The ratio of natural science to social science majors would be:

$$\text{Ratio} = \frac{f_1}{f_2} = \frac{2799}{1884} = 1.49$$

For every social science major, there are 1.49 natural science majors at College B.

included 1,370 Protestant families and 930 Catholic families. To find the ratio of Protestants (f_1) to Catholics (f_2), divide 1,370 by 930:

$$\text{Ratio} = \frac{f_1}{f_2} = \frac{1370}{930} = 1.47$$

The ratio of 1.47 means there are 1.47 Protestant families for every Catholic family.

Ratios can be very economical ways of expressing the relative size of two categories. In our example, it is obvious from the raw data that Protestants outnumber Catholics, but ratios tell us exactly how much one category outnumbers the other.

Ratios are often multiplied by some power of 10 to eliminate decimal points. For example, the ratio just computed might be multiplied by 100 and reported as 147 instead of 1.47. This would mean that there are 147 Protestant families for every 100 Catholic families in the community. To ensure clarity, the comparison units for the ratio are also often expressed. Based on a unit of ones, the ratio of Protestants to Catholics would be expressed as 1.47:1. Based on hundreds, the same statistic might be expressed as 147:100. *(For practice in computing and interpreting ratios, see problems 2.1 and 2.2.)*

Rates

Rates provide still another way of summarizing the distribution of a single variable. Rates are defined as the number of actual occurrences of some phenomenon divided by the number of possible occurrences per some unit of time. Rates are usually multiplied by some power of 10 to eliminate decimal points. For example, the crude death rate for a population is defined as the number of deaths in that population (actual occurrences) divided by the number of people in the population (possible occurrences) per year. This quantity is then multiplied by 1,000. The formula for the crude death rate can be expressed as:

$$\text{Crude death rate} = \frac{\text{Number of deaths}}{\text{Total population}} \times 1{,}000$$

Applying Statistics 2.3 Rates

In 2010, there were 2,500 births in a city of 167,000. In 1970, when the population of the city was only 133,000, there were 2,700 births. Is the birthrate rising or falling? Although this question can be answered from the preceding information, the trends will be much more obvious if we compute birthrates for both years. Like crude death rates, crude birthrates are usually multiplied by 1,000 to eliminate decimal points. For 1970:

$$\text{Crude birthrate} = \frac{2{,}700}{133{,}000} \times 1{,}000 = 20.30$$

In 1970, there were 20.30 births for every 1,000 people in the city. For 2010:

$$\text{Crude birthrate} = \frac{2{,}500}{167{,}000} \times 1{,}000 = 14.97$$

In 2010, there were 14.97 births for every 1,000 people in the city. With the help of these statistics, the decline in the birthrate is clearly expressed.

If there were 100 deaths during a given year in a town of 7,000, the crude death rate for that year would be:

$$\text{Crude death rate} = \frac{100}{7{,}000} \times 1{,}000 = (0.01429) \times 1{,}000 = 14.29$$

Or, for every 1,000 people, there were 14.29 deaths during this particular year. In the same way, if a city of 237,000 people experienced 120 auto thefts during a particular year, the auto theft rate would be:

$$\text{Auto theft rate} = \frac{120}{237{,}000} \times 100{,}000 = (0.0005063) \times 100{,}000 = 50.63$$

Or, for every 100,000 people, there were 50.63 auto thefts during the year in question. *(For practice in computing and interpreting rates, see problems 2.3 and 2.4a.)*

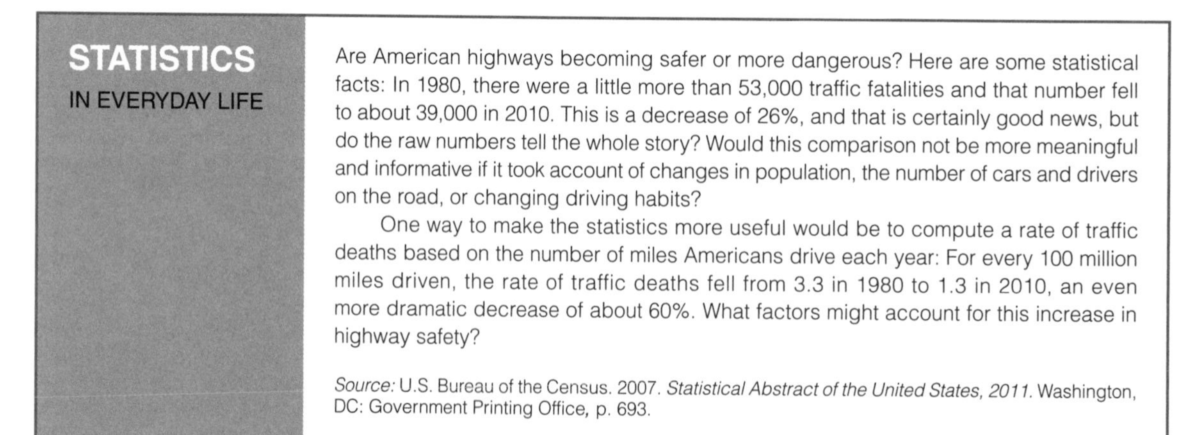

STATISTICS IN EVERYDAY LIFE

Are American highways becoming safer or more dangerous? Here are some statistical facts: In 1980, there were a little more than 53,000 traffic fatalities and that number fell to about 39,000 in 2010. This is a decrease of 26%, and that is certainly good news, but do the raw numbers tell the whole story? Would this comparison not be more meaningful and informative if it took account of changes in population, the number of cars and drivers on the road, or changing driving habits?

One way to make the statistics more useful would be to compute a rate of traffic deaths based on the number of miles Americans drive each year: For every 100 million miles driven, the rate of traffic deaths fell from 3.3 in 1980 to 1.3 in 2010, an even more dramatic decrease of about 60%. What factors might account for this increase in highway safety?

Source: U.S. Bureau of the Census. 2007. *Statistical Abstract of the United States, 2011*. Washington, DC: Government Printing Office, p. 693.

Percentage Change

Measuring social change, in all its variety, is an important task for all social sciences. One very useful statistic for this purpose is the **percentage change**, which tells us how much a variable has increased or decreased over a certain span of time.

To compute this statistic, we need the scores of a variable at two different points in time. The scores could be in the form of frequencies, rates, or percentages. The percentage change will tell us how much the score has changed at the later time relative to the earlier time. Using death rates as an example once again, imagine a society suffering from a devastating outbreak of disease in which the death rate rose from 16 deaths per 1,000 population in 1995 to 24 deaths per 1,000 in 2005. Clearly, the death rate is higher in 2005 but by how much relative to 1995?

The formula for the percent change is:

FORMULA 2.4

$$\text{Percent change} = \left(\frac{f_2 - f_1}{f_1}\right) \times 100$$

Where f_1 = first score, frequency, or value
f_2 = second score, frequency, or value

In our example, f_1 is the death rate in 1995 ($f_1 = 16$) and f_2 is the death rate in 2005 ($f_2 = 24$). The formula tells us to subtract the earlier score from the later and then divide by the earlier score. The value that results expresses the size of the change in scores ($f_2 - f_1$) relative to the score at the earlier time (f_1). The value is then multiplied by 100 to express the change in the form of a percentage:

$$\text{Percent change} = \left(\frac{24 - 16}{16}\right) \times 100 = \left(\frac{8}{16}\right) \times 100 = (.50) \times 100 = 50\%$$

Applying Statistics 2.4 Percent Change

The American family has been changing rapidly over the past several decades. One major change has been an increase in the number of married women and mothers with jobs outside the home. For example, in 1970, 30.3% of married women with children under the age of 6 worked outside the home. In 2009, this percentage had risen to 61.6.* How large has this change been?

It is obvious that the 2009 percentage is much higher, and calculating the percentage change will give us an exact idea of the magnitude of the change. The 1970 percentage is f_1 and the 2009 figure is f_2. Thus:

$$\text{Percent Change} = \left(\frac{f_2 - f_1}{f_1}\right) \times 100$$

$$\text{Percent Change} = \left(\frac{61.6 - 30.3}{30.3}\right) \times 100$$

$$\text{Percent Change} = \left(\frac{31.3}{30.3}\right) \times 100$$

$$\text{Percent Change} = (1.033) \times 100$$

$$\text{Percent Change} = 103.30$$

In the almost 40-year period between 1970 and 2009, the percentage of women with children younger than 6 who worked outside the home increased by 103.30%. This is an extremely large change—more than double the earlier percentage—in a short time frame and signals major changes in this social institution.

**Source:* U.S. Bureau of the Census. 2007. *Statistical Abstract of the United States, 2007*. Washington, DC: Government Printing Office. p. 380.

TABLE 2.4 Projected Population Growth for Six Nations, 1990–2020

Nation	Population, 1990 (f_1)	Population, 2020 (f_2)	Increase/ Decrease ($f_2 - f_1$)	Percent Change
China	1,148,364,000	1,384,545,000	236,181,000	20.57
United States	249,623,000	341,387,000	91,764,000	36.76
Canada	27,791,000	36,387,000	8,596,000	30.93
Hungary	10,372,000	9,772,000	−600,000	−5.78
Japan	123,537,000	121,633,000	−1,904,000	−1.54
Nigeria	96,604,000	182,344,000	85,740,000	88.75

Source: U.S. Bureau of the Census 2011. *Statistical Abstract of the United States.* 2011: http://www.census.gov/prod/2011pubs/11statab/intlstat.pdf

The death rate in 2005 is 50% higher than in 1995. This means that the 2005 rate was equal to the 1995 rate *plus* half of the earlier score. If the rate had risen to 32 deaths per 1,000, the percent change would have been 100% (the rate would have doubled), and if the death rate had fallen to 8 per 1,000, the percent change would have been −50%. Note the negative sign: It means that the death rate has decreased by 50%. The 2005 rate would have been half the size of the 1995 rate.

An additional example should make the computation and interpretation of the percentage change clearer. Suppose we wanted to compare the projected population growth rates for various nations. Table 2.4 shows the actual population for each nation in 1990 and the projected population for 2020. The Increase/Decrease column shows the estimate of how many people will be added or lost over the 30-year time span.

Casual inspection will give us some information about population trends. For example, compare the Increase/Decrease column for Nigeria and the United States. These societies are projected to add roughly similar numbers of people (about 86 million for Nigeria; a little more for the United States), but because Nigeria's 1990 population is less than half the size of the population of the United States, its percent change will be much higher (almost 90% vs. about 37%).

The percent change statistic makes comparisons more precise. The right-hand column of Table 2.4 shows the percent change in projected population for each nation. These values were computed by subtracting the 1990 population (f_1) from the 2020 population (f_2), dividing by the 1990 population, and multiplying by 100.

China will add the most people by far, but because it is the largest nation of these six (and the largest in the world), its actual growth rate will be lower than the United States or Canada. Hungary and Japan will actually lose population over the time span (note the negative growth rates). Nigeria has by far the highest growth rate and its population will increase by almost 90%. This means that in 2020, the population of Nigeria will be almost twice its 1990 size. *(For practice in computing and interpreting percent change, see problem 2.4b.)*

ONE STEP AT A TIME **Finding Ratios, Rates, and Percent Change**

Step	*Operation*
To Find Ratios	
1.	Determine the values for f_1 and f_2. The value for f_1 will be the number of cases in the first category (e.g., the number of males on your campus), and the value for f_2 will be the number of cases in the second category (e.g., the number of females on your campus).
2.	Divide the value of f_1 by the value of f_2.
3.	You may multiply the ratio by some power of 10 to eliminate decimal points and increase clarity.
To Find Rates	
1.	Determine the number of actual occurrences (e.g., births, deaths, homicides, assaults). This value will be the numerator.
2.	Determine the number of possible occurrences. This value will usually be the total population for the area in question.
3.	Divide the number of actual occurrences (step 1) by the number of possible occurrences (step 2).
4.	Multiply the value you calculated in step 3 by some power of 10. Conventionally, birthrates and death rates are multiplied by 1,000 and crime rates are multiplied by 100,000.
To Find Percent Change	
1.	Determine the values for f_1 and f_2. The former will be the score at time 1 (the earlier time) and the latter will be the score at time 2 (the later time).
2.	Subtract f_1 from f_2.
3.	Divide the quantity you found in step 2 by f_1.
4.	Multiply the quantity you found in step 3 by 100.

Frequency Distributions: Introduction

Frequency distributions are tables that report the number of cases in each category of a variable. They are very helpful and commonly used ways of organizing and working with data. In fact, the construction of frequency distributions is almost always the first step in any statistical analysis.

To illustrate the usefulness of frequency distributions and to provide some data for examples, assume that the counseling center at a university is assessing its effectiveness. Any realistic evaluation research would collect a variety of information from a large group of students, but for the sake of this example, we will confine our attention to just four variables and 20 students. The data are reported in Table 2.5.

Table 2.5 includes an unrealistically low number of cases, but it is still difficult to discern patterns. For example, try to ascertain the general level of satisfaction from this table. You may be able to do so with just 20 cases, but it will take some time and effort. Imagine the difficulty with 50 cases or 100 cases presented in this fashion. Clearly, the data need to be organized so the researcher (and his or her audience) can easily understand the distribution of the variables.

One general rule that applies to all frequency distributions, regardless of the level of measurement of the variable, is that the categories of the

TABLE 2.5 Data from Counseling Center Survey

Student	Sex	Marital Status	Satisfaction with Services	Age
A	Male	Single	4	18
B	Male	Married	2	19
C	Female	Single	4	18
D	Female	Single	2	19
E	Male	Married	1	20
F	Male	Single	3	20
G	Female	Married	4	18
H	Female	Single	3	21
I	Male	Single	3	19
J	Female	Divorced	3	23
K	Female	Single	3	24
L	Male	Married	3	18
M	Female	Single	1	22
N	Female	Married	3	26
O	Male	Single	3	18
P	Male	Married	4	19
Q	Female	Married	2	19
R	Male	Divorced	1	19
S	Female	Divorced	3	21
T	Male	Single	2	20

frequency distribution must be exhaustive and mutually exclusive. In other words, the categories must be stated so each case can be sorted into one and only one category.

Beyond this rule, there are only guidelines to help you construct useful frequency distributions. The researcher has a fair amount of discretion in stating the categories of the frequency distribution (especially with variables measured at the interval-ratio level). I will identify the issues, but ultimately, the guidelines are aids for decision making and helpful suggestions. As always, the researcher has the final responsibility for making sensible decisions and presenting his or her data in a meaningful way.

Frequency Distributions for Variables Measured at the Nominal Level

Constructing frequency distributions for nominal-level variables is typically very straightforward. Count the number of times each category or score of the variable occurred and then display the frequencies in table format. Table 2.6

displays a frequency distribution for the variable "*sex*" from the counseling center survey. For purposes of illustration, a column for tallies has been included in this table to illustrate how the score could be counted. This column would *not* be included in the final form of the frequency distribution. Note that the table has a title, clearly labeled categories (male and female), and reports the total number of cases (N) at the bottom of the frequency column. These items must be included in *all* frequency distributions. The meaning of the table is quite clear. There are 10 males and 10 females in the sample—a fact that is much easier to comprehend from the frequency distribution than from the unorganized data presented in Table 2.5.

For some nominal variables, the researcher might have to make some choices about the number of categories included. For example, Table 2.7 presents the distribution of the variable "*marital status*" by using the categories listed in Table 2.5, but, if the researcher wants to focus only on married and unmarried students, categories could be collapsed like in Table 2.8. Notice that, when categories are grouped, information and detail will be lost. Table 2.8 would not allow the researcher to distinguish between the two unmarried states.

TABLE 2.6 Sex of Respondents, Counseling Center Survey

Sex	Tallies	Frequency (f)
Male	~~||||~~ ~~||||~~	10
Female	~~||||~~ ~~||||~~	10
		$N = 20$

TABLE 2.7 Marital Status of Respondents, Counseling Center Survey

Status	Frequency (f)
Single	10
Married	7
Divorced	3
	$N = 20$

TABLE 2.8 Marital Status of Respondents, Counseling Center Survey

Status	Frequency (f)
Married	7
Not married	13
	$N = 20$

Frequency Distributions for Variables Measured at the Ordinal Level

Frequency distributions for ordinal-level variables are constructed in the same way as for nominal-level variables. Table 2.9 reports the frequency distribution of "satisfaction" from the counseling center survey. Note that a column of percentages by category has been added to this table. Such columns heighten the clarity of the table (especially with larger samples) and are common adjuncts to the basic frequency distribution for variables measured at all levels. This table reports that most students were either satisfied or very satisfied with the services of the counseling center. The most common response (nearly half the sample) was "satisfied." If the researcher wanted to emphasize this major trend, the categories could be collapsed like in Table 2.10. Again, the price paid for this increased compactness is that some information (in this case, the exact breakdown of degrees of satisfaction and dissatisfaction) is lost. *(For practice in constructing and interpreting frequency distributions for nominal- and ordinal-level variables, see problem 2.5.)*

Frequency Distributions for Variables Measured at the Interval-Ratio Level

In general, the construction of frequency distributions for variables measured at the interval-ratio level is more complex than for nominal and ordinal variables. Interval-ratio variables usually have a wide range of scores, and this means that the researcher must collapse or group categories to produce reasonably compact

TABLE 2.9 Satisfaction with Services, Counseling Center Survey

Satisfaction	Frequency (*f*)	Percentage (%)
(4) Very satisfied	4	20
(3) Satisfied	9	45
(2) Dissatisfied	4	20
(1) Very dissatisfied	3	15
	$N = 20$	100%

TABLE 2.10 Satisfaction with Services, Counseling Center Survey

Satisfaction	Frequency (*f*)	Percentage (%)
Satisfied	13	65
Dissatisfied	7	35
	$N = 20$	100%

tables. To construct frequency distributions for interval-ratio-level variables, you must decide how many categories to use and how wide these categories should be.

For example, suppose you wished to report the distribution of the variable "*age*" for a sample drawn from a community. Unlike the college data reported in Table 2.5, a community sample would have a very broad range of ages. If you simply reported the number of times that each year of age (or score) occurred, you could easily wind up with a frequency distribution that contained 80, 90, or even more categories and would be very difficult to read. The scores (years) must be grouped into larger categories to increase the ease of comprehension. How large should these categories be? How many categories should be included in the table? Although there are no hard-and-fast rules, these decisions always involve a trade-off between more detail (a greater number of narrow categories) or more compactness (a smaller number of wide categories).

Constructing the Frequency Distribution

We will construct a frequency distribution for the ages of the students in the counseling center survey to demonstrate the decision-making processes for frequency distributions for interval-ratio-level variables. Because of the narrow age range of a group of college students, we can use categories of only one year (these categories are often called **class intervals** when working with interval-ratio data). The frequency distribution is constructed by listing the ages in order, counting the number of times each score (year of age) occurs, and then totaling the number of scores for each category. Table 2.11 presents the information and reveals a concentration or clustering of scores in the 18 and 19 class intervals.

Even though this table is fairly clear, assume for the sake of illustration that you desire a more compact (less detailed) summary. To do this, you will have to group scores into wider class intervals. Increasing the interval width (say, to two years) will reduce the number of intervals and achieve a more compact expression. The grouping of scores in Table 2.12 clearly emphasizes the predominance

TABLE 2.11 Age of Respondents, Counseling Center Survey (interval width = one year of age)

Class Intervals	Frequency (f)
18	5
19	6
20	3
21	2
22	1
23	1
24	1
25	0
26	1
	$N = 20$

TABLE 2.12 Age of Respondents, Counseling Center Survey (interval width = two years of age)

Class Intervals	Frequency (*f*)	Percentage (%)
18–19	11	55
20–21	5	25
22–23	2	10
24–25	1	5
26–27	1	5
	$N = 20$	100%

of younger respondents. This trend in the data can be stressed even more by the addition of a column displaying the percentage of cases in each category.

Stated Limits

Note that the class intervals in Table 2.12 have been stated with an apparent gap between them (that is, the **stated class limits** are separated by a distance of one unit). At first glance, these gaps may appear to violate the principle of exhaustiveness, but because age has been measured in whole numbers, the gaps are actually not a problem. Given the level of precision of the measurement (in whole years, as opposed to, say, tenths of a year), no case could have a score falling between these class intervals. For these data, the set of class intervals in Table 2.12 is exhaustive and mutually exclusive. Each of the 20 respondents in the sample can be sorted into one and only one age category.

However, consider the potential difficulties if age had been measured with greater precision. If age had been measured in tenths of a year, into which class interval in Table 2.12 would a 19.4-year-old subject be placed? You can avoid this ambiguity by always stating the limits of the class intervals at the same level of precision as the data. Thus, if age were being measured in tenths of a year, the limits of the class intervals in Table 2.12 would be stated in tenths of a year. For example:

17.0–18.9
19.0–20.9
21.0–22.9
23.0–24.9
25.0–26.9

To maintain mutual exclusivity between categories, do not overlap the class intervals. If you state the limits of the class intervals at the same level of precision as the data (which might be in whole numbers, tenths, hundredths, etc.) and maintain a gap between intervals, you will always produce a frequency distribution in which each case can be assigned to one and only one category.

Midpoints

On occasion for example, when constructing certain graphs—you will need to work with the **midpoints** of the class intervals. A midpoint is the point exactly

ONE STEP AT A TIME **Finding Midpoints**

Step	*Operation*
1.	Find the upper and lower limits of the lowest interval in the frequency distribution. For any interval, the upper limit is the highest score included in the interval and the lower limit is the lowest score included in the interval. For example, for the top set of intervals in Table 2.13, the lowest interval (0–2) includes scores of 0, 1, and 2. The upper limit of this interval is 2 and the lower limit is 0.
2.	Add the upper and lower limits and then divide by 2. For the interval 0–2: (0 + 2)/2 = 1. The midpoint for this interval is 1.
3.	Midpoints for other intervals can be found by repeating steps 1 and 2 for each interval. As an alternative, you can find the midpoint for any interval by adding the value of the interval width to the midpoint of the next lower interval. For example, the lowest interval in Table 2.13 is 0–2 and the midpoint is 1. Intervals are 3 units wide (that is, they each include three scores), so the midpoint for the next higher interval (3–5) is 1 + 3, or 4. The midpoint for the interval 6–8 is 4 + 3, or 7, and so forth.

halfway between the upper and lower limits of a class interval and can be found by dividing the sum of the upper and lower limits by 2. Table 2.13 displays midpoints for two different sets of class intervals. *(For practice in finding midpoints, see problems 2.8b and 2.9b.)*

Cumulative Frequency and Cumulative Percentage

Two commonly used adjuncts to the basic frequency distribution for interval-ratio data are the **cumulative frequency** and **cumulative percentage** columns. These additional columns allow us to tell at a glance how many cases fall below a given score or class interval in the distribution. To add these columns, follow

TABLE 2.13 Midpoints

Class Interval Width = 3	
Class Intervals	Midpoints
0–2	1.0
3–5	4.0
6–8	7.0
9–11	10.0
Class Interval Width = 6	
Class Intervals	Midpoints
100–105	102.5
106–111	108.5
112–117	114.5
118–123	120.5

ONE STEP AT A TIME **Adding Cumulative Frequency and Percentage Columns to Frequency Distributions**

Step ***Operation***

To Add the Cumulative Frequency Column

1. Begin with the lowest class interval (i.e., the class interval with the lowest scores). The entry in the cumulative frequency columns for that interval will be the same as the number of cases in the interval.
2. Go to the next class interval. The cumulative frequency for this interval is the number of cases in the interval plus all the cases in the lower interval.
3. Continue adding (or accumulating) cases from interval to interval until you reach the interval with the highest scores, which will have a cumulative frequency equal to *N*.

To Add the Cumulative Percentage Column

1. Compute the percentage of cases in each category one at a time, and then follow the pattern for cumulative frequencies. The cumulative percentage for the lowest class interval will be the same as the percentage of cases in the interval.
2. For the next-higher interval, the cumulative percentage is the percentage of cases in the interval plus the percentage of cases in the lower interval.
3. Continue adding (or accumulating) percentages from interval to interval until you reach the interval with the highest scores, which will have a cumulative percentage of 100%.

TABLE 2.14 Age of Respondents, Counseling Center Survey

Class Intervals	Frequency (*f*)	Cumulative Frequency
18–19	11	11
20–21	5	16
22–23	2	18
24–25	1	19
26–27	1	20
	$N = 20$	

the instructions in the "One Step at a Time" box. Table 2.14 shows a cumulative frequency column added to Table 2.12, and Table 2.15 adds a cumulative percentage column to the same distribution.

These cumulative columns are quite useful in situations where the researcher wants to make a point about how cases are spread across the range of scores. For example, Tables 2.14 and 2.15 clearly show that most students in the counseling center survey are less than 21 years of age. If the researcher wishes to stress this fact, then these cumulative columns are quite handy. Most realistic research situations will be concerned with many more than 20 cases and/or many more categories than our tables have, and the cumulative percentage column would usually be preferred to the cumulative frequencies column.

TABLE 2.15 Age of Respondents, Counseling Center Survey

Class Intervals	Frequency (*f*)	Cumulative Frequency	Percentage	Cumulative Percentage
18–19	11	11	55%	55%
20–21	5	16	25%	80%
22–23	2	18	10%	90%
24–25	1	19	5%	95%
26–27	1	20	5%	100%
	N = 20		100%	

Unequal Class Intervals

As a general rule, the class intervals of frequency distributions should be equal in size in order to maximize clarity and ease of comprehension. For example, note that all the class intervals in Tables 2.14 and 2.15 are the same width (2 years). However, there are several other possibilities for stating class interval, and we will examine each situation separately.

Open-Ended Intervals. What would happen to the frequency distribution in Table 2.14 if we added one more student who was 47 years of age? We would now have 21 cases, and there would be a large gap between the oldest respondent (now 47) and the second oldest (age 26). If we simply added the older student to the frequency distribution, we would have to include nine new class intervals (28–29, 30–31, 32–33, etc.) with zero cases in them before we got to the 46–47 interval. This would waste space and probably be unclear and confusing. An alternative way to handle the situation where we have a few very high or low scores would be to add an open-ended interval to the frequency distribution, as in Table 2.16.

The open-ended interval in Table 2.16 allows us to present the information more compactly than listing all the empty intervals from 28–29 through 46–47. We could handle an extremely low score by adding an open-ended interval as the lowest class interval (e.g., "17 and younger"). There is a small price to pay for this efficiency (there is no information in Table 2.16 about the exact scores included in the open-ended interval), so this technique should not be used indiscriminately.

TABLE 2.16 Age of Respondents, Counseling Center Survey (*N* = 21)

Class Intervals	Frequency (*f*)	Cumulative Frequency
18–19	11	11
20–21	5	16
22–23	2	18
24–25	1	19
26–27	1	20
28 and older	1	21
	N = 21	

Intervals of Unequal Size. For some variables, most scores are tightly clustered together but others are strewn across a broad range of scores. As an example, consider the distribution of income in the United States. In 2010, most households (almost 56%) reported annual incomes between $25,000 and $100,000 and a sizable grouping (about 23%) earned less than that. The problem (from a statistical viewpoint) comes with more affluent households with incomes above $100,000. The number of these households is quite small, of course, but we must still account for these cases.

If we tried to use a frequency distribution with equal intervals of, say, $10,000, we would need 30 or 40 or more intervals to include all of the more affluent households, and many of our intervals in the higher income ranges—especially those over $150,000—would have few or zero cases. In situations such as this, we can use intervals of unequal size to summarize the variable more efficiently, as in Table 2.17.

Some of the intervals in Table 2.17 are $10,000 wide, others are $25,000 or $50,000 wide, and two (the lowest and highest intervals) are open ended. Tables that use intervals of mixed widths might be a little confusing for the reader, but the trade-off in compactness and efficiency can be considerable. *(For practice in constructing and interpreting frequency distributions for interval-ratio-level variables, see problems 2.5 to 2.9.)*

TABLE 2.17 Distribution of Income by Household, United States, 2009

Income Categories	Number of Households (Frequencies)	Number of Households (Percent)
Less than $10,000	8,329,488	7.40
$10,000 to $14,999	6,305,311	5.60
$15,000 to $19,999	6,024,160	5.35
$20,000 to $24,999	6,147,899	5.46
$25,000 to $29,999	6,024,956	5.35
$30,000 to $34,999	5,960,273	5.29
$35,000 to $39,999	5,601,938	4.97
$40,000 to $44,999	5,501,951	4.89
$45,000 to $49,999	4,960,432	4.40
$50,000 to $59,999	9,385,944	8.33
$60,000 to $74,999	11,853,169	10.36
$75,000 to $99,999	13,853,787	12.30
$100,000 to $124,999	8,639,394	7.67
$125,000 to $149,999	4,939,327	4.39
$150,000 to $199,999	4,724,616	4.20
$200,000 or more	4,544,384	4.04
	112,611,029	100.00

Source: U.S. Census Bureau. *American Community Survey, 2005–2009.* Retrieved June 25, 2010, from http://factfinder.census.gov/servlet/DTTable?_bm=y&-geo_id=01000US&-ds_name=ACS_2009_5YR_G00_&-_lang=en&-mt_name=ACS_2009_5YR_G2000_B19001&-format=&-CONTEXT=dt.

Frequency Distributions for Interval-Ratio-Level Variables: An Example

We covered a lot of ground in the preceding section, so let us pause and review these principles by considering a specific research situation. The following data represent the numbers of visits received over the past year by 90 residents of a retirement community.

0	52	21	20	21	24	1	12	16	12
16	50	40	28	36	12	47	1	20	7
9	26	46	52	27	10	3	0	24	50
24	19	22	26	26	50	23	12	22	26
23	51	18	22	17	24	17	8	28	52
20	50	25	50	18	52	46	47	27	0
32	0	24	12	0	35	48	50	27	12
28	20	30	0	16	49	42	6	28	2
16	24	33	12	15	23	18	6	16	50

Listed in this format, the data are a hopeless jumble from which no one could derive much meaning. The function of the frequency distribution is to arrange and organize these data so their meanings will be made obvious.

First, we must decide how many class intervals to use in the frequency distribution. Following the guidelines presented in the "One Step at a Time: Constructing Frequency Distributions for Interval-Ratio Variables" box, let us use 10 intervals ($k = 10$). By inspecting the data, we can see that the lowest score is 0 and the highest is 52. The range of these scores (R) is $52 - 0$, or 52. To find the approximate interval size (i), divide the range (52) by the number of intervals (10). Because $52/10 = 5.2$, we can set the interval size at 5.

The lowest score is 0, so the lowest class interval will be 0–4. The highest class interval will be 50–54, which will include the high score of 52. All that remains is to state the intervals in table format, count the number of scores that fall into each interval, and report the totals in a frequency column. The results of these steps are shown in Table 2.18, which also includes columns for the percentages and cumulative percentages. Note that this table is the product of several relatively arbitrary decisions. The researcher should remain aware of this fact and inspect the frequency distribution carefully. If the table is unsatisfactory for any reason, it can be reconstructed with a different number of categories and interval sizes.

Now, with the aid of the frequency distribution, some patterns in the data can be discerned. There are three distinct groupings of scores in the table. Ten residents were visited rarely—if at all (the 0–4 visits per year interval). The single-largest interval, with 18 cases, is 20–24. Combined with the intervals immediately above and below, this represents quite a sizeable grouping of cases (42 out of 90, or 46.66% of all cases) and suggests that the dominant visiting rate is about twice a month, or approximately 24 visits per year. The third grouping, in the 50–54 class interval (12 cases), reflects a visiting rate of about once a week. The cumulative percentage column indicates that the majority of the residents (58.89%) were visited 24 or fewer times a year.

TABLE 2.18 Number of Visits Per Year, 90 Retirement Community Residents

Class Intervals	Frequency (*f*)	Cumulative Frequency	Percentage (%)	Cumulative Percentage
0–4	10	10	11.11	11.11
5–9	5	15	5.56	16.67
10–14	8	23	8.89	25.56
15–19	12	35	13.33	38.89
20–24	18	53	20.00	58.89
25–29	12	65	13.33	72.22
30–34	3	68	3.33	75.55
35–39	2	70	2.22	77.77
40–44	2	72	2.22	79.99
45–49	6	78	6.67	86.66
50–54	12	90	13.33	99.99
	$N = 90$		99.99%*	

*Percentage columns will occasionally fail to total 100% because of a rounding error. If the total is between 99.90% and 100.10%, ignore the discrepancy. Discrepancies of greater than ±0.10% may indicate mathematical errors, and the entire column should be computed again.

Applying Statistics 2.5 Frequency Distributions

The following list shows the ages of 50 prisoners enrolled in a work-release program. Is this group young or old? A frequency distribution will provide an accurate picture of the overall age structure.

18	60	57	27	19
20	32	62	26	20
25	35	75	25	21
30	45	67	41	30
37	47	65	42	25
18	51	22	52	30
22	18	27	53	38
27	23	32	35	42
32	37	32	40	45
55	42	45	50	47

We will use about 10 intervals to display this data. By inspection, we see that the youngest prisoner is 18 and the oldest is 75. The range is thus 57. Interval size will be 57/10, or 5.7, which we can round off to either 5 or 6. Let us use a six-year interval beginning at 18. The limits of the lowest interval will be 18–23. Now we must state the limits of all other intervals, count the number of cases in each interval, and display these counts in a frequency distribution. Columns may be added for percentages, cumulative percentages, and/or cumulative frequency. The complete distribution, with a column added for percentages, is:

Ages	Frequency	Percentages
18–23	10	20%
24–29	7	14%
30–35	9	18%
36–41	5	10%
42–47	8	16%
48–53	4	8%
54–59	2	4%
60–65	3	6%
66–71	1	2%
72–77	1	2%
	$N = 50$	100%

The prisoners seem to be fairly evenly spread across the age groups—up to the 48–53 interval. There is a noticeable lack of prisoners in the oldest age groups and a concentration of prisoners in their 20s and 30s.

ONE STEP AT A TIME **Constructing Frequency Distributions for Interval-Ratio Variables**

Step	***Operation***
1.	Decide how many class intervals (k) you wish to use. One reasonable convention suggests that the number of intervals should be about 10. Many research situations may require fewer than 10 intervals ($k < 10$), and it is common to find frequency distributions with as many as 15 intervals. Only rarely will more than 15 intervals be used because the resultant frequency distribution would be too large for easy comprehension.
2.	Find the range (R) of the scores by subtracting the low score from the high score.
3.	Find the size of the class intervals (i) by dividing R (from step 2) by k (from step 1): $$i = R/k$$ Round the value of i to a convenient whole number. This will be the interval size or width.
4.	State the lowest interval so its lower limit is equal to or less than the lowest score. By the same token, your highest interval will be the one that includes the highest score. Generally, intervals should be equal in size, but unequal and open-ended intervals may be used when convenient.
5.	State the limits of the class intervals at the same level of precision as you have used to measure the data. Do not overlap intervals. You will thereby define the class intervals so each case can be sorted into one and only one category.
6.	Count the number of cases in each class interval, and report these subtotals in a column labeled '*Frequency*. Report the total number of cases (N) at the bottom of this column. The table may also include a column for percentages, cumulative frequencies, and cumulative percentages.
7.	Inspect the frequency distribution carefully. Has too much detail been lost? If so, reconstruct the table with a greater number of class intervals (or use a smaller interval size). Is the table too detailed? If so, reconstruct the table with fewer class intervals (or use wider intervals). Are there too many intervals with no cases in them? If so, consider using open-ended intervals or intervals of unequal size. Remember that the frequency distribution results from a number of decisions you make in a rather arbitrary manner. If the appearance of the table seems less than optimal given the purpose of the research, redo the table until you are satisfied that you have struck the best balance between detail and conciseness.
8.	Give your table a clear, concise title, and number the table if your report contains more than one. All categories and columns must also be clearly labeled.

Graphic Presentations of Data

Researchers frequently use charts and graphs to present their data in ways that are visually more dramatic than frequency distributions. These devices are particularly useful for conveying an impression of the overall shape of a distribution and for highlighting any clustering of cases in a particular range of scores. Many graphing techniques are available, but we will examine just four. The first two—pie and bar charts—are appropriate for variables at any level of measurement that have only a few categories. The last two—histograms and line charts (or frequency polygons)—are primarily used with interval-ratio variables with many scores.

The sections that follow explain how to construct graphs and charts "by hand." However, these days, computer programs are almost always used to produce graphic displays. Graphing software is sophisticated and flexible but also relatively easy to use; if such programs are available to you, you should familiarize yourself with them. The effort required to learn these programs will be repaid in the quality of the final product.

Pie Charts

To construct a **pie chart**, begin by computing the percentage of all cases that fall into each category of the variable. Then, divide a circle (the pie) into segments (slices) proportional to the percentage distribution. Be sure that the chart and all segments are clearly labeled.

Figure 2.1 is a pie chart that displays the distribution of marital status from the counseling center survey. The frequency distribution (Table 2.7) is reproduced as Table 2.19, with a column added for the percentage distribution. Because a circle's circumference is 360°, we will apportion 180° (or 50%) for the first category, 126° (35%) for the second, and 54° (15%) for the last category. The pie chart visually reinforces the relative preponderance of single respondents and the relative absence of divorced students in the counseling center survey.

Bar Charts

Like pie charts, **bar charts** are relatively straightforward. Conventionally, the categories of the variable are arrayed along the horizontal axis (or *abscissa*) and frequencies—or percentages, if you prefer—are arrayed along the vertical axis (or *ordinate*). For each category of the variable, construct (or draw) a rectangle of constant width and with a height that corresponds to the number of cases in the category. The bar chart in Figure 2.2 reproduces the marital status data from Figure 2.1 and Table 2.19.

Figure 2.2 would be interpreted in exactly the same way as Figure 2.1, and researchers are free to choose between these two methods of displaying data.

FIGURE 2.1 Marital Status of Respondents, Counseling Center Survey (N = 20)

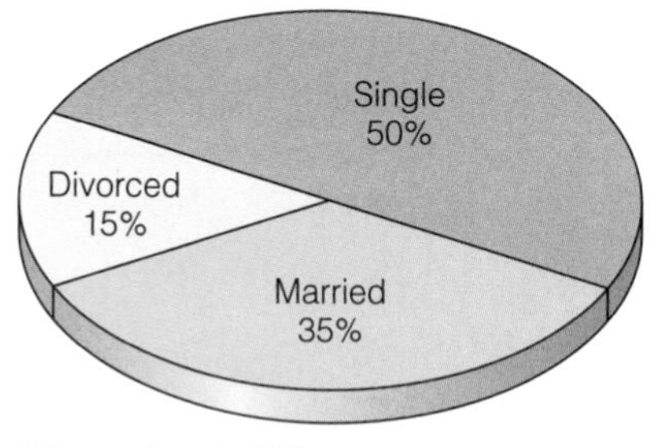

TABLE 2.19 Marital Status of Respondents, Counseling Center Survey

Status	Frequency (f)	Percentage (%)
Single	10	50
Married	7	35
Divorced	3	15
	N = 20	100%

However, if a variable has more than four or five categories, the bar chart would be preferred. With too many categories, the pie chart gets very crowded and loses its visual clarity. To illustrate, Figure 2.3 uses a bar chart to display the data on visiting rates for the retirement community presented in Table 2.18. A pie chart for this same data would have had 11 different slices—a more complex or busier picture than that presented by the bar chart. In Figure 2.3, the clustering of scores in the 20–24 range (approximately two visits a month) is readily apparent, as are the groupings in the 0–4 and 50–54 ranges.

Bar charts are particularly effective ways to display the relative frequencies for two or more categories of a variable when you want to emphasize some comparisons. For example, suppose you want to make a point about the changing rates of homicide victimization for white males and females since 1955. Figure 2.4 displays the data in a dramatic and easily comprehended way. The bar chart shows that rates for males are higher than rates for females, that rates for both sexes were highest in 1975, and that rates declined after that time. *(For practice in constructing and interpreting pie and bar charts, see problems 2.5b and 2.10.)*

FIGURE 2.2 Marital Status of Respondents, Counseling Center Survey, $N = 20$

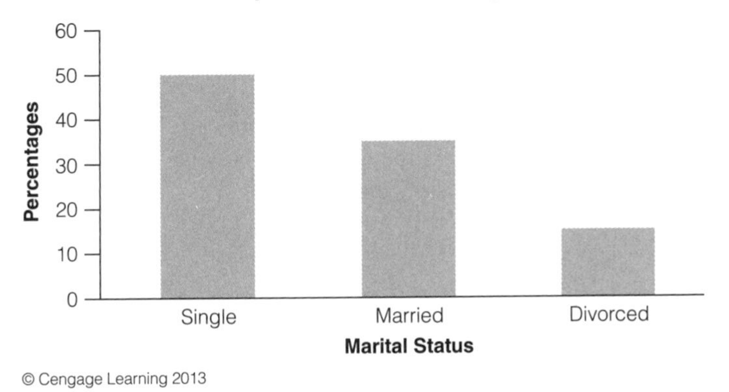

FIGURE 2.3 Visits per Year, Retirement Community Respondents, $N = 90$

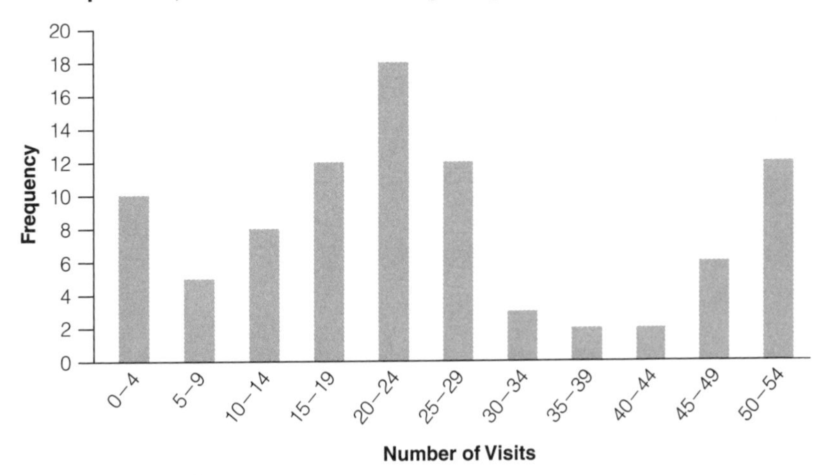

FIGURE 2.4 Homicide Victimization Rates for Males and Females, Selected Years (Whites Only)

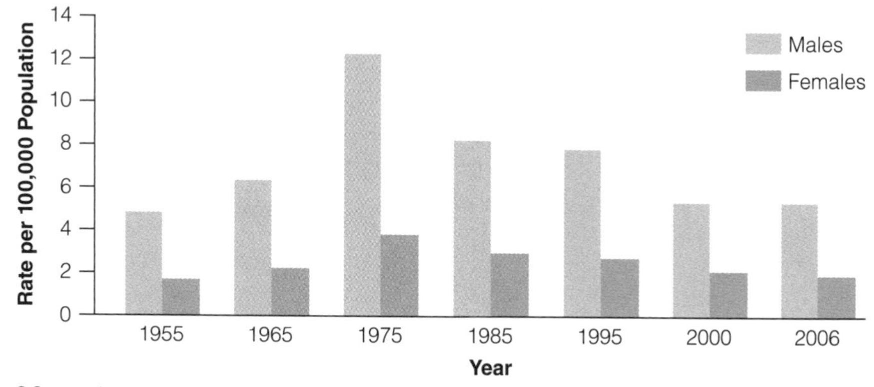

Histograms

Histograms look a lot like bar charts and, in fact, are constructed in much the same way. However, the bars in a histogram touch each other in a continuous series from the lowest to the highest scores. These graphs are most appropriate for interval-ratio-level variables that have many scores covering a wide range. To construct a histogram from a frequency distribution, follow the steps in the One Step at a Time box.

As an example, Figure 2.5 uses a histogram to display the distribution of incomes for U.S. households in 2009. This same information is presented in table format in Table 2.17. Note how the graph makes it easier to detect basic patterns. We can see a noticeable grouping of cases in the lowest income category, a relatively even number of households in the next higher income groups, and a very large grouping in the $50,000 to $100,000 range, we might call these middle-income Americans. Finally, note the gradual decline of cases in the highest income brackets.

Line Charts

Line Charts. The construction of a **line chart** (or **frequency polygon**) is similar to the construction of a histogram. However, instead of using bars to represent the frequencies, use a dot at the midpoint of each interval. Straight lines then connect the dots. Figure 2.6 displays a line chart for the visiting data previously displayed in the bar chart in Figure 2.3.

ONE STEP AT A TIME Constructing Histograms

Step	*Operation*
1.	Array the scores along the horizontal axis (abscissa).
2.	Array frequencies along the vertical axis (ordinate).
3.	For each score or class interval in the frequency distribution, construct a bar with the height corresponding to the number of cases in the category.
4.	Label each axis of the graph.
5.	Title the graph.

FIGURE 2.5 Distribution of Income by Household, United States, 2009

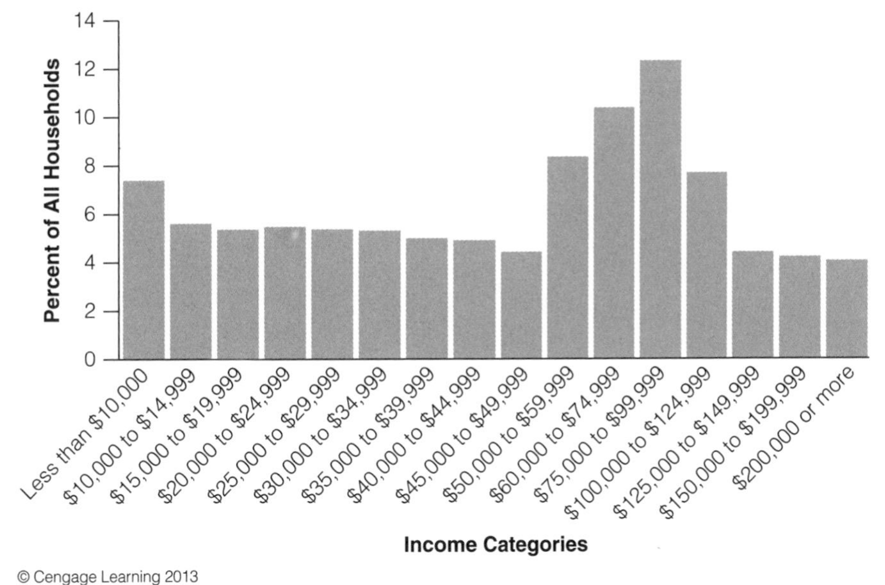

FIGURE 2.6 Number of Visits per Year. Retirement Community Residents

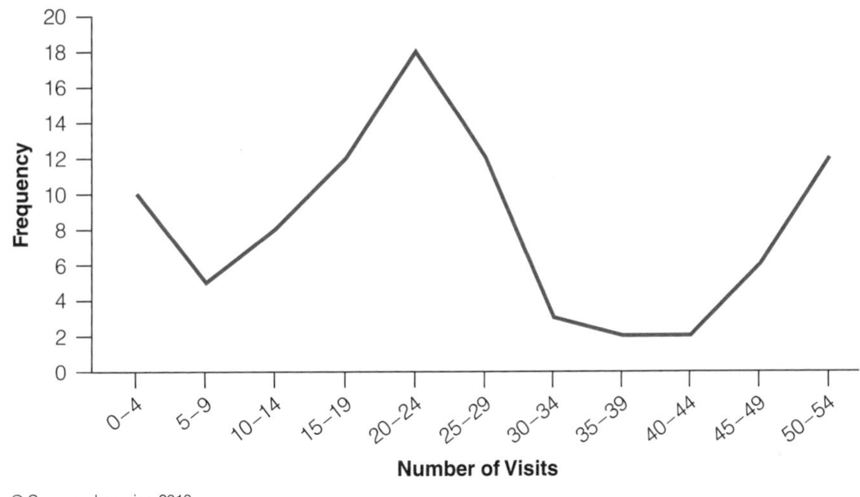

Line charts can also be used to display trends across time. Figure 2.7 shows the total number of newspapers published in the United States from 1940 through 2009 and displays a dramatic decline in print journalism over this time period. Note that the vertical axis begins at 1,000, not zero. This makes the graph more compact by eliminating empty white space at the bottom.

The decline in newspapers is quite obvious in Figure 2.7. The numbers fell from almost 1,900 to less than 1,400. Remember that the population of the United States increased dramatically over this same time period. What would the chart look like if we plotted number of newspapers per capita rather than the raw numbers?

FIGURE 2.7 Number of Newspapers Published in the United States, 1940–2009

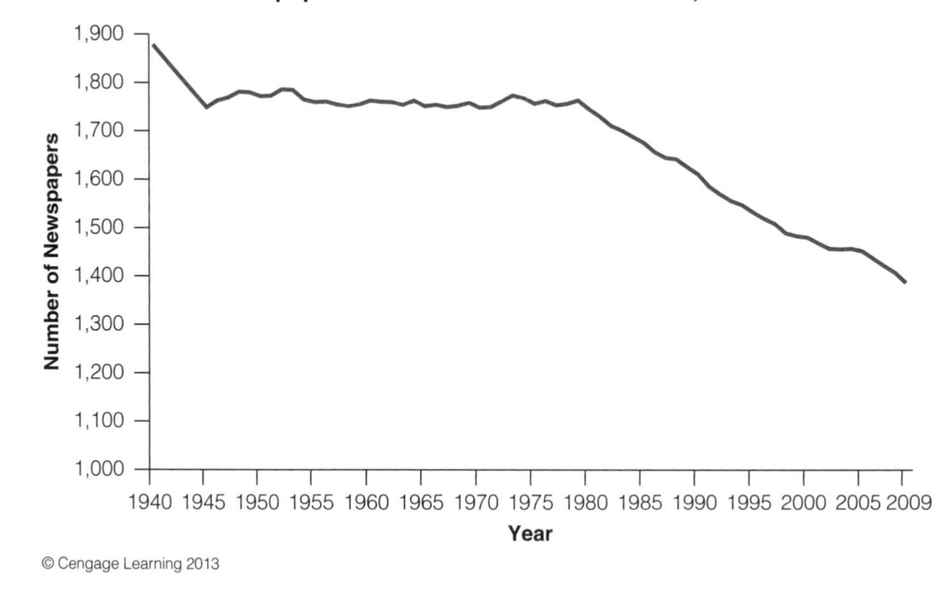

STATISTICS IN EVERYDAY LIFE

There is a great deal of concern in U.S. society about the future of the American family and, more specifically, our high divorce rate. The U.S. divorce rate is higher than that of other Western industrialized nations (and almost twice the Canadian rate) but has been falling since the early 1980s. Why? One reason is that the marriage rate has also been falling—you cannot get divorced unless you get married! Figure 2.8 displays both trends.

FIGURE 2.8 U.S. Rates of Marriage and Divorce, 1950–2008 (rates per 100,000 population)

This information could have been presented in a table. What are the advantages of the picture presented in a graph? For example, can you see a difference in the rate of decline of the two lines? Is one rate falling faster than the other? What are the implications of this difference? What else does the graph imply about family life in the United States? What sociological variables might help to account for these changes?

Histograms and frequency polygons are alternative ways of displaying essentially the same message. Thus, the choice between the two techniques is left to the aesthetic pleasures of the researcher. *(For practice in constructing and interpreting histograms and line charts, see problems 2.7b, 2.8d, 2.9d, 2.11, and 2.12.)*

BECOMING A CRITICAL CONSUMER: Urban Legends, Road Rage, and Context

The statistics covered in this chapter may seem simple and even humble. However, as with any tool, they can be misunderstood and applied inappropriately. Here, we will examine some ways in which these statistics can be misused and abused and also reinforce some points about their usefulness in communicating information. We will finish by examining the ways in which percentages and rates are used in the professional research literature.

First of all, by themselves, statistics guarantee nothing about the accuracy or validity of a statement. False information, such as so-called "urban legends," can be expressed statistically, and this may enhance their credibility in the eyes of many people. For example, consider the legend that domestic violence increases on Super Bowl Sunday, the day the championship game in American football is played. You may have heard a variation of this report that used specific percentages (for example: "Admissions to domestic abuse shelters rise by 40% in the city of the team that loses the Super Bowl"). The credibility of reports like these stems partly from the close association between football "macho" values, aggression, and violence. Also, people celebrate the Super Bowl with parties and gatherings, during which large quantities of alcohol and other substances may be consumed. It seems quite reasonable that this heady mixture of macho values and alcohol would lead to higher rates of domestic violence.

The problem is that there is no evidence of a connection between the Super Bowl and spouse abuse. Two different studies, conducted at different times and locations, found no increase in spouse abuse on Super Bowl Sunday.[1] Of course, a connection may still exist at some level or in some form between football and domestic violence, maybe we just have not found it. My point, though, is that the mere presence of seemingly exact percentages (or any other statistic) is no guarantee of accuracy or validity. Incorrect and even outrageously wrong information can (and does) seep into everyday conversation and become part of what "everyone knows" to be true.

Of course, even "true" statistics and solid evidence can be misused. This brings up a second point: the need to carefully examine the context in which the statistic is reported. Sometimes, exactly the same statistical fact can be made to sound alarming and scary or trivial and uninteresting simply by changing the context in which the information is imbedded. To illustrate, consider the phenomena of "road rage" or aggressive driving. Angry drivers have been around since the invention of the automobile (and maybe since the invention of the wheel), but the term "road rage" entered the language in the mid-1990s, sparked by several violent incidents on the nation's highways and a frenzy of media coverage. We will follow sociologist Barry Glassner's (1999)[2] analysis of road rage and look first at what the media reported at that time and then look at context.

Beginning in the mid-1990s, the media began to characterize road rage as a "growing American danger," an "exploding phenomenon," and a "plague" (Glassner, 1999, pp. 3–5). One widely cited statistic was that incidents of road rage rose almost 60% between 1990 and 1996. This percentage change was based on two numbers: There were 1,129 "road rage" incidents in 1990 and 1,800 in 1996. Thus so, the increase is:

$$\text{Percent Change} = \left(\frac{f_2 - f_1}{f_1}\right) \times 100$$

$$\text{Percent Change} = \left(\frac{1{,}800 - 1{,}129}{1{,}129}\right) \times 100$$

$$\text{Percent Change} = \left(\frac{671}{1{,}129}\right) \times 100$$

$$\text{Percent Change} = 59.43\%$$

(continued next page)

BECOMING A CRITICAL CONSUMER: (Continued)

The media reported the percentage increase—not the frequency of incidents—and a 60% increase certainly seems to justify the characterization of road rage as "an exploding phenomenon." However, in this case, it is the raw frequencies that are actually the crucial pieces of information.

Note how the perception of the percentage increase in road rage changes when it is framed in a broader context:

- Between 1990 and 1996, there were 20 million injuries from traffic accidents and about 250,000 fatalities on U.S. roadways.
- In this same period, the total number of road rage incidents was 11,000.
- Alcohol is involved in about 1 out of 2 traffic fatalities; road rage in about one in 1,000.

In the context of the total volume of traffic mayhem, injury, and death (and alcohol-related incidents), is it reasonable to label road rage a "plague"?

As Professor Glassner says, "big percentages don't always have big numbers behind them" (Glassner, 1999, p. 5). Road rage represents a miniscule danger compared to drunk driving. In fact, concern about road rage may actually be harmful if it deflects attention from more serious problems. Considered in isolation, the increase in road rage seems very alarming. When viewed against the total volume of traffic injury and death, the problem fades in significance.

In a related point, you should consider the time frame used to report changes in statistical trends. Consider that the homicide rate (number of homicides per 100,000 population) in the United States went up by 3.6% between 2000 and 2006—a change that could cause concern and increased fear in the general public. However, over a different time frame—from 1980 to 2010—the homicide rate actually declined almost 50%.[3] Thus, different time frames can lead to different conclusions about the dangers of living in the United States.

Finally, let us consider the proper use of these statistics as devices for communicating facts clearly and simply. We will use an example from professional social science research to make this final point. Social scientists rely heavily on the U.S. census for information about the characteristics and trends of change in American society, including age composition, birth and death rates, residential patterns, educational levels, and a host of other variables. Census data is readily available (at www.census.gov), but because the census [to fix dangling modifier] presents information about the entire population (over 300 million people), the numbers are often large, cumbersome, and awkward to use or understand. Percentages and rates are extremely useful ways of analyzing and presenting census information.

For example, suppose a report on voter turnout in the United States included this statement:

> In 1996, 104,985,400 of the 193,700,000 eligible voters actually turned out to cast their ballots in the national election. In 2008, on the other hand, 131,241,000 of the 225,500,000 Americans over 18 voted[4].

Can you distill any meaningful understandings about American politics from these sentences? Raw information simply does not speak for itself and these facts have to be organized or placed in some context to reveal their meaning. Thus, social scientists almost always use percentages or rates to present this kind of information so that they can understand it themselves, assess the meaning, and convey their interpretations to others.

In contrast with the raw information above, consider this

- The percentage of eligible voters in national elections who actually went to the polls increased from 54.2% in 1996 to 58.2% in 2008.

The second paragraph actually contains less information—because it omits the raw numbers and these are very important—but is much easier to comprehend.

Finally, remember that most research projects analyze interrelationships among many variables. Because the statistics covered in this chapter summarize variables one at a time, they are unlikely to be included in such research reports (or perhaps, included only as background information). Even when they are not reported, you can be sure that the research began with an inspection of percentages and frequency distributions for each variable.

NOTES:

1. See Oths, K., and Robertson, T. (2007). "Give Me Shelter: Temporal Patterns of Women Fleeing Domestic Abuse." *Human Organization* 66: 249–260; and Sachs, C., and Chu, L. 2000. "The Association Between Professional Football Games and Domestic Violence in Los Angeles County." *Journal of Interpersonal Violence* 15: 1192–1201. See also http://www.snopes.com/crime/statistics/superbowl.asp
2. Glassner, Barry (1999). *The Culture of Fear: Why Americans Are Afraid of the Wrong Things.* Basic Books: New York.
3. U.S. Bureau of the Census (2011). *Statistical Abstract of the United States, 2011*, p. 194.
4. U.S. Bureau of the Census (2011), *Statistical Abstract of the United States, 2011*, p. 260.

SUMMARY

1. We considered several different ways of summarizing the distribution of a single variable and, more generally, reporting the results of our research. Our emphasis throughout was on the need to communicate our results clearly and concisely. As you strive to communicate statistical information to others, you will often find that the meanings of the information will also become clearer to you.
2. Percentages and proportions, ratios, rates, and percentage change represent several different ways to enhance clarity by expressing our results in terms of relative frequency. Percentages and proportions report the relative occurrence of some category of a variable compared with the distribution as a whole. Ratios compare two categories with each other, and rates report the actual occurrences of some phenomenon compared with the number of possible occurrences per some unit of time. Percentage change shows the relative increase or decrease in a variable over time.
3. Frequency distributions are tables that summarize the entire distribution of some variable. Statistical analysis almost always starts with the construction and review of these tables for each variable. Columns for percentages, cumulative frequency, and/or cumulative percentages often enhance the readability of frequency distributions.
4. Pie and bar charts, histograms, and line charts (or frequency polygons) are graphs used to express the basic information contained in the frequency distribution in a compact and visually dramatic way.

SUMMARY OF FORMULAS

FORMULA 2.1 Percentages $\% = \left(\frac{f}{N}\right) \times 100$

FORMULA 2.2 Proportions $p = \frac{f}{N}$

FORMULA 2.3 Ratios $\text{Ratio} = \frac{f_1}{f_2}$

FORMULA 2.4 Percent Change $\text{Percent Change} = \left(\frac{f_2 - f_1}{f_1}\right) \times 100$

GLOSSARY

Bar chart. A graph for nominal- and ordinal-level variables. Categories are represented by bars of equal width, the height of each corresponding to the number (or percentage) of cases in the category.

Class intervals. The categories used in the frequency distributions for interval-ratio variables.

Cumulative frequency. An optional column in a frequency distribution that displays the number of cases within an interval and all preceding intervals.

Cumulative percentage. An optional column in a frequency distribution that displays the percentage of cases within an interval and all preceding intervals.

Frequency distribution. A table that displays the number of cases in each category of a variable.

Frequency polygon. A graph for interval-ratio variables. Class intervals are represented by dots placed over the midpoints—the height of each corresponding to the number (or percentage) of cases in the interval. All dots are connected by straight lines. Same as a **line chart**.

Histogram. A graph for interval-ratio variables. Class intervals are represented by contiguous bars of equal width—the height of each corresponding to the number (or percentage) of cases in the interval.

Line chart. See **Frequency polygon**.

Midpoint. The point exactly halfway between the upper and lower limits of a class interval.

Percentage. The number of cases in a category of a variable divided by the number of cases in all categories of the variable, and then multiplied by 100.

Percent change. A statistic that expresses the magnitude of change in a variable from time 1 to time 2.

Pie chart. A graph for nominal- and ordinal-level variables with only a few categories. A circle (the pie) is divided into segments proportional in size to the percentage of cases in each category of the variable.

Proportion. The number of cases in one category of a variable divided by the number of cases in all categories of the variable.

Rate. The number of actual occurrences of some phenomenon or trait divided by the number of possible occurrences per some unit of time.

Ratio. The number of cases in one category divided by the number of cases in some other category.

Stated class limits. The class intervals of a frequency distribution when stated as discrete categories.

PROBLEMS

Problems are labeled with the social science discipline from which they are drawn: SOC is for sociology, SW is for social work, PS is for Political Science, CJ is for Criminal Justice, PA is for Public Administration, and GER is for Gerontology.

2.1 SOC The tables that follow report the marital status of 20 respondents in two different apartment complexes.

Status	Complex A	Complex B
Married	5	10
Unmarried ("living together")	8	2
Single	4	6
Separated	2	1
Widowed	0	1
Divorced	1	0
	20	20

Read each of the following problems carefully before constructing the fraction and solving for the answer. (*HINT: Make sure you have the correct numbers in the numerator and denominator before solving these problems. For example, problem 2.1a asks which "percentage of the respondents at each complex are married" and the denominators will be 20 for these two fractions. On the other hand, problem 2.1d, asks which "percentage of the single respondents live in Complex B" and the denominator for this fraction will be 4 + 6, or 10.*)

a. What percentage of the respondents at each complex are married?

b. What is the ratio of single to married respondents at each complex?

c. What proportion of each sample are widowed?

d. What percentage of the single respondents live in Complex B?

e. What is the ratio of the unmarried/living together to the married at each complex?

2.2 SOC At St. Algebra College, the numbers of males and females in the various major fields of study are:

Major	Males	Females	Totals
Humanities	117	83	200
Social sciences	97	132	229
Natural sciences	72	20	92
Business	156	139	295
Nursing	3	35	38
Education	30	15	45
Totals	475	424	899

Read each of the following problems carefully before constructing the fraction and solving for the answer. *(HINT: Be sure you place the proper number in the denominator of the fractions. For example, some problems use the total number of males or females as the denominator, whereas others use the total number of majors.)*

a. What percentage of social science majors are male?
b. What proportion of business majors are female?
c. For the humanities, what is the ratio of males to females?
d. What percentage of the total student body are males?
e. What is the ratio of males to females for the entire sample?
f. What proportion of the nursing majors are male?
g. What percentage of the sample are social science majors?
h. What is the ratio of humanities majors to business majors?
i. What is the ratio of female business majors to female nursing majors?
j. What proportion of the males are education majors?

2.3 CJ The town of Shinbone, Kansas, has a population of 211,732 and experienced 47 bank robberies, 13 murders, and 23 auto thefts during the past year. Compute a rate for each type of crime per 100,000 population. (*HINT: Make sure you set up the fraction with size of population in the denominator.*)

2.4 CJ The numbers of homicides in five states and five Canadian provinces for the years 1997 and 2009 were:

	1997		2009	
State	Homicides	Population	Homicides	Population
New Jersey	338	8,053,000	319	8,708,000
Iowa	52	2,852,000	34	3,008,000
Alabama	426	4,139,000	323	4,709,000
Texas	1327	19,439,000	1328	24,782,000
California	2579	32,268,000	1972	36,962,000

Source: Federal Bureau of Investigation. *Uniform Crime Reports.* http://www2.fbi.gov/ucr/cius2009/data/table_04.html.

	1997		2009	
Province	Homicides	Population	Homicides	Population
Nova Scotia	24	936,100	15	942,500
Quebec	132	7,323,600	88	7,907,400
Ontario	178	11,387,400	178	13,310,700
Manitoba	31	1,137,900	57	1,235,400
British Columbia	116	3,997,100	118	4,531,000

Source: Statistics Canada. http://www.statcan.gc.ca

a. Calculate the homicide rate per 100,000 population for each state and each province for each year. Relatively speaking, which state and which province had the highest homicide rates in each year? Which nation seems to have the higher homicide rate? Write a paragraph describing these results.

b. Using the rates you calculated in part a, calculate the percent change between 1997 and 2009 for each state and each province. Which states and provinces had the largest increase and decrease? Which nation seems to have the largest change in homicide rates? Summarize your results in a paragraph.

2.5 SOC The scores of 15 respondents on four variables are as reported below. These scores were taken from a public opinion survey called the General Social Survey, or the GSS. This data set is used for the computer exercises in this text. Small subsamples from the GSS will be used throughout the text to provide "real" data for problems. For the actual questions and other details, see Appendix G. The numerical codes for the variables are:

Sex	Support for Gun Control	Level of Education	Age
1 = Male	1 = In favor	0 = Less than HS	Actual years
2 = Female	2 = Opposed	1 = HS	
		2 = Jr. college	
		3 = Bachelor's	
		4 = Graduate	

Case Number	Sex	Support for Gun Control	Level of Education	Age
1	2	1	1	45
2	1	2	1	48
3	2	1	3	55
4	1	1	2	32
5	2	1	3	33
6	1	1	1	28
7	2	2	0	77
8	1	1	1	50
9	1	2	0	43
10	2	1	1	48
11	1	1	4	33
12	1	1	4	35
13	1	1	0	39
14	2	1	1	25
15	1	1	1	23

a. Construct a frequency distribution for each variable. Include a column for percentages.

b. Construct pie and bar charts to display the distributions of sex, support for gun control, and level of education.

2.6 SW A local youth service agency has begun a sex education program for teenage girls who have been referred by the juvenile courts. The girls were given a 20-item test for general knowledge about sex, contraception, and anatomy and physiology on admission to the program and again after completing the program. The scores of the first 15 girls to complete the program are:

	Pretest	Posttest
A	8	12
B	7	13
C	10	12
D	15	19
E	10	8
F	10	17
G	3	12
H	10	11
I	5	7
J	15	12
K	13	20
L	4	5
M	10	15
N	8	11
O	12	20

Construct frequency distributions for the pretest and posttest scores. Include a column for percentages. (*HINT: There were 20 items on the test, so the maximum range for these scores is 20. If you use 10 class intervals to display these scores, the interval size will be 2. Because there are no scores of 0 or 1 for either test, you may state the first interval as 2–3. To make comparisons easier, both frequency distributions should have the same intervals.*)

2.7 SOC Sixteen high school students completed a class to prepare them for the college boards. Their scores were:

420	345	560	650
459	499	500	657
467	480	505	555
480	520	530	589

These same 16 students were given a test of math and verbal ability to measure their readiness for college-level work. Scores are reported here in terms of the percentage of correct answers for each test.

Math Test			
67	45	68	70
72	85	90	99
50	73	77	78
52	66	89	75

Verbal Test			
89	90	78	77
75	70	56	60
77	78	80	92
98	72	77	82

a. Display each of these variables in a frequency distribution with columns for percentages and cumulative percentages.

b. Construct a histogram and frequency polygon for this data.

2.8 GER The reported number of times 25 residents of a community for senior citizens left their homes for any reason during the past week:

0	2	1	7	3
7	0	2	3	17
14	15	5	0	7
5	21	4	7	6
2	0	10	5	7

a. Construct a frequency distribution to display these data.

b. What are the midpoints of the class intervals?

c. Add columns to the table to display the percentage distribution, cumulative frequency, and cumulative percentage.

d. Construct a histogram and a frequency polygon to display this distribution.

e. Write a paragraph summarizing this distribution of scores.

2.9 SOC Twenty-five students completed a questionnaire that measured their attitudes toward interpersonal violence. Respondents who scored high believed that a person could legitimately use physical force against another person in many situations. Respondents who scored low believed that the use of violence could be justified in few or no situations.

52	47	17	8	92
53	23	28	9	90
17	63	17	17	23
19	66	10	20	47
20	66	5	25	17

a. Construct a frequency distribution to display this data.

b. What are the midpoints of the class intervals?

c. Add columns to the table to display the percentage distribution, cumulative frequency, and cumulative percentage.

d. Construct a histogram and a frequency polygon to display these data.

e. Write a paragraph summarizing this distribution of scores.

2.10 PA/CJ As part of an evaluation of the efficiency of your local police force, you have gathered the following data on police response time to calls for assistance during two different years. (Response times were rounded off to whole minutes.) Convert both frequency distributions into percentages, and construct pie charts and bar charts to display the data. Write a paragraph comparing the changes in response time between the two years.

Response Time, 2000	Frequency (*f*)
21 minutes or more	35
16–20 minutes	75
11–15 minutes	180
6–10 minutes	375
Less than 6 minutes	210
	875

Response Time, 2010	Frequency (*f*)
21 minutes or more	45
16–20 minutes	95
11–15 minutes	155
6–10 minutes	350
Less than 6 minutes	250
	895

2.11 SOC Figures 2.9 through 2.11 display trends in crime in the United States over the last two decades. Write a paragraph describing each of these graphs. What similarities and differences can you observe among the three graphs? (For example, do crime rates always change in the same direction?) Note the differences in the vertical axes from chart to chart; for homicide, the axis ranges from 0 to 12, while for burglary and auto theft, the range is from 0 to 1,600. The latter crimes are far more common, and a scale with smaller intervals is needed to display the rates.

FIGURE 2.9 U.S. Homicide Rate, 1984–2009 (per 100,000 Population)

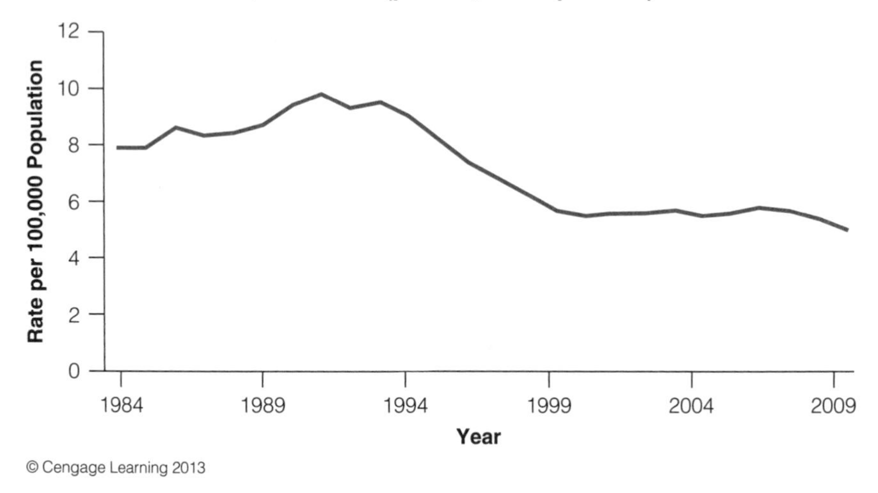

FIGURE 2.10 U.S. Robbery and Aggravated Assault Rates, 1984–2009 (per 100,000 Population)

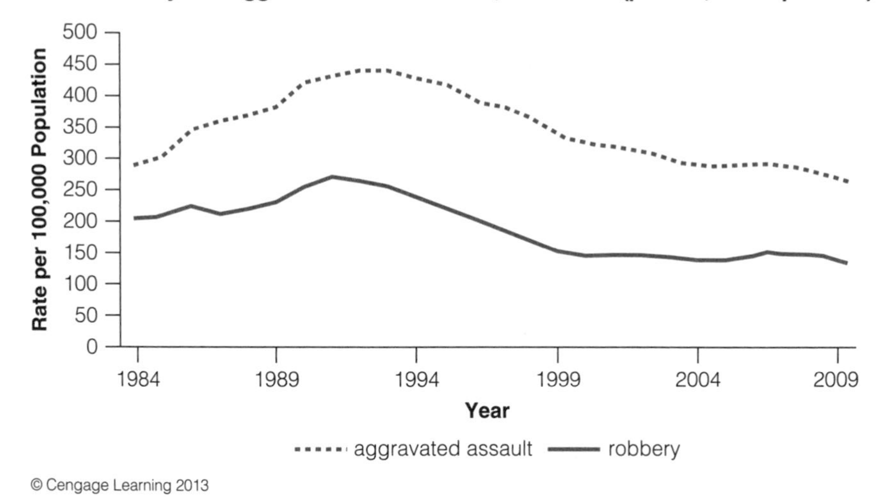

FIGURE 2.11 U.S. Burglary and Car Theft Rates, 1984–2009 (per 100,000 Population)

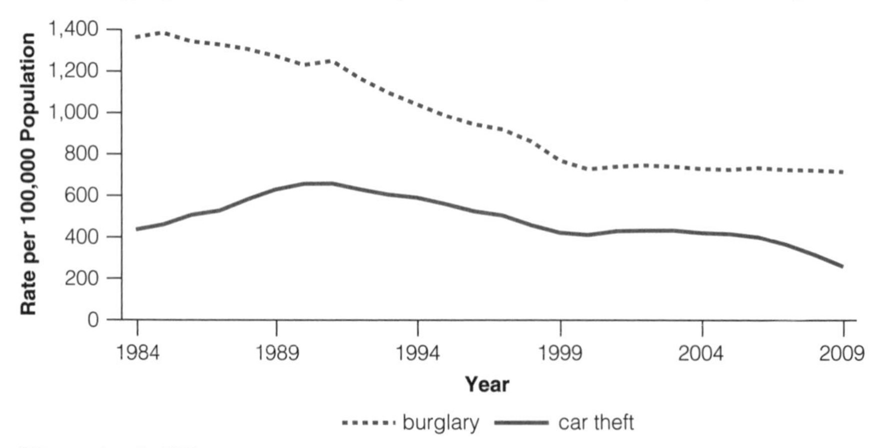

2.12 PA A city's Department of Transportation has been keeping track of accidents on a particularly dangerous stretch of highway. Early in the year, the city lowered the speed limit on this highway and increased police patrols. Data on the number of accidents before and after the changes are presented here. Did the changes work? Is the highway safer? Construct a line chart to display these two sets of data (use graphics software if available), and write a paragraph describing the changes.

Month	12 Months Before	12 Months After
January	23	25
February	25	21
March	20	18
April	19	12
May	15	9
June	17	10
July	24	11
August	28	15
September	23	17
October	20	14
November	21	18
December	22	20

YOU ARE THE RESEARCHER

Is there a "culture war" in the United States?

One of the early steps in a research project is to produce frequency distributions and use them to inspect the variables. If nothing else, these tables provide excellent background information, and sometimes, you can use them to begin to answer research questions. In this installment of "You Are the Researcher," you will use SPSS to produce summary tables for several variables that measure attitudes about controversial issues in U.S. society and that may map the battlefronts in what many call the American culture war.

The United States seems to be deeply divided on a number of religious, political, and cultural issues and values. We might characterize the opposing sides in terms of liberal vs. conservative, modern vs. traditional, or progressive vs. old school. Some of the bitterest debates concern abortion, gay marriage, and gun control. As you know, debates over issues like these can be intense, rancorous, and even violent; adherents of one position may view their opponents with utter contempt, blast them with insults, demonize them, and dismiss their arguments. How deep is this fault line in U.S. society? How divided are the American people?

We can begin to investigate these questions by choosing three variables from the 2010 General Social Survey (GSS) that seem to differentiate the sides in the American culture war. Before doing this, let us take a moment to consider the process of picking variables. Technically, selecting a variable to represent a concept such as "culture war"—is called *operationalization* and this can be one of the most challenging steps in a research project. Recall that we discussed operationalization in Chapter 1 when we reviewed the research process. When we move from the theory stage of research to the hypothesis stage (see Figure 1.1), we identify specific variables or measures (such as responses to survey items) that match our concept (such as prejudice). This can be difficult because our concepts are usually quite abstract and subject to a variety of perspectives. What exactly is a culture war, and what (and why) are some positions liberal or traditional, conservative or progressive? In order to do research, we must use concrete, specific variables to represent our abstract and general concepts, but which variables relate to which concepts?

Any pairing we make between variables and concepts is bound to be at least a little arbitrary. In many cases, the best strategy is to use several variables to represent the concept; if our operationalizations are reasonable, our selected variables will behave similarly, and each will behave as the abstract concept would if we could measure it directly. This is why I ask you to select three different variables—not just one—to represent the culture war. Each researcher may select different variables, but if everyone makes reasonable decisions, the chosen variables should be close representations of the concept.

Begin by starting SPSS and then opening the 2010 GSS. See Appendix F for instructions on opening SPSS. Once you see the SPSS screen, find the **File** command at the far left and then click **File → Open → Data**. Find the 2010 GSS file and then click on the file name to open it. You may have to change the drive specification to locate the 2010 GSS data supplied with this text (probably named **GSS2010.sav**).

Next, select three variables that seem relevant to the culture war. You can browse through the available variables by consulting Appendix G or you can click **Utilities → Variables** on the menu bar of the Data Editor Window of SPSS for a list of variables. After you have made your selections, complete the following steps.

STEP 1: IDENTIFY YOUR THREE VARIABLES

In the second column, enter the name of the variable as it appears in the database (for example: *abany*, *marital*, or *sex*). To "explain what the variable measures," look at the wording of the survey items in Appendix G or use the abbreviated statements in the **Utilities → Variables** window.

Variable	SPSS name	Explain exactly what this variable measures
1		
2		
3		

STEP 2: OPERATIONALIZATION

Explain how each variable you selected represents an issue in the culture wars. Which value or response of the variable indicates that the respondent is liberal or progressive, and which denotes "conservative or traditional"?

Variable	SPSS name	Explain how this variable relates to the culture war	Identify which value (e.g, "agree" or "support") is	
			Liberal	Conservative

STEP 3: USING SPSS FOR WINDOWS TO PRODUCE FREQUENCY DISTRIBUTIONS

Now we are ready to generate output and get some background on the nature of disagreements over values and issues among Americans.

Generating Frequency Distributions

We produced and examined a frequency distribution for the variable *sex* in Appendix F. Use the same procedures to produce frequency distributions for the three variables you used to represent the American culture wars. From the menu bar, click **Analyze**. From the menu that drops down, click **Descriptive Statistics** and then **Frequencies**. The **Frequencies** window appears with the variables listed in alphabetical order in the left-hand box. The window may display variables by name (e.g, *abany*, *abpoor*) or by label (e.g., ABORTION IF WOMAN WANTS FOR ANY REASON). If labels are displayed, you may switch to variable names by clicking **Edit**, clicking **Options**, and then making the appropriate selections on the "General" tab. See Appendix F and Table F.2 for further information.

Find the first of your three variables, click on its name to highlight it, and then click the arrow button in the middle of the screen to move it to the right-hand window. Find your other two variables and then follow the same procedure to move their names to the right-hand window. SPSS will process all variables listed in the right-hand box together. Click **OK** at the bottom of the **Frequencies** window, and SPSS will rush off to create the frequency distributions you requested.

The tables will be in the **Viewer** window that will now be "closest" to you on the screen. The tables, along with other information, will be on the right-hand side of the window. To change the size of this window, click the middle symbol (shaped like either a square or two intersecting squares) in the upper-right-hand corner of the **Output** window.

Reading Frequency Distributions

I will illustrate how to decipher the SPSS output by using the variable *marital*, which measures current marital status. I chose *marital* so as not to duplicate any of the variables you selected in step 1 of this exercise. The output looks like this:

MARITAL STATUS

		Frequency	Percent	Valid Percent	Cumulative Percent
Valid	Married	647	43.3	43.3	43.3
	Widowed	140	9.4	9.4	52.7
	Divorced	241	16.1	16.1	68.9
	Separated	47	3.1	3.1	72.0
	Never Married	418	28.0	28.0	100.0
	Total	1493	99.9	100.0	
Missing	NA	1	.1		
Total		1494	100.0		

Let us examine the elements of this table. The variable name and label are printed at the top of the output ("MARITAL STATUS"). The various categories are printed on the left. Moving one column to the right, we find the actual frequencies, or the number of times each score of the variable occurred. We see that 647 of the respondents were married, 140 were widowed, and so forth. Next are two columns that report percentages. The entries in the "Percent" column are based on all respondents who were asked this question and may include the scores NA (No Answer), DK (Don't Know), or IAP (Inapplicable). The Valid Percent column eliminates all cases with missing values. Because we almost always ignore missing values, we will pay attention only to the Valid Percent column (even though, in this case, only one respondent did not supply this information and the columns are virtually identical). The final column is a cumulative percentage column (see Table 2.15). For nominal-level variables like *marital*, this information is not meaningful because the order in which the categories are stated is arbitrary.

The three frequency distributions you generated will have the same format and can be read in the same way. Use these tables—especially the "Valid Percents" column—to complete step 4.

STEP 4: INTERPRETING RESULTS

Characterize your results by reporting the percentage (not the frequencies) of respondents who endorsed each response. How large are the divisions in American values? Is there consensus on the issue measured by your variable (do the great majority endorse the same response) or is there considerable disagreement? The lower the consensus, the greater the opportunity for the issue to be included in the culture war.

SPSS Name of variable 1: ___________

Summarize the frequency distribution in terms of the percentage of respondents who endorsed each position:

Are these results consistent with the idea that there is a "war" over Americans' values? How?

SPSS Name of variable 2: ___________

Summarize the frequency distribution in terms of the percentage of respondents who endorsed each position:

Are these results consistent with the idea that there is a "war" over Americans' values? How?

SPSS Name of variable 3: ___________

Summarize the frequency distribution in terms of the percentage of respondents who endorsed each position:

Are these results consistent with the idea that there is a "war" over Americans' values? How?

3 Measures of Central Tendency

LEARNING OBJECTIVES By the time you finish this chapter, you will be able to:

1. Explain the purposes of measures of central tendency and interpret the information they convey.
2. Calculate, explain, and compare and contrast the mode, median, and mean.
3. Explain the mathematical characteristics of the mean.
4. Select an appropriate measure of central tendency according to level of measurement and skew.

USING STATISTICS

The statistical techniques presented in this chapter are used to find the typical case or average score on a single variable. They can be used to:

- Identify the most popular political party in a community.
- Compare the average cost of living or house or gas prices in different cities or states. For example, in 2010, the typical house in San Francisco sold for $525,000, while the typical house in Buffalo, New York, sold for $121,000.
- Measure the "gender gap" in income by comparing the average wages for males and females.
- Track changes in American family life by reporting changes in the average number of children or typical age at first marriage over time.

Introduction

One clear benefit of frequency distributions, graphs, and charts is that they summarize the overall shape of a distribution of scores in a way that can be quickly comprehended. However, you will almost always need to report more detailed information about the distribution. Specifically, two additional kinds of statistics are extremely useful: some idea of the typical or average case in the distribution (for example, "The average starting salary for social workers is $39,000 per year") and some idea of how much variety or heterogeneity there is in the distribution ("In this state, starting salaries for social workers range from $31,000 per year to $42,000 per year"). The first kind of statistic—called **measures of central tendency**—is the subject of this chapter. The second kind of statistic—measures of dispersion—is presented in Chapter 4.

The three commonly used measures of central tendency—mode, median, and mean—are all probably familiar to you. All three summarize an entire distribution of scores by describing the most common score (the mode), the score of the middle case (the median), or the average score (the mean) of that distribution. These statistics are powerful because they can reduce huge arrays of data to a single, easily understood number. Remember that the central purpose of descriptive statistics is to summarize or "reduce" data.

Even though they share a common purpose, the three measures of central tendency are different statistics, and they will have the same value only under certain conditions. As we shall see, they vary in terms of level of measurement, and perhaps more importantly, they also vary in terms of how they define central tendency; they will not necessarily identify the same score or case as "typical." Thus, your choice of an appropriate measure of central tendency will depend in part on the way you measure the variable and in part on the purpose of the research.

The Mode

The **mode** of any distribution of scores is the value that occurs most frequently. For example, in the set of scores 58, 82, 82, 90, 98, the mode is 82 because it occurs twice and the other scores occur only once.

The mode is a simple statistic—most useful when you are interested in the most common score and when you are working with nominal-level variables. In fact, the mode is the only measure of central tendency for nominal-level variables. For example, Table 3.1 reports the religious affiliations of adult Americans. The mode of this distribution—the single-largest category—is Protestant.

You should note that the mode has several limitations. First, some distributions have no mode at all (see Table 2.6). Other distributions have so many modes that the statistic becomes meaningless. For example, consider the distribution of test scores for a class of 100 students presented in Table 3.2. The distribution has modes at 55, 66, 78, 82, 90, and 97. Would reporting all six of these modes actually convey any useful information about central tendency in the distribution?

Second, the modal score of ordinal or interval-ratio variables may not be central to the distribution as a whole. That is, *most common* does not necessarily mean "typical" in the sense of identifying the center of the distribution. For example,

TABLE 3.1 Religious Preference of Adult Americans

Denomination	Frequency	Percentages
Protestant	116,203,000	53.7
Catholic	57,199,000	26.4
Jewish	2,680,000	1.2
Muslim	1,349,000	0.6
None	34,169,000	15.8
Other	4,767,000	2.2
	N = 216,367,000	99.9%

Source: U.S. Bureau of the Census. 2011. *Statistical Abstract of the United States, 2011,* p. 61. http://www.census.gov/prod/2011pubs/12statab/pop.pdf.

TABLE 3.2 A Distribution of Test Scores

Scores (% Correct)	Frequency
97	10
91	7
90	10
86	8
82	10
80	3
78	10
77	6
75	9
70	7
66	10
55	10
	$N = 100$

consider the rather unusual (but not impossible) distribution of test scores in Table 3.3. The mode of the distribution is 93. Is this score very close to the majority of the scores? If the instructor summarized this distribution by reporting only the modal score, would he or she be conveying an accurate picture of the distribution as a whole? *(For practice in finding and interpreting the mode, see problems 3.1 to 3.7.)*

TABLE 3.3 A Distribution of Test Scores

Scores (% correct)	Frequency
93	5
70	1
69	1
68	1
67	4
66	3
64	2
62	3
60	2
58	2
	$N = 24$

STATISTICS IN EVERYDAY LIFE

As we saw in Table 3.1, the most common (or modal) religious affiliation in the United States, by far, is Protestant. However, the nation is changing rapidly in this area of life, as in so many others. For example, since 1990, the percentage of adult Americans who identify as "Protestant" dropped from 61% to 54% and the percentage with no religious affiliation (including atheists and agnostics) has doubled from 8% to almost 16%.* Do these trends suggest that American society is growing stronger (more secular, more rational) or weaker (less moral, less guided by spiritual considerations)? Why? What additional information would you like to have before coming to a conclusion on this issue?

*U.S. Bureau of the Census. 2011. *Statistical Abstract of the United States, 2011,* p. 61. http://www.census.gov/prod/2011pubs/12statab/pop.pdf.

The Median

Unlike the mode, the **median (Md)** is always at the exact center of a distribution of scores. The median is the score of the case that is at the middle of a distribution: Half the cases have scores higher than the median and half the cases have scores lower than the median. Thus, if the median family income for a community is $45,000, half the families earn more than $45,000 and half earn less.

Before finding the median, the cases must be placed in order from the highest to the lowest score—or from the lowest to the highest. Then, you can determine the median by locating the case that divides the distribution into two equal halves. The median is the score of the middle case. If five students received grades of 93, 87, 80, 75, and 61 on a test, the median would be 80—the score that splits the distribution into two equal halves.

When the number of cases (*N*) is odd, the value of the median is unambiguous because there will always be a middle case. However, with an even number of cases, there will be two middle cases, and in this situation, the median is defined as the score exactly halfway between the scores of the two middle cases.

To illustrate, assume that seven students were asked to indicate their level of support for the intercollegiate athletic program at their universities on a scale ranging from 10 (indicating great support) to 0 (no support). After arranging their responses from high to low, you can find the median by locating the case that divides the distribution into two equal halves. With a total of seven cases, the middle case would be the fourth case because there will be three cases above and three cases below this case. Table 3.4 lists the cases in order and identifies the median. With seven cases, the median is the score of the fourth case.

Now suppose we added one more student to the sample, whose support for athletics was measured as a "1." This would make *N* an even number (8) and we would no longer have a single middle case. Table 3.5 presents the new distribution of scores, and as you can see, any value between 7 and 5 would technically satisfy the definition of a median (that is, it would split the distribution into two equal halves of four cases each). We resolve this ambiguity by defining the median as the average of the scores of the two middle cases. In this example, the median would be defined as $(7 + 5)/2$, or 6.

Remember that we follow different procedures to find the median depending on whether *N* is odd or even. The procedures are stated in general terms in the "One Step at a Time" box.

TABLE 3.4 Finding the Median with Seven Cases (*N* is odd)

Case	Score	
1	10	
2	10	
3	8	
4	7	← Md
5	5	
6	4	
7	2	

TABLE 3.5 Finding the Median with Eight Cases (N is even)

Case	Score	
1	10	
2	10	
3	8	
4	7	
		Md = (7 + 5) / 2 = 6
5	5	
6	4	
7	2	
8	1	

ONE STEP AT A TIME Finding the Median

Step	*Operation*
1.	Array the scores in order from high score to low score.
2.	Count the number of scores to see if N is odd or even.

IF N is ODD ↓

3. The median will be the score of the middle case.
4. To find the middle case, add 1 to N and then divide by 2.
5. The value you calculated in step 4 is the number of the middle case. The median is the score of this case. For example, if $N = 13$, the median will be the score of the $(13 + 1)/2$, or seventh, case.

IF N is EVEN ↓

3. The median is halfway between the scores of the two middle cases.
4. To find the first middle case, divide N by 2.
5. To find the second middle case, increase the value you computed in step 4 by 1.
6. Find the scores of the two middle cases. Add the scores together and then divide by 2. The result is the median. For example, if $N = 14$, the median is the score halfway between the scores of the seventh and eighth cases. If the middle cases have the same score, that score is defined as the median.

The median cannot be calculated for variables measured at the nominal level because it requires that scores be ranked from high to low. Remember that the scores of nominal level variables cannot be ordered or ranked: The scores are different from each other but do not form a mathematical scale of any sort. The median can be found for either ordinal or interval-ratio data but is generally more appropriate for the former. *(The median may be found for any problem at the end of this chapter.)*

The Mean

The mean, or arithmetic average, is by far the most commonly used measure of central tendency. It reports the average score of a distribution and is calculated by dividing the total sum of the scores by the number of scores (*N*). In this text, we will use two different symbols for the mean: If we are referring to a sample, we will use the symbol $\overline{X}$, which can be read as "X-bar." For a population mean, we will use μ, the Greek letter Mu (pronounced "mew"). In this chapter, we will use only the first symbol, but the distinction between sample and population means will become extremely important in Part II of this text.

We will illustrate the computation of the mean with a specific example. A birth control clinic administered a 20-item test of general knowledge about contraception and safe sex practices to 10 clients. The number of correct responses was 2, 10, 15, 11, 9, 16, 18, 10, 11, and 7. To find the mean of this distribution, add the scores (total = 109) and then divide by the number of scores (10). The result (10.9) is the average score on the test.

The mathematical formula for the mean:

FORMULA 3.1

$$\overline{X} = \frac{\Sigma(X_i)}{N}$$

Where: $\overline{X}$ = the sample mean
$\Sigma(X_i)$ = the summation of the scores
N = the number of cases

Let us take a moment to consider the new symbols introduced in this formula. First, the symbol Σ (uppercase Greek letter sigma) is a mathematical operator just like the plus sign $(+)$ or divide sign $(\div)$. It stands for "the summation of" and directs us to add whatever quantities are stated immediately following it. The second new symbol is X_i ("*X*-sub-*i*"), which refers to any single score—the "*i*th" score. If we wished to refer to the score of a particular case, the number of the case could replace the subscript. Thus, X_1 would refer to the score of the first case, X_2 to the score of the second case, X_{26} to the 26th case, and so forth.

The operation of adding all the scores is symbolized as $\Sigma(X_i)$. This combination of symbols directs us to sum the scores—beginning with the first score and ending with the last score in the distribution. Thus, Formula 3.1 states in symbols what has already been stated in words (to calculate the mean, add the scores and divide by the number of scores) but in a succinct and precise way. *(For practice in computing the mean, see any of the problems at the end of this chapter.)*

Because computation of the mean requires addition and division, it should be used with variables measured at the interval-ratio level. However, researchers do calculate the mean for variables measured at the ordinal level because the mean is much more flexible than the median and is a central feature of many interesting and powerful advanced statistical techniques. Thus, if the researcher plans to do any more than merely describe his or her data, the mean will probably be the preferred measure of central tendency even for ordinal-level variables.

ONE STEP AT A TIME	Finding the Mean
Step	***Operation***
1.	Add up the scores (X_i).
2.	Divide the quantity you found in step 1 (ΣX_i) by N.

Three Characteristics of the Mean

The mean is the most commonly used measure of central tendency, and we will consider its mathematical and statistical characteristics in some detail.

The Mean Balances All the Scores. First, the mean is an excellent measure of central tendency because it acts like a fulcrum that "balances" all the scores; the mean is the point around which all of the scores cancel out. We can express this property symbolically:

$$\Sigma(X_i - \overline{X}) = 0$$

This expression says that if we subtract the mean $(\overline{X})$ from each score (X_i) in a distribution and then sum the differences, the result will always be zero.

To illustrate, consider the test scores presented in Table 3.6. The mean of these five scores is 390/5, or 78. The difference between each score and the mean is listed in the right-hand column $(X_i - \overline{X})$, and the sum of these differences is zero. The total of the negative differences (-19) is exactly equal to the total of the positive differences $(+19)$, as will always be the case. Thus, the mean "balances" the scores and is at the center of the distribution.

The Mean Minimizes the Variation of the Scores. A second characteristic of the mean is called the "least squares" principle—a characteristic that is expressed in this statement:

$$\Sigma(X_i - \overline{X})^2 = \text{minimum}$$

That is, the mean is the point in a distribution around which the variation of the scores (as indicated by the squared differences) is minimized. If the differences

TABLE 3.6 A Demonstration Showing That All Scores Cancel Out Around the Mean

X_i	$(X_i - X)$
65	65 − 78 = −13
73	73 − 78 = −5
77	77 − 78 = −1
85	85 − 78 = 7
90	90 − 78 = 12
$\Sigma X_i = 390$	$\Sigma(X_i - \overline{X}) = 0$
$\overline{X} = 390/5 = 78$	

Table 3.7 A Demonstration Showing that the Mean is the Point of Minimized Variation

X_i	$(X_i - \overline{X})$	$(X_i - \overline{X})^2$	$(X_i - 77)^2$
65	$65 - 78 = -13$	$(-13)^2 = 169$	$65 - 77 = (-12)^2 = 144$
73	$73 - 78 = -5$	$(-5)^2 = 25$	$73 - 77 = (-4)^2 = 16$
77	$77 - 78 = -1$	$(-1)^2 = 1$	$77 - 77 = (0)^2 = 0$
85	$85 - 78 = 7$	$(7)^2 = 49$	$85 - 77 = (8)^2 = 64$
90	$90 - 78 = 12$	$(12)^2 = 144$	$90 - 77 = (13)^2 = 169$
$\Sigma(X_i) = 390$	$\Sigma(X_i - \overline{X}) = 0$	$\Sigma(X_i - \overline{X})^2 = 388$	$\Sigma(X_i - 77)^2 = 393$

between the scores and the mean are squared and then added, the resultant sum will be less than the sum of the squared differences between the scores and any other point in the distribution.

To illustrate this principle, consider the distribution of five scores mentioned earlier: 65, 73, 77, 85, and 90. The differences between the scores and the mean have already been found. As illustrated in Table 3.7, if we square and sum these differences, we would get a total of 388. If we performed those same mathematical operations with any number other than the mean—say, the value 77—the resultant sum would be greater than 388. Table 3.7 illustrates this point by showing that the sum of the squared differences around 77 is 393—a value greater than 388.

This least-squares principle underlines the fact that the mean is closer to all the scores than the other measures of central tendency. Also, this characteristic of the mean is important for the statistical techniques of correlation and regression—topics we take up toward the end of this text.

The Mean Can Be Misleading If the Distribution Is Skewed. The final important characteristic of the mean is that every score in the distribution affects it. The mode (which is only the most common score) and the median (which deals only with the score of the middle case or cases) are not so affected. This quality is an advantage and a disadvantage. On one hand, the mean utilizes all the available information; every score in the distribution is included in the calculation of the mean. On the other hand, when a distribution has a few very high or very low scores, the mean may be a very misleading measure of centrality.

To illustrate, consider Table 3.8. The five scores listed in column 1 have a mean and median of 25 (see column 2). In column 3, the scores are listed again with one score changed: 35 is changed to 3,500. Look in column 4 and you will see that this change has no effect on the median; it remains 25. This is because the median is based *only* on the score of the middle case and is not affected by changes in the scores of other cases in the distribution.

The mean, in contrast, is very much affected by the change because it takes *all* scores into account. The mean changes from 25 to 718 solely because of the one extreme score of 3,500. Note also that the mean in column 4 is very different from four of the five scores listed in column 3. In this case, is the mean or the median a better representation of the scores? For distributions that have a few very high or very low scores, the mean may present a very misleading picture of the typical or central score. In these cases, the median may be the preferred measure of central tendency for interval-ratio variables. *(For practice in dealing with the effects of extreme scores on means and medians, see problems 3.11, 3.13, 3.14, and 3.15.)*

TABLE 3.8 A Demonstration Showing That the Mean Is Affected by Every Score

1 Scores	2 Measures of Central Tendency	3 Scores	4 Measures of Central Tendency
15		15	Mean = 3590 / 5 = 718
20	Mean = 125 / 5 = 25	20	
25		25	
30	Median = 25	30	Median = 25
35		3500	
Σ = 125		Σ = 3590	

The general principle to remember is that when there are a few very high or very low scores, the mean will be pulled in the direction of the extreme scores (relative to the median). The mean and median will have the same value when and only when a distribution is symmetrical. When a distribution has some extremely high scores (this is called a *positive skew*), the mean will always have a greater numerical value than the median. If the distribution has some very low scores (a *negative skew*), the mean will be lower in value than the median. Figures 3.1 to 3.3 depict three different frequency polygons that demonstrate these relationships.

These relationships between medians and means also have a practical value. For one thing, a quick comparison of the median and mean will always tell you if a distribution is skewed and the direction of the skew. If the mean is less than the median, the distribution has a negative skew. If the mean is greater than the median, the distribution has a positive skew.

Second, these characteristics of the mean and median also provide a simple and effective way to "lie" with statistics. For example, if you want to maximize

FIGURE 3.1 A Positively Skewed Distribution

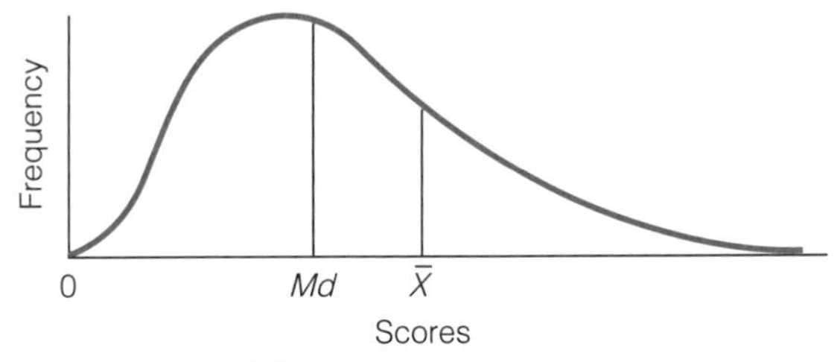

FIGURE 3.2 A Negatively Skewed Distribution

FIGURE 3.3 An Unskewed, Symmetrical Distribution

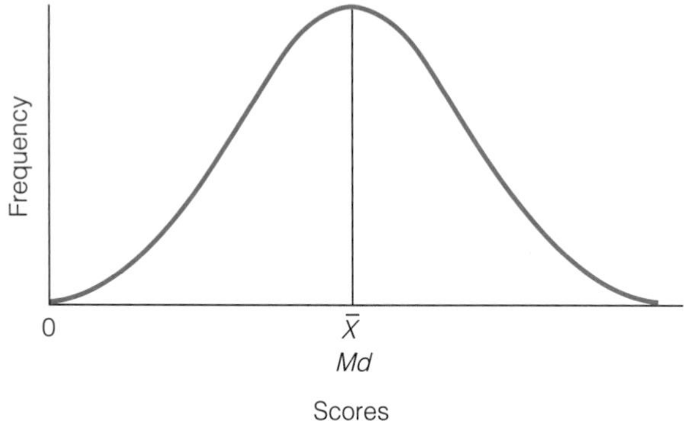

the average score of a positively skewed distribution, report the mean. Income data usually have a positive skew (there are only a few very wealthy people). If you want to impress someone with the general affluence of a mixed-income community, report the mean. If you want a lower figure, report the median.

Which measure is most appropriate for skewed distributions? This will depend on what point the researcher wishes to make, but as a rule, either both statistics or the median alone should be reported when the distribution has a few extreme scores.

Choosing a Measure of Central Tendency

You should consider two main criteria when choosing a measure of central tendency. First, make sure you know the level of measurement of the variable in question. This will generally tell you whether you should report the mode, median, or mean. Table 3.9 shows the relationship between the level of measurement and measures of central tendency. The capitalized, boldface "**YES**" identifies the most appropriate measure of central tendency for each level of measurement, and the nonboldface "Yes" indicates the levels of measurement for which the measure is also permitted. An entry of "No" in the table means that the statistic cannot be computed for that level of measurement. Finally, the "Yes (?)" entry in the

STATISTICS IN EVERYDAY LIFE

The global reach of McDonald's is well known but the near ubiquity of American fast food also provides a way of comparing the costs of living and workers' purchasing power around the globe. According to research conducted by UBS (an international financial services corporation) in 73 cities around the globe, the amount of time required for the typical worker to purchase a Big Mac varies from 12 minutes in Tokyo and most North American and Western European cities to over 2½ hours in Nairobi, Kenya. The average amount of time for all 73 cities was 37 minutes and the median was 27 minutes. Note the positive skew in this distribution: the higher value of the mean reflects the very high scores of a few cities.

Source: UBS. *Prices and Earnings: A Comparison of Purchasing Power Around the Globe.* Available at www.ubs.com/1/ShowMedia/ubs_ch/wealth_mgmt_ch?contentId=103982&name=eng.pdf.

TABLE 3.9 The Relationship Between Level of Measure and Measures of Central Tendency

	Level of Measurement		
Measure of Central Tendency:	Nominal	Ordinal	Interval-Ratio
Mode	**YES**	Yes	Yes
Median	No	**YES**	Yes
Mean	No	Yes (?)	**YES**

bottom row indicates that the mean is often used with ordinal-level variables even though, strictly speaking, this practice violates level of measurement guidelines.

Second, consider the definitions of the three measures of central tendency and remember that they provide different types of information. They will have the same value only under certain specific conditions (that is, for symmetrical distributions with one mode), and each has its own message to report. In many circumstances, you might want to report all three. The guidelines in Table 3.10 stress both selection criteria and may be helpful when choosing a specific measure of central tendency.

Applying Statistics 3.1 The Mean and Median

Ten students have been asked how many hours they spent in the college library during the past week. What is the average library time for these students? The hours are reported in the following list, and we will find the mode, the median, and the mean for these scores.

By scanning the scores, we can see that two scores—5 and 14—occurred twice and no other score occurred more than once. This distribution has two modes: 5 and 14.

Student	Hours Spent in the Library Last Week (X_i)
1	0
2	2
3	5
4	5
5	7
6	10
7	14
8	14
9	20
10	30
	$\Sigma(X_i) = 107$

Because the number of cases is even, the median will be the average of the two middle cases after all cases have been ranked in order. With 10 cases, the first middle case will be the (N/2), or (10/2), or fifth case. The second middle case is the (N/2) + 1, or (10/2) + 1, or sixth case. The median will be the score halfway between the scores of the fifth and sixth cases. Counting down from the top, we find that the score of the fifth case is 7 and the score of the sixth case is 10. The median for these data is (7 + 10)/2, or (17/2), or 8.5. The median—the score that divides this distribution in half—is 8.5.

The mean is found by first adding all the scores and then dividing by the number of scores. The sum of the scores is 107, so the mean is:

$$\bar{X} = \frac{\Sigma(X_i)}{N} = \frac{107}{10} = 10.7$$

These 10 students spent an average of 10.7 hours in the library during the week in question.

Note that the mean is higher than the median. This indicates a positive skew in the distribution (a few extremely high scores). By inspection, we can see that the positive skew is caused by the two students who spent many more hours (20 hours and 30 hours) in the library than the other eight students.

TABLE 3.10 Choosing a Measure of Central Tendency

Use the mode when	1. The variable is measured at the nominal level. 2. You want a quick and easy measure for ordinal and interval-ratio variables. 3. You want to report the most common score.
Use the median when	1. The variable is measured at the ordinal level. 2. A variable measured at the interval-ratio level has a highly skewed distribution. 3. You want to report the central score. The median always lies at the exact center of a distribution.
Use the mean when	1. The variable is measured at the interval-ratio level (except when the variable is highly skewed). 2. You want to report the typical score. The mean is "the fulcrum that exactly balances all the scores." 3. You anticipate additional statistical analysis.

BECOMING A CRITICAL CONSUMER: Using an Appropriate Measure of Central Tendency

Consider these recent headlines:

"Median Price of Housing Falls"
"Average Price of Gas at Record High"
"Nonwhites Will Soon Become Majority in Area"

By now, you recognize that each of these reports cites a different measure of central tendency, and we can analyze how appropriately these choices were made.

Our first concern is the level of measurement of the variable being summarized. The first two variables—the cost of houses and the price of gas—are interval-ratio. Why does one report use the median and the other the mean or average? Consider the nature of the variables. Housing costs, like almost any variable related to money, is positively skewed (see Figure 3.1). Why? Any community will have a wide range of housing available. Some houses will be tumble-down shacks ("handyman specials"), many will be modest in quality and price, some will be upscale, and a few will cost millions. That is, most cases (houses) will cluster in the low to middle range, but there will be a few with extremely high scores, or prices. When a variable is positively skewed, the mean will have a higher value than the median and may not present an accurate picture of what is "typical" or average. When a variable is highly skewed, the median is the preferred measure of central tendency—even for interval-ratio-level variables. Responsible descriptions of the central tendency of any variable related to money or price generally will use the median, and you should be immediately suspicious of reports that do not.

This raises an issue with the second report: Why are gas prices reported with the mean or average? Like housing prices or income, gas prices fluctuate over time and across the country (see www.gasbuddy.com/gb_gastemperaturemap.aspx for a visual representation) and a few service stations will have very high prices. However, the range of variation in price per gallon is small—at least compared to that of housing prices. While there can be a difference of millions of dollars in the price of two houses, the maximum difference around the country at any one time in the cost of a gallon of regular is only about a dollar. The smaller range means that gas prices cannot be very skewed (relatively speaking), and it is reasonable to use the mean to report the average in this case.

The final headline uses the mode to report information about the changing racial and ethnic composition of U.S. society. Race and ethnicity are nominal-level variables, and the mode is the only measure of central tendency available for these variables. The majority category of any variable is its mode, or most common score.

The fact that there are three different measures of central tendency in common use can lead to some confusion and misunderstanding, especially because the word "average" can be used as a synonym for any of the three (although "average" refers only to the mean, strictly speaking). Always check to see which of the three is being used when an average is reported, and use extra caution when the exact statistic is not identified.

SUMMARY

1. The three measures of central tendency presented in this chapter share a common purpose. Each reports some information about the most typical or representative value in a distribution. Appropriate use of these statistics permits the researcher to report important information about an entire distribution of scores in a single, easily understood number.
2. The mode reports the most common score and is used most appropriately with nominally measured variables.
3. The median (Md) reports the score that is the exact center of the distribution. It is most appropriately used with variables measured at the ordinal level and with variables measured at the interval-ratio level when the distribution is skewed.
4. The mean $(\bar{X})$, the most frequently used of the three measures, reports the most typical score. It is used most appropriately with variables measured at the interval-ratio level (except when the distribution is highly skewed).
5. The mean has a number of mathematical characteristics that are significant for statisticians. First, it is the point in a distribution of scores around which all other scores cancel out. Second, the mean is the point of minimized variation. Last, as distinct from the mode or median, the mean is affected by every score in the distribution and is therefore pulled in the direction of extreme scores.

SUMMARY OF FORMULAS

FORMULA 3.1 Mean $\bar{X} = \frac{\Sigma X_i}{N}$

GLOSSARY

Least Squares Principle. The mean is the point of minimum variation of a set of scores

Mean. The arithmetic average of the scores. $\bar{X}$ represents the mean of a sample and μ is the mean of a population.

Measures of central tendency. Statistics that summarize a distribution of scores by reporting the most typical or representative value of the distribution.

Median (Md). The point in a distribution of scores above and below which exactly half of the cases fall.

Mode. The most common value in a distribution or the largest category of a variable.

Σ (uppercase Greek letter sigma). "The summation of."

Skew. The extent to which a distribution of scores has a few scores that are extremely high (positive skew) or extremely low (negative skew).

X_i ("X sub i"). Any score in a distribution.

PROBLEMS

3.1 SOC A variety of information has been gathered from a sample of college freshmen and seniors, including their region of birth, the extent to which they support legalization of marijuana (measured on a scale on which 7 = strong support, 4 = neutral, and 1 = strong opposition), the amount of money they spend each week out of pocket for food, drinks, and entertainment, how many movies they watched in their dorm rooms last week, their opinion of cafeteria food (10 = excellent, 0 = very bad), and their religious affiliation. Some results are presented here. Find the *most appropriate* measure of central tendency for each variable for freshmen and then for seniors. Report the measure you selected as well as its value for each variable (e.g., "Mode = 3" or "Median = 3.5"). *(HINT: Determine the level of measurement for each variable first. In general, this will tell you which measure of central tendency is appropriate. See Tables 3.9 and 3.10 to review the relationship between measure of central tendency and level of measurement. Also, remember that the mode is the* most common *score and especially remember to array scores from high to low before finding the median.)*

FRESHMEN						
Student	Region of Birth	Legalization	Out-of-Pocket Expenses	Movies	Cafeteria Food	Religion
A	North	7	33	0	10	Protestant
B	North	4	39	14	7	Protestant
C	South	3	45	10	2	Catholic
D	Midwest	2	47	7	1	None
E	North	3	62	5	8	Protestant
F	North	5	48	1	6	Jewish
G	South	1	52	0	10	Protestant
H	South	4	65	14	0	Other
I	Midwest	1	51	3	5	Other
J	West	2	43	4	6	Catholic

SENIORS						
Student	Region of Birth	Legalization	Out-of-Pocket Expenses	Movies	Cafeteria Food	Religion
K	North	7	65	0	1	None
L	Midwest	6	62	5	2	Protestant
M	North	7	60	11	8	Protestant
N	North	5	90	3	4	Catholic
O	South	1	62	4	3	Protestant
P	South	5	57	14	6	Protestant
Q	West	6	40	0	2	Catholic
R	West	7	49	7	9	None
S	North	3	45	5	4	None
T	West	5	85	3	7	Other
U	North	4	78	5	4	None

3.2 SOC A variety of information have been collected for all district high schools. Find the most appropriate measure of central tendency for each variable and summarize this information in a paragraph. *(HINT: The level of measurement of the variable will generally tell you which measure of central tendency is appropriate. Remember to organize the scores from high to low before finding the median.)*

High School	Enrollment	Largest Racial/ Ethnic Group	Percent College-Bound	Most Popular Sport	Condition of Physical Plant (scale of 1–10, with 10 high)
1	1400	White	25	Football	10
2	1223	White	77	Baseball	7
3	876	Black	52	Football	5
4	1567	Hispanic	29	Football	8
5	778	White	43	Basketball	4
6	1690	Black	35	Basketball	5
7	1250	White	66	Soccer	6
8	970	White	54	Football	9

3.3 PS You have been observing the local Democratic Party in a large city and have compiled some information about a small sample of party regulars. Find the appropriate measure of central tendency for each variable.

Respondent	Sex	Social Class	No. of Years in Party	Education	Marital Status	Number of Children
A	M	High	32	High school	Married	5
B	M	Medium	17	High school	Married	0
C	M	Low	32	High school	Single	0
D	M	Low	50	Eighth grade	Widowed	7
E	M	Low	25	Fourth grade	Married	4
F	M	Medium	25	High school	Divorced	3
G	F	High	12	College	Divorced	3
H	F	High	10	College	Separated	2
I	F	Medium	21	College	Married	1
J	F	Medium	33	College	Married	5
K	M	Low	37	High school	Single	0
L	F	Low	15	High school	Divorced	0
M	F	Low	31	Eighth grade	Widowed	1

3.4 SOC You have compiled the following information on each of the graduates voted "most likely to succeed" by a local high school for a 10-year period. For each variable, find the appropriate measure of central tendency.

Case	Present Income	Marital Status	Owns a BMW?	Years of Education Post–High School
A	24,000	Single	No	8
B	48,000	Divorced	No	4
C	54,000	Married	Yes	4
D	45,000	Married	No	4
E	30,000	Single	No	4
F	35,000	Separated	Yes	8
G	30,000	Married	No	3
H	17,000	Married	No	1
I	33,000	Married	Yes	6
J	48,000	Single	Yes	4

3.5 SOC For 15 respondents, data have been gathered on four variables (see the following table). Find and report the appropriate measure of central tendency for each variable.

Respondent	Marital Status	Racial/Ethnic Group	Age	Attitude on Abortion Scale (high score = strong opposition)
A	Single	White	18	10
B	Single	Hispanic	20	9
C	Widowed	White	21	8
D	Married	White	30	10
E	Married	Hispanic	25	7
F	Married	White	26	7
G	Divorced	Black	19	9
H	Widowed	White	29	6
I	Divorced	White	31	10
J	Married	Black	55	5
K	Widowed	Asian American	32	4
L	Married	American Indian	28	3
M	Divorced	White	23	2
N	Married	White	24	1
O	Divorced	Black	32	9

3.6 SOC Following are four variables for 30 cases from the General Social Survey. Age is reported in years. The variable *happiness* consists of answers to the question "Taken all together, would you say that you are (1) very happy (2) pretty happy, or (3) not too happy?" Respondents were asked how many sex partners they had over the past five years. Responses were measured on the following scale: 0–4 = actual numbers; 5 = 5–10 partners; 6 = 11–20 partners; 7 = 21–100 partners; 8 = more than 100. For each variable, find the appropriate measure of central tendency and then write a sentence reporting this statistical information as you would in a research report.

Respondent	Age	Happiness	Number of Partners	Religion
1	20	1	2	Protestant
2	32	1	1	Protestant
3	31	1	1	Catholic
4	34	2	5	Protestant
5	34	2	3	Protestant
6	31	3	0	Jewish
7	35	1	4	None
8	42	1	3	Protestant
9	48	1	1	Catholic
10	27	2	1	None
11	41	1	1	Protestant
12	42	2	0	Other
13	29	1	8	None
14	28	1	1	Jewish
15	47	2	1	Protestant

Respondent	Age	Happiness	Number of Partners	Religion
16	69	2	2	Catholic
17	44	1	4	Other
18	21	3	1	Protestant
19	33	2	1	None
20	56	1	2	Protestant
21	73	2	0	Catholic
22	31	1	1	Catholic
23	53	2	3	None
24	78	1	0	Protestant
25	47	2	3	Protestant
26	88	3	0	Catholic
27	43	1	2	Protestant
28	24	1	1	None
29	24	2	3	None
30	60	1	1	Protestant

3.7 SOC Find the appropriate measure of central tendency for each variable displayed in Problem 2.5. Report each statistic as you would in a formal research report.

3.8 SOC The following table lists the median family incomes for 13 Canadian provinces and territories in 2000 and 2006. Compute the mean and median for each year and then compare the two measures of central tendency. Which measure of central tendency is greater for each year? Are the distributions skewed? In which direction?

Province or Territory	2000	2006
Newfoundland and Labrador	38,800	50,500
Prince Edward Island	44,200	56,100
Nova Scotia	44,500	56,400
New Brunswick	43,200	54,000
Quebec	47,700	59,000
Ontario	55,700	66,600
Manitoba	47,300	58,700
Saskatchewan	45,800	60,500
Alberta	55,200	78,400
British Columbia	49,100	62,600
Yukon	56,000	76,000
Northwest Territories	61,000	88,800
Nunavut	37,600	54,300

3.9 SOC The administration is considering a total ban on student automobiles. You have conducted a poll on this issue of 20 fellow students and 20 of the neighbors who live around the campus and have calculated scores for your respondents. On the scale you used, a high score indicates strong opposition to the proposed ban. The scores are presented here for both groups. Calculate an appropriate measure of central tendency and then compare the two groups in a sentence or two.

Students		Neighbors	
10	11	0	7
10	9	1	6
10	8	0	0
10	11	1	3
9	8	7	4
10	11	11	0
9	7	0	0
5	1	1	10
5	2	10	9
0	10	10	0

3.10 SW As the head of a social services agency, you believe your staff of 20 social workers is very much overworked compared to 10 years ago. The caseloads for each worker are reported for each of the two years in question. Has the average caseload increased? What measure of central tendency is most appropriate to answer this question? Why?

2000		2010	
52	55	42	82
50	49	75	50
57	50	69	52
49	52	65	50
45	59	58	55
65	60	64	65
60	65	69	60
55	68	60	60
42	60	50	60
50	42	60	60

3.11 SOC The following table lists the price of a liter of regular gas for 20 nations. Compute the mean and median for these data. *NOTE: The costs are listed in order. This will make the median easy to find.* Which statistic is greater in value? Is there a positive skew in the data? How do you know?

Nation	Cost of a Liter of Gasoline (U.S. Dollars)
U.S.	0.56
Nigeria	0.59
Australia	0.74
Mexico	0.74
Canada	0.76
Argentina	0.78
South Africa	0.87
Ukraine	0.88
China	0.99
Colombia	1.04
India	1.09
Kenya	1.20
Spain	1.23
Brazil	1.26
Sweden	1.38
United Kingdom	1.44
Germany	1.56
Ireland	1.56
Italy	1.57
Japan	1.74

Source: http://www.nationmaster.com

3.12 SW For the test scores first presented in problem 2.6 and reproduced here, compute a median and mean for the pretest and posttest and then interpret these statistics.

Case	Pretest	Posttest
A	8	12
B	7	13
C	10	12
D	15	19
E	10	8
F	10	17
G	3	12
H	10	11
I	5	7
J	15	12
K	13	20
L	4	5
M	10	15
N	8	11
O	12	20

3.13 SOC A sample of 25 freshmen at a major university completed a survey that measured their degree of racial prejudice (the higher the score, the greater the prejudice).

a. Compute the median and mean scores for these data.

10	43	30	30	45
40	12	40	42	35
45	25	10	33	50
42	32	38	11	47
22	26	37	38	10

b. These same 25 students completed the same survey during their senior year. Compute the median and mean for this second set of scores and then compare them to the earlier set. What happened?

10	45	35	27	50
35	10	50	40	30
40	10	10	37	10
40	15	30	20	43
23	25	30	40	10

3.14 PA The following table presents the annual person-hours of time lost due to traffic congestion for a group of cities for 2007. This statistic is a measure of traffic congestion.

City	Annual Person-Hours of Time Lost to Traffic Congestion per Year per Person
Baltimore	25
Boston	22
Buffalo	5
Chicago	22
Cleveland	7
Dallas	32
Detroit	29
Houston	32
Kansas City	8
Los Angeles	38
Miami	27
Minneapolis	22
New Orleans	10
New York	21
Philadelphia	21
Pittsburgh	8
Phoenix	23
San Antonio	21
San Diego	29
San Francisco	29
Seattle	24
Washington, DC	31

Source: U.S. Bureau of the Census. 2011. *Statistical Abstract of the United States, 2011.* Table 1098. Available at http://www.census.gov/prod/2011pubs/11statab/trans.pdf.

a. Calculate the mean and median of this distribution.
b. Compare the mean and median. Which is the higher value? Why?
c. If you removed Los Angeles from this distribution and recalculated, what would happen to the mean? To the median? Why?
d. Report the mean and median as you would in a formal research report.

3.15 CJ The table below presents the homicide rates for the 50 states for two different years.

a. Calculate the mean and median rates for both years. Remember to rank-order the states before finding the median.
b. Compare the mean and median for each year separately. Which is the higher value? Why?
c. If you removed Louisiana (the state with the highest rate in both years) from the distributions and recalculated, what would happen to the mean? To the median? Why?
d. If you removed North Dakota (the state with the lowest rate) from the 1994 distribution and New Hampshire (the state with the lowest rate) from the 2009 distribution and recalculated, what would happen to the means and medians for both years? Why?
e. Compare the means and medians for the two different years and then write a brief paragraph summarizing the changes over time.

Homicide Rates (Number of Homicides per 100,000 Population) for 1994 and 2009

	1994	2009
Alabama	11.9	6.9
Alaska	6.3	3.1
Arizona	10.5	5.4
Arkansas	12.0	6.2
California	11.8	5.3
Colorado	5.4	3.5
Connecticut	6.6	3.0
Delaware	4.7	4.6
Florida	8.3	5.5
Georgia	10.0	5.8
Hawaii	4.2	1.7
Idaho	3.5	1.4
Illinois	11.7	6.0
Indiana	7.9	4.8
Iowa	1.7	1.1
Kansas	6.7	4.2
Kentucky	6.4	4.1
Louisiana	19.8	11.8
Maine	2.3	2.0
Maryland	11.6	7.7
Massachusetts	3.5	2.6
Michigan	9.8	6.3
Minnesota	3.2	1.4
Mississippi	15.3	6.4
Missouri	10.5	6.4
Montana	3.3	2.9
Nebraska	3.1	2.2
Nevada	11.7	5.9
New Hampshire	1.4	0.8
New Jersey	5.0	3.7
New Mexico	10.7	8.7
New York	11.1	4.0

Homicide Rates (Number of Homicides per 100,000 Population) for 1994 and 2009		
North Carolina	10.9	5.3
North Dakota	0.2	1.5
Ohio	6.0	4.5
Oklahoma	6.9	6.2
Oregon	4.9	2.2
Pennsylvania	5.9	5.2
Rhode Island	4.1	2.9
South Carolina	9.6	6.3
South Dakota	1.4	2.6
Tennessee	9.3	7.3
Texas	11.0	5.4
Utah	2.9	1.3
Vermont	1.0	1.1
Virginia	8.7	4.4
Washington	5.5	2.7
West Virginia	5.4	4.6
Wisconsin	4.5	2.5
Wyoming	3.4	2.4

3.16 Professional athletes are threatening to strike because they claim they are underpaid. The team owners have released a statement that says, in part, "The average salary for players was $1.2 million last year." The players counter by issuing their own statement that says, in part, "The average player earned only $753,000 last year." Is either side necessarily lying? If you were a sports reporter and had just read Chapter 3 of this text, what questions would you ask about these statistics?

YOU ARE THE RESEARCHER: The Typical American

Is there such a thing as a "typical" American? In this exercise, you will develop a profile of the average American based on measures of central tendency for 10 variables chosen by you from the 2010 General Social Survey. Choose variables that you think are important in defining what it means to be a member of this society and then choose an appropriate measure of central tendency for each variable. Use this information to write a description of the typical American. We will also use this opportunity to introduce a new SPSS program.

STEP 1: Choosing Your Variables

Scroll through the list of available variables either in Appendix G or by using the **Utilities → Variables** command in SPSS to get the online codebook. Select 10 variables that, in your view, are central to defining or describing the "typical American" and then list them in the table below. *Select at least one variable from each of the three levels of measurement.*

Variable	SPSS Name	Explain Exactly What This Variable Measures	Level of Measurement
1			
2			
3			
4			
5			
6			
7			
8			
9			
10			

STEP 2: Getting the Statistics

Start *SPSS for Windows* by clicking the SPSS icon on your monitor screen. Load the 2010 GSS, and when you see the message "SPSS Processor is Ready" on the bottom of the "closest" screen, you are ready to proceed.

Using the "Frequencies" Procedure for the Mode and Median

The only procedure in SPSS that will produce all three commonly used measures of central tendency (mode, median, and mean) is **Frequencies**. We used this procedure to produce frequency distributions in Chapter 2 and in Appendix F. Here, you will use **Frequencies** to get modes and medians for any nominal and ordinal-level variables you selected in step 1. We will process the interval-ratio-level variables in the next step.

Begin by clicking **Analyze** from the menu bar and then click **Descriptive Statistics** and **Frequencies**. In the **Frequencies** dialog box, find the names of your nominal- and ordinal-level variables in the list on the left and then click the arrow button in the middle of the screen to move the names to the **Variables** box on the right.

To request specific statistics, click the **Statistics** button in the **Frequencies** dialog box, and the **Frequencies: Statistics** dialog box will open. Find the **Central Tendency** box on the right and then click **Median** and **Mode**. Click **Continue**, and you will be returned to the **Frequencies** dialog box, where you might want to click the **Display Frequency Tables** box. When this box is *not* checked, SPSS will *not* produce frequency distribution tables and only the statistics we request (mode and median) will appear in the **Output** window. Click **OK**, and SPSS will produce your output.

Record your results in the table below. Report the mode for all nominal-level variables and the median for ordinal-level variables and use as many lines as needed.

Variable	SPSS Name	Level of Measurement	Mode	Median
1				
2				
3				
4				
5				
6				
7				
8				
9				

Using the "Descriptives" Procedure for the Mean

The **Descriptives** command in *SPSS for Windows* is designed to provide summary statistics for interval-ratio-level variables. By default (i.e., unless you tell it otherwise), **Descriptives** produces the mean, the minimum and maximum scores (i.e., the lowest and highest scores, which can be used to compute the range), and the standard deviation. We will introduce the range and standard deviation in the next chapter.

To use **Descriptives**, click **Analyze**, **Descriptive Statistics**, and **Descriptives**. The **Descriptives** dialog box will open. This dialog box looks just like the **Frequencies** dialog box and works in the same way. Find the names of your interval-ratio-level variables in the list on the left, and once they are highlighted, click the arrow button in the middle of the screen to transfer them to the **Variables** box on the right. Click **OK** to produce your output and then record your results in the table below, using as many lines as necessary.

Variable	SPSS Name	Level of Measurement	Mean
1			
2			
3			
4			
5			
6			
7			
8			
9			

STEP 3: Interpreting Results

Examine the two tables displaying your results and then write a summary paragraph or two describing the typical American. Be sure to report all 10 variables and, as appropriate, describe the most common case (the mode), the typical case (the median), or the typical score (mean). Write as if you are reporting in a newspaper; your goal should be clarity and accuracy. As an example, you might report that "The typical American is Protestant and married."

4 Measures of Dispersion

LEARNING OBJECTIVES By the end of this chapter, you will be able to:

1. Explain the purpose of measures of dispersion and the information they convey.
2. Compute and explain range (R), interquartile range (Q), the standard deviation (s), and variance (s^2).
3. Select an appropriate measure of dispersion and then correctly calculate and interpret the statistic.
4. Describe and explain the mathematical characteristics of the standard deviation.

USING STATISTICS

The statistical techniques presented in this chapter are used to describe the variability or diversity in a set of scores. They can be used to describe:

- Variability in a student's grades. Some students get the same grade virtually all the time, while others vary from high to low grades.
- Diversity of lifestyles in different social settings. Larger cities tend to support a greater variety of lifestyles than small towns.
- Differences in racial and ethnic diversity from state to state in the United States. Some states (California and New York, for example) are home to many different racial, cultural, and language groups, while others (such as Iowa and Maine) are much less variable.
- Variations in income inequality from nation to nation or from time to time. Some nations feature vast differences in wealth and income from the very richest to the poorest members, while the differences might be much smaller in other nations.

Introduction

Chapters 2 and 3 presented a variety of ways of describing a variable, including frequency distributions, graphs, and measures of central tendency. For a complete description of a distribution of scores, we must combine these with **measures of dispersion**—the subject of this chapter. While measures of central tendency describe the typical, average, or central score, measures of dispersion describe variety, diversity, or heterogeneity of a distribution of scores.

The importance of the concept of **dispersion** might be easier to grasp if we consider a brief example. Suppose the director of public safety wants to evaluate two ambulance services that have contracted with the city to provide emergency medical aid. As a part of the investigation, she has collected data on the response time of both services to calls for assistance. Data collected for the past year show that the mean response time is 7.4 minutes for Service A and 7.6 minutes for Service B. These averages, or means (calculated by adding up the response times to all calls and dividing by the number of calls), are very similar and provide no basis for judging one service as more or less efficient than the other. However, measures of dispersion can reveal substantial differences between distributions even when the measures of central tendency are the same. For example, consider Figure 4.1, which displays in line charts the distribution of response times for the two services (see Chapter 2).

Compare the shapes of these two figures. Note that the chart for Service B is much flatter than that for Service A. This is because the scores for Service B are more spread out, or more diverse, than the scores for Service A. In other words, Service B was much more variable in response time and had more scores in the high *and* low ranges and fewer in the middle. Service A was more consistent in its response time, and its scores are more clustered, or grouped, around the mean. Both distributions have essentially the same *average* response time, but there is considerably more *variation*, or dispersion, in the response times for Service B.

If you were the director of public safety, would you be more likely to select an ambulance service that was always on the scene of an emergency in about the same amount of time (Service A) or one that was sometimes very slow and sometimes very quick to respond (Service B)? Note that if you had not considered

FIGURE 4.1 Response Time for Two Ambulance Services

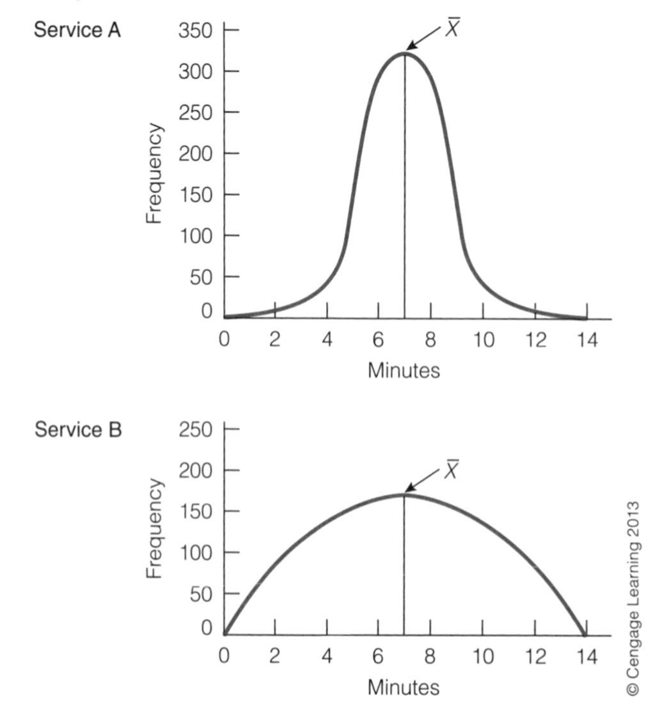

dispersion, a possibly important difference in the performance of the two ambulance services might have gone unnoticed.

Keep the two shapes in Figure 4.1 in mind as visual representations of the concept of dispersion. The greater clustering of scores around the mean in the distribution for Service A indicates *less* dispersion, and the flatter curve of the distribution for Service B indicates *more* dispersion, or variety. Measures of dispersion decrease in value as the scores become less dispersed and the distribution becomes more peaked (that is, as the distribution looks more and more like Service A's) and increase in value as the scores become more dispersed and the distribution becomes flatter (that is, as the distribution looks more and more like Service B's).

These ideas and Figure 4.1 may give you a general notion of what is meant by dispersion, but the concept is not easily described in words alone. In this chapter, we introduce some of the more common measures of dispersion, each of which provides a quantitative indication of the variety in a set of scores. We devote most of our attention to the standard deviation—the most important and widely used measure of dispersion.

STATISTICS IN EVERYDAY LIFE

The United States is growing more diverse on a number of dimensions. As noted in Chapter 3, Americans espouse an increasing variety of religious affiliations. As we saw, the numerical predominance of Protestants declined from 61% in 1990 to 54% in 2010. Over the same time period, some religions (Islam, for example) increased in size, as did the percentage of the population that has no religious affiliation. Overall, the religious profile of the United States is becoming more "spread out," or variable. A similar increase in diversity is occurring in terms of race, ethnicity, culture, and language. For example, in 1990, over 75% of Americans identified with a single racial category: white. By 2050, this percentage will decline to less than 50% and the United States will be majority minority. What are the implications of this increasing religious, racial, and ethnic diversity? Will social life in the United States become more interesting and richer or more discordant and quarrelsome?

The Range (*R*) and Interquartile Range (*Q*)

The Range. The **range (*R*)** is defined as the distance between the highest and lowest scores in a distribution:

FORMULA 4.1

$$R = \text{High score} - \text{Low score}$$

The range is quite easy to calculate and is perhaps most useful to gain a quick and general notion of variability while scanning many distributions. This statistic is also easy to interpret: the greater the value of the range, the greater the distance from high to low score and the greater the dispersion in the distribution.

Unfortunately, the range has some important limitations related to the fact that it is based on only two scores (the highest and lowest scores). First, almost any sizable distribution will contain some scores that are unusually high and/or low compared to most of the scores (for example, see Table 3.3). Thus, *R* might exaggerate the amount of dispersion for most of the scores in the distribution. Also, *R* yields no information about the variation of the scores between the highest and lowest scores.

The Interquartile Range. The **interquartile range (Q)** is a kind of range. It avoids some of the problems associated with R by considering only the middle 50% of the cases in a distribution. To find Q, arrange the scores from highest to lowest and then divide the distribution into quarters (as distinct from halves when locating the median). The first quartile (Q_1) is the point below which 25% of the cases fall and above which 75% of the cases fall. The second quartile (Q_2) divides the distribution into halves (thus, Q_2 is equal in value to the median). The third quartile (Q_3) is the point below which 75% of the cases fall and above which 25% of the cases fall. Thus, if line LH represents a distribution of scores, the quartiles are located as shown:

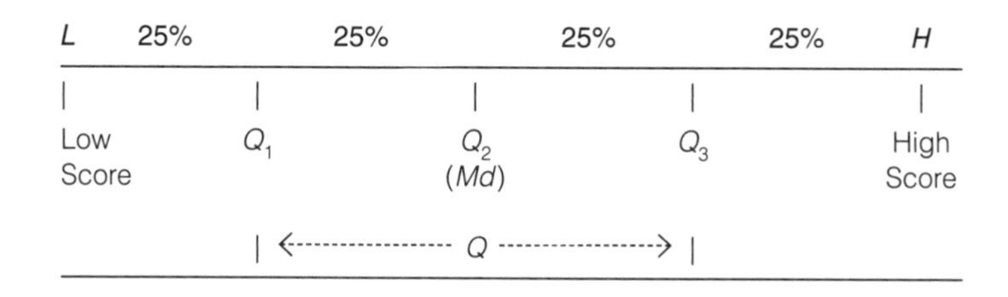

The interquartile range is defined as the distance from the third to the first quartile, as stated in Formula 4.2:

FORMULA 4.2

$$Q = Q_3 - Q_1$$

The interquartile range essentially extracts the middle 50% of the distribution and, like R, is based on only two scores. Unlike the range, Q avoids the problem of being based on the most extreme scores, but it also fails to yield any information about the variation of the scores other than the two on which it is based.

Computing the Range and Interquartile Range[1]

Table 4.1 presents a measure of educational attainment for 20 states. What are the range and interquartile range of these data? Note that the scores have already been ordered from high to low. This makes the range easy to calculate and is necessary for finding the interquartile range. Of these 20 states, Connecticut ranked the highest (35.6% of the population had a college degree) and West Virginia ranked the lowest (17.1%). The range is therefore 35.6 − 17.1, or 18.5 (R = 18.5).

To find Q, we must locate the first and third quartiles (Q_1 and Q_3). We will define these points in terms of the scores associated with certain cases, as we did when finding the median. Q_1 is determined by multiplying N by (.25). Because (20) × (.25) is 5, Q_1 is the score associated with the fifth case, counting up from the lowest score. The fifth case is Indiana, with a score of 22.9. Thus, Q_1 = 22.9. The case that lies at the third quartile (Q_3) is given by multiplying N by (.75): (20) × (.75) = 15th case.

[1]This section is optional.

TABLE 4.1 Percent of Population Aged 25 and Older in Twenty States With a College Degree, 2010

Rank	State	Percent With College Degree
20 (highest)	Connecticut	35.6
19	New Hampshire	33.3
18	New York	31.9
17	California	29.6
16	Kansas	29.6
15	Montana	27.1
14	Pennsylvania	26.3
13	North Carolina	26.1
12	Florida	25.8
11	Maine	25.4
10	Texas	25.3
9	Arizona	25.1
8	Michigan	24.7
7	Ohio	24.1
6	Idaho	24.0
5	Indiana	22.9
4	Alabama	22.0
3	Louisiana	20.3
2	Mississippi	19.4
1 (lowest)	West Virginia	17.1

Source: U.S. Bureau of the Census. 2011. *Statistical Abstract of the United States, 2011*, p. 151. Available at http://www.census.gov/prod/2011pubs/11statab/educ.pdf.

The 15th case, again counting up from the lowest score, is Montana, with a score of 27.1 ($Q_3 = 27.1$). Therefore:

$$Q = Q_3 - Q_1$$
$$Q = 27.1 - 22.9$$
$$Q = 4.2$$

In most situations, the locations of Q_1 and Q_3 will not be as obvious as they are when $N = 20$. For example, if N had been 157, then Q_1 would be (157)(.25)—or the score associated with the 39.25th case—and Q_3 would be (157)(.75)—or the score associated with the 117.75th case. Because fractions of cases are impossible, these numbers present some problems. The easy solution to this difficulty is to round off and take the score of the closest case to the numbers that mark the quartiles. Thus, Q_1 would be defined as the score of the 39th case and Q_3 as the score of the 118th case.

The more accurate solution would be to take the fractions of cases into account. For example, Q_1 could be defined as the score that is one-quarter of the distance between the scores of the 39th and 40th cases and Q_3 could be defined as the score that is three-quarters of the distance between the scores of the 117th and 118th cases. (This procedure could be analogous to defining the median—which is also Q_2—as halfway between the two middle scores when N is even.) In most cases, the differences in the values of Q for these two methods would be quite small. *(For practice in finding and interpreting Q, see problem 4.5. The range may be found for any of the problems at the end of this chapter.)*

ONE STEP AT A TIME **Finding the Interquartile Range (Q)**

Step	*Operation*
1.	Arrange the cases in order.
2.	Find the case that lies at the first quartile (Q_1) by multiplying N by 0.25. If the result is not a whole number, round off to the nearest whole number. This is the number of the case that marks the first quartile. Note the score of this case.
3.	Find the case that lies at the third quartile (Q_3) by multiplying N by 0.75. If the result is not a whole number, round off to the nearest whole number. This is the number of the case that marks the third quartile. Note the score of this case.
4.	Subtract the score of the case at the first quartile (Q_1)—see step 2—from the score of the case at the third quartile (Q_3)—see step 3. The result is the interquartile range (Q).

The Standard Deviation and Variance

A basic limitation of Q and R is that they are based on only two scores. They do not use all the scores or all the available information. Also, neither statistic provides any information on how far the scores are from each other or from some central point, such as the mean. How can we design a measure of dispersion that would correct these faults? We can begin with some specifications. A good measure of dispersion should:

1. Use all the scores in the distribution. The statistic should use all the information available.
2. Describe the average or typical deviation of the scores. The statistic should give us an idea about how far the scores are from each other or from the center of the distribution.
3. Increase in value as the scores became more diverse. This would be a very handy feature because it would permit us to tell at a glance which distribution was more variable: the higher the numerical value of the statistic, the greater the dispersion.

One way to develop a statistic to meet these criteria would be to start with the distances between each score and the mean, or **deviations** $(X_i - \overline{X})$. The value of the deviations will increase as the differences between the scores and the mean increase. If the scores are more clustered around the mean (remember the graph for Service A in Figure 4.1), the deviations will be smaller. If the scores are more spread out or more varied (like the scores for Service B in Figure 4.1), the deviations will be larger. How can we use the deviations to develop a useful statistic?

One course of action would be to use the sum of the deviations—$\Sigma(X_i - \overline{X})$—as the basis for a statistic, but as we saw in Section 3.6, the sum of the deviations will always be zero. To illustrate, consider the distribution of five scores presented in Table 4.2. If we sum the deviations of the scores from the mean, we will always wind up with a total of zero, regardless of the amount of variety in the scores.

Still, the sum of the deviations is a logical basis for a measure of dispersion, and statisticians have developed a way around the fact that the positive deviations always equal the negative deviations. If we square each of the deviations, all values will be positive because a negative number multiplied by itself becomes positive.

TABLE 4.2 A Demonstration That the Sum of the Deviations of the Scores Around the Mean Will Always Total Zero

Scores (X_i)	Deviations ($X_i - \bar{X}$)
10	10 − 30 = −20
20	20 − 30 = −10
30	30 − 30 = 0
40	40 − 30 = 10
50	50 − 30 = 20
$\Sigma(X_i) = 150$	$\Sigma(X_i - \bar{X}) = 0$
$\bar{X} = 150/5 = 30$	

For example: (–20) × (−20) = +400. In our example, the sum of the squared deviations would be (400 + 100 + 0 + 100 + 400), or 1,000. Thus, a statistic based on the sum of the squared deviations will have the properties we want in a good measure of dispersion.

Before we finish designing our measure of dispersion, we must deal with another problem. The sum of the squared deviations will increase with the sample size: the larger the *number* of scores, the greater the value of the measure. This would make it very difficult to compare the relative variability of distributions based on samples of different size. We can solve this problem by dividing the sum of the squared deviations by N (sample size) and thus standardizing for samples of different sizes.

These procedures yield a statistic known as the **variance**, which is symbolized as s^2. The variance is used primarily in inferential statistics, although it is a central concept in the design of some measures of association. For purposes of describing the dispersion of a distribution, a closely related statistic—called the **standard deviation**, or s—is used, and this will be our focus for the remainder of this chapter.

The formulas for the variance and standard deviation are:

FORMULA 4.3

$$s^2 = \frac{\Sigma(X_i - \bar{X})^2}{N}$$

FORMULA 4.4

$$s = \sqrt{\frac{\Sigma(X_i - \bar{X})^2}{N}}$$

Strictly speaking, Formulas 4.3 and 4.4 are for the variance and standard deviation of a population. Slightly different formulas—with $N - 1$ instead of N in the denominator—should be used when we are working with random samples rather than entire populations. This is an important point because many of the electronic calculators and statistical software packages you might be using (including SPSS) use $N - 1$ in the denominator and, thus, produce results that are at least slightly different from these formulas. The size of the difference will decrease as the sample size increases, but the problems and examples in this chapter use small samples, and the differences between using N and $N - 1$ in the denominator can be considerable in such cases. Some calculators offer the choice of $N - 1$ or N in the denominator. If you use the latter, the values calculated for the standard deviation should match the values in this text.

TABLE 4.3 Computing the Standard Deviation

Scores (X_i)	Deviations ($X_i - \bar{X}$)	Deviations Squared ($X_i - \bar{X}$)
10	$10 - 30 = -20$	$(-20)^2 = 400$
20	$20 - 30 = -10$	$(-10)^2 = 100$
30	$30 - 30 = 0$	$(0)^2 = 0$
40	$40 - 30 = 10$	$(10)^2 = 100$
50	$50 - 30 = 20$	$(20)^2 = 400$
$\Sigma(X_i) = 150$	$\Sigma(X_i - \bar{X}) = 0$	$\Sigma(X_i - \bar{X})^2 = 1{,}000$

To compute the standard deviation, it is advisable to construct a table like Table 4.3 to organize computations. The five scores used in the previous example are listed in the left-hand column, the deviations are in the middle column, and the squared deviations are in the right-hand column.

The total of the last column in Table 4.3 is the sum of the squared deviations and should be substituted into the numerator of Formula 4.4 . To finish solving the formula, divide the sum of the squared deviations by N and take the square root of the result.

$$s = \sqrt{\frac{\Sigma(X_i - \bar{X})^2}{N}}$$

$$s = \sqrt{\frac{1{,}000}{5}}$$

$$s = \sqrt{200}$$

$$s = 14.14$$

To find the variance, square the standard deviation. For our example problem, the variance is $s^2 = (14.14)^2 = 200$:

ONE STEP AT A TIME Finding the Standard Deviation (s) and the Variance (s^2)

Step ***Operation***

1. Construct a computing table like Table 4.3, with columns for the scores (X_i), the deviations ($X_i - \bar{X}$), and the deviations squared [$(X_i - \bar{X})^2$].
2. List the scores (X_i) in the left-hand column. Add up the scores and then divide by N to find the mean.
3. Find the deviations ($X_i - \bar{X}$) by subtracting the mean from each score—one at a time. List the deviations in the second column.
4. Add up the deviations. The sum must equal zero (within rounding error). If the sum of the deviations does not equal zero, you have made a computational error and need to repeat steps 2 and 3.
5. Square each deviation and list the result in the third column.
6. Add up the squared deviations listed in the third column and then transfer this sum to the numerator in Formula 4.4.
7. Divide the sum of the squared deviations (see step 6) by N.
8. Take the square root of the quantity you computed in step 7. This is the standard deviation.

To Find the Variance (s^2):

1. Square the value of the standard deviation (s) or see step 7.

Computing the Standard Deviation: An Additional Example

An additional example will help to clarify the procedures for computing and interpreting the standard deviation. A researcher is comparing the student bodies of two campuses. One college is located in a small town, and almost all the students reside on campus. The other is located in a large city, and the students are almost all part-time commuters. The researcher wishes to compare the age structure of the two campuses and has compiled the information presented in Table 4.4. Which student body is older and more diverse on age? (Needless to say, these very small groups are much too small for serious research and are used here only to simplify computations.)

We see from the means that the students from the residential campus are quite a bit younger than the students from the urban campus (19 vs. 23 years of age). Which group is more diverse in age? Computing the standard deviation will answer this question. To solve Formula 4.4 , substitute the sum of the right-hand column ("Deviations Squared") in the numerator and N (5 in this case) in the denominator:

Residential Campus:

$$s = \sqrt{\frac{\Sigma(X_i - \overline{X})^2}{N}} = \sqrt{\frac{4}{5}} = \sqrt{.8} = .89$$

Urban Campus:

$$s = \sqrt{\frac{\Sigma(X_i - \overline{X})^2}{N}} = \sqrt{\frac{88}{5}} = \sqrt{17.6} = 4.20$$

TABLE 4.4 Computing the Standard Deviation in Students' Ages for Two Campuses

Residential Campus		
Scores (X_i)	Deviations ($X_i - \overline{X}$)	Deviations Squared $(X_i - \overline{X})^2$
18	$18 - 19 = -1$	$(-1)^2 = 1$
19	$19 - 19 = 0$	$(0)^2 = 0$
20	$20 - 19 = 1$	$(-1)^2 = 1$
18	$18 - 19 = -1$	$(-1)^2 = 1$
20	$20 - 19 = 1$	$(1)^2 = 1$
$\sum(X_i) = 95$	$\sum(X_i - \overline{X}) = 0$	$\sum(X_i - \overline{X})^2 = 4$
$\overline{X} = \frac{\Sigma(X_i)}{N} = \frac{95}{5} = 19$		

Urban Campus		
Scores (X_i)	Deviations ($X_i - \overline{X}$)	Deviations Squared $(X_i - \overline{X})^2$
20	$20 - 23 = -3$	$(-3)^2 = 9$
22	$22 - 23 = -1$	$(1)^2 = 1$
18	$18 - 23 = -5$	$(-5)^2 = 25$
25	$25 - 23 = 2$	$(2)^2 = 4$
30	$30 - 23 = 7$	$(7)^2 = 49$
$\sum(X_i) = 115$	$\sum(X_i - \overline{X}) = 0$	$\sum(X_i - \overline{X})^2 = 88$
$\overline{X} = \frac{\Sigma(X_i)}{N} = \frac{115}{5} = 23$		

STATISTICS
IN EVERYDAY LIFE

The citizens of modern industrial societies can routinely expect to live into their 70s, 80s, or even older. This is not the case, of course, for nations with lower standards of living, nutrition, and health care. The table below presents some statistical information on life expectancy for a group of 95 nations at three different income levels. Note that average life expectancy increases as income level increases but that the measures of dispersion (the range and the standard deviation) do just the opposite. Why are low-income nations more diverse on this variable? ***(Hint: Are people in low-income nations more vulnerable at all ages?)***

Summary Statistics on Life Expectancy for 92 Nations by Income Level, 2010

Income Level	Mean	Standard Deviation	High	Low	Range	*N*
High	79.9	1.80	83	75	8	25
Middle	70.1	6.41	76	55	21	42
Low	53.7	7.01	66	43	23	25

The higher value of the standard deviation for the urban campus means that it is more diverse. As you can see by scanning the scores, the students at the residential college are within a narrow age range ($R = 20 - 18 = 2$), whereas the students at the urban campus are more mixed and include students of ages 25 and 30 ($R = 30 - 18 = 12$). *(For practice in computing and interpreting the standard deviation, see any of the problems at the end of this chapter. Problems with smaller data sets, such as 4.1 to 4.2, are recommended for practicing computations until you are comfortable with these procedures.)*

Interpreting the Standard Deviation

It is very possible that the meaning of the standard deviation is not completely obvious to you at this point. You might be asking: "Once I've gone to the trouble of calculating the standard deviation, what do I have?" The meaning of this measure of dispersion can be expressed in three ways. The first and most important involves the normal curve, and we will defer this interpretation until the next chapter.

A second way of thinking about the standard deviation is as an index of variability that increases in value as the distribution becomes more variable. In other words, the standard deviation is higher for more diverse distributions and lower for less diverse distributions. The lowest value the standard deviation can have is zero, and this would occur for distributions with no dispersion or distributions in which every single case had exactly the same score. Thus, 0 is the lowest value possible for the standard deviation (although there is no upper limit).

A third way to get a feel for the meaning of the standard deviation is by comparing one distribution with another. We already did this when comparing the two ambulance services in Figure 4.1 and the residential and urban campuses in Table 4.4. You might also do this when comparing one group against another (e.g., men vs. women, blacks vs. whites) or the same variable at two different times. For example, suppose we found that the ages of the students on a particular campus had changed over time, as indicated by the following summary statistics:

2000	2010
$\bar{X} = 21$	$\bar{X} = 25$
$s = 1$	$s = 3$

In 2000, students were, on average, 21 years old. By 2010, the average age had risen to 25. Clearly, the student body has grown older, and according to the standard deviation, it has also grown more diverse in age. The lower standard deviation for 2000 shows that ages would be more clustered around the mean in that year (remember the distribution for Service A in Figure 4.1). In 2010, in contrast, the larger standard deviation means that the distribution is flatter and more spread out, like the distribution for Service B in Figure 4.1. In other words, the students became more diverse in age and less clustered in a narrower age range. The standard deviation is extremely useful for making comparisons of this sort between distributions of scores.

Applying Statistics 4.1 The Standard Deviation

At a local preschool, 10 children were observed for 1 hour, and the number of aggressive acts committed by each was recorded in the following list. What is the standard deviation of this distribution? We will use Formula 4.4 to compute the standard deviation.

NUMBER OF AGGRESSIVE ACTS

Scores (X_i)	Deviations $(X_i - \bar{X})$	Deviations Squared $(X_i - \bar{X})^2$
1	$1 - 4 = -3$	$(-3)^2 = 9$
3	$3 - 4 = -1$	$(\ 1)^2 = 1$
5	$5 - 4 = 1$	$(\ 1)^2 = 1$
2	$2 - 4 = -2$	$(-2)^2 = 4$
7	$7 - 4 = 3$	$(\ 3)^2 = 9$
11	$11 - 4 = 7$	$(\ 7)^2 = 49$
1	$1 - 4 = 3$	$(-3)^2 = 9$
8	$8 - 4 = 4$	$(\ 4)^2 = 16$
2	$2 - 4 = -2$	$(-2)^2 = 4$
0	$0 - 4 = -4$	$(-4)^2 = 16$
$\sum(X_i) = 40$	$\sum(X_i - \bar{X}) = 0$	$\sum(X_i - \bar{X})^2 = 118$

$$\bar{X} = \frac{\sum(X_i)}{N} = \frac{40}{10} = 4.0$$

Substituting into Formula 4.4, we have:

$$s = \sqrt{\frac{\sum(X_i - \bar{X})^2}{N}} = \sqrt{\frac{118}{10}} = \sqrt{11.8} = 3.44$$

The standard deviation for these data is 3.44.

Applying Statistics 4.2 Describing Dispersion

The homicide rates of five western and five New England states are listed in the table below. Which group of states has the highest rate? Which is most variable?

HOMICIDE RATE FOR FIVE NEW ENGLAND STATES, 2010 (HOMICIDES PER 100,000 POPULATION)

State	Homicides (X_i)	Deviations ($X_i - \bar{X}$)	Deviations Squared $(X_i - \bar{X})^2$
Connecticut	3.80	1.14	1.30
Massachusetts	2.60	−0.06	0.00
Vermont	2.80	0.14	0.02
Rhode Island	3.00	0.34	0.12
New Hampshire	1.10	−1.56	2.43
	$\sum(X_i) = 13.30$	$\sum(X_i - \bar{X}) = 0.00$	$\sum(X_i - \bar{X})^2 = 3.87$

$$\bar{X} = \frac{\Sigma X_i}{N} = \frac{13.30}{5} = 2.66$$

$$s = \sqrt{\frac{\Sigma(X_i - \bar{X})}{N}} = \sqrt{\frac{3.87}{5}} = \sqrt{0.77} = 0.88$$

HOMICIDE RATE FOR FIVE WESTERN STATES, 2010 (HOMICIDES PER 100,000 POPULATION)

State	Homicides (X_i)	Deviations ($X_i - \bar{X}$)	Deviations Squared $(X_i - \bar{X})^2$
Nevada	6.30	2.30	5.29
California	5.80	1.80	3.24
Wyoming	2.30	−1.70	2.89
Oregon	2.30	−1.70	2.89
Montana	3.30	−0.70	0.49
	$\sum(X_i) = 20.00$	$\sum(X_i - \bar{X}) = 0.00$	$\sum(X_i - \bar{X})^2 = 14.80$

Source: U.S. Bureau of the Census. 2011. *Statistical Abstract of the United States, 2011.* U.S. Government Printing Office: Washington, D.C., p. 195.

$$\bar{X} = \frac{\Sigma X_i}{N} = \frac{20.00}{5} = 4.00$$

$$s = \sqrt{\frac{\Sigma(X_i - \bar{X})}{N}} = \sqrt{\frac{14.80}{5}} = \sqrt{2.96} = 1.72$$

With such small groups, you can tell by inspection that the western states have higher homicide rates. This impression is confirmed by the median, which is 2.80 for the New England states (Vermont is the middle case) and 3.3 for the western states (Montana is the middle case), and the mean (2.66 for the New England states and 4.00 for the western states). For the western states, the mean is greater than the median, indicating a positive skew. For the eastern states, the mean is lower than the median and this indicates a negative skew.

The five western states are also much more variable. The range for the western states is 4 ($R = 6.3 - 2.3 = 4$)—higher than the range for the New England states of 2.7 ($R = 3.8 - 1.1 = 2.7$). Similarly, the standard deviation for the western states (1.72) is twice the value of the standard deviation for the eastern states (0.88). The greater dispersion in the western states is also largely a reflection of Nevada's high score.

In summary, the five western states average higher homicide rates and are also more variable than the five eastern states.

Application 4.3 Finding The Perfect Place to Live

If you could live anywhere in the United States, where would it be? What criteria would you use to make your selection? Suppose you place a very high value on climate and consider mild temperatures (around 70° F.) to be especially desirable and decide to use this criterion to select your next place of residence.

So far, so good, but your decision about the "nicest place to live" could be sadly mistaken if you consider *only* the average temperature. What if the cities you examined had roughly the same average temperature of 70° but differed in the range or in the extremes of temperature they experienced? What if some cities had temperatures that ranged from blistering (a high of 103°) to freezing (a low of –10°) but others had nothing but mild, balmy temperatures year-round? Clearly, you would want to consider dispersion as well as central tendency in choosing your residence.

What American city comes closest to the ideal of a constant 70° temperature year-round? The table below presents data for five cities:

AVERAGE MONTHLY TEMPERATURE DATA FOR FIVE CITIES

	Miami	Chicago	New York City	Dallas	San Diego
Mean	82.79	58.63	62.32	76.28	70.79
Std. Dev.	5.12	19.90	17.31	15.21	4.55
Range	13.80	54.70	47.60	42.50	11.90

Compared to your ideal average temperature of 70°, Chicago and New York City are too cold and Dallas and Miami are too hot. This leaves San Diego, which matches your ideal average daily temperature almost exactly.

Considering dispersion, temperature fluctuations are greatest for Chicago (with a standard deviation of almost 20°) and are also considerable for New York City and Dallas. These cities experience all four seasons, and this is not what you want in your ideal residence. Miami and San Diego have much smaller temperature swings throughout the year, and the latter city, with its 71° average temperature and small standard deviation, comes closest to matching your ideal climate. Although there are many more places to investigate in the search for the perfect place to live (and, no doubt, other criteria to consider besides climate), San Diego is the clear winner among these five cities.

BECOMING A CRITICAL CONSUMER: Getting the Whole Picture

As we have seen, measures of dispersion provide important information about variables. In fact, we can say that in order to completely understand a variable, we need three different kinds of information: overall shape of a table or graph (see Chapter 2), central tendency (see Chapter 3), and dispersion or variability (the subject of this chapter). Reporting only one type of information can be misleading and lead to completely incorrect impressions. This is particularly a problem in the popular media, which typically reports only central tendency (see the "headlines" about housing and gas prices in the installment of "Becoming a Critical Consumer" in Chapter 3). Here, we will consider several examples of problems stemming from incomplete reporting and take a look at the reporting styles you will find in the professional research literature.

As our first example, consider the following scenario: Assume that the real estate market in a particular city is doing poorly and that every single

(*continued next page*)

BECOMING A CRITICAL CONSUMER: *(continued)*

house is losing value.[2] You would naturally expect that the average sale price of housing would fall, right? Maybe not: It is possible that the median or mean price of a house will be rising even during a recession. How? The answer relates to the overall shape of the distribution, not just the central point identified by the mean or median. For example, the measures of central tendency may rise if expensive houses—say, houses costing $2 million—are still selling, even though at a lower price (say $1.6 million), while houses in the lower and middle ranges (houses costing less than $250,000) are not selling at all. This pattern would exaggerate the positive skew in the distribution of sale prices, and measures of central tendency could rise even when the housing market is suffering.

The table below presents a simplified picture of how this might work. The left-hand side of the table is a "balanced" market where houses in all price ranges are selling and the right-hand side is the "skewed" market where only high-end houses are selling (but for less than before). As you can see, the two measures of central tendency rise from time 1 to time 2, giving the false impression that the housing market is prospering. On the other hand, the measures of dispersion decline over the time period. This indicates that the distribution is becoming less variable and diverse: The sales are concentrated in one sector (the high end) of the market. Taken together, we can see what these statistics are telling us: The rising mean and median reflect the unbalanced nature of sales, not an increase in housing values. The point, of course, is that partial information about a variable can be dangerously misleading.

Let us consider another example—this time using real data. It is well known that males have higher average incomes than females. Because income is positively skewed, the median is the most useful measure of central tendency, and it shows that the income gap between the genders has shrunk over the decades. In 1955, women earned about 65% of what men earned, but the gap shrunk to the point where women earned 76% of men's earnings in 2009. This trend would be heartening to those who advocate gender equality, but it does not present the full picture. The line chart below shows the overall shape of the distribution of incomes by gender for 2009. The figure shows that women are more concentrated in the lower and middle income brackets, while men are concentrated in higher incomes. In fact, in the three highest income

Time 1		***Time 2***	
Price	Number of Sales	Price	Number of Sales
2,000,000	1	1,600,000	2
750,000	1	650,000	1
500,000	1	500,000	0
250,000	1	250,000	0
100,000	1	100,000	0
Mean Sale Price = 720,000		Mean Sale Price = 1,316,667	
Median Sale Price = 500,000		Median Sale Price = 1,600,00	
Range of Sale Prices = 1,900,00		Range of Sale Prices = 950,000	
Standard Deviation of Sale Prices = 677,200.10		Standard Deviation of Sale Prices = 449,073.84	

[2]We are grateful to Felix Salmon at http://seekingalpha.com/article/81663-lies-damn-lies-and-median-house-prices for this example.

categories, men outnumber women by 2 to 1. Looking only at the trends in median income may lead to a rosier impression of the extent of income inequality than a consideration of the entire distribution would merit.

Reading the Professional Literature. Our point about the need to describe variables fully—their shape, central tendency, and dispersion—may not be honored in the professional research literature of the social sciences. These projects typically include many variables and space in journals is expensive, so there may not be room to describe each variable fully. Furthermore, virtually all research reports focus on *relationships* between variables rather than the distribution of single variables. In this sense, univariate descriptive statistics are irrelevant to the main focus of the report.

This does not mean, of course, that univariate descriptive statistics are irrelevant to the researchers. They will calculate and interpret these statistics for virtually every variable in virtually every research project. However, these statistics are less likely to be included in final research reports than the statistics found by the more analytical techniques to be presented in the remainder of this text.

When included in research reports, measures of central tendency and dispersion will most often be presented in summary form—often as a table. To see an example of how this is done, we will take a brief look at a project conducted by professors Virginia Chang, Amy Hillier, and Neil Mehta. They were concerned with the effects of racial residential segregation and neighborhood disorder on health in general and obesity in particular. They hypothesized that black Americans living in racially isolated, heavily segregated neighborhoods would tend to have a higher body mass index, especially when the neighborhood was highly disordered in terms of physical and social characteristics.

As is typical, the researchers use means and standard deviations as a way of describing the overall characteristics of their sample. The table below displays these statistics for some of

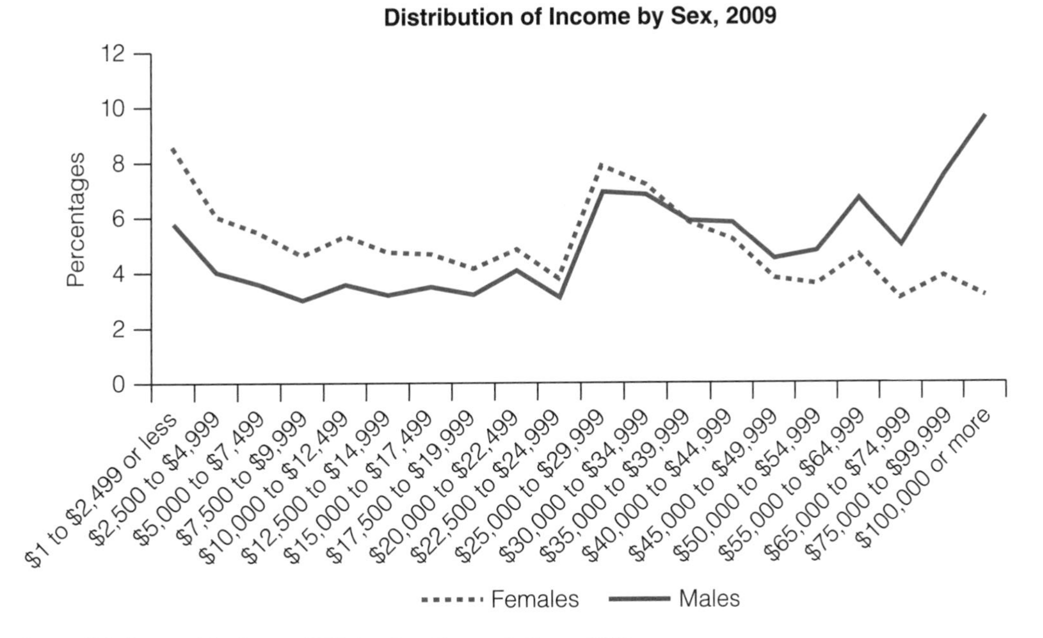

Source: U.S. Bureau of the Census, 2011. American Community Survey, 2009

(*continued next page*)

BECOMING A CRITICAL CONSUMER: *(continued)*

the neighborhood variables. Researchers report variables such as these to provide a context for their more analytical findings and to supply the reader with information about basic statistical characteristics. The actual tests of hypotheses are reported in separate sections.

Want to find out more? The reference is given below.

Variable	Mean	Standard Deviation
% of Population below the Poverty Line	22.80	14.70
Number of Vacant Lots	7.94	10.25
Number of Housing Code Violations	18.45	23.94
Serious crime against Persons (rate per 100 population)	1.37	1.12
Serious crimes against property (rate per 100 population)	6.45	6.91
Number of Supermarkets	1.68	3.57

Source: Chang, Virginia, Hillier, Amy, and Mehta, Neil. 2009. "Neighborhood Racial Isolation, Disorder, and Obesity." *Social Forces* 87: 2063–2092.

SUMMARY

1. Measures of dispersion summarize information about the heterogeneity, or variety, in a distribution of scores. When combined with an appropriate measure of central tendency, these statistics convey a large volume of information in just a few numbers. While measures of central tendency locate the central points of the distribution, measures of dispersion indicate the amount of diversity in the distribution.
2. The range (R) is the distance from the highest to the lowest score in the distribution. The interquartile range (Q) is the distance from the third to the first quartile (the "range" of the middle 50% of the scores). These two ranges can be used with variables measured at either the ordinal or interval-ratio level.
3. The standard deviation (s) is the most important measure of dispersion because of its key role in many more advanced statistical applications. The standard deviation has a minimum value of zero (indicating no variation in the distribution) and increases in value as the variability of the distribution increases. It is used most appropriately with variables measured at the interval-ratio level.

SUMMARY OF FORMULAS

FORMULA 4.1	Range	$R = \text{High Score} - \text{Low Score}$
FORMULA 4.2	Interquartile Range	$Q = Q_3 - Q_1$

FORMULA 4.3 Variance

$$s^2 = \frac{\Sigma(X_i - \overline{X})^2}{N}$$

FORMULA 4.4 Standard Deviation

$$s = \sqrt{\frac{\Sigma(X_i - \overline{X})^2}{N}}$$

GLOSSARY

Deviations. The distances between the scores and the mean.

Dispersion. The amount of variety, or heterogeneity, in a distribution of scores.

Interquartile range (*Q*). The distance from the third quartile to the first quartile.

Measures of dispersion. Statistics that indicate the amount of variety, or heterogeneity, in a distribution of scores.

Range (*R*). The highest score minus the lowest score.

Standard deviation. The square root of the sum of the squared deviations of the scores around the mean, divided by *N*. The most important and useful descriptive measure of dispersion.

Variance. The sum of the squared deviations of the scores around the mean divided by *N*. A measure of dispersion used primarily in inferential statistics and also in correlation and regression techniques.

PROBLEMS

4.1 Compute the range and standard deviation of the following 10 scores. *(HINT: It will be helpful to organize your computations as in Table 4.3.)*

10, 12, 15, 20, 25, 30, 32, 35, 40, 50

4.2 Compute the range and standard deviation of the following 10 test scores.

77, 83, 69, 72, 85, 90, 95, 75, 55, 45

4.3 In problem 3.1 at the end of Chapter 3, you calculated measures of central tendency for six variables for freshman and seniors. Three of those variables are reproduced here. Calculate the mean (if necessary), the range, and the standard deviation for each variable. What information is added by the measures of dispersion? Write a paragraph summarizing the differences between freshmen and seniors.

Out-of-Pocket Expenses		Number of Movies		Rating of Cafeteria Food	
Freshmen	Seniors	Freshmen	Seniors	Freshmen	Seniors
33	65	0	0	10	1
39	62	14	5	7	2
45	60	10	11	2	8
47	90	7	3	1	4
62	62	5	4	8	3
48	57	1	14	6	6
52	40	0	0	10	2
65	49	14	7	0	9
51	45	3	5	5	4
43	85	4	3	6	7
	78		5		4

4.4 SOC In problem 3.5 at the end of Chapter 3, you calculated measures of central tendency for four variables for 15 respondents. Two of those variables are reproduced here. Calculate the mean (if necessary), the range, and the standard deviation for each variable. What information is added by the measures of dispersion?

Respondent	Age	Attitude on Abortion (High Score = Strong Opposition)
A	18	10
B	20	9
C	21	8
D	30	10
E	25	7
F	26	7
G	19	9
H	29	6
I	31	10
J	55	5
K	32	4
L	28	3
M	23	2
N	24	1
O	32	9

4.5 SOC In problem 3.8, you computed mean and median income for 13 Canadian provinces and territories in two separate years. Now compute the standard deviation and range for each year, and taking account of the two measures of central tendency and the two measures of dispersion, write a paragraph summarizing the distributions. What do the measures of dispersion add to what you already knew about central tendency? Did the provinces become more or less variable over the period? The scores are reproduced here.

Province or Territory	2000	2006
Newfoundland and Labrador	38,800	50,500
Prince Edward Island	44,200	56,100
Nova Scotia	44,500	56,400
New Brunswick	43,200	54,000
Quebec	47,700	59,000
Ontario	55,700	66,600
Manitoba	47,300	58,700
Saskatchewan	45,800	60,500
Alberta	55,200	78,400
British Columbia	49,100	62,600
Yukon	56,000	76,000
Northwest Territories	61,000	88,800
Nunavut	37,600	54,300

4.6 SOC Data on several variables measuring overall heath and well-being for five nations are reported here for 2005, with projections to 2020. Are nations becoming more or less diverse on these variables? Calculate the mean, range, and standard deviation for each year for each variable. Summarize the results in a paragraph.

	Life Expectancy (years)		Infant Mortality Rate*		Fertility Rate #	
	2005	2020	2005	2020	2005	2020
Canada	80	82	4.8	4.4	1.6	1.6
U. S.	78	80	6.5	5.4	2.1	2.1
Mexico	75	78	20.9	13.2	2.5	2.1
Columbia	72	77	21.0	12.3	2.6	1.9
Japan	81	83	3.3	2.7	1.4	1.3
China	72	76	24.2	12.6	1.7	1.6
Sudan	59	64	62.5	38.9	4.9	4.0
Kenya	53	62	60.7	42.4	5.0	2.6
Italy	80	81	5.9	4.6	1.3	1.4
Germany	79	81	4.2	3.6	1.4	1.5

Sources: For 2005: U.S. Bureau of the Census. *Statistical Abstract of the United States, 2007*, p. 837. Available at http://www.census.gov/compendia/statab. For 2020: U.S. Bureau of the Census. *Statistical Abstract of the United States, 2011*, p. 842. Available at http://www.census.gov/prod/2011pubs/11statab/intlstat.pdf.

* Number of deaths of children under one year of age per 1,000 live births.

Average number of children per female.

4.7 SOC Labor force participation rates (percent employed), percent high school graduates, and mean income for males and females in 10 states are reported here. Calculate a mean and a standard deviation for both groups for each variable and then describe the differences. Are males and females unequal on any of these variables? How great is the gender inequality?

	Labor Force Participation		% High School Graduates		Mean Income	
State	Male	Female	Male	Female	Male	Female
A	74	54	65	67	35,623	27,345
B	81	63	57	60	32,345	28,134
C	81	59	72	76	35,789	30,546
D	77	60	77	75	38,907	31,788
E	80	61	75	74	42,023	35,560
F	74	52	70	72	34,000	35,980
G	74	51	68	66	25,800	19,001
H	78	55	70	71	29,000	26,603
I	77	54	66	66	31,145	30,550
J	80	75	72	75	34,334	29,117

4.8 SW Compute the standard deviation for the pretest and posttest scores that were used in problem 3.12. The scores are reproduced here. Taking into account all of the information you have on these variables, write a paragraph describing how the sample changed from test to test. What does the standard deviation add to the information you already had?

Case	Pretest	Posttest
A	8	12
B	7	13
C	10	12
D	15	19
E	10	8
F	10	17
G	3	12
H	10	11
I	5	7
J	15	12
K	13	20
L	4	5
M	10	15
N	8	11
O	12	20

4.9 SOC In problem 3.11, you computed measures of central tendency for the price of a liter of regular gas for 20 nations. The scores are reproduced here. Compute the standard deviation for this variable and then write a paragraph summarizing the mean, median, and standard deviation.

Nation	Cost of a Liter of Gasoline (U.S. Dollars)
United States	0.56
Nigeria	0.59
Australia	0.74
Mexico	0.74
Canada	0.76
Argentina	0.78
South Africa	0.87
Ukraine	0.88
China	0.99
Colombia	1.04
India	1.09
Kenya	1.20
Spain	1.23
Brazil	1.26
Sweden	1.38
United Kingdom	1.44
Germany	1.56
Ireland	1.56
Italy	1.57
Japan	1.74

Source: http://www.nationmaster.com.

4.10 CJ/SW Per capita expenditures for police protection for 20 cities are reported here for 2000 and 2010. Compute a mean and standard deviation for each year and then describe the differences in expenditures for the five-year period.

City	2000	2010
A	180	210
B	95	110
C	87	124
D	101	131
E	52	197
F	117	200
G	115	119
H	88	87
I	85	125
J	100	150
K	167	225
L	101	209
M	120	201
N	78	141
O	107	94
P	55	248
Q	78	140
R	92	131
S	99	152
T	103	178

4.11 SOC Compute the range and standard deviation for the data presented in problem 3.14. The data are reproduced here. What would happen to the value of the standard deviation if you removed Los Angeles from this distribution and recalculated? Why?

City	Annual Person-Hours of Time Lost to Traffic Congestion per Year per Person
Baltimore	25
Boston	22
Buffalo	5
Chicago	22
Cleveland	7
Dallas	32
Detroit	29
Houston	32
Kansas City	8
Los Angeles	38
Miami	27
Minneapolis	22
New Orleans	10
New York	21
Philadelphia	21
Pittsburgh	8
Phoenix	23
San Antonio	21
San Diego	29
San Francisco	29
Seattle	24
Washington, D.C.	31

Source: U.S. Bureau of the Census. *Statistical Abstract of the United States, 2009*. Table 1063. (Available at http://www.census.gov/prod/2009pubs/10statab/trans.pdf.)

4.12 SOC Listed here are the rates of abortion per 100,000 women for 20 states in 1973 and 1975. Describe what happened to these distributions over the two-year period. Did the average rate increase or decrease? What happened to the dispersion of this distribution? What happened between 1973 and 1975 that might explain these changes in central tendency and dispersion? *(Hint: It was a Supreme Court decision.)*

State	1973	1975
Maine	3.5	9.5
Massachusetts	10.0	25.7
New York	53.5	40.7
Pennsylvania	12.1	18.5
Ohio	7.3	17.9
Michigan	18.7	20.3
Iowa	8.8	14.7
Nebraska	7.3	14.3
Virginia	7.8	18.0
South Carolina	3.8	10.3
Florida	15.8	30.5
Tennessee	4.2	19.2
Mississippi	0.2	0.6
Arkansas	2.9	6.3
Texas	6.8	19.1
Montana	3.1	9.9
Colorado	14.4	24.6
Arizona	6.9	15.8
California	30.8	33.6
Hawaii	26.3	31.6

Source: United States Bureau of the Census. *Statistical* Abstracts *of the United States, 1977* (98th edition). Washington, D.C., 1977.

4.13 SW One of your goals as the new chief administrator of a large social service bureau is to equalize workloads within the various divisions of the agency. You have gathered data on caseloads per worker within each division. Which division comes closest to the ideal of an equalized workload? Which is farthest away?

A	B	C	D
50	60	60	75
51	59	61	80
55	58	58	74
60	55	59	70
68	56	59	69
59	61	60	82
60	62	61	85
57	63	60	83
50	60	59	65
55	59	58	60

4.14 SOC Compute the standard deviation for both sets of data presented in problem 3.13 and reproduced here. Compare the standard deviation computed for freshmen with the standard deviation computed for seniors. What happened? Why? Does this change relate at all to what happened to the mean over the four-year period? How? What happened to the shapes of the underlying distributions?

Freshmen				
10	43	30	30	45
40	12	40	42	35
45	25	10	33	50
42	32	38	11	47
22	26	37	38	10

Seniors				
10	45	35	27	50
35	10	50	40	30
40	10	10	37	10
40	15	30	20	43
23	25	30	40	10

4.15 SOC/CJ Calculate the range and standard deviation for the homicide rates presented in problem 3.15 and reproduced here. Using all the information you have on central tendency and dispersion, write a paragraph describing the changes in this variable between 1994 and 2009.

Homicide Rates (Number of Homicides per 100,000 Population) for 1994 and 2009		
	1994	2009
Alabama	11.9	6.9
Alaska	6.3	3.1
Arizona	10.5	5.4
Arkansas	12.0	6.2
California	11.8	5.3
Colorado	5.4	3.5
Connecticut	6.6	3.0
Delaware	4.7	4.6
Florida	8.3	5.5
Georgia	10.0	5.8
Hawaii	4.2	1.7
Idaho	3.5	1.4
Illinois	11.7	6.0
Indiana	7.9	4.8
Iowa	1.7	1.1
Kansas	6.7	4.2
Kentucky	6.4	4.1
Louisiana	19.8	11.8
Maine	2.3	2.0
Maryland	11.6	7.7
Massachusetts	3.5	2.6
Michigan	9.8	6.3
Minnesota	3.2	1.4
Mississippi	15.3	6.4
Missouri	10.5	6.4
Montana	3.3	2.9
Nebraska	3.1	2.2
Nevada	11.7	5.9
New Hampshire	1.4	0.8
New Jersey	5.0	3.7
New Mexico	10.7	8.7
New York	11.1	4.0
North Carolina	10.9	5.3
North Dakota	0.2	1.5
Ohio	6.0	4.5
Oklahoma	6.9	6.2
Oregon	4.9	2.2
Pennsylvania	5.9	5.2
Rhode Island	4.1	2.9
South Carolina	9.6	6.3
South Dakota	1.4	2.6
Tennessee	9.3	7.3
Texas	11.0	5.4
Utah	2.9	1.3
Vermont	1.0	1.1
Virginia	8.7	4.4
Washington	5.5	2.7
West Virginia	5.4	4.6
Wisconsin	4.5	2.5
Wyoming	3.4	2.4

4.16 At St. Algebra College, the math department ran some special sections of the freshman math course by using a variety of innovative teaching techniques. Students were randomly assigned to either the traditional sections or the experimental sections, and all students were given the same final exam. The results of the final are summarized here. What was the effect of the experimental course?

Traditional	Experimental
$\bar{X} = 77.8$	$\bar{X} = 76.8$
$s = 12.3$	$s = 6.2$
$N = 478$	$N = 465$

4.17 You are the governor of the state and must decide which of four metropolitan police departments will win the annual award for efficiency. The performance of each department is summarized in monthly arrest statistics, as reported here. Which department will win the award? Why?

	Departments			
	A	B	C	D
$\bar{X} =$	601.30	633.17	592.70	599.99
$s =$	2.30	27.32	40.17	60.23

YOU ARE THE RESEARCHER

Two projects are presented below. The first is a follow-up on the project presented at the end of Chapter 3 and the second presents a new SPSS command. You are urged to complete both.

PROJECT 1: The Typical American

In Chapter 3, you described the typical American by using 10 variables selected from the 2010 GSS. Now you will examine variation or dispersion on some of the variables you selected.

Step 1: Choosing the Variables

Select at least five of the ordinal and interval-ratio variables you used in Chapter 3. Add more variables if you had fewer than five variables at this level of measurement.

Step 2: Getting the Statistics

Use **Descriptives** to find the range and standard deviation for each of your selected variables. With the 2010 GSS loaded, click **Analyze**, **Descriptive Statistics**, and **Descriptives** from the Main Menu of SPSS. The **Descriptives** dialog box will open. Find the names of your variables in the list on the left and then click the right arrow button to transfer them to the **Variables** box. Click **OK** and SPSS will produce the same output you analyzed in Chapter 3. Now, however, we will consider dispersion rather than central tendency.

The standard deviation for each variable is reported in the column labeled "Std. Deviation," and the range can be computed from the values given in the "Minimum" and "Maximum" columns. At this point, the range is probably easier to understand and interpret than the standard deviation. As we have seen, the latter is more meaningful when we have a point of comparison. For example, suppose we were interested in the variable ***tvhours*** and how television-viewing habits have changed over the years. The **Descriptives** output for 2010 shows that people watched an average of 3.09 hours a day, with a standard deviation of 2.90. Suppose that data from 1980 showed an average of 3.70 hours of television viewing a day, with a standard deviation of 1.10. You could conclude that television watching had, on average, decreased over the 30-year period but that Americans had also become much more diverse in their viewing habits.

Record your results in the table below:

Variable	SPSS Name	Mean	Range	Standard Deviation
1				
2				
3				
4				
5				

Step 3: Interpreting Results

Start with the descriptions of the variables you wrote in Chapter 3 and then add information about dispersion, referring to the range and the standard deviation. Your job now is to describe the "typical" American *and* to suggest the amount of variation around this central point.

PROJECT 2: The Culture War Revisited

In Chapter 2, you examined some of the dimensions of the culture war in U.S. society. In this project, you will re-examine the topic and learn how to use a new SPSS command to combine existing variables into a new summary variable. This computed variable can be used to summarize feelings and attitudes in general and to explore new dimensions of the issue.

Step 1: Creating a Scale to Summarize Attitudes About Abortion

One of the most controversial issues in the American culture war is abortion: Under what conditions, if any, should abortion be legal? The GSS data set supplied with this text includes two variables that measure support for legal abortion. The variables differ in context: One asks specifically if an abortion should be available if the family is poor and wants no more children (*abpoor*). The other asks if an abortion should be available when the woman wants it "for any reason" (*abany*). Because these two situations are distinct, each item should be analyzed in its own right. However, suppose you wanted to create a summary scale that indicated a person's *overall* feelings about abortion.

One way to do this would be to add the scores on the two variables together. This would create a new variable, which we will call *abscale,* with three possible scores. If a respondent was consistently pro-abortion and answered "yes" (coded as "1") to both items, the respondent's score on the summary variable would be 1 + 1 or 2. A score of 3 would occur when a respondent answered "yes" (1) to one item and "no" (2) to the other. This might be labeled an "intermediate" or "moderate" position. The final possibility would be a score of 4 if the respondent answered "no" (coded as "2") to both items. This would be a consistent anti-abortion position. The table below summarizes the scoring possibilities.

If response on *abany* is:	and response on *abpoor* is:	score on *abscale* will be:
1 (Yes)	1 (Yes)	2 (pro-abortion)
1 (Yes)	2 (No)	3 (moderate)
2 (No)	1 (Yes)	3 (moderate)
2 (No)	2 (No)	4 (anti-abortion)

The new variable—*abscale*—summarizes each respondent's overall position on the issue. Once created, *abscale* could be analyzed, transformed, and manipulated exactly like a variable actually recorded in the data file.

Using Compute

To use the **Compute** command, click **Transform** and then **Compute Variable** from the main menu. The **Compute Variable** window will appear. Find the **Target Variable** box in the upper left-hand corner of this window. The first thing we need to do is assign a name (*abscale*) to the new variable we are about to compute by typing that name in this box. Next, we need to tell SPSS how to compute the new variable. In this case, *abscale* will be computed by adding the scores of *abany* and *abpoor.* Find *abany* in the variable list on the left and then click the arrow button in the middle of the screen to transfer the variable name to the **Numeric Expression** box. Next, click the plus sign (+) on the calculator pad under the **Numeric Expression** box and the sign will appear next to *abany.* Finally, highlight *abpoor* in the variable list and then click the arrow button to transfer the variable name to the **Numeric Expression** box.

The expression in the **Numeric Expression** box should now read:

abany + *abpoor*

Click **OK,** and ***abscale*** will be created and added to the data set. If you want to keep this new variable permanently, click **Save** from the **File** menu, and the updated data set with ***abscale*** added will be saved to disk. (If you are using the student version of SPSS, remember that your data set is limited to 50 variables and you will not be allowed to save the data set with ***abscale*** added.)

Examining the Variables

We now have three variables that measure attitudes about abortion—two items referring to specific situations and a more general summary item. It is always a good idea to check the frequency distribution for computed variables to make sure that the computations were carried out as we intended. Use the **Frequencies** procedure (click **Analyze**, **Descriptive Statistics**, and then **Frequencies**) to get tables for *abany, abpoor,* and *abscale.* Your output will look like this:

Abortion If Woman Wants One for Any Reason					
		Frequency	Percent	Valid Percent	Cumulative Percent
Valid	YES	390	26.1	43.2	43.2
	NO	513	34.3	56.8	100.0
	Total	903	60.4	100.0	
Missing	IAP	554	37.1		
	DK	26	1.7		
	NA	11	.7		
	Total	591	39.6		
Total		1,494	100.0		

Low Income—Can't Afford More Children					
		Frequency	Percent	Valid Percent	Cumulative Percent
Valid	YES	408	27.3	45.0	45.0
	NO	498	33.3	55.0	100.0
	Total	906	60.6	100.0	
Missing	IAP	554	37.1		
	DK	23	1.5		
	NA	11	.7		
	Total	588	39.4		
Total		1,494	100.0		

abscale					
		Frequency	Percent	Valid Percent	Cumulative Percent
Valid	2.00	331	22.2	37.1	37.1
	3.00	128	8.6	14.3	51.5
	4.00	433	29.0	48.5	100.0
	Total	892	59.7	100.0	
Missing	System	602	40.3		
Total		1,494	100.0		

Missing Cases

Note that a little more that 900 of the 1,494 total respondents to the 2010 GSS answered the that two specific abortion items. Remember that no respondent is given the entire GSS and that the vast majority of the "missing cases" received a form of the GSS that did not include these two items. These cases are coded as "IAP" in the tables above.

Now look at *abscale* and note that even fewer cases (892) are included in the summary scale than in the two original items. When SPSS executes a **Compute** statement, it automatically eliminates any cases that are missing scores on any of the constituent items. If these cases were not eliminated, a variety of errors and misclassifications could result. For example, if cases with missing scores were included, a person who scored a 2 ("anti-abortion") on *abany* and then did not respond to *abpoor* would have a total score of 2 on *abscale*. Thus, this case would be treated as pro-abortion when the only information we have indicates that this respondent is anti-abortion. Cases with missing scores are deleted to avoid this type of error.

Step 2: Interpreting the Results

Compare the distributions of these three variables with each other and then write a report in which you answer these questions:

1. How does the level of approval for abortion differ from one specific situation to another?
2. What percent of the respondents approved of abortion in both circumstances and what percent disapproved of abortion in both circumstances? What percentage approved in one situation but not the other? (Use the distribution of abscale to answer this question.)
3. What do these patterns reveal about value consensus in American society? Are Americans generally in agreement about the issue of abortion?

5 The Normal Curve

LEARNING OBJECTIVES

By the end of this chapter, you will be able to:

1. Define and explain the concept of the normal curve.
2. Convert empirical scores to *Z* scores and then use *Z* scores and the normal curve table (Appendix A) to find areas above, below, and between points on the curve.
3. Express areas under the curve in terms of probabilities.

USING STATISTICS

The statistical techniques presented in this chapter can be used to:

- Describe the position of scores on certain tests: "John scored higher than 75% of the students who took the test" or "Mary is in the 98th percentile in math ability."
- Estimate the probability that certain events will occur. For example: "The probability that a randomly selected fourth-grade student will have a score more than 123 is 23 out of 100."

The **normal curve** is a concept of great importance in statistics. In combination with the mean and standard deviation, it is used to make precise descriptive statements about empirical distributions. Also, the normal curve is central to the theory that underlies inferential statistics. Thus, this chapter concludes our treatment of Part I (Descriptive Statistics) and lays important groundwork for Part II (Inferential Statistics).

The normal curve is sometimes used to assign grades on tests, and you may be familiar with this application. In this guise, the normal curve is often called the "bell curve," and grading "on the curve" means the instructor wants the scores to follow a specific pattern: The modal grade is a C and there will be equal numbers of As and Fs, Bs and Ds. In other words, the distribution of grades should look like a bell or a mound.

Properties of the Normal Curve

The normal curve is a theoretical model—a kind of frequency polygon or line chart—that is unimodal (i.e., has a single mode, or peak), perfectly smooth, and symmetrical (unskewed), thus, its mean, median, and mode are all exactly the same value. It is bell shaped, and its tails extend infinitely in both directions. Of course, no empirical distribution matches this ideal model perfectly, but some variables (e.g., test results from large classes or standardized scores on tests such as the GRE) are close enough to permit the assumption of normality. In turn, this assumption makes possible one of the most important uses of the normal curve—the description of empirical distributions based on our knowledge of the theoretical normal curve.

The crucial point about the normal curve is that distances along the horizontal axis—when measured in standard deviations from the mean—always encompass the same proportion of the total area under the curve. In other words, the distance from any point to the mean—when measured in standard deviations—will cut off exactly the same proportion of the area under the curve.

To illustrate, Figures 5.1 and 5.2 present two hypothetical distributions of IQ scores for fictional groups of males and females—normally distributed—such that:

Males	Females
$\bar{X} = 100$	$\bar{X} = 100$
$s = 20$	$s = 10$
$N = 1{,}000$	$N = 1{,}000$

Figures 5.1 and 5.2 are drawn with two scales on the horizontal axis of the graph. The upper scale is stated in IQ units and the lower scale in standard deviations from the mean. These scales are interchangeable, and we can easily shift from one to the other. For example, for the males, an IQ score of 120 is one standard deviation (remember that for the male group, $s = 20$) above the mean and an IQ of 140 is two standard deviations above (to the right of) the mean. Scores to the left of the mean are marked as negative values because they are less than the mean. An IQ of 80 is one standard deviation below the mean, an IQ score of 60 is two standard deviations less than the mean, and so forth. Figure 5.2 is marked in a similar way, except that because its standard deviation is a different value ($s = 10$), the markings occur at different points. For the female sample, one standard deviation above the mean is an IQ of 110, one standard deviation below the mean is an IQ of 90, and so forth.

Recall that on any normal curve, distances along the horizontal axis—when measured in standard deviations—always encompass exactly the same proportion of the total area under the curve. Specifically, the distance between one

FIGURE 5.1 IQ Scores for a Group of Males

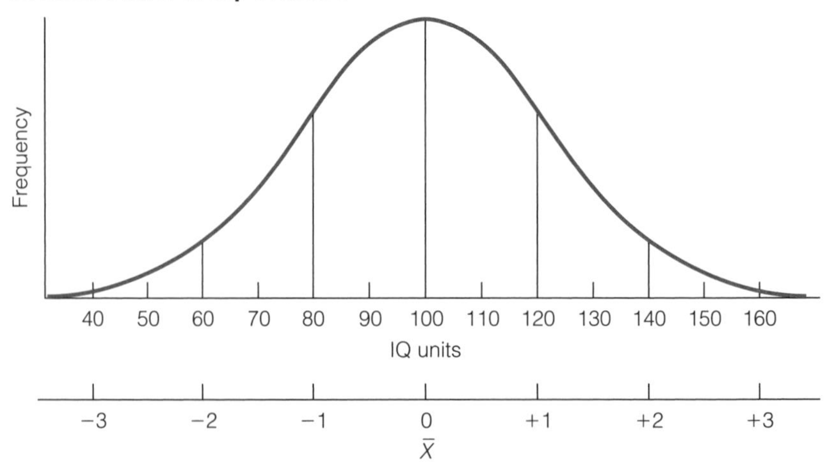

FIGURE 5.2 IQ Scores for a Group of Females

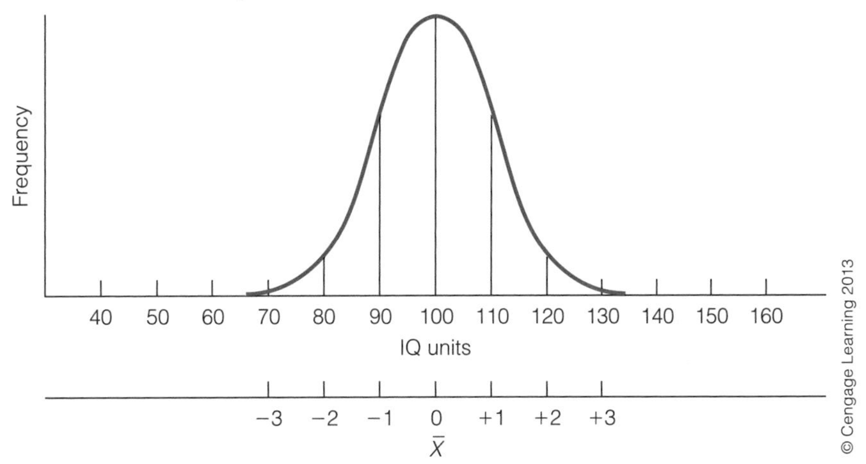

standard deviation above the mean and one standard deviation below the mean (or ±1 standard deviation) encompasses exactly 68.26% of the total area under the curve. This means that in Figure 5.1, 68.26% of the total area lies between the score of 80 (−1 standard deviation) and 120 (+1 standard deviation). The standard deviation for females is 10, so the same percentage of the area (68.26%) lies between the scores of 90 and 110. On any normal distribution, 68.26% of the total area will always fall between ±1 standard deviation, regardless of the trait being measured and the number values of the mean and standard deviation.

It will be useful to familiarize yourself with the following relationships between distances from the mean and areas under the curve:

Between	% of the Area
±1 standard deviation lies	68.26
±2 standard deviations lies	95.44
±3 standard deviations lies	99.72

These relationships are displayed graphically in Figure 5.3.

We can describe empirical distributions that are at least approximately normal by using these relationships between distance from the mean and area.

FIGURE 5.3 Areas Under the Theoretical Normal Curve

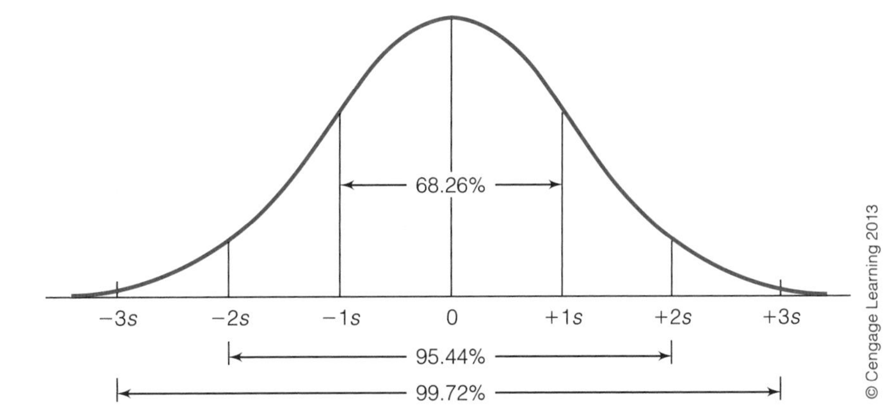

Tests of intelligence are designed to generate distributions of scores that are approximately normal. That is, the tests mix easier and harder questions so that the mean will be about 100 and the final distribution of scores will be bell shaped, with equal numbers of high and low scores.

Intelligence testing can be controversial, and although we cannot go into detail here, you should be aware that there is considerable debate about the meaning of IQ scores. Tests of intelligence may have cultural and other biases, and there is considerable disagreement about whether the tests actually measure native mental ability. In particular, you should be aware that the more or less normal distribution of IQ scores does not necessarily mean that the underlying and elusive quality called intelligence—however that might be defined—is also normally distributed.

The position of individual scores can be described with respect to the mean, the distribution as a whole, or any other score in the distribution.

The areas between scores can also be expressed in the number of cases rather than the percentage of total area. For example, a normal distribution of 1,000 cases will contain about 683 cases (68.26% of 1,000 cases) between ± 1 standard deviation of the mean, about 954 between ± 2 standard deviations, and about 997 between ± 3 standard deviations. Thus, for any normal distribution, only a few cases will be farther away from the mean than ± 3 standard deviations.

Using the Normal Curve

We have seen that we can find areas under the normal curve for scores that are 1, 2, or 3 standard deviations above or below the mean. To work with values that are not exact multiples of the standard deviation, we must express the original scores in units of the standard deviation or convert them into **Z scores**. The original scores could be in any unit of measurement (feet, IQ, dollars), but Z scores always have the same values for their mean (0) and standard deviation (1).

Computing Z Scores

Think of converting the original scores into Z scores as a process of changing value scales—similar to changing from meters to yards, kilometers to miles, or gallons to liters. These units are different but equally valid ways of expressing length, distance, or volume. For example, a mile is equal to 1.61 kilometers, so two towns that are 10 miles apart are also 16.1 kilometers apart and a 5K race covers about 3.10 miles. Although you may be more familiar with miles than kilometers, either unit works perfectly well as a way of expressing distance.

In the same way, the original (or "raw") scores and Z scores are two equally valid but different ways of measuring distances under the normal curve. In Figure 5.1, for example, we could describe a particular score in terms of IQ units ("John's score was 120") or standard deviations ("John scored one standard deviation above the mean").

When we compute Z scores, we convert the original units of measurement (IQ scores, inches, dollars, etc.) to Z scores and, thus, "standardize" the normal curve to a distribution that has a mean of 0 and a standard deviation of 1. The mean of the

ONE STEP AT A TIME	Finding Z Scores
Step	***Operation***
1.	Subtract the value of the mean ($\overline{X}$) from the value of the score (X_i).
2.	Divide the quantity found in step 1 by the value of the standard deviation (s). The result is the Z-score equivalent for this raw score.

empirical normal distribution will be converted to 0 and its standard deviation to 1, and all values will be expressed in Z-score form. The formula for computing Z scores is

FORMULA 5.1

$$Z = \frac{X_i - \overline{X}}{s}$$

This formula will convert any score (X_i) from an empirical normal distribution into the equivalent Z score. To illustrate with the men's IQ data (Figure 5.1), the Z-score equivalent of a raw score of 120 would be:

$$Z = \frac{120 - 100}{20} = +1.00$$

The Z score of positive 1.00 indicates that the original score lies one standard deviation unit above (to the right of) the mean. A negative score would fall below (to the left of) the mean. *(For practice in computing Z scores, see any of the problems at the end of this chapter.)*

The Normal Curve Table

The theoretical normal curve has been very thoroughly described by statisticians. The areas related to any Z score have been precisely determined and organized into a table format. This **normal curve table**, or Z-score table, is presented as Appendix A in this text; for purposes of illustration, a small portion of it is reproduced here as Table 5.1.

TABLE 5.1 An Illustration of How to Find Areas Under the Normal Curve by Using Appendix A

(a) Z	(b) Area Between Mean and Z	(c) Area Beyond Z
0.00	0.0000	0.5000
0.01	0.0040	0.4960
0.02	0.0080	0.4920
0.03	0.0120	0.4880
⋮	⋮	⋮
1.00	0.3413	0.1587
1.01	0.3438	0.1562
1.02	0.3461	0.1539
1.03	0.3485	0.1515
⋮	⋮	⋮
1.50	0.4332	0.0668
1.51	0.4345	0.0655
1.52	0.4357	0.0643
1.53	0.4370	0.0630
⋮	⋮	⋮

The normal curve table consists of three columns, with Z scores in the left-hand column a, areas between the Z score and the mean in the middle column b, and areas beyond the Z score in the right-hand column c. To find the area between any Z score and the mean, go down the column labeled "Z" until you find the Z score. For example, go down column a either in Appendix A or in Table 5.1 until you find a Z score of $+1.00$. The entry in column b shows that the "area between mean and Z" is 0.3413. The table presents all areas in the form of proportions, but we can easily translate these into percentages by multiplying them by 100 (see Chapter 2). We could say either "a proportion of 0.3413 of the total area under the curve lies between a Z score of 1.00 and the mean" or "34.13% of the total area lies between a score of 1.00 and the mean."

To illustrate further, find the Z score of 1.50 either in column a of Appendix A or the abbreviated table presented in Table 5.1. This score is 1.5 standard deviations to the right of the mean and corresponds to an IQ of 130 for the men's IQ data. The area in column b for this score is 0.4332. This means that a proportion of 0.4332—or a percentage of 43.32%—of all the area under the curve lies between this score and the mean.

The third column in the table presents "Areas Beyond Z." These are areas above positive scores or below negative scores. This column will be used when we want to find an area above or below certain Z scores—an application that will be explained later in this chapter.

To conserve space, the normal curve table in Appendix A includes only positive Z scores. However, because the normal curve is perfectly symmetrical, the area between the score and the mean—column b—for a negative score will be exactly the same as that for a positive score of the same numerical value. For example, the area between a Z score of -1.00 and the mean will also be 34.13%—exactly the same as the area we found previously for a score of $+1.00$. However, as will be repeatedly demonstrated later, the sign of the Z score is extremely important and should be carefully noted.

For practice in using Appendix A to describe areas under an empirical normal curve, verify that the following Z scores and areas are correct for the men's IQ distribution. For each IQ score, the equivalent Z score is computed by using Formula 5.1 and then Appendix A is consulted to find the area between the score and the mean. ($\bar{X} = 100$, $s = 20$ throughout.)

IQ Score	*Z* Score	Area Between *Z* and the Mean
110	+0.50	19.15%
125	+1.25	39.44%
133	+1.65	45.05%
138	+1.90	47.13%

The same procedures apply when the Z-score equivalent of an actual score happens to be a minus value (that is, when the raw score lies below the mean).

IQ Score	*Z* Score	Area Between *Z* and the Mean
93	−0.35	13.68%
85	−0.75	27.34%
67	−1.65	45.05%
62	−1.90	47.13%

Remember that the areas in Appendix A will be the same for Z scores of the same numerical value regardless of sign. The area between the score of 138 (+1.90) and the mean is the same as the area between 62 (−1.90) and the mean. *(For practice in using the normal curve table, see any of the problems at the end of this chapter.)*

Finding the Total Area Above and Below a Score

To this point, we have seen how the normal curve table can be used to find areas between a Z score and the mean. The information presented in the table can also be used to find other kinds of areas in empirical distributions that are at least approximately normal in shape. For example, suppose you need to determine the total area below the scores of two male subjects in the distribution described in Figure 5.1. The first subject has a score of 117 ($X_1 = 117$), which is equivalent to a Z score of +0.85:

$$Z_1 = \frac{X_i - \overline{X}}{s} = \frac{117 - 100}{20} = \frac{17}{20} = +0.85$$

The plus sign of the Z score indicates that the score should be placed above (to the right of) the mean. To find the area below a positive Z score, the area between the score and the mean—given in column b—must be added to the area below the mean. As we noted earlier, the normal curve is symmetrical (unskewed) and its mean will be equal to its median. Therefore, the area below the mean (just like the median) will be 50%. Study Figure 5.4 carefully. We are interested in the shaded area.

By consulting the normal curve table, we find that the area between the score and the mean in column b is 30.23% of the total area. The area below a Z score of +0.85 is therefore 80.23% (50.00% + 30.23%). This subject scored higher than 80.23% of the persons tested.

The second subject has an IQ score of 73 ($X_2 = 73$), which is equivalent to a Z score of −1.35:

$$Z_2 = \frac{X_i - \overline{X}}{s} = \frac{73 - 100}{20} = -\frac{27}{20} = -1.35$$

To find the area below a negative score, we use column c, labeled "Area Beyond Z." The area of interest is depicted in Figure 5.5, and we must determine

FIGURE 5.4 Finding the Area Below a Positive Z Score

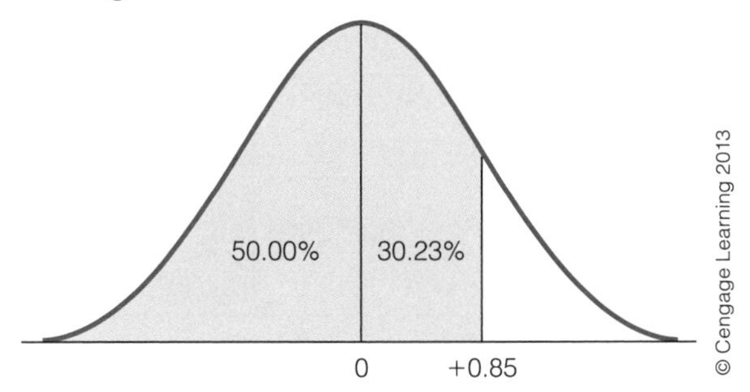

FIGURE 5.5 Finding the Area Below a Negative Z Score

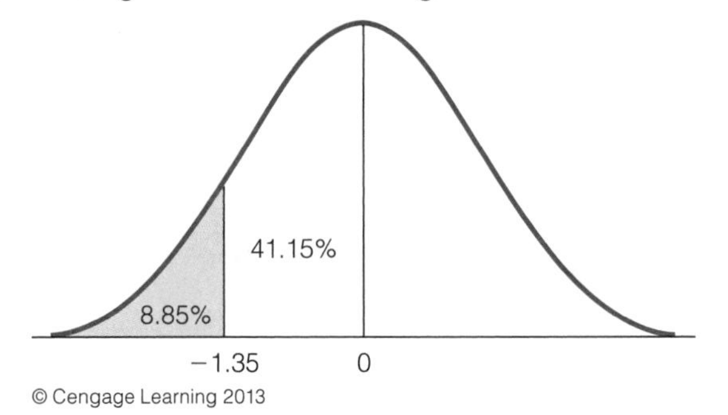

the size of the shaded area. The area beyond a score of −1.35 is given as 0.0885, which we can express as 8.85%. The second subject ($X_2 = 73$) scored higher than 8.85% of the tested group.

In the foregoing examples, we found the area below a score. Essentially, the same techniques are used to find the area above a score. For example, if we need to determine the area above an IQ score of 108, we would first convert to a Z score:

$$Z = \frac{X_i - \overline{X}}{s} = \frac{108 - 100}{20} = \frac{8}{20} = +0.40$$

Then, we proceed to Appendix A. The shaded area in Figure 5.6 represents the area in which we are interested. The area above a positive score is found in the "Area Beyond Z" column, and in this case, the area is 0.3446, or 34.46%.

These procedures are summarized in Table 5.2 and in the "One Step at a Time" box. They might be confusing at first, and you should *always* draw the curve and shade in the areas in which you are interested. *(For practice in finding areas above or below Z scores, see problems 5.1 to 5.7.)*

Finding Areas Between Two Scores

On occasion, you will need to determine the area between two scores. When the scores are on opposite sides of the mean, the area between them can be found by adding the areas between each score and the mean. Using the men's

FIGURE 5.6 Finding the Area Above a Positive Z Score

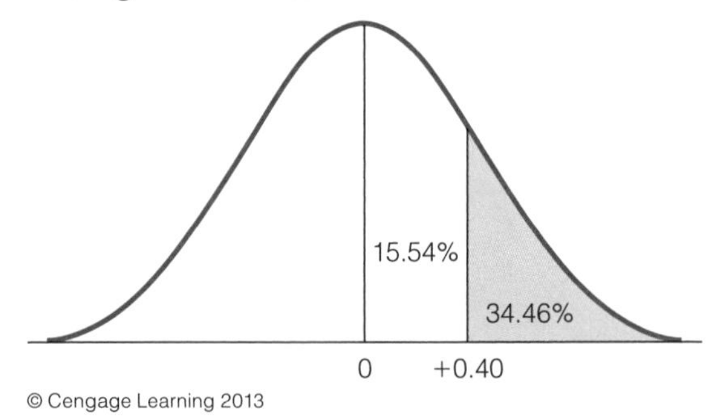

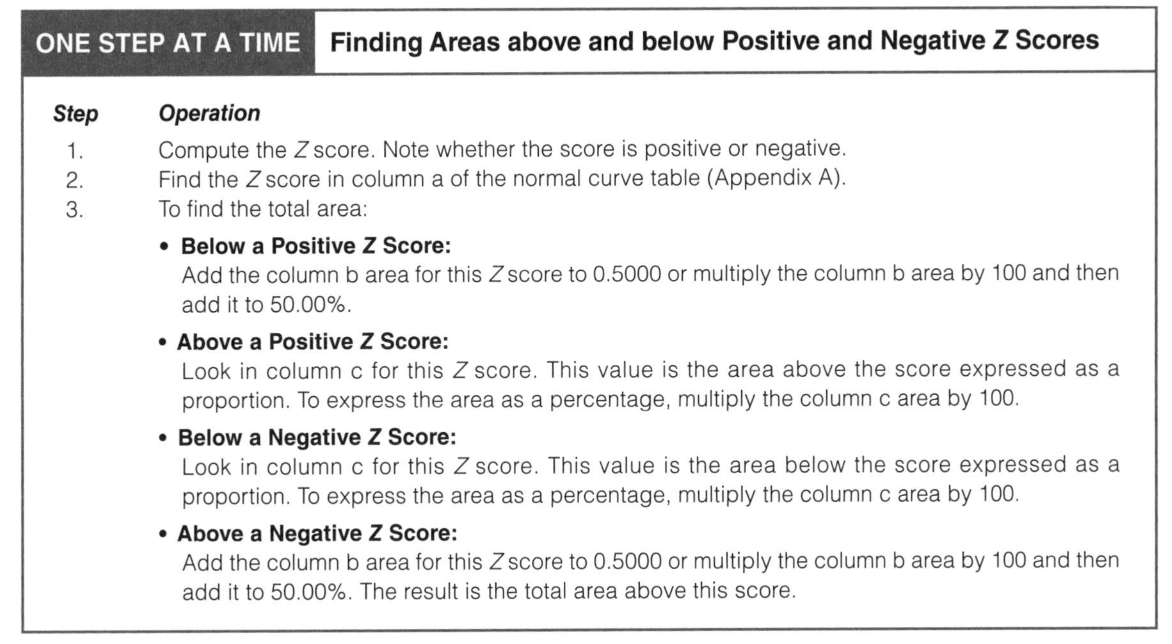

ONE STEP AT A TIME **Finding Areas above and below Positive and Negative *Z* Scores**

Step	*Operation*
1.	Compute the *Z* score. Note whether the score is positive or negative.
2.	Find the *Z* score in column a of the normal curve table (Appendix A).
3.	To find the total area:

- **Below a Positive *Z* Score:**
  Add the column b area for this *Z* score to 0.5000 or multiply the column b area by 100 and then add it to 50.00%.
- **Above a Positive *Z* Score:**
  Look in column c for this *Z* score. This value is the area above the score expressed as a proportion. To express the area as a percentage, multiply the column c area by 100.
- **Below a Negative *Z* Score:**
  Look in column c for this *Z* score. This value is the area below the score expressed as a proportion. To express the area as a percentage, multiply the column c area by 100.
- **Above a Negative *Z* Score:**
  Add the column b area for this *Z* score to 0.5000 or multiply the column b area by 100 and then add it to 50.00%. The result is the total area above this score.

IQ data as an example, if we wished to know the area between the IQ scores of 93 and 112, we would convert both scores to *Z* scores, find the area between each score and the mean in column b of Appendix A, and then add these two areas together. The first IQ score of 93 converts to a *Z* score of −0.35:

$$Z_1 = \frac{X_i - \overline{X}}{s} = \frac{93 - 100}{20} = -\frac{7}{20} = -0.35$$

The second IQ score (112) converts to +0.60:

$$Z_2 = \frac{X_i - \overline{X}}{s} = \frac{112 - 100}{20} = \frac{12}{20} = 0.60$$

Both scores are placed on Figure 5.7. We are interested in the total shaded area. The total area between these two scores is 13.68% + 22.57%, or 36.25%. Therefore, 36.25% of the total area (or about 363 of the 1,000 cases) lies between the IQ scores of 93 and 112.

TABLE 5.2 Finding Areas Above and Below Positive and Negative Scores

	When the *Z* Score Is	
To Find Area	Positive	Negative
Above *Z*	Look in column c.	Add column b area to 0.5000 or change column b area to a percentage and add it to 50.00%.
Below *Z*	Add column b area to 0.5000 or change column b area to a percentage and add it to 50.00%.	Look in column c.

FIGURE 5.7 Finding the Area Between Two Scores

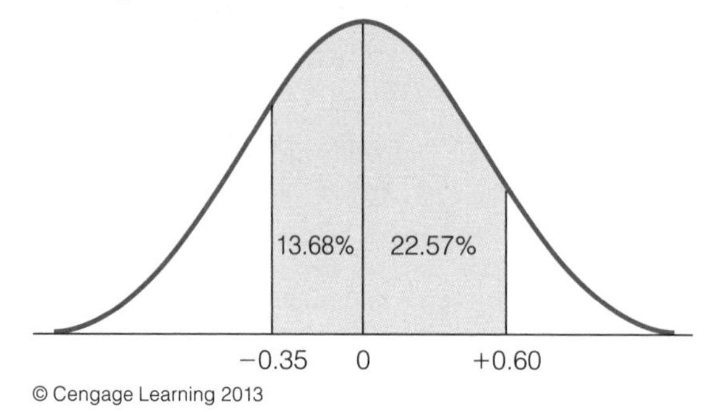

When the scores of interest are on the same side of the mean, a different procedure must be followed to determine the area between them. For example, if we were interested in the area between the scores of 113 and 121, we would begin by converting these scores into Z scores:

$$Z_1 = \frac{X_i - \overline{X}}{s} = \frac{113 - 100}{20} = \frac{13}{20} = +0.65$$

$$Z_2 = \frac{X_i - \overline{X}}{s} = \frac{121 - 100}{20} = \frac{21}{20} = +1.05$$

The scores are noted in Figure 5.8; we are interested in the shadowed area. To find the area between two scores on the same side of the mean, find the area between each score and the mean (given in column b of Appendix A) and then subtract the smaller area from the larger. Between the Z score of +0.65 and the mean lies 24.22% of the total area. Between +1.05 and the mean lies 35.31% of the total area. Therefore, the area between these two scores is 35.31% − 24.22%, or 11.09% of the total area. The same technique would be followed if both scores had been below the mean. The procedures for finding areas between two scores are summarized in Table 5.3 and in the "One Step at a Time" box. *(For practice in finding areas between two scores, see problems 5.3, 5.4, and 5.6 to 5.9.)*

FIGURE 5.8 Finding the Area Between Two Scores

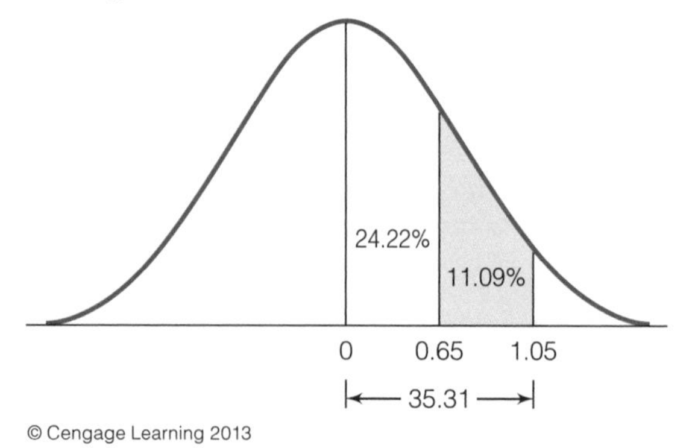

ONE STEP AT A TIME **Finding Areas Between Z Scores**

Step	***Operation***
1.	Compute the *Z* scores for both raw scores. Note whether the scores are positive or negative.
2.	Find the areas between each score and the mean in column b.

If the scores are on the same side of the mean:

3.	Subtract the smaller area from the larger area. Multiply this value by 100 to express it as a percentage.

If the scores are on opposite sides of the mean:

4.	Add the two areas together to get the total area between the scores. Multiply this value by 100 to express it as a percentage.

Applying Statistics 5.1 Finding *Z* Scores and Areas

You have just received your score on a test of intelligence. If your score was 78 and you know that the mean score on the test was 67 with a standard deviation of 5, how does your score compare with the distribution of all test scores?

If you can assume that the test scores are normally distributed, you can compute a *Z* score and then find the area below or above your score. The *Z*-score equivalent of your raw score would be:

$$Z = \frac{X_i - \bar{X}}{s} = \frac{78 - 67}{5} = \frac{11}{5} = +2.20$$

Turning to Appendix A, we find that the "Area Between Mean and *Z*" for a *Z* score of 2.20 is 0.4861, which could also be expressed as 48.61%. Because this is a positive *Z* score, we need to add this area to 50.00% to find the total area below. Your score is higher than 48.61 + 50.00, or 98.61%, of all the test scores. You did pretty well!

Applying Statistics 5.2 Finding *Z* Scores and Areas

All sections of Biology 101 at a large university were given the same final exam. Test scores were distributed normally, with a mean of 72 and a standard deviation of 8. What percentage of students scored between 60 and 69 (a grade of D) and what percentage scored between 70 and 79 (a grade of C)? The first two scores are below the mean. Using Table 5.3 as a guide, we must first compute *Z* scores, find areas between each score and the mean, and then subtract the smaller area from the larger:

$$Z_1 = \frac{X_i - \bar{X}}{s} = \frac{60 - 72}{8} = -\frac{12}{8} = -1.50$$

$$Z_2 = \frac{X_i - \bar{X}}{s} = \frac{69 - 72}{8} = -\frac{3}{8} = -0.38$$

Using column b, we see that the area between $z = -1.50$ and the mean is .4332, and the area between $z = -0.38$ and the mean is .1480. Subtracting the smaller from the larger (.4332 − .1480) gives .2852. Changing to percentage format, we can say that 28.52% of the students earned a D on the test.

To find the percentage of students who earned a C, we must add column b areas together because the scores (70 and 79) are on opposite sides of the mean (see Table 5.3):

$$Z_1 = \frac{X_i - \bar{X}}{s} = \frac{70 - 72}{8} = -\frac{2}{8} = -0.25$$

$$Z_2 = \frac{X_i - \bar{X}}{s} = \frac{79 - 72}{8} = \frac{7}{8} = 0.88$$

Using column b, we see that the area between $Z = -0.25$ and the mean is .0987, and the area between $Z = 0.88$ and the mean is .3106. Therefore, the total area between these two scores is .0987 + .3106, or 0.4093. Translating to percentages again, we can say that 40.93% of the students earned a C on this test.

TABLE 5.3 Finding Areas Between Scores

Situation	Procedure
Scores are on the same side of the mean.	Find areas between each score and the mean in column b. Subtract the smaller area from the larger area.
Scores are on opposite sides of the mean.	Find areas between each score and the mean in column b. Add the two areas together.

Using the Normal Curve to Estimate Probabilities

To this point, we have thought of the theoretical normal curve as a way of describing the percentage of total area above, below, and between scores in an empirical distribution. We have also seen that these areas can be converted into the number of cases above, below, and between scores. In this section, we introduce the idea that the theoretical normal curve may also be thought of as a distribution of probabilities. Specifically, we may use the properties of the theoretical normal curve (Appendix A) to estimate the probability that a randomly selected case will have a score that falls in a certain range. In terms of techniques, these probabilities will be found in exactly the same way as areas were found. However, before we consider these mechanics, let us examine what is meant by the concept of probability.

Although we are rarely systematic or rigorous about it, we all attempt to deal with probabilities every day; indeed, we base our behavior on our estimates of the likelihood that certain events will occur. We constantly ask (and answer) questions such as: What is the probability of rain? Of drawing an inside straight in poker? Of the worn-out tires on my car going flat? Of passing a test if I don't study?

To estimate the probability of an event, we must first be able to define what would constitute a "success." The preceding examples contain several different definitions of a success (that is, rain, drawing a certain card, flat tires,

STATISTICS IN EVERYDAY LIFE

How normal is the normal curve? How common is it in everyday life? Mathematically speaking, the scores of a variable are most likely to be normal in shape if they represent random deviations around a mean. For example, imagine a manufacturing process designed to produce parts of a uniform size. Any errors or mistakes will be equally likely to be too large or too small, and the number of errors will decrease as we move further away from the desired size. In other words, the errors will be distributed around the mean in a bell-shaped curve: Minor errors will be common and larger errors will be rare.

In the social world, the variables that are of most concern to us are either known *not* to be normal (e.g., income, the distribution of which is positively skewed) or they have shapes that are unknown (e.g., the degree of support for gay marriage). In other words, we usually do not find normal distributions when we examine the variables in our research projects. As you will see, the importance of the normal curve relates more to the concepts and logic that underlie hypothesis testing—the subject of the next part of this text—not to the descriptions of our variables.

and passing grades). To determine a probability, a fraction must be established, with the numerator equaling the number of events that would constitute a success and the denominator equaling the total number of possible events where a success could theoretically occur:

$$\text{Probability} = \frac{\#\ successes}{\#\ events}$$

To illustrate, assume we wish to know the probability of selecting a specific card—say, the king of hearts—in one draw from a well-shuffled deck of cards. Our definition of a success is specific (drawing the king of hearts), and with the information given, we can establish a fraction. Only one card satisfies our definition of success, so the number of events that would constitute a success is 1; this value will be the numerator of the fraction. There are 52 possible events (that is, 52 cards in the deck), so the denominator will be 52. The fraction is thus 1/52, which represents the probability of selecting the king of hearts on one draw from a well-shuffled deck of cards. Our probability of success is 1 out of 52.

We can leave this fraction as it is or we can express it in several other ways. For example, we can express it as an odds ratio by inverting the fraction, showing that the odds of selecting the king of hearts on a single draw are 52:1 (or 52 to 1). We can express the fraction as a proportion by dividing the numerator by the denominator. For our example, the corresponding proportion is .0192, which is the proportion of all possible events that would satisfy our definition of a success. In the social sciences, probabilities are usually expressed as proportions, and we will follow this convention throughout the remainder of this section. Using p to represent "probability," the probability of drawing the king of hearts (or any specific card) can be expressed as:

$$p\,(\text{king of hearts}) = \frac{\#\ successes}{\#\ events} = \frac{1}{52} = .0192$$

As conceptualized here, probabilities have an exact meaning: Over the long run, the events that we define as successes will bear a certain proportional relationship to the total number of events. The probability of .0192 for selecting the king of hearts in a single draw really means that over thousands of selections of one card at a time from a full deck of 52 cards, the proportion of successful draws would be .0192. Or for every 10,000 draws, 192 would be the king of hearts and the remaining 9,808 selections would be other cards. Thus, when we say that the probability of drawing the king of hearts in one draw is .0192, we are essentially applying to a single draw our knowledge of what would happen over thousands of draws.

Like proportions, probabilities range from 0.00 (meaning that the event has absolutely no chance of occurrence) to 1.00 (a certainty). As the value of the probability increases, the likelihood that the defined event will occur also increases. A probability of .0192 is close to zero, and this means that the event (drawing the king of hearts) is unlikely or improbable.

These techniques can be used to establish simple probabilities in any situation in which we can specify the number of successes and the total number of events.

For example, a single die has six sides, or faces, each with a different value, ranging from 1 to 6. The probability of getting any specific number (say, a 4) in a single roll of a die is:

$$p\,(\text{rolling a } 4) = \frac{1}{6} = .1667$$

Combining this way of thinking about probability with our knowledge of the theoretical normal curve allows us to estimate the likelihood of selecting a case that has a score within a certain range. For example, suppose we wished to estimate the probability that a randomly chosen subject from the distribution of men's IQ scores would have an IQ score between 95 and the mean score of 100. Our definition of a success here would be the selection of any subject with a score in the specified range. Normally, we would next establish a fraction with the numerator equal to the number of subjects with scores in the defined range and the denominator equal to the total number of subjects. However, if the empirical distribution is normal in form, we can skip this step because the probabilities—in proportion form—are already stated in Appendix A. That is, the areas in Appendix A can be interpreted as probabilities.

To determine the probability that a randomly selected case will have a score between 95 and the mean, we would convert the original score to a Z score:

$$Z = \frac{X_i - \overline{X}}{s} = \frac{95 - 100}{20} = -\frac{5}{20} = -0.25$$

Using Appendix A, we see that the area between this score and the mean is 0.0987. This is the probability we are seeking. The probability that a randomly selected case will have a score between 95 and 100 is 0.0987 (or, rounded off, 0.1, or 1 out of 10). In the same fashion, the probability of selecting a subject from any range of scores can be estimated. Note that the techniques for estimating probabilities are exactly the same as those for finding areas. The only new information introduced in this section is the idea that the areas in the normal curve table can also be thought of as probabilities.

To consider an additional example, what is the probability that a randomly selected male will have an IQ less than 123? We will find probabilities in exactly the same way we found areas. The score (X_i) is above the mean, and following the directions in Table 5.2, we will find the probability we are seeking by adding the area in column b to 0.5000. First, we find the Z score:

$$Z = \frac{X_i - \overline{X}}{s} = \frac{123 - 100}{20} = \frac{23}{20} = +1.15$$

Next, look in column b of Appendix A to find the area between this score and the mean. Then, add the area (0.3749) to 0.5000. The probability of selecting a male with an IQ of less than 123 is 0.3749 + 0.5000, or 0.8749. Rounding this value to .88, we can say that the odds are .88 (very high) that we will select a male with an IQ score in this range. Technically, remember that this probability expresses what would happen over the long run: For every 100 males selected from this group over an infinite number of trials, 88 would have IQ scores less than 123 and 12 would not.

Let us close by stressing a very important point about probabilities and the normal curve. The probability is very high that any case randomly selected from a normal distribution will have a score close in value to that of the mean. The shape of the normal curve is such that most cases are clustered around the mean and decline in frequency as we move farther away—either to the right or to the left—from the mean value. In fact, given what we know about the normal curve, the probability that a randomly selected case will have a score within ± 1 standard deviation of the mean is 0.6826. Rounding off, we can say that 68 out of 100 cases—or about two-thirds of all cases—selected over the long run will have a score between ± 1 standard deviation, or Z score, of the mean. The probabilities are that any randomly selected case will have a score close in value to the mean.

In contrast, the probability of the case having a score beyond three standard deviations from the mean is very small. Look in column c ("Area Beyond Z") for a Z score of 3.00 and you will find the value .0014. Adding the areas in the upper tail (beyond $+3.00$) to the area in the lower tail (beyond -3.00) gives us $.0014 + .0014$, for a total of .0028. The probability of selecting a case with a very high score or a very low score is .0028. If we randomly select cases from a normally distributed variable, we would select cases with Z scores beyond ± 3.00 only 28 times out of every 10,000 trials.

The general point to remember is that cases with scores close to the mean are common and that cases that have scores far above or below the mean are rare. This relationship is central for an understanding of inferential statistics in Part II. *(For practice in using the normal curve table to find probabilities, see problems 5.8 to 5.10 and 5.13.)*

Applying Statistics 5.3 Finding Probabilities

The distribution of scores on a biology final exam used in Applying Statistics 5.2 had a mean of 72 and a standard deviation of 8. What is the probability that a student selected at random will have a score less than 61? More than 80? Less than 98? To answer these questions, we must first calculate Z scores and then consult Appendix A. We are looking for probabilities, so we will leave the areas in proportion form. The Z score for a score of 61 is

$$Z_1 = \frac{X_i - \overline{X}}{s} = \frac{61 - 72}{8} = -\frac{11}{8} = -1.38$$

This score is a negative value (below, or to the left of, the mean) and we are looking for the area below. Using Table 5.2 as a guide, we see that we must use column c to find the area below a negative score. This area is .0838. Rounding off, we can say that the odds of selecting a student with a score less than 61 are only 8 out of 100. This low value tells us this would be an unlikely event.

The Z score for the score of 80 is:

$$Z_1 = \frac{X_i - \overline{X}}{s} = \frac{80 - 72}{8} = \frac{8}{8} = 1.00$$

The Z score is positive, and to find the area above (greater than) 80, we look in column c (see Table 5.2). This value is .1587. The odds of selecting a student with a score greater than 80 is roughly 16 out of 100—about twice as likely as selecting a student with a score of less than 61.

The Z score for the score of 98 is:

$$Z_1 = \frac{X_i - \overline{X}}{s} = \frac{98 - 72}{8} = \frac{26}{8} = 3.25$$

To find the area below a positive Z score, we add the area between the score and the mean (column b) to .5000 (see Table 5.2). This value is $.4994 + .5000$, or .9994. It is extremely likely that a randomly selected student will have a score less than 98. Remember that scores more than ± 3 standard deviations from the mean are very rare.

ONE STEP AT A TIME **Finding Probabilities**

Step	*Operation*
1.	Compute the *Z* score (or scores). Note whether the score is positive or negative.
2.	Find the *Z* score (or scores) in column a of the normal curve table (Appendix A).
3.	Find the area above or below the score (or between the scores) as you would normally (see the three previous "One Step at a Time" boxes in this chapter) and then express the result as a proportion. Typically, probabilities are expressed as a value between 0.00 and 1.00 rounded to two digits beyond the decimal point.

STATISTICS IN EVERYDAY LIFE

If you enjoy games of chance—bingo, card games, the state lottery, or board games such as Parcheesi—you can use your knowledge of the laws of probability to improve your performance and your likelihood of winning. This point was illustrated in the recent best seller *Bringing Down the House*, later made into a movie called *21*. The book describes the adventures of a group of MIT students who used their understanding of probability to develop a system that almost guaranteed they could beat the house in the game of blackjack. They won hundreds of thousands of dollars before they were finally stopped by casino security.

SUMMARY

1. The normal curve, in combination with the mean and standard deviation, can be used to construct precise descriptive statements about empirical distributions that are normally distributed. This chapter also lays some important groundwork for Part II.
2. To work with the theoretical normal curve, raw scores must be transformed into their equivalent *Z* scores. *Z* scores allow us to find areas under the theoretical normal curve (Appendix A).
3. We considered three uses of the theoretical normal curve: finding total areas above and below a score, finding areas between two scores, and expressing these areas as probabilities. This last use of the normal curve is especially germane because inferential statistics are centrally concerned with estimating the probabilities of defined events in a fashion very similar to the process introduced in this chapter.

SUMMARY OF FORMULA

FORMULA 5.1

$$Z = \frac{X_i - \overline{X}}{s}$$

GLOSSARY

Normal curve. A theoretical distribution of scores that is symmetrical, unimodal, and bell shaped. The standard normal curve always has a mean of 0 and a standard deviation of 1.

Normal curve table. Appendix A; a detailed description of the area between a *Z* score and the mean of any standardized normal distribution.

***Z* scores.** Standard scores; the way scores are expressed after they have been standardized to the theoretical normal curve.

PROBLEMS

5.1 Scores on a quiz were normally distributed and had a mean of 10 and a standard deviation of 3. For each of the following scores, find the *Z* score and the percentage of area above and below the score.

X_i	*Z* Score	% Area Above	% Area Below
5			
6			
7			
8			
9			
11			
12			
14			
15			
16			
18			

5.2 Assume that the distribution of a college entrance exam is normal, with a mean of 500 and a standard deviation of 100. For each of the following scores, find the equivalent *Z* score, the percentage of the area above the score, and the percentage of the area below the score.

X_i	*Z* Score	% Area Above	% Area Below
650			
400			
375			
586			
437			
526			
621			
498			
517			
398			

5.3 A senior class has been given a comprehensive examination to assess educational experience. The mean on the test was 74 and the standard deviation was 10. What percentage of the students had scores

a. between 75 and 85? ________
b. between 80 and 85? ________
c. above 80? ________
d. above 83? ________
e. between 80 and 70? ________
f. between 75 and 70? ________
g. below 75? ________
h. below 77? ________
i. below 80? ________
j. below 85? ________

5.4 For a normal distribution where the mean is 50 and the standard deviation is 10, what percentage of the area is

a. between the scores of 40 and 47? ________
b. above a score of 47? ________
c. below a score of 53? ________
d. between the scores of 35 and 65? ________
e. above a score of 72? ________
f. below a score of 31 and above a score of 69? ________
g. between the scores of 55 and 62? ________
h. between the scores of 32 and 47? ________

5.5 At St. Algebra College, the 200 freshmen enrolled in introductory biology took a final exam on which their mean score was 72 and the standard deviation was 6. The following table presents the grades of 10 students. Convert each into a *Z* score and determine the number of people who scored higher or lower than each of the 10 students. *(Hint: Multiply the appropriate proportion by N and round the result.)*

X_i	*Z* Score	Number of Students Above	Number of Students Below
60			
57			
55			
67			
70			
72			
78			
82			
90			
95			

5.6 If a distribution of test scores is normal, with a mean of 78 and a standard deviation of 11, what percentage of the area lies

a. below 60? ________
b. below 70? ________
c. below 80? ________
d. below 90? ________
e. between 60 and 65? ________
f. between 65 and 79? ________
g. between 70 and 95? ________
h. between 80 and 90? ________
i. above 99? ________
j. above 89? ________
k. above 75? ________
l. above 65? ________

5.7 A scale measuring prejudice has been administered to a large sample of respondents. The distribution of scores is approximately normal, with a mean of 31 and a standard deviation of 5. What percentage of the sample had scores

a. below 20? ________
b. below 40? ________
c. between 30 and 40? ________
d. between 35 and 45? ________
e. above 25? ________
f. above 35? ________

5.8 SOC On the scale mentioned in problem 5.7, if a score of 40 or more is considered "highly prejudiced," what is the probability that a person selected at random will have a score in that range?

5.9 For a math test on which the mean was 59 and the standard deviation was 4, what is the probability that a student randomly selected from this class will have a score

a. between 55 and 65? ________
b. between 60 and 65? ________
c. above 65? ________
d. between 60 and 50? ________
e. between 55 and 50? ________
f. below 55? ________

5.10 The average number of dropouts for a school district has been 305 per year, with a standard deviation of 50. What is the probability that next year the number of dropouts will be

a. less than 250? ________
b. less than 300? ________
c. more than 350? ________
d. more than 400? ________
e. between 250 and 350? ________
f. between 300 and 350? ________
g. between 350 and 375? ________

5.11 CJ The local police force gives all applicants an entrance exam and accepts only those applicants who score in the top 15% on this test. If the mean score this year is 87 and the standard deviation is 8, would an individual with a score of 110 be accepted?

5.12 To be accepted into an honor society, students must have GPAs in the top 10% of the school. If the mean GPA is 2.78 and the standard deviation is .33, which of the following GPAs would qualify?

3.20, 3.21, 3.25, 3.30, 3.35

5.13 In a distribution of scores with a mean of 35 and a standard deviation of 4, which event is more likely: that a randomly selected score will be between 29 and 31 or that a randomly selected score will be between 40 and 42?

5.14 SOC After taking the state merit examinations for the positions of school counselor and social worker, you receive the following information on the tests and on your performance. On which of the tests did you do better?

School Counselor	Social Worker
$\bar{X} = 118$	$\bar{X} = 27$
$s = 17$	$s = 3$
Your score = 127	Your score = 29

5.15 A large university administers entrance exams to all entering students. The test is administered again as an exit exam for all graduates. The test results for one group of students are:

Freshman Year	Senior Year
$\bar{X} = 53$	$\bar{X} = 92$
$s = 7$	$s = 4$

a. The scores for five students on both tests are listed below. For each student, calculate Z scores and then determine whether they performed better as freshmen or as seniors.

Student	Score: Freshman Year	Score: Senior Year
A	57	97
B	51	94
C	45	82
D	73	101
E	62	98

b. Determine the probabilities that randomly selected students will have scores in each of these ranges:

Freshman Year	
Score	Probability
Less than 52	
Less than 57	
Between 40 and 50	
More than 51	
More than 62	

Senior Year	
Score	Probability
Less than 88	
Less than 98	
Between 70 and 100	
More than 97	
More than 85	

Part II Inferential Statistics

The five chapters in this part cover the techniques and concepts of inferential statistics. Generally speaking, these applications allow us to learn about large groups (populations) from smaller, carefully selected subgroups (samples). These statistical techniques are powerful and extremely useful. They are used to poll public opinion, to research the potential market for new consumer products, to project the winners of elections, to test the effects of new drugs, and in hundreds of other ways inside and outside the social sciences.

Chapter 6 includes a brief description of sampling, but the most important part of this chapter concerns the sampling distribution—the single-most important concept in inferential statistics. The sampling distribution is normal in shape, and it is the key link between populations and samples. This chapter also covers estimation—the first of two main applications in inferential statistics. In this section, you will learn how to use means and proportions computed from samples to estimate the characteristics of a population. This technique is most commonly used in public opinion polling and election projection.

Chapters 7 through 10 cover the second application of inferential statistics: hypothesis testing. Most of the relevant concepts for this material are introduced in Chapter 7, and each chapter covers a different situation in which hypothesis testing is done. For example, Chapter 8 presents the techniques used when we compare information from two different samples or groups (e.g., men vs. women), while Chapter 9 covers applications involving more than two groups or samples (e.g., Republicans vs. Democrats vs. Independents).

Hypothesis testing is one of the more challenging aspects of statistics for beginning students, and we have included an abundance of learning aids to ease the chore of assimilating this material. Hypothesis testing is also one of the most common and important statistical applications to be found in social science research. Mastery of this material is essential for developing your ability to read the professional literature.

6 Introduction to Inferential Statistics, the Sampling Distribution, and Estimation

LEARNING OBJECTIVES

By the end of this chapter, you will be able to:

1. Explain the purpose of inferential statistics in terms of generalizing from a sample to a population.
2. Explain the principle of random sampling and these key terms: population, sample, parameter, statistic, representative, EPSEM.
3. Differentiate between the sampling distribution, the sample, and the population.
4. Explain the two theorems presented.
5. Explain the logic of estimation and the role of the sample, sampling distribution, and population.
6. Define and explain the concepts of bias and efficiency.
7. Construct and interpret confidence intervals for sample means and sample proportions.

USING STATISTICS

The statistical techniques presented in this chapter can be used to estimate the characteristics of large populations from small samples. They can be used to estimate:

- Changes in values and attitudes (e.g., support for capital punishment or gay marriage) in the United States over the years.
- Levels of happiness and well-being in nations throughout the world.
- The effectiveness of drugs or other therapies for the treatment of disease.
- The appeal of candidates for political office in various segments of the voting population (e.g., among women or Catholics or Southerners).
- Immediate public reactions to controversial issues, such as new laws on immigration or health care.

One of the goals of social science research is to test our theories and hypotheses by using many different people, groups, societies, and historical eras. Obviously, we can have the greatest confidence in theories that have stood up to testing against the greatest variety of cases and social settings. However, a major problem in social science research is that the populations in which we are interested are too large to test. For example, a theory concerning political party preference among U.S. citizens would be most suitably tested by using the entire electorate, but it is impossible to interview every voting member of this population (over 230 million people). Indeed, even for theories that could be reasonably tested with smaller populations—such as a local community or the student body at a university—the logistics of gathering data from every single case (entire populations) are staggering to contemplate.

If it is too difficult or expensive to do research with entire populations, how can we test our theories? To deal with this problem, social scientists select samples, or subsets of cases, from the populations of interest. Our goal in inferential statistics is to learn about the characteristics (or **parameters**) of a population based on what we can learn from our samples. Two applications of inferential statistics are covered in this text. In estimation procedures, covered in this chapter, we make an "educated guess" about the population parameter based on what is known about the sample. In hypothesis testing, covered in Chapters 7 through 10, the validity of a hypothesis about the population is tested against sample outcomes. However, before we can address these applications, we need to consider sampling (the techniques for selecting cases for a sample) and a key concept in inferential statistics: the sampling distribution.

Probability Sampling: Basic Concepts

Social scientists have developed a variety of sampling techniques. In this section, we will review the basic procedure for selecting probability samples—the only type of sample that fully supports the use of inferential statistical techniques to generalize to populations. These types of samples are often described as "random," and you may be more familiar with this terminology. Because of its greater familiarity, we will often use the phrase "random sample" in the following chapters. However, the term "probability sample" is preferred because in everyday language, "random" is often used to mean "by coincidence" or to give a connotation of unpredictability. To the contrary, probability samples are selected by techniques that are careful and methodical and leave no room for haphazardness. Interviewing the people you happen to meet in a mall one afternoon may be "random" in some sense, but this technique will not result in a sample that could support inferential statistics or justify a generalization to a population.

Before considering probability sampling, let me point out that social scientists often use nonprobability samples. For example, researchers who are studying small-group dynamics or the structure of attitudes or personal values might use the students enrolled in their classes as subjects. Such "convenience" samples are very useful for a number of purposes (e.g., exploring ideas or pretesting survey forms before embarking on a more ambitious project) and are typically less costly and easier to assemble. The major limitation of these samples is that results cannot be generalized beyond the group being tested. For example, if a theory of prejudice has been tested only on the students who happen to have been enrolled in a particular section of an introductory sociology class at a particular university in a particular year, then the researcher cannot generalize the findings to other types of people in other social locations or at other times. Therefore, we cannot place a lot of confidence in the generalizability of theories tested on nonprobability samples only, even when the evidence is very strong.

Selecting Representative Samples. When constructing a probability sample, our goal is to select cases so the final sample is **representative** of the population from which it was drawn. A sample is representative if it reproduces the important characteristics of the population. For example, if the population consists of 60% females and 40% males, the sample should contain essentially the same proportions. In other words, a representative sample is very much like the population—only smaller. How can we be sure that our samples are representative? Unfortunately, it is not possible to guarantee that samples are representative. However, we can maximize the chances of drawing a representative sample by following **EPSEM** (the **E**qual **P**robability of **SE**lection **M**ethod), the fundamental principle of probability sampling. To follow EPSEM and maximize the probability that our sample will be representative, we select the sample so every element or case in the population has an equal probability of being selected for the sample. Our goal is to select a representative sample, and the technique we use to maximize the chance of achieving that goal is to follow the rule of EPSEM.

Remember that EPSEM and representativeness are two different things. In other words, selecting a sample by EPSEM does not guarantee that it will be an exact representation of the population. The probability that an EPSEM sample is representative is very high, but an EPSEM sample will occasionally present an inaccurate picture of the population—just as a perfectly honest coin will sometimes show 10 heads in a row when flipped. One of the great strengths of inferential statistics is that they allow the researcher to estimate the probability of this type of error and interpret results accordingly.

Simple Random Samples. The most basic EPSEM sampling technique produces a **simple random sample**. There are numerous variations and refinements on this technique, but in this text, we will consider only the most straightforward application. To draw a simple random sample, we need a list of all cases in the population and a system for selecting cases that ensures that every case has an equal chance of being selected for the sample. The selection process could be based on a number of different kinds of operations (for example, drawing cards from a well-shuffled deck, flipping coins, throwing dice, drawing numbers from a hat, and so on). Cases are often selected by using a list of random numbers that has been generated by a computer program or that is presented in a table of random numbers. In either case, the numbers are random in the sense that they have no order or pattern: Each number in the list is just as likely as any other number. An example of a table of random numbers is available at the website for this text.

STATISTICS IN EVERYDAY LIFE

Advertisers sometimes use language that might seem scientific. Claims such as "80% of the 2,000 people surveyed prefer Brand X" might sound convincing, and a sample size of 2,000 certainly seems impressive. What if the sample consisted solely of people leaving a promotional event for Brand X during which free samples of the product were distributed? Does the claim still have legitimacy? Only properly selected probability samples can be used to make conclusions about larger populations.

STATISTICS IN EVERYDAY LIFE

EPSEM sampling techniques are also used for commercial purposes and for mass marketing. The next time you are annoyed by a telemarketer calling during dinnertime, remember that your phone number may have been selected by a sophisticated, complex application of random sampling. This probably will not lessen your annoyance, but it will remind you that sampling technology is a part of daily life, including many areas outside of the social sciences.

To use a table of random numbers to select a simple random sample, first assign each case on the population list a unique identification number. Then, select a case for the sample when its identification number corresponds to the number chosen from the table. This procedure will produce an EPSEM sample because the numbers in the table are in random order, and any number is just as likely as any other number. Stop selecting cases when you have reached your desired sample size, and if an identification number is selected more than once, ignore the repeats.[1]

Remember that the purpose of inferential statistics is to acquire knowledge about populations based on the information derived from samples of that population. Each of the applications of inferential statistics to be presented in this text requires that samples be selected according to EPSEM. While even the most painstaking and sophisticated sampling techniques will not guarantee representativeness, the probability is high that EPSEM samples will be representative of the populations from which they are selected.

The Sampling Distribution

Once we have selected a probability sample, what do we know? On one hand, we can gather a great deal of information about the cases in the sample. On the other hand, we know nothing about the population. Indeed, if we had information about the population, we probably would not need the sample. Remember that we use inferential statistics to learn more about populations, and information from the sample is important primarily insofar as it allows us to generalize to the population.

When we use inferential statistics, we generally measure some variable (e.g., age, political party preference, or opinions about abortion) in the sample and then use that information to learn more about that variable in the population. In Part I of this text, you learned that three types of information are generally necessary to adequately characterize a variable: (1) the shape of its distribution, (2) some measure of central tendency, and (3) some measure of dispersion. Clearly, we can gather all three kinds of information about the cases in the sample. Just as clearly, none of the information is available for the population. The means and standard

---

[1]Ignoring identification numbers when they are repeated is called "sampling without replacement." Technically, this practice compromises the randomness of the selection process. However, if the sample is a small fraction of the total population, we will be unlikely to select the same case twice, and ignoring repeats will not bias our conclusions.

deviations of variables in the population as well as the shapes of their distributions are unknown. Let me remind you that if we had this information for the population, inferential statistics would be unnecessary.

In statistics, we link information from the sample to the population with the **sampling distribution**: the theoretical, probabilistic distribution of a statistic for all possible samples of a certain sample size (*N*). That is, the sampling distribution is the distribution of a statistic (e.g., a mean or a proportion) for every conceivable combination of cases from the population. A crucial point about the sampling distribution is that its characteristics are based on the laws of probability, not on empirical information, and are very well known. In fact, the sampling distribution is the central concept in inferential statistics, and a prolonged examination of its characteristics is certainly in order.

The Three Distributions Used in Inferential Statistics. As illustrated by Figure 6.1, we move between the sample and population by means of the sampling distribution. Thus, three separate and distinct distributions are involved in every application of inferential statistics:

FIGURE 6.1 The Relationships Between the Sample, Sampling Distribution, and Population

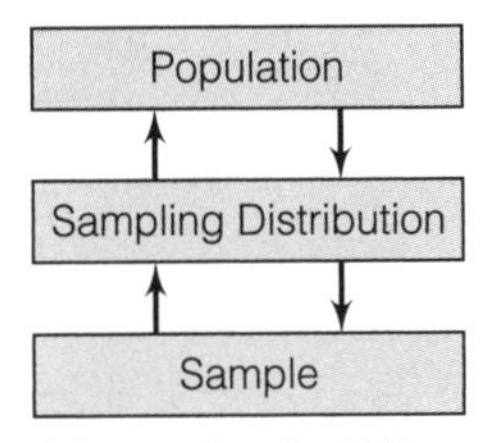

1. The population distribution, which, of course, is unknown. Amassing information about or making inferences to the population is the sole purpose of inferential statistics.
2. The sampling distribution, which is nonempirical or theoretical. Because of the laws of probability, a great deal is known about this distribution. Specifically, the shape, central tendency, and dispersion of the distribution can be deduced; therefore, the distribution can be adequately characterized.
3. The sample distribution of the variable, which is empirical (i.e., it exists in reality) and is known. The shape, central tendency, and dispersion of any variable can be ascertained for the sample. Remember that the information from the sample is important primarily insofar as it allows the researcher to learn about the population.

The utility of the sampling distribution is implied by its definition. Because it encompasses all possible sample outcomes, the sampling distribution enables us to estimate the probability of any particular sample outcome—a process that will occupy our attention for the remainder of this chapter and the next four chapters.

Constructing a Sampling Distribution. The sampling distribution is theoretical, which means that it is never obtained in reality. However, to better understand the structure and function of the distribution, let us consider an example of how one might be constructed. Suppose we wanted to gather some information about the age of a particular community of 10,000 individuals. We draw an EPSEM sample of 100 residents, ask all 100 respondents their age, and use those individual scores to compute a mean age of 27. This score is noted on the graph in Figure 6.2. Note that this sample is one of countless possible combinations of 100 people taken from this population of 10,000 and that the mean of 27 is one of millions of possible sample outcomes.

FIGURE 6.2 Constructing a Sample Distribution

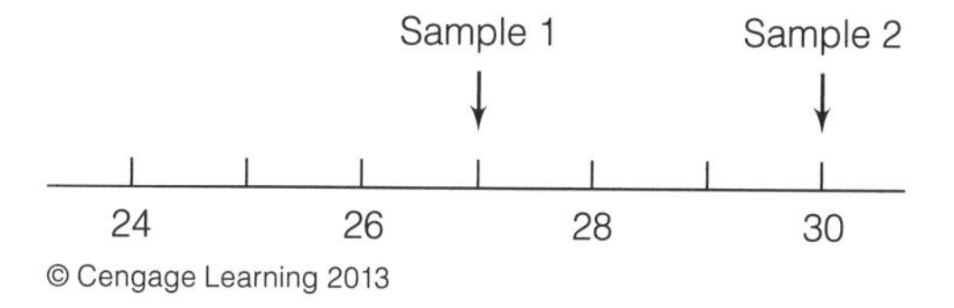

Now replace the 100 respondents in the first sample and then draw another sample of the same size ($N = 100$) and again compute the average age. Assume that the mean for the second sample is 30 and then note this sample outcome on Figure 6.2. This second sample is another of the countless possible combinations of 100 people taken from this population of 10,000, and the sample mean of 30 is another of the millions of possible sample outcomes. Replace the respondents from the second sample and then draw still another sample, calculate and note the mean, replace this third sample, and draw a fourth sample, continuing these operations an infinite number of times, calculating and noting the mean of each sample. Now try to imagine what Figure 6.2 would look like after tens of thousands of individual samples had been collected and the mean had been computed for each sample. What shape, mean, and standard deviation would this distribution of sample means have after we had collected all possible combinations of 100 respondents from the population of 10,000?

For one thing, we know that each sample will be at least slightly different from every other sample because it is very unlikely that we will sample exactly the same 100 people twice. Because each sample will almost certainly be a unique combination of individuals, each sample mean will be at least slightly different in value. We also know that even though the samples are chosen according to EPSEM, they will not be representative of the population in every single case. For example, if we continue taking samples of 100 people long enough, we will eventually choose a sample that includes only the very youngest residents of the community. The mean of this very young sample would be much lower than the true population mean. Likewise, by random chance alone, some of our samples will include only senior citizens and will have means that are much higher than the population mean. However, common sense suggests that such nonrepresentative samples will be rare and that most sample means will cluster around the true population value.

To illustrate further, assume we somehow come to know that the true mean age of the population is 30. As we have seen, if the population mean is 30, most of the sample means will also be approximately 30 and the sampling distribution of these sample means should peak at 30. Some of the sample means will be much too low or much too high, but the frequency of such misses should decline as we get farther away from 30. That is, the distribution should slope to the base as we get farther away from the population value; sample means of 29 or 31 might be common, but means of 20 or 40 would be rare. Because the samples are random, the means should miss an equal number of times on either side of the population value and the distribution itself should therefore be roughly symmetrical. In other words, the sampling distribution of all possible sample means should be approximately normal and will resemble the distribution presented in Figure 6.3. Recall from Chapter 5 that on any normal curve, cases close to the mean (say, within ± 1 standard deviation) are common and cases far away from the mean (say, beyond ± 3 standard deviations) are rare.

FIGURE 6.3 A Sampling Distribution of Sample Means

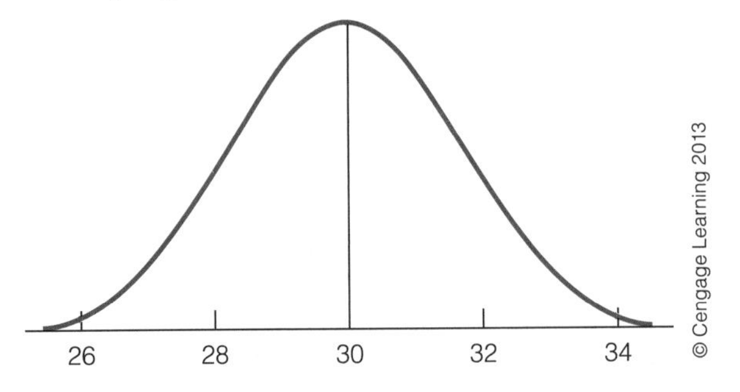

Two Theorems. These commonsense notions about the shape of the sampling distribution and other very important information about central tendency and dispersion are stated in two theorems. Before examining the theorems, we need to review some of the symbols we will use. Recall that the symbols for the means of samples and populations are $\overline{X}$ and μ, respectively. For measures of dispersion, we will use s to symbolize the standard deviation of a sample and σ (the lowercase Greek letter sigma) to refer to the standard deviation of a population.

The first theorem states:

> If repeated random samples of size N are drawn from a normal population with mean μ and standard deviation σ, then the sampling distribution of sample means will be normal, with a mean μ and a standard deviation of $\sigma/\sqrt{N}$.

To translate: If we begin with a trait that is normally distributed across a population (IQ, height, or weight, for example) and take an infinite number of equally sized random samples from that population, then the sampling distribution of sample means will be normal. If it is known that the variable is distributed normally in the population, it can be assumed that the sampling distribution will be normal.

However, the theorem tells us more than the shape of the sampling distribution of all possible sample means. It also defines its mean and standard deviation. In fact, it says that the mean of the sampling distribution will be exactly the same value as the mean of the population. That is, if we know that the mean IQ of the entire population is 100, then we know that the mean of any sampling distribution of sample mean IQs will also be 100. Exactly why this should be so is not a matter that can be fully explained at this level. However, recall that most sample means will cluster around the population value over the long run. Thus, the fact that these two values are equal should have intuitive appeal. As for dispersion, the theorem says that the standard deviation of the sampling distribution—also called the **standard error of the mean**—will be equal to the standard deviation of the population divided by the square root of N (symbolically: $\sigma/\sqrt{N}$).

If the mean and standard deviation of a normally distributed population are known, the theorem allows us to compute the mean and standard deviation of the sampling distribution.[2] Thus, we will know exactly as much about the sampling

[2]In the typical research situation, the values of the population mean and standard deviation are, of course, unknown. However, these values can be estimated from sample statistics, as we shall see in the chapters that follow.

distribution (shape, central tendency, and dispersion) as we ever knew about any empirical distribution.

The first theorem requires a normal population distribution. What happens when the distribution of the variable in question is unknown or is known not to be normal in shape (such as income, which always has a positive skew)? These eventualities (very common, in fact) are covered by a second theorem called the **central limit theorem**:

> If repeated random samples of size N are drawn from any population, with mean μ and standard deviation σ, then as N becomes large, the sampling distribution of sample means will approach normality, with mean μ and standard deviation $\sigma/\sqrt{N}$.

To translate: For *any* trait or variable, even those that are not normally distributed in the population, as the sample size grows larger, the sampling distribution of sample means will become normal in shape. When N is large, the mean of the sampling distribution will equal the population mean and its standard deviation (or the standard error of the mean) will be equal to $\sigma/\sqrt{N}$.

The importance of the central limit theorem is that it removes the constraint of normality in the population. Whenever the sample size is large, we can assume that the sampling distribution is normal, with a mean equal to the population mean and a standard deviation equal to $\sigma/\sqrt{N}$, regardless of the shape of the variable in the population. Thus, even if we are working with a variable that is known to have a skewed distribution (such as income), we can still assume a normal sampling distribution.

The issue remaining, of course, is to define what is meant by a large sample. A good rule of thumb is that if the sample size (N) is 100 or more, the central limit theorem applies and you can assume that the sampling distribution is normal in shape. When N is less than 100, you must have good evidence of a normal population distribution before you can assume that the sampling distribution is normal. Thus, a normal sampling distribution can be ensured by the expedient of using fairly large samples.

The Sampling Distribution: An Additional Example

Developing an understanding of the sampling distribution—what it is and why it is important—can be a challenging task. It may be helpful to briefly list the most important points about the sampling distribution:

1. *Its definition:* The sampling distribution is the distribution of a statistic (like a mean or a proportion) for all possible sample outcomes of a certain size.
2. *Its shape:* It is normal (see Chapter 5 and Appendix A).
3. *Its central tendency and dispersion:* The mean of the sampling distribution is the same value as the mean of the population. The standard deviation of the sampling distribution—or the standard error—is equal to the population

standard deviation divided by the square root of N. (See the two theorems presented in the previous section of this chapter).

4. *The role of the sampling distribution in inferential statistics:* It links the sample with the population (see Figure 6.1).

To reinforce these points, let us consider an additional example using the General Social Survey (GSS), the database used for SPSS exercises in this text. The GSS has been administered to randomly selected samples of adult Americans since 1972 and explores a broad range of characteristics and issues, including confidence in the Supreme Court, attitudes about assisted suicide, one's number of siblings, and level of education. The GSS has its limits, of course, but it is a very valuable resource for testing theory and for learning more about American society. Focusing on the General Social Survey, let us review the roles played by the population, the sample, and the sampling distribution when we use this database.

We will start with the population or the group we want to learn more about. In the case of the GSS, the population consists of all adult (older than 18) Americans, which includes more than 230 million people. Clearly, we can never interview all these people and learn what they are like or what they are thinking about abortion, gun control, sex education in the public schools, or any other issue. We should also note that this information is worth having. It could help inform public debates, provide some basis in fact for the discussion of many controversial issues (e.g., the polls consistently show that the majority of Americans favor some form of gun control), and assist people in clarifying their personal beliefs. If the information is valuable, how can we learn more about this huge population?

This brings us to the sample—a carefully chosen subset of the population. The GSS is administered to several thousand people, each of whom is chosen by a sophisticated technology based on the principle of EPSEM. A key point to remember is that samples chosen by this method are very likely to be representative of the populations from which they were selected. Whatever is true of the sample will also be true of the population (with some limits and qualifications, of course).

The respondents are contacted at home and asked for background information (religion, gender, years of education, and so on) as well as their opinions and attitudes. When all this information is collated, the GSS database includes information (shape, central tendency, dispersion) on hundreds of variables (age, level of prejudice, marital status) for the people in the sample. Thus, we have a lot of information about the variables for the *sample* (the people who actually responded to the survey) but no information about these variables for the *population* (the more than 230 million adult Americans). How do we get from the known sample to the unknown population? This is the central question of inferential statistics; the answer, as you hopefully realize by now, is "by using the sampling distribution."

Remember that unlike the sample and the population, the sampling distribution is theoretical, and because of the theorems presented earlier in this chapter, we know its shape, central tendency, and dispersion. For any variable from the GSS, the theorems tell us that:

- The sampling distribution will be normal in shape because the sample is "large" (N is much greater than 100). This will be true regardless of the shape of the variable in the population.

STATISTICS IN EVERYDAY LIFE

The GSS has been administered since 1972 and this permits us to track change in our society over time. As one illustration, data from the GSS demonstrate a considerable softening in the tendency of Americans to condemn homosexuality. In the 1970s, about 72% of respondents felt that homosexuality was "always wrong." In the most recent administration of the GSS in 2010, that percentage had fallen to about 45%.

- The mean of the sampling distribution will be the same value as the mean of the population. If *all* adult Americans have completed an average of 13.5 years of schooling ($\mu = 13.5$), the mean of the sampling distribution will also be 13.5.
- The standard deviation (or standard error) of the sampling distribution is equal to the population standard deviation (σ) divided by the square root of N.

Thus, the theorems tell us the statistical characteristics of this distribution (shape, central tendency, and dispersion), and this information allows us to link the sample to the population.

How does the sampling distribution link the sample to the population? The fact that the sampling distribution will be normal when N is large is crucial. This means that more than two-thirds (68%) of all samples will be within ± 1 Z score of the mean of the sampling distribution (which is the same value as the mean of the population), about 95% are within ± 2 Z scores, and so forth. We do not (and cannot) know the actual value of the mean of the sampling distribution, but we do know that the probabilities are very high that our sample statistic is approximately equal to this parameter. Similarly, the theorems give us crucial information about the mean and standard error of the sampling distribution that we can use—as you will see in the remainder of this chapter and the other chapters in this part—to link information from the sample to the population.

To summarize, our goal is to infer information about the population (in the case of the GSS, all adult Americans). When populations are too large to test, we use information from randomly selected samples—carefully drawn from the population of interest—to estimate the characteristics of the population. In the case of the GSS, the full sample for 2010 consists of about 2,000 adult Americans who have responded to the questions on the survey. The sampling distribution—the theoretical distribution whose characteristics are defined by the theorems—links the known sample to the unknown population.

Symbols and Terminology

In the following chapters, we will be working with three entirely different distributions. Furthermore, we will be concerned with several different kinds of sampling distributions—including the sampling distribution of sample means and the sampling distribution of sample proportions.

TABLE 6.1 Symbols for Means and Standard Deviations of Three Distributions

	Mean	Standard Deviation	Proportion
1. Samples	$\bar{X}$	s	P_s
2. Populations	μ	σ	P_u
3. Sampling distributions			
Of means	$\mu_{\bar{X}}$	$\sigma_{\bar{X}}$	
Of proportions	μ_p	σ_p	

To distinguish among these various distributions, we will often use symbols; for quick reference. Table 6.1 presents the symbols that will be used for the sampling distribution. Basically, the sampling distribution is denoted with Greek letters that are subscripted according to the sample statistic of interest.

Note that the mean and standard deviation of a sample are denoted with English letters ($\bar{X}$ and s), while the mean and standard deviation of a population are denoted with the Greek-letter equivalents (μ and σ). Proportions calculated on samples are symbolized as *P*-sub-*s* (*s* for sample), while population proportions are denoted as *P*-sub-*u* (*u* for "universe" or population). The symbols for the sampling distribution are Greek letters with English-letter subscripts. The mean and standard deviation of a sampling distribution of sample means are $\mu_{\bar{X}}$ ("mu-sub-*x*-bar") and $\sigma_{\bar{X}}$ ("sigma-sub-*x*-bar"). The mean and standard deviation of a sampling distribution of sample proportions are μ_p ("mu-sub-*p*") and σ_p ("sigma-sub-*p*").

Introduction to Estimation

The object of this branch of inferential statistics is to estimate population values or parameters from statistics computed from samples. Although the mathematical techniques may be new to you, you are certainly familiar with their most common applications. Polls and surveys on every conceivable issue—from the sublime to the trivial—have become a staple of mass media and popular culture. The techniques you will learn here are essentially the same as those used by the most reputable, sophisticated, and scientific pollsters.

The standard procedure for estimating population values is to construct **confidence intervals**—a mathematical statement that says that the parameter lies within a certain interval or range of values. For example, a confidence interval estimate might say "68% ±3%—or between 65% and 71%—of Americans approve of the death penalty."[3] In the media, the central value of the interval (68% in this case) is usually stressed, but it is important to realize that the population parameter (the percentage of *all* Americans who support capital punishment) could be anywhere between 65% and 71%.

[3]This estimate is based on the General Social Survey GSS, 2010, which was administered to a representative sample of adult residents of the United States.

Estimation Selection Criteria

Estimation procedures are based on sample statistics. Which of the many available sample statistics should be used? Estimators can be selected according to two criteria: **bias** and **efficiency**. Estimates should be based on sample statistics that are unbiased and relatively efficient. We cover each of these criteria separately.

Bias. An estimator is unbiased *if the mean of its sampling distribution is equal to the population value of interest.* We know from the theorems presented earlier in this chapter that sample means conform to this criterion. The mean of the sampling distribution of sample means (which we will note symbolically as $\mu_{\bar{X}}$) is the same as the population mean (μ).

Sample proportions (P_s) are also unbiased. That is, if we calculate sample proportions from repeated random samples of size N and then array them in a line chart, the sampling distribution of sample proportions will have a mean (μ_p) equal to the population proportion (P_u). Thus, if we are concerned with coin flips and sample honest coins 10 at a time ($N = 10$), the sampling distribution will have a mean equal to 0.5, which is the probability that an honest coin will be heads (or tails) when flipped. All statistics other than sample means and sample proportions are biased (that is, have sampling distributions with means not equal to the population value).[4]

Because sample means and proportions are unbiased, we can determine the probability that they lie within a given distance of the population values we are trying to estimate. To illustrate, consider a specific problem. Assume that we wish to estimate the average income of a community. A random sample of 500 households is taken ($N = 500$) and a sample mean of $45,000 is computed. In this example, the population mean is the average income of *all* households in the community and the sample mean is the average income for the 500 households that happened to be selected for our sample. Note that we do not know the value of the population mean (μ)—if we did, we would not need the sample—but it is μ that we are interested in. The sample mean of $45,000 is important primarily insofar as it can give us information about the population mean.

The two theorems presented earlier in this chapter give us a great deal of information about the sampling distribution of all possible sample means in this situation. Because N is large ($N > 100$), we know that the sampling distribution is normal and that its mean is equal to the population mean. We also know that all normal curves contain about 68% of the cases (the cases here are sample means) within ± 1 Z, 95% of the cases within ± 2 Zs, and more than 99% of the cases within ± 3 Zs of the mean. Remember that we are discussing the sampling distribution here—the distribution of all possible sample outcomes or, in this instance, sample means. Thus, the probabilities are very good (approximately 68 out of 100 chances) that our sample mean of $45,000 is within ± 1 Z, excellent (95 out of 100) that it is within ± 2 Zs, and overwhelming (99 out of 100) that it is within ± 3 Zs

[4]In particular, the sample standard deviation (s) is a biased estimator of the population standard deviation (σ). As you might expect, there is less dispersion in a sample than in a population, and as a consequence, s will underestimate σ. However, as we shall see, sample standard deviation can be corrected for this bias and still serve as an estimate of the population standard deviation for large samples.

FIGURE 6.4 Areas Under the Sampling Distribution of Sample Means

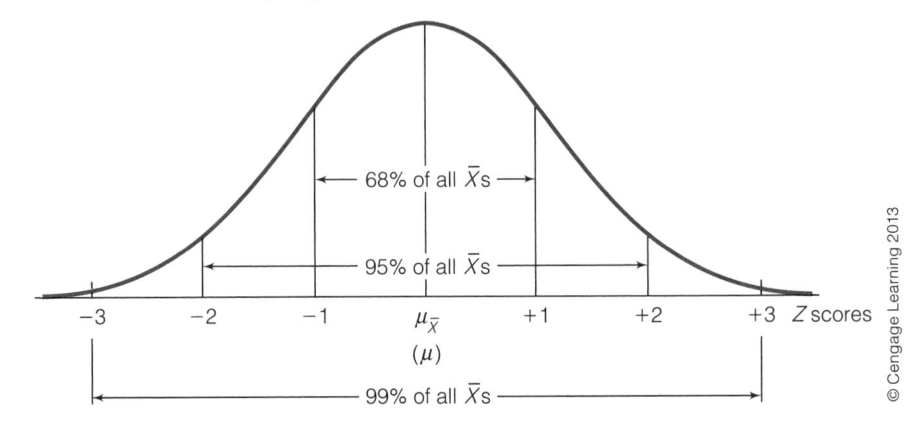

of the mean of the sampling distribution (which is the same value as the population mean). These relationships are graphically depicted in Figure 6.4.

If an estimator is unbiased, it is almost certainly an accurate estimate of the population parameter (μ in this case). However, in less than 1% of the cases, a sample mean will be more than ± 3 *Z*s away from the mean of the sampling distribution (very inaccurate) by random chance alone. We literally have no idea if our particular sample mean of \$45,000 is in this small minority. However, we do know that the odds are high that our sample mean is considerably closer than ± 3 *Z*s to the mean of the sampling distribution and, thus, to the population mean.

Efficiency. The second desirable characteristic of an estimator is efficiency, which is the extent to which the sampling distribution is clustered about its mean. Efficiency or clustering is essentially a matter of dispersion—the topic of Chapter 4 (see Figure 4.1). The smaller the standard deviation of a sampling distribution, the greater the clustering and the higher the efficiency. Remember that the standard deviation of the sampling distribution of sample means—or the standard error of the mean—is equal to the population standard deviation divided by the square root of *N*. Therefore, the standard deviation of the sampling distribution is an inverse function of *N* ($\sigma_{\bar{X}} = \sigma/\sqrt{N}$). As the sample size increases, $\sigma_{\bar{X}}$ will decrease. We can improve the efficiency (or decrease the standard deviation of the sampling distribution) for any estimator by increasing the sample size.

An example should make this clearer. Consider two samples of different sizes:

Sample 1	Sample 2
$\bar{X} = \$45{,}000$	$\bar{X} = \$45{,}000$
$N_1 = 100$	$N_2 = 1{,}000$

Both sample means are unbiased, but which is the more efficient estimator? Consider Sample 1 and assume for the sake of illustration that the population

FIGURE 6.5 A Sampling Distribution with N = 100 and $\sigma_{\bar{X}}$ = \$50.00

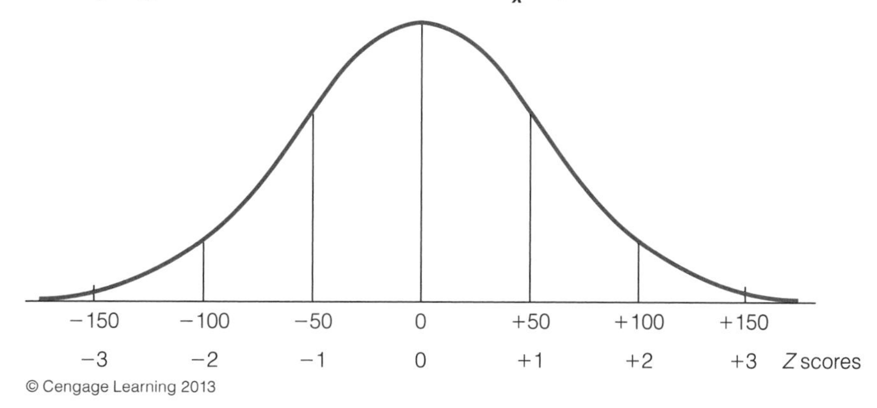

standard deviation (σ) is \$500.[5] In this case, the standard deviation of the sampling distribution of all possible sample means with an N of 100 would be $\sigma/\sqrt{N}$ or $500/\sqrt{100}$ or \$50.00. For Sample 2, the standard deviation of all possible sample means with an N of 1,000 would be much smaller. Specifically, it would be equal to $500/\sqrt{1{,}000}$, or \$15.81.

Sampling distribution 2 is much more clustered than sampling distribution 1. In fact, distribution 2 contains 68% of all possible sample means within ± 15.81 of μ, while distribution 1 requires a much broader interval of ± 50.00 to do the same. The estimate based on a sample with 1,000 cases is much more likely to be close to the population parameter than is the estimate based on a sample of 100 cases. Figures 6.5 and 6.6 illustrate these relationships graphically.

The key point to remember is that the standard deviation of all sampling distributions is an inverse function of N: the larger the sample, the greater the clustering and the higher the efficiency. In part, these relationships between the sample size and the standard deviation of the sampling distribution do nothing more than underscore our commonsense notion that much more confidence can be placed in large samples than in small (as long as both have been randomly selected).

FIGURE 6.6 A Sampling Distribution with N = 1,000 and $\sigma_{\bar{X}}$ = \$15.81

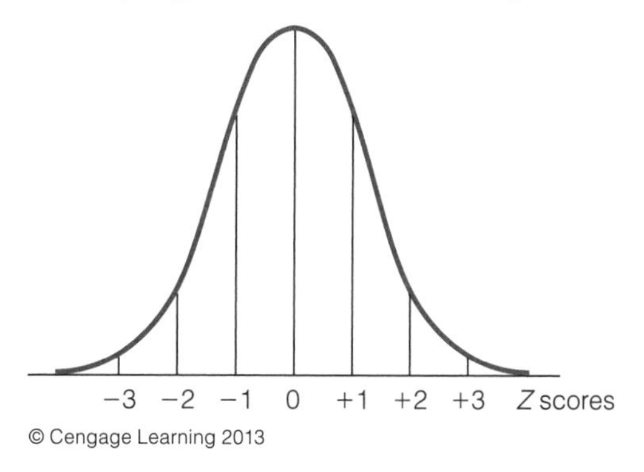

[5]In reality, of course, the value of σ would be unknown.

Interval Estimation Procedures

We are now ready to estimate population values based on sample statistics. We will do this by constructing confidence intervals or statements that say that the parameter is within a certain range of values. Confidence intervals are constructed in three steps.

The first step is to decide on the risk that you are willing to take of being wrong. An estimate is wrong if it does not include the population parameter. This probability of error is called **alpha** (symbolized as α). The exact value of alpha will depend on the nature of the research situation, but a 0.05 probability is commonly used. Setting alpha equal to 0.05—also called using the 95% **confidence level**—means that over the long run, the researcher is willing to be wrong only 5% of the time. Or to put it another way, if an infinite number of intervals were constructed at this alpha level (and with all other things being equal), 95% of them would contain the population value and 5% would not. In reality, of course, only one interval is constructed, and by setting the probability of error very low, we are setting the odds in our favor that the interval will include the population value.

The second step is to picture the sampling distribution, divide the probability of error equally into the upper and lower tails of the distribution, and then find the corresponding *Z* score. For example, if we decided to set alpha equal to 0.05, we would place half (0.025) of this probability in the lower tail and half in the upper tail of the distribution. The sampling distribution would thus be divided as illustrated in Figure 6.7.

We need to find the *Z* score that marks the beginnings of the shaded areas in Figure 6.7. In Chapter 5, we learned how to first calculate a *Z* score and then find an area under the normal curve. Here, we will reverse that process. We need to find the *Z* score beyond which lies a proportion of .0250 of the total area. To do this, go down column c of Appendix A until you find this proportional value (.0250). The associated *Z* score is 1.96. Because the curve is symmetrical and we are interested in the upper and lower tails, we designate the *Z* score that corresponds to an alpha of .05 as ± 1.96 (see Figure 6.8).

We now know that 95% of all possible sample outcomes fall within ± 1.96 *Z* score units of the population value. In reality, of course, there is only one sample

FIGURE 6.7 The Sampling Distribution With Alpha (α) Equal to 0.05

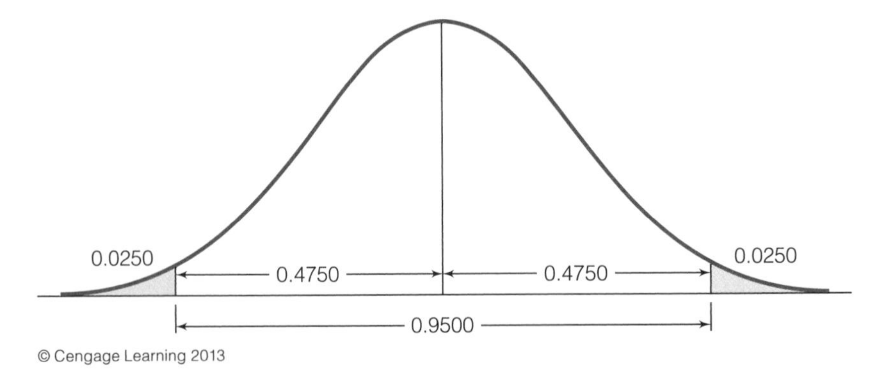

FIGURE 6.8 Finding the *Z* Score That Corresponds to an Alpha (α) of 0.05

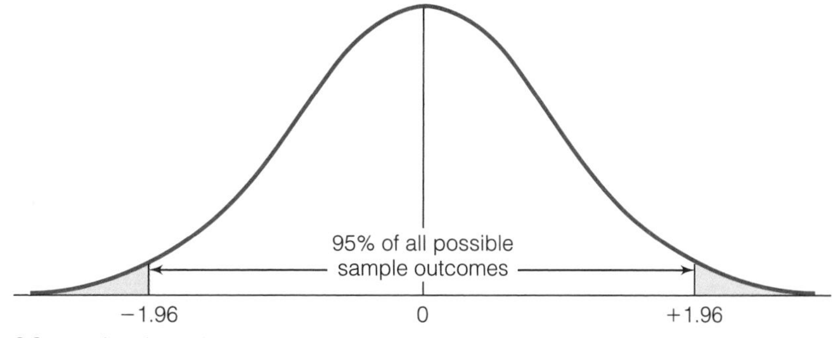

outcome, but if we construct an interval estimate based on ±1.96 *Z*s, the probability is that 95% of all such intervals will trap the population value. Thus, we can be 95% confident that our interval contains the population value.

Besides the 95% level, there are four other commonly used confidence levels—all listed in the left-hand column of Table 6.2, along with the 95% level. Note the relationship between the confidence levels (expressed as percentages) and the alpha levels (expressed as proportions); they are different ways of saying the same thing. The alpha level expresses the probability that the confidence interval will be *wrong* and *will not* include the population value. Confidence levels express our confidence that the interval is *correct* and *will* include the parameter in which we are interested. To find the corresponding *Z* scores for any levels, follow the procedures outlined above for an alpha of 0.05. Table 6.2 summarizes all the information you will need.

You should turn to Appendix A and confirm for yourself that the *Z* scores in Table 6.2 do indeed correspond to these alpha levels. As you do, note that in the cases where alpha is set at 0.10 and 0.01, the precise areas we seek do not appear in the table. For example, with an alpha of 0.10, we would look in column c ("Area Beyond *Z*") for the area 0.0500. Instead, we find an area of 0.0505 ($Z = \pm 1.64$) and an area of 0.0495 ($Z = \pm 1.65$). The *Z* score we are seeking is somewhere between these two other scores. When this condition occurs, take the larger of the two scores as *Z*. This will make the interval as wide as possible under the circumstances and is thus the most conservative course of action. In the case of an alpha of 0.01, we encounter the same problem (the exact area .0050 is not in the table); resolve it the same way and take the larger score as *Z*. For the alpha of 0.001, we take the largest of the several Z scores listed for the area as our

TABLE 6.2 *Z* Scores for Various Levels of Alpha (α)

Confidence Level	Alpha (α)	$\alpha/2$	*Z* Score
90%	0.10	0.0500	±1.65
95%	0.05	0.0250	±1.96
99%	0.01	0.0050	±2.58
99.9%	0.001	0.0005	±3.32
99.99%	0.0001	0.00005	±3.90

Z score. Finally, for the lowest alpha of 0.0001, the table is not detailed enough to show exact areas, and we will use ± 3.90 as our Z score. *(For practice in finding Z scores for various levels of confidence, see problem 6.3.)*

The third step is to actually construct the confidence interval. In the sections that follow, we illustrate how to construct an interval estimate—first with sample means and then with sample proportions.

Interval Estimation Procedures for Sample Means (Large Samples)

The formula for constructing a confidence interval based on sample means is given in Formula 6.1:

FORMULA 6.1

$$c.i. = \overline{X} \pm Z\left(\frac{\sigma}{\sqrt{N}}\right)$$

Where: $c.i.$ = confidence interval
$\overline{X}$ = the sample mean
Z = the Z score as determined by the alpha level
$\frac{\sigma}{\sqrt{N}}$ = the standard deviation of the sampling distribution or the standard error of the mean

As an example, suppose you wanted to estimate the average IQ of a community and had randomly selected a sample of 200 residents, with a sample mean IQ of 105. Assume that the population standard deviation for IQ scores is about 15, so we can set σ equal to 15. If we are willing to run a 5% chance of being wrong and set alpha at 0.05, the corresponding Z score will be 1.96. These values can be directly substituted into Formula 6.1 and an interval can be constructed:

$$c.i. = \overline{X} \pm Z\left(\frac{\sigma}{\sqrt{N}}\right)$$

$$c.i. = 105 \pm 1.96\left(\frac{15}{\sqrt{200}}\right)$$

$$c.i. = 105 \pm 1.96\left(\frac{15}{14.14}\right)$$

$$c.i. = 105 \pm (1.96)(1.06)$$

$$c.i. = 105 \pm 2.08$$

That is, our estimate is that the average IQ for the population in question is somewhere between 102.92 (105 − 2.08) and 107.08 (105 + 2.08). Because 95% of all possible sample means are within ± 1.96 Zs (or 2.08 IQ units in this case) of the mean of the sampling distribution, the odds are very high that our interval will contain the population mean. In fact, even if the sample mean is as far off as ± 1.96 Zs (which is unlikely), our interval will still contain $\mu_{\overline{X}}$ and, thus, μ. Only if our sample mean is one of the few that is more than ± 1.96 Zs from the mean of the sampling distribution will we have failed to include the population mean.

Note that in this example the value of the population standard deviation was supplied. Needless to say, it is unusual to have such information about a population. In the great majority of cases, we will have no knowledge of σ. However, in such cases, we can estimate σ with s, the sample standard deviation. Unfortunately, s is a biased estimator of σ, and the formula must be changed slightly to correct for the bias. For larger samples, the bias of s will not affect the interval very much. The revised formula for cases in which σ is unknown is:

FORMULA 6.2

$$c.i. = \overline{X} \pm Z\left(\frac{s}{\sqrt{N-1}}\right)$$

In comparing this formula with 6.1, note that there are two changes. First, σ is replaced by s, and second, the denominator of the last term is the square root of $N - 1$ rather than the square root of N. The latter change is the correction for the fact that s is biased.

Let me stress here that the substitution of s for σ is permitted only for large samples (that is, samples with 100 or more cases). For smaller samples, when the value of the population standard deviation is unknown, the standardized normal distribution summarized in Appendix A cannot be used in the estimation process. To construct confidence intervals from sample means with samples smaller than 100, we must use a different theoretical distribution, called the Student's t distribution, to find areas under the sampling distribution. We will defer the presentation of the t distribution until Chapter 7 and confine our attention here to estimation procedures for large samples only.

We will close this section by working through a sample problem with Formula 6.2. Average income for a random sample of a particular community is \$45,000, with a standard deviation of \$200. What is the 95% interval estimate of the population mean, μ?

Given that

$$\overline{X} = \$45{,}000$$

$$s = \$200$$

$$N = 500$$

and using an alpha of 0.05, the interval can be constructed:

$$c.i. = \overline{X} \pm Z\left(\frac{s}{\sqrt{N-1}}\right)$$

$$c.i. = 45{,}000 \pm 1.96\left(\frac{200}{\sqrt{499}}\right)$$

$$c.i. = 45{,}000 \pm 1.96\left(\frac{200}{22.34}\right)$$

$$c.i. = 45{,}000 \pm (1.96)(8.95)$$

$$c.i. = 45{,}000 \pm 17.55$$

ONE STEP AT A TIME Constructing Confidence Intervals for Sample Means by Using Formula 6.2

Step	*Operation*
1.	Select an alpha level and then find the associated *Z* score in Table 6.1. If you use the conventional alpha level of 0.05, the *Z* score is ±1.96.
2.	Substitute the sample values into Formula 6.2.

To Solve Formula 6.2

1. Find the square root of $N - 1$.
2. Divide the value you found in step 1 into *s*, the sample standard deviation.
3. Multiply the value you found in step 2 by the value of *Z*.
4. The value you found in step 3 is the width of the confidence interval. To find the lower and upper limits of the interval, subtract and add this value to the sample mean.

Interpreting the Confidence Interval

1. Express the confidence interval in a sentence or two that identifies each of these elements:
   a. The sample statistic (a mean in this case)
   b. The confidence interval
   c. The sample size (*N*)
   d. The population for which you are estimating
   e. The confidence level (e.g., 95%)

The confidence interval we constructed in this section could be expressed as: "The average income for this community is $45,000 ± $17.55. This estimate is based on a sample of 500 respondents, and we can be 95% confident that the interval estimate is correct."

Applying Statistics 6.1 Estimating a Population Mean

A study of the leisure activities of Americans was conducted on a sample of 1,000 households. The respondents identified television viewing as a major form of recreation. If the sample reported an average of 6.2 hours of television viewing a day, what is the estimate of the population mean? The information from the sample is:

$$\bar{X} = 6.2$$

$$s = 0.7$$

$$N = 1{,}000$$

If we set alpha at 0.05, the corresponding *Z* score will be ±1.96 and the 95% confidence interval will be:

$$c.i. = \bar{X} \pm Z\left(\frac{s}{\sqrt{N-1}}\right)$$

$$c.i. = 6.2 \pm 1.96\left(\frac{0.7}{\sqrt{1{,}000-1}}\right)$$

$$c.i. = 6.2 \pm 1.96\left(\frac{0.7}{31.61}\right)$$

$$c.i. = 6.2 \pm (1.96)(0.02)$$

$$c.i. = 6.2 \pm 0.04$$

Based on this result, we would estimate that the population spends an average of 6.2 ± .04 hours per day viewing television. The lower limit of our interval estimate (6.2 − 0.04) is 6.16 and the upper limit (6.2 + 0.04) is 6.24. Thus, another way to state the interval would be:

$$6.16 \leq \mu \geq 6.24$$

The population mean is greater than or equal to 6.16 and less than or equal to 6.24. Because alpha was set at the .05 level, this estimate has a 5% chance of being wrong (that is, of not containing the population mean).

The average income for the community as a whole is between $44,982.45 (45,000 − 17.55) and $45,017.55 (45,000 + 17.55). Remember that this interval has only a 5% chance of being wrong (that is, of not containing the population mean). See the "One Step at a Time" box for instructions on reporting the confidence interval clearly and completely. *(For practice in constructing and expressing confidence intervals for sample means, see problems 6.1, 6.4–6.7, and 6.18a–6.18c.)*

Interval Estimation Procedures for Sample Proportions (Large Samples)

Estimation procedures for sample proportions are essentially the same as those for sample means. The major difference is that because proportions are different statistics, we must use a different sampling distribution. In fact, again based on the central limit theorem, we know that sample proportions have sampling distributions that are normal in shape, with means (μ_p) equal to the population value (P_u) and standard deviations (σ_p) equal to $\sqrt{\frac{P_u(1 - P_u)}{N}}$. The formula for constructing confidence intervals based on sample proportions is:

FORMULA 6.3

$$c.i. = P_s \pm Z\sqrt{\frac{P_u(1 - P_u)}{N}}$$

The values for P_s and N come directly from the sample and the value of Z is determined by the confidence level, as was the case with sample means. This leaves one unknown in the formula: P_u—the same value we are trying to estimate. This dilemma can be resolved by setting the value of P_u at 0.5. Because the second term in the numerator under the radical $(1 - P_u)$ is the reciprocal of P_u, the entire expression will always have a value of 0.5 × 0.5, or 0.25, which is the maximum value this expression can attain. That is, if we set P_u at any value other than 0.5, the expression $P_u(1 - P_u)$ will decrease in value. For example, if we set P_u at 0.4, the second term $(1 - P_u)$ would be 0.6 and the value of the entire expression would decrease to 0.24. Setting P_u at 0.5 ensures that the expression $P_u(1 - P_u)$ will be at its maximum possible value, and consequently, the interval will be at maximum width. This is the most conservative solution possible to the dilemma posed by having to assign a value to P_u in the estimation equation.

To illustrate these procedures, assume you wish to estimate the proportion of students at your university who missed at least one day of classes because of illness last semester. Out of a random sample of 200 students, 60 reported that they had been sick enough to miss classes at least once during the previous semester. The sample proportion on which we will base our

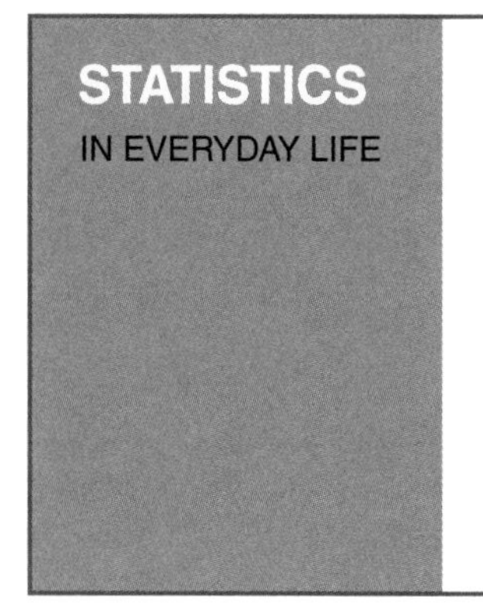

What are the happiest days of the year for Americans? The Gallup poll has been tracking people's happiness since October 2009 and found that during that period, the days on which the most Americans felt "happiness and enjoyment without stress and worry" were Christmas 2010 (65%), Easter 2011 (64%), Mother's Day 2010 (63%), and the 4th of July 2010 (63%). These results are based on daily samples of about 1,000 Americans, are accurate to within ±3%, and have a 95% confidence level.

How could these results be explained? Why would these particular holidays be the most enjoyable? What additional information would you like to have to begin to answer these questions?

The complete data set is available at http://www.gallup.com/poll/106915/Gallup-Daily-US-Mood.aspx.

estimate is thus 60/200, or 0.30. At the 95% level, the interval estimate will be

$$c.i. = P_s \pm Z\sqrt{\frac{P_u(1 - P_u)}{N}}$$

$$c.i. = 0.30 \pm 1.96\sqrt{\frac{(0.5)(0.5)}{200}}$$

$$c.i. = 0.30 \pm 1.96\sqrt{\frac{0.25}{200}}$$

$$c.i. = 0.30 \pm 1.96\sqrt{0.00125}$$

$$c.i. = 0.30 \pm (1.96)(0.035)$$

$$c.i. = 0.30 \pm 0.07$$

Based on the sample proportion of 0.30, you would estimate that the proportion of students who missed at least one day of classes because of illness was between 0.23 and 0.37. The estimate could, of course, also be phrased in percentages by reporting that between 23% and 37% of the student body was affected by illness at least once during the past semester.

As was the case with sample means, the final step in the process is to express the confidence interval in a way that is easy to understand and that includes all relevant information. See the "One Step at a Time" box for guidelines and an example. *(For practice with confidence intervals for sample proportions, see problems 6.2, 6.8–6.12, 6.16–6.17, and 6.18d–6.18g.)*

ONE STEP AT A TIME **Constructing Confidence Intervals for Sample Proportions**

Step	*Operation*
1.	Select an alpha level and then find the associated Z score in Table 6.1. If you use the conventional alpha level of 0.05, the Z score is ±1.96.

(continued next page)

ONE STEP AT A TIME ***(continued)***

Step	***Operation***
	To Solve Formula 6.3
1.	Substitute the value 0.25 for the expression $P_u(1 - P_u)$ in the numerator of the fraction under the square root sign.
2.	Divide N into 0.25.
3.	Find the square root of the value you found in step 2.
4.	Multiply the value you found in step 3 by the value of Z.
5.	The value you found in step 4 is the width of the confidence interval. To find the lower and upper limits of the interval, subtract and add this value to the sample proportion.
	Interpreting the Confidence Interval
1.	Express the confidence interval in a sentence or two that identifies each of these elements: a. The sample statistic (a proportion in this case) b. The confidence interval c. The sample size (N) d. The population for which you are estimating e. The confidence level (e.g., 95%)

The confidence interval we constructed in this section could be expressed as: "On this campus, 30% ± 7% of students were sick enough to miss class at least once during the semester. This estimate is based on a sample of 200 respondents, and we can be 95% confident that the interval estimate is correct."

Applying Statistics 6.2 Estimating Population Proportions

A total of 2,164 adult Canadians and 1,249 residents of the United States were randomly selected to participate in a study of attitudes and values, including their ideas about marriage and the family. One item asked if they agreed or disagreed with the idea that "marriage is an outdated institution." The results, expressed in proportions, are:

	Canada	U.S.
N	2,164	1,249
Proportion Agreeing	0.22	0.13

For Canadians, the confidence interval estimate to the population at the 95% confidence level is:

$$c.i. = P_s \pm Z\sqrt{\frac{P_u(1 - P_u)}{N}}$$

$$c.i. = 0.22 \pm 1.96\sqrt{\frac{0.50(1 - 0.50)}{2164}}$$

$$c.i. = 0.22 \pm 1.96\sqrt{\frac{0.25}{2164}}$$

$$c.i. = 0.22 \pm 1.96\sqrt{0.0001}$$

$$c.i. = 0.22 \pm (1.96)(0.01)$$

$$c.i. = 0.22 \pm 0.02$$

Expressing these results in terms of percentages, we can conclude that between 20% and 24% of adult Canadians agree that marriage is an outdated institution. This estimate is based on a sample of 2,164 and is constructed at the 95% confidence level.

For residents of the U.S., the confidence interval estimate to the population at the 95% confidence level is:

$$c.i. = P_s \pm Z\sqrt{\frac{P_u(1 - P_u)}{N}}$$

$$c.i. = 0.13 \pm 1.96\sqrt{\frac{0.50(1 - 0.50)}{1249}}$$

Applying Statistics 6.2 *(continued)*

$$c.i. = 0.13 \pm 1.96\sqrt{\frac{0.25}{1249}}$$

$$c.i. = 0.13 \pm 1.96\sqrt{0.0002}$$

$$c.i. = 0.13 \pm (1.96)(0.014)$$

$$c.i. = 0.13 \pm 0.03$$

Again, expressing results in terms of percentages, we can conclude that between 10% and 16% of adult Americans agree that marriage is an outdated institution. This estimate is based on a sample of 1,249 and is constructed at the 95% confidence level.

Source: World Values Survey, http://worldvaluessurvey.com.

Computing Confidence Intervals: A Summary

To this point, we have covered the construction of confidence intervals for sample means and sample proportions. In both cases, the procedures assume large samples (N greater than 100). The procedures for constructing confidence intervals for small samples are not covered in this text. Table 6.3 presents the three formulas for confidence intervals organized by the situations in which they are used. For sample means, when the population standard deviation is known, use Formula 6.1. When the population standard deviation is unknown (which is the usual case), use Formula 6.2. For sample proportions, always use Formula 6.3.

Controlling the Width of Interval Estimates

The width of a confidence interval for either sample means or sample proportions can be partly controlled by manipulating two terms in the equation. First, the confidence level can be raised or lowered, and second, the interval can be widened or narrowed by gathering samples of different size. The researcher alone determines the risk he or she is willing to take of being wrong (that is, of not including the population value in the interval estimate). The exact confidence level (or alpha level) will depend, in part, on the purpose of the research. For example, if potentially harmful drugs were being tested, the researcher would naturally demand very high levels of confidence (99.99%

TABLE 6.3 Choosing Formulas for Confidence Intervals

If the sample statistic is a	and	use formula	
mean	the population standard deviation is known	6.1	$c.i. = \bar{X} \pm Z\left(\frac{\sigma}{\sqrt{N}}\right)$
mean	the population standard deviation is unknown	6.2	$c.i. = \bar{X} \pm Z\left(\frac{s}{\sqrt{N-1}}\right)$
proportion		6.3	$c.i. = P_s \pm Z\sqrt{\frac{P_u(1-P_u)}{N}}$

or even 99.999%). On the other hand, if intervals are being constructed only for loose "guesstimates," then much lower confidence levels can be tolerated (such as 90%).

The relationship between interval size and confidence level is that intervals widen as confidence levels increase. This relationship should make intuitive sense. Wider intervals are more likely to trap the population value; hence, more confidence can be placed in them.

To illustrate this relationship, let us return to the example where we estimated the average income for a community. In this problem, we were working with a sample of 500 residents, and the average income for this sample was $45,000, with a standard deviation of $200. We constructed the 95% confidence interval and found that it extended 17.55 around the sample mean (that is, the interval was $45,000 ± 17.55).

If we had constructed the 90% confidence interval for these sample data (a lower confidence level), the Z score in the formula would have decreased to ± 1.65 and the interval would have been narrower:

$$c.i. = \overline{X} \pm Z\left(\frac{s}{\sqrt{N-1}}\right)$$

$$c.i. = 45{,}000 \pm 1.65\left(\frac{200}{\sqrt{499}}\right)$$

$$c.i. = 45{,}000 \pm 1.65(8.95)$$

$$c.i. = 45{,}000 \pm 14.77$$

On the other hand, if we had constructed the 99% confidence interval, the Z score would have increased to ± 2.58 and the interval would have been wider:

$$c.i. = \overline{X} \pm Z\left(\frac{s}{\sqrt{N-1}}\right)$$

$$c.i. = 45{,}000 \pm 2.58\left(\frac{200}{\sqrt{499}}\right)$$

$$c.i. = 45{,}000 \pm 2.58(8.95)$$

$$c.i. = 45{,}000 \pm 23.09$$

At the 99.9% confidence level, the Z score would be ± 3.32 and the interval would be wider still:

$$c.i. = \overline{X} \pm Z\left(\frac{s}{\sqrt{N-1}}\right)$$

$$c.i. = 45{,}000 \pm 3.32\left(\frac{200}{\sqrt{499}}\right)$$

$$c.i. = 45{,}000 \pm 3.32(8.95)$$

$$c.i. = 45{,}000 \pm 29.71$$

These four intervals are grouped together in Table 6.4 and the increase in interval size can be readily observed. Although sample means have been used to illustrate the relationship between interval width and confidence level, exactly the same relationships apply to sample proportions. (*To further explore the relationship between alpha and interval width, see problem 6.13.*)

TABLE 6.4 Interval Estimates for Four Confidence Levels ($\bar{X}$ = \$45,000, s = \$200, N = 500 Throughout)

Alpha	Confidence Level	Interval	Interval Width
0.10	90%	\$45,000 ± 14.77	\$29.54
0.05	95%	\$45,000 ± 17.55	\$35.10
0.01	99%	\$45,000 ± 23.09	\$46.18
0.001	99.9%	\$45,000 ± 29.71	\$59.42

The sample size bears the opposite relationship to interval width. As the sample size increases, interval width decreases. Larger samples give more precise (narrower) estimates. Again, an example should make this clearer. In Table 6.5, confidence intervals for four samples of various sizes are constructed and then grouped together for purposes of comparison. The sample data are the same as in Table 6.4 and the confidence level is 95% throughout. The relationships illustrated in Table 6.5 also hold true, of course, for sample proportions. *(To further explore the relationship between the sample size and the interval width, see problem 6.14.)*

Notice that the decrease in interval width (or increase in precision) does not bear a constant or linear relationship with the sample size. For example, sample 2 is five times larger than sample 1, but the interval constructed with the larger sample size is not five times as narrow. This is an important relationship because it means that N might have to be increased many times over to appreciably improve the accuracy of an estimate. Because the cost of a research project is directly related to the sample size, this relationship implies a point of diminishing returns in estimation procedures. A sample of 10,000 will cost about twice as much as a sample of 5,000, but estimates based on the larger sample will not be twice as precise.

TABLE 6.5 Interval Estimates for Four Different Samples ($\bar{X}$ = \$45,000, s = \$200, Alpha = 0.05 Throughout)

Sample 1 (N = 100)	Sample 2 (N = 500)
$c.i. = 45{,}000 \pm 1.96\left(\frac{200}{\sqrt{99}}\right)$	$c.i. = 45{,}000 \pm 1.96\left(\frac{200}{\sqrt{499}}\right)$
$c.i. = 45{,}000 \pm 39.40$	$c.i. = 45{,}000 \pm 17.55$

Sample 3 (N = 1,000)	Sample 4 (N = 10,000)
$c.i. = 45{,}000 \pm 1.96\left(\frac{200}{\sqrt{999}}\right)$	$c.i. = 45{,}000 \pm 1.96\left(\frac{200}{\sqrt{9{,}999}}\right)$
$c.i. = 45{,}000 \pm 12.40$	$c.i. = 45{,}000 \pm 3.92$

Sample	N	Interval Width
1	100	\$78.80
2	500	\$35.10
3	1,000	\$24.80
4	10,000	\$ 7.84

BECOMING A CRITICAL CONSUMER: Public Opinion Polls, Election Projections, and Surveys

Public opinion polls have become a part of everyday life in the United States and in many other societies, and statements such as those below are routinely found in the press, on the Internet, and in mass media:

- *55% of drivers have changed their driving habits as a result of the high cost of fuel.*
- *40% of voters are likely to vote for Candidate X.*
- *The president's approval rating stands at 62%.*
- *17% of Americans have watched an X-rated movie in the past 6 months.*

Can reports like these be trusted? How much credence should be accorded to statements such as these? How many times have you said (or heard someone else say) "Where do they get these numbers? They aren't talking to anyone I know." How can you evaluate these claims?

Three Cautions. First of all, whenever you encounter an attempt to characterize how "the public" feels or thinks, you need to examine the source of the statement. Generally, you can place more trust in reports that come from reputable polling firms (e.g., Gallup) or a national news source (CBS News, *USA Today*) and very little (if any) in polls commissioned for partisan purposes (that is, by organizations that represent a particular viewpoint, such as political parties or advocacy groups).

Secondly, you should examine how the information is reported. Professional polling firms use interval estimates, and responsible reporting by the media will usually emphasize the estimate itself (for example, "In a survey of the American public, 47% approved of gay marriage") but will also report the width of the interval ("This estimate is accurate to within ±3%" or "Figures from this poll are subject to a sampling error of ±3%"), the alpha level (usually as the confidence level of 95%), and the size of the sample ("1,458 households were surveyed"). Your suspicions should be raised if any of this information is missing.

Third, you should examine the sample for not only adequate size but also for representativeness. In particular, you should greatly discount reports based on the folksy, "man in the street" approach used in many news programs and "comments from our readers or viewers" sometimes found in mass media outlets. These may be interesting and even useful, but because they are not based on EPSEM samples, the results *cannot* be generalized or used to characterize the opinions of anyone other than the actual respondents.

Election Projections and Presidential Approval Ratings. In politics, the very same estimation techniques presented in this chapter are used to track public sentiment, measure how citizens perceive the performance of our leaders, and project the likely winners of upcoming elections.

We should note that these applications are controversial. Many people wonder if the easy availability of polls makes our political leaders too sensitive to the whims of public sentiment. Also, there is concern that election projections could work against people's willingness to participate in the political process and cast their votes on Election Day. These are serious concerns, but we can do little more than acknowledge them in this text and hope that you will have the opportunity to pursue them fully in other contexts.

Here, we will examine the accuracy of election projections for the 2004 and 2008 presidential election, and we will also examine the polls that have measured the approval ratings of incumbent U.S. presidents since the middle of the 20th century. Both kinds of polls use the same formulas introduced in this chapter to construct confidence intervals (although the random samples were assembled according to a complex and sophisticated technology that is beyond the scope of this text).

"Too Close to Call." Pollsters have become very accurate in predicting the outcomes of presidential elections, but remember that confidence intervals based on sample sizes of 1,000 or 2,000 respondents are accurate only to ±3 percentage points (assuming that the conventional alpha level of 0.05 is being used). This means that we cannot identify the likely winner in very close races. The 2004 election was

(*continued next page*)

BECOMING A CRITICAL CONSUMER: *(continued)*

very close, and the polls indicated a statistical dead heat right up to the end of the campaigns. The table below shows CNN's "poll of polls"—or averages of polls from various sources—for the final weeks of the 2004 campaign and the final breakdown of votes for the two major candidates. The polls included by CNN were based on sample sizes of about 1,000.

Polling Results and Actual Vote, 2004 Presidential Election

Actual Vote		
	BUSH	**KERRY**
	51%	48%

Election Projections:

	Percentage of Electorate Estimated to Vote for:	
Date of Poll	**BUSH**	**KERRY**
November 1	48	46
October 25	49	46
October 18	50	45

The final polls in October and November show that the race was very close and that the difference between the candidates was so small that a winner could not be projected. For example, in the November 1 poll, Bush's support could have been as low as 45% (48% − 3%) and Kerry's could have been as high as 49% (46% + 3%). When the confidence intervals overlap, the race is said to be "too close to call" and "a statistical dead heat."

The 2008 presidential election was not nearly so close. Democratic candidate and eventual winner Barack Obama enjoyed a comfortable margin over Republican candidate John McCain, as indicated by the three polls cited in the table below and by the final margin of victory:

Polling Results and Actual Vote, 2008 Presidential Election

Actual Vote		
	Obama	**McCain**
	55%	44%

Election Projections:

	Percentage of Sample Estimated to Vote for:	
Poll, Date, and Margin of Error	**Obama**	**McCain**
Gallup, Nov. 2, 2008, ±2%	53%	42%
NBC News, Nov. 1, 2008, ±3%	51%	43%
CBS News, Nov. 1, 2008, ±4%	54%	41%

Note that two of the polls were within their margin of error for both candidates. The confidence interval of the NBC poll was one percentage point too low in predicting Obama's final share of the vote but still showed a comfortable lead for the Democratic candidate.

The Ups and Downs of Presidential Popularity. Once you get to be president, you are not free of polls and confidence intervals. Since the middle of the 20th century, pollsters have tracked the president's popularity by asking randomly selected samples of adult Americans if they approve or disapprove of the way the president is handling his job.

Ratings for presidents Bush and Obama are presented below. For purposes of clarity, only the sample percentages are used for this graph, but you should remember that the confidence interval estimate would range about ±3% (at the 95% confidence level) around these points.

Perhaps the single-most dramatic feature of the graph is the huge increase in the approval of President Bush that followed the 9/11 terrorist attacks on the World Trade Center and the Pentagon in 2001. This burst of support reflects the increased solidarity, strong emotions, and high levels of patriotism with which Americans responded to the attacks. The president's approval rating reached an astounding 90% shortly after the attacks but then, inevitably, began to trend down as society (and politics) gradually returned to the daily routine of everyday business. President Bush's approval received another boost with the invasion of Iraq in March 2003—an echo of the "rally" effect that followed 9/11—but then began to sink to prewar levels and below, reflecting the substantial reluctance of many Americans to support this war effort.

Americans greeted President Obama's inauguration with great enthusiasm and high levels

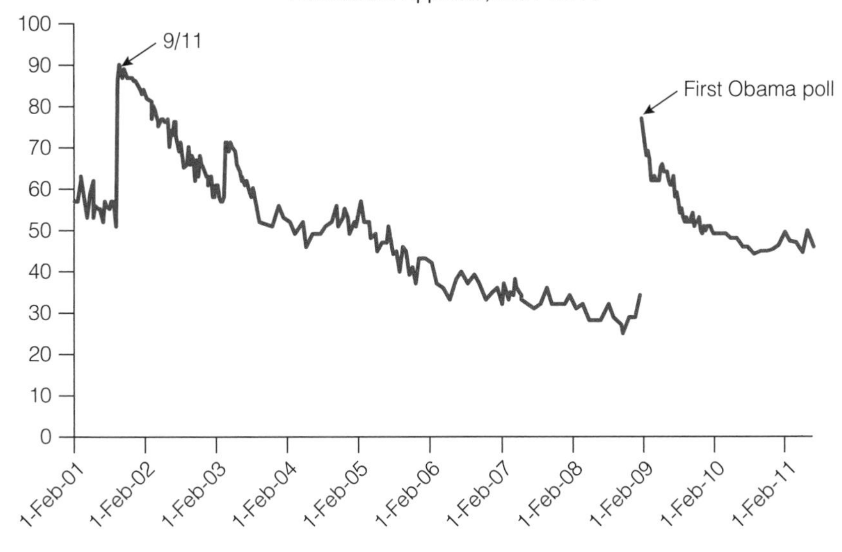

of approval. His initial ratings were in the high 70s—much higher than President Bush's first ratings and more than double his last—but then gradually declined as the nation's economic woes continued and the administration was caught up in a variety of acrimonious issues, including health care and the conduct of the wars in Iraq and Afghanistan.

Returning to the controversies surrounding public opinion polls, many would argue that it is harmful to judge the president so continuously. To some extent, these approval ratings expose the president (and other leaders and politicians whose performance is similarly measured) to the whims of public sentiment and, at some level, tend to pressure him or her to cater to popular opinion and shape ("spin") his or her image to maintain support. On the other hand, this information supplies some interesting and useful insights into the affairs of the society as a whole, not only of the political institution. For example, the approval ratings provide convincing evidence that Americans responded to the attacks of 9/11 with strong solidarity and cohesion, contrary to the expectations of some. The numbers also show the debilitating effects of controversial wars and the sputtering economy on public morale and confidence in the leadership of the nation.

Surveys in the Professional Literature. For the social sciences, probably the single-most important consequence of the growth in opinion polling is that many nationally representative databases are now available for research purposes. These high-quality databases are often available for free or for a nominal fee, and they make it possible to conduct "state of the art" research without the expense and difficulty of collecting data yourself. This is an important development because we can now test our theories against very high-quality data, and our conclusions will therefore have a stronger empirical basis. Our research efforts will have greater credibility with our colleagues, with policymakers, and with the public at large.

One of the more important and widely used databases of this sort is called the General Social Survey, or the GSS. Since 1972, the National Opinion Research Council has questioned a nationally representative sample of Americans about a wide variety of issues and concerns. Because many of the questions are asked every year, the GSS offers a longitudinal record of American sentiment and opinion about a large variety of topics for almost four decades. Each year, new topics of current concern are added and explored, and the variety of information available continues to expand. Like other nationally representative samples, the GSS sample is chosen by a complex probability design. Sample size varies from 1,400 to over 4,000, and estimates based on samples this large will be accurate to within about ±3%. The computer exercises in this text are based on the 2010 GSS, and this database is described more fully in Appendix G.

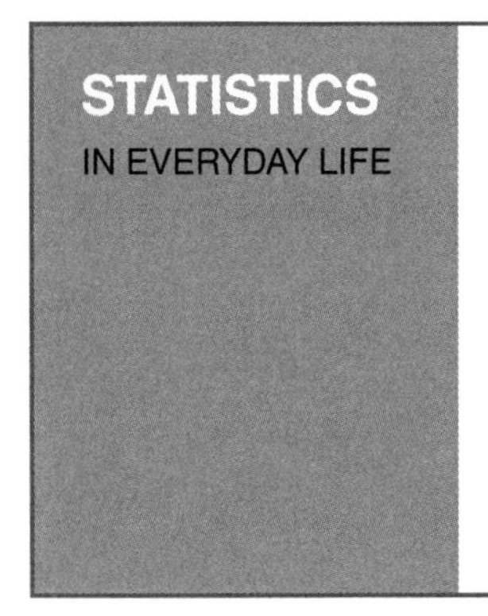

"Dewey Defeats Truman" read the banner headline of the morning edition of the *Chicago Tribune* the day after the 1948 presidential election. The headline became famous because the final results were just the reverse: Not only had Truman won his bid for re-election, but he won by a handy margin. What happened? Because the political "experts" were certain that Truman would lose, the pollsters had stopped taking polls early in the fall of 1948 and missed a late surge of support for the incumbent president. This major embarrassment for the *Chicago Tribune* (and many other newspapers) resulted in the current practice of polling voters right up until election day.

What kinds of opinions or attitudes would be more likely to change rapidly, as did support for President Truman? Can you think of attitudes or opinions that would be less likely to fluctuate?

SUMMARY

1. Because populations are almost always too large to test, a fundamental strategy of social science research is to select a sample from the defined population and then use information from the sample to generalize to the population. This is done either by estimation or by hypothesis testing.
2. Simple random samples are created by selecting cases from a list of the population following the rule of EPSEM (each case has an equal probability of being selected). Samples selected by the rule of EPSEM have a very high probability of being representative.
3. The sampling distribution—the central concept in inferential statistics—is a theoretical distribution of all possible sample outcomes. Because its overall shape, mean, and standard deviation are known (under the conditions specified in the two theorems), the sampling distribution can be adequately characterized and utilized by researchers.
4. The two theorems that were introduced in this chapter state that when the variable of interest is normally distributed in the population or when the sample size is large, the sampling distribution will be normal in shape, its mean will be equal to the population mean, and its standard deviation (or standard error) will be equal to the population standard deviation divided by the square root of N.
5. Population values can be estimated with sample values. With confidence intervals, which can be based on either proportions or means, we estimate that the population value falls within a certain range of values. The width of the interval is a function of the risk we are willing to take of being wrong (the alpha level) and the sample size. The interval widens as our probability of being wrong decreases and as the sample size decreases.
6. Estimates based on sample statistics must be unbiased and relatively efficient. Of all the sample statistics, only means and proportions are unbiased. The means of the sampling distributions of these statistics are equal to the respective population values. Efficiency is largely a matter of the sample size. The greater the sample size, the lower the value of the standard deviation of the sampling distribution, the more tightly clustered the sample outcomes will be around the mean of the sampling distribution, and the more efficient the estimate.

SUMMARY OF FORMULAS

FORMULA 6.1 Confidence interval for a sample mean (large sample, population standard deviation known):

$$c.i. = \bar{X} \pm Z\left(\frac{\sigma}{\sqrt{N}}\right)$$

FORMULA 6.2 Confidence interval for a sample mean (large sample, population standard deviation unknown):

$$c.i. = \bar{X} \pm Z\left(\frac{s}{\sqrt{N-1}}\right)$$

FORMULA 6.3 Confidence interval for a sample proportion (large sample)

$$c.i. = P_s \pm Z\sqrt{\frac{P_u(1 - P_u)}{N}}$$

GLOSSARY

Alpha (α). The probability of error or the probability that a confidence interval does not contain the population value. Commonly used alpha levels are 0.10, 0.05, 0.01, 0.001, and 0.0001.

Bias. A criterion used to select sample statistics as estimators. A statistic is unbiased if the mean of its sampling distribution is equal to the population value of interest.

Central limit theorem. A theorem that specifies the mean, standard deviation, and shape of the sampling distribution, given that the sample is large.

Confidence interval. An estimate of a population value in which a range of values is specified.

Confidence level. A frequently used alternative way of expressing alpha—the probability that an interval estimate will not contain the population value. Confidence levels of 90%, 95%, 99%, 99.9%, and 99.99% correspond to alphas of 0.10, 0.05, 0.01, 0.001, and 0.0001, respectively.

Efficiency. The extent to which the sample outcomes are clustered around the mean of the sampling distribution.

EPSEM. The **E**qual **P**robability of **SE**lection **M**ethod for selecting samples. Every element or case in the population must have an equal probability of selection for the sample.

μ. The mean of a population.

$\mu_{\overline{X}}$. The mean of a sampling distribution of sample means.

μ_P. The mean of a sampling distribution of sample proportions.

Parameter. A characteristic of a population.

P_s. (*P*-sub-*s*) Any sample proportion.

P_u. (*P*-sub-*u*) Any population proportion.

Representative. The quality a sample is said to have if it reproduces the major characteristics of the population from which it was drawn.

Sampling distribution. The distribution of a statistic for all possible sample outcomes of a certain size. Under conditions specified in two theorems, the sampling distribution will be normal in shape with a mean equal to the population value and a standard deviation equal to the population standard deviation divided by the square root of *N*.

Simple random sample. A method for choosing cases from a population by which every case and every combination of cases have an equal chance of being included.

Standard error of the mean. The standard deviation of a sampling distribution of sample means.

PROBLEMS

6.1 For each set of sample outcomes below, construct the 95% confidence interval for estimating μ, the population mean.

a. $\overline{X} = 5.2$, $s = .7$, $N = 157$
b. $\overline{X} = 100$, $s = 9$, $N = 620$
c. $\overline{X} = 20$, $s = 3$, $N = 220$
d. $\overline{X} = 1{,}020$, $s = 50$, $N = 329$
e. $\overline{X} = 7.3$, $s = 1.2$, $N = 105$
f. $\overline{X} = 33$, $s = 6$, $N = 220$

6.2 For each set of sample outcomes below, construct the 99% confidence interval for estimating P_u.

a. $P_s = .14$, $N = 100$
b. $P_s = .37$, $N = 522$
c. $P_s = .79$, $N = 121$
d. $P_s = .43$, $N = 1{,}049$
e. $P_s = .40$, $N = 548$
f. $P_s = .63$, $N = 300$

6.3 For each confidence level below, determine the corresponding *Z* score.

Confidence Level	Alpha	Area Beyond *Z*	*Z* Score
95%	.05	.0250	±1.96
94%			
92%			
97%			
98%			
99.9%			

6.4 SW You have developed a series of questions to measure job satisfaction for bus drivers in New York City. A random sample of 100 drivers has an average score of 10.6, with a standard deviation of 2.8. What is your estimate of the average job satisfaction score for the population as a whole? Use the 95% confidence level.

6.5 SOC A researcher has gathered information from a random sample of 178 households. For each variable below, construct confidence intervals to estimate the population mean. Use the 90% level.

a. An average of 2.3 people resides in each household. Standard deviation is .35.

b. There was an average of 2.1 television sets ($s = .10$) and .78 telephones ($s = .55$) per household.

c. The households averaged 6.0 hours of television viewing per day ($s = 3.0$).

6.6 SOC A random sample of 100 television programs contained an average of 2.37 acts of physical violence per program. At the 99% level, what is your estimate of the population value?

$$\bar{X} = 2.37$$

$$s = 0.30$$

$$N = 100$$

6.7 SOC A random sample of 429 college students was interviewed about a number of matters.

a. They reported that they had spent an average of \$478.23 on textbooks during the previous semester. If the sample standard deviation for these data is \$15.78, construct an estimate of the population mean at the 99% level.

b. They also reported that they had visited the health clinic an average of 1.5 times a semester. If the sample standard deviation is 0.3, construct an estimate of the population mean at the 99% level.

c. On average, the sample had missed 2.8 days of classes per semester because of illness. If the sample standard deviation is 1.0, construct an estimate of the population mean at the 99% level.

d. On average, the sample had missed 3.5 days of classes per semester for reasons other than illness. If the sample standard deviation is 1.5, construct an estimate of the population mean at the 99% level.

6.8 CJ A random sample of 100 inmates at a maximum security prison shows that exactly 10 of the respondents had been the victims of violent crime during their incarceration. Estimate the proportion of victims for the population as a whole, using the 90% confidence level. *(Hint: Calculate the sample proportion P_s before using Formula 6.3. Remember that proportions are equal to frequency divided by N.)*

6.9 SOC The survey mentioned in problem 6.5 found that 25 of the 178 households consisted of unmarried couples who were living together. What is your estimate of the population proportion? Use the 95% level.

6.10 PA A random sample of 324 residents of a community revealed that 30% were very satisfied with the quality of trash collection. At the 99% level, what is your estimate of the population value?

6.11 SOC A random sample of 1,496 respondents of a major metropolitan area was questioned about a number of issues. Construct estimates for the population at the 95% level for each of the results reported below. Express the final confidence interval in percentages (e.g., "between 40 and 45% agreed that premarital sex was always wrong").

a. When asked to agree or disagree with the statement "Explicit sexual books and magazines lead to rape and other sex crimes," 823 agreed.

b. When asked to agree or disagree with the statement "Handguns should be outlawed," 650 agreed.

c. 375 of the sample agreed that marijuana should be legalized.

d. 1,023 of the sample said that they had attended church or synagogue at least once within the past month.

e. 800 agreed that public elementary schools should have sex education programs starting in the fifth grade.

6.12 SW A random sample of 100 patients treated in a program for alcoholism and drug dependency over the past 10 years was selected. It was determined that 53 of the patients had been readmitted to the program at least once. At the 95% level, construct an estimate for the population proportion.

6.13 For the sample data below, construct four different interval estimates of the population mean—one each for the 90%, 95%, 99%, and 99.9% levels. What happens to the interval width as the confidence level increases? Why?

$$\bar{X} = 100$$

$$s = 10$$

$$N = 500$$

6.14 For each of the three sample sizes below, construct the 95% confidence interval. Use a sample proportion of 0.40 throughout. What happens to the interval width as the sample size increases? Why?

$$P_s = 0.40$$

Sample A: $N = 100$

Sample B: $N = 1{,}000$

Sample C: $N = 10{,}000$

6.15 PS Two individuals are running for mayor of Shinbone, Kansas. You conduct an election survey a week before the election and find that 51% of the respondents prefer candidate A. Can you predict a winner? Use the 99% level. *(Hint: In a two-candidate race, what percentage of the vote would the winner need? Does the confidence interval indicate that candidate A has a sure margin of victory? Remember that while the population parameter is probably (α = .01) in the confidence interval, it may be anywhere in the interval.)*

$$P_s = 0.51$$

$$N = 578$$

6.16 SOC The World Values Survey (http://www.worldvaluessurvey.org) is administered periodically to random samples from societies around the globe. Listed below are the number of respondents in each nation who said they are "very happy." Compute sample proportions and then construct confidence interval estimates for each nation at the 95% level.

Nation	Year	Number "Very Happy"	Sample Size	Confidence Interval
Great Britain	2006	528	1041	
Japan	2005	311	1096	
Brazil	2006	510	1500	
China	2007	424	2015	
Malaysia	2006	428	1201	
Mali	2007	598	1534	
Russia	2006	214	2033	

6.17 SOC The fraternities and sororities at St. Algebra College have been plagued by declining membership over the past several years and want to know if the incoming freshman class will be a fertile recruiting ground. Not having enough money to survey all 1,600 freshmen, they commission you to survey the interests of a random sample. You find that 35 of your 150 respondents are "extremely" interested in social clubs. At the 95% level, what is your estimate of the number of freshmen who would be extremely interested? *(Hint: The high and low values of your final confidence interval are proportions. How can proportions also be expressed as numbers?)*

6.18 SOC The results listed below are from a survey given to a random sample of the American public. For each sample statistic, construct a confidence interval estimate of the population parameter at the 95% confidence level. The sample size (*N*) is 2,987 throughout.

a. The average occupational prestige score was 43.87, with a standard deviation of 13.52.

b. The respondents reported watching an average of 2.86 hours of TV per day, with a standard deviation of 2.20.

c. The average number of children was 1.81, with a standard deviation of 1.67.

d. Of the 2,987 respondents, 876 identified themselves as Catholic.

e. 535 of the respondents said that they had never married.

f. The proportion of respondents who said they voted for McCain in the 2008 presidential election was 0.43.

g. When asked about capital punishment, 2,425 of the respondents said they favored the death penalty for murder.

YOU ARE THE RESEARCHER: Estimating the Characteristics of the Typical American

SPSS does not provide a program specifically for constructing confidence intervals, although some of the procedures we will cover in future chapters do include confidence intervals as part of the output. Rather than make use of these programs, we will use SPSS to produce sample statistics from the 2010 GSS, which you can then use as the basis for interval estimates to the population (U.S. society as of 2010).

Step 1: Choosing the Variables

As a framework for this exercise, we will return to the task of describing the "typical American" begun in Chapter 4. Select four of the variables you used in the earlier exercises and add four more that you did *not* use earlier. Make sure you have at least one variable from the nominal level and at least one measured at the ordinal or interval-ratio level. List the variables here along with level of measurement.

Variable	SPSS Name	Explain Exactly What This Variable Measures	Level of Measurement
1			
2			
3			
4			
5			
6			
7			
8			

Step 2: Getting Sample Statistics

For interval-ratio variables and ordinal level variables with four or more scores, use **Descriptives** to get sample means, standard deviations, and sample sizes. For nominal-level variables and for ordinal-level variables with three or fewer scores, use **Frequencies** to produce frequency distributions. For either procedure, click **Analyze → Descriptive Statistics** and then select either **Frequencies** or **Descriptives**. Select your variables from the box on the left and then click the arrow to move the variable name to the window on the right. SPSS will process all variables listed at the same time.

Step 3: Constructing Confidence Intervals

Once you have the SPSS output, use the results to construct 95% confidence intervals around the sample statistics. For nominal-level variables and ordinal-level variables with three or fewer scores, select one category (e.g., female for *sex* or Catholic for *relig*) and then look in the **Valid Percent** column of the frequency distribution to get the percentage of cases in the sample in that category. Change this value to a proportion (divide by 100). This value is P_s and can be substituted directly into Formula 6.3. Remember to estimate P_u at 0.5. After you have found the confidence interval, change your proportion back to a percentage and then record your results in the table below.

For ordinal-level variables with four or more scores and for interval-ratio variables, find the sample mean, the standard deviation, and the sample size in the output of the **Descriptives** procedure. Substitute these values into Formula 6.2 and then record your results in the table below.

Step 4: Recording Results

Use the table to summarize your confidence intervals:

Variable	SPSS Name	Sample Statistic ($\overline{X}$ or P_s)	95% Confidence Interval	N
1				
2				
3				
4				
5				
6				
7				
8				

Step 5: Reporting Results

For each statistic, express the confidence interval in words, as if you were reporting results in a newspaper story. Be sure to include the sample statistic, the sample size, the upper and lower limits of the confidence interval, and the confidence level (95%) and then identify the population. A sample sentence might read "I estimate that 45% of all Americans support candidate X for president. This estimate is accurate to within ±3%, has a 95% chance of being accurate, and is based on a sample of 1,362 adult Americans."

7 Hypothesis Testing I
The One-Sample Case

LEARNING OBJECTIVES

By the end of this chapter, you will be able to:

1. Explain the logic of hypothesis testing.
2. Define and explain the conceptual elements involved in hypothesis testing, especially the null hypothesis, the sampling distribution, the alpha level, and the test statistic.
3. Explain what it means to "reject the null hypothesis" or "fail to reject the null hypothesis."
4. Identify and cite examples of situations in which one-sample tests of hypotheses are appropriate.
5. Test the significance of single-sample means and proportions by using the five-step model and then correctly interpret the results.
6. Explain the difference between one- and two-tailed tests and specify when each is appropriate.
7. Define and explain Type I and Type II errors and then relate each to the selection of an alpha level.

USING STATISTICS

The statistical techniques presented in this chapter can be used to answer research questions in situations where we want to compare a sample to a population. Examples of this situation include:

1. A researcher working for a senior citizens advocacy group has asked a sample of 789 older residents of a particular state if they have been victimized by crime over the past year. He knows the percentage of the entire population of the state that was victimized and wants to test his hypothesis that senior citizens, as represented by this sample, are more likely to be victimized than the population in general.
2. The board of trustees of a small college wants to expand the school's athletic program but is concerned about the academic progress of student athletes. They have asked the dean to investigate. Among other tests, he will compare the GPAs of a random sample of 235 student athletes with the GPA of all students at the university.
3. A sociologist is assessing the effectiveness of a rehabilitation program for alcoholics in her city. The program serves a large area, so she cannot test every single client. Instead, she draws a random sample of 127 people from the list of all clients and questions them on a variety of issues. She notices

that, on average, the people in her sample miss fewer days of work each year than workers in the city as a whole. On the basis of this unexpected finding, she decides to conduct a test to see if the workers in her sample really are more reliable than workers in the community as a whole.

In Chapter 6, you were introduced to inferential statistics techniques for estimating population parameters from sample statistics. In Chapters 7 through 10, we investigate a second application of inferential statistics called **hypothesis testing** or **significance testing**. In this chapter, the techniques for hypothesis testing in the one-sample case are introduced. We can use these techniques when we have a randomly selected sample that we want to compare to a population. Note that we are not interested in the sample per se but in the larger group from which it was selected. We want to know if the group represented by the sample differs from a population parameter on a specific characteristic.

Of course, it would be better if we could test all members of the group in which we are interested rather than a small sample. However, as we have seen, researchers usually do not have the resources to test everyone in a large group and must use random samples instead. In these situations, conclusions will be based on a comparison of a single sample (representing the larger group) and the population. For example, if we found that the rate of criminal victimization for a sample of senior citizens was higher than the rate in the state as a whole, we might conclude that "senior citizens are significantly more likely to be crime victims." The word *significantly* is a key word: It means that the difference between the sample's victimization rate and the population's rate is very unlikely to be caused by random chance alone. In other words, it is very likely that *all* senior citizens (not just those in the sample) have a higher victimization rate than the state as a whole. On the other hand, if we found little difference between the GPAs of a sample of athletes and the whole student body, we might conclude that athletes (*all* athletes, not just those in the sample) are essentially the same as other students in terms of academic achievement.

Thus, we can use samples to represent larger groups (senior citizens, athletes, or treated alcoholics), compare and contrast the characteristics of the sample with those of the population, and be extremely confident in our conclusions. However, remember that there is no guarantee that random samples will be representative; thus, there will always be a small amount of uncertainty in our conclusions. One of the great advantages of inferential statistics is that we will be able to estimate the probability of error and evaluate our decisions accordingly.

An Overview of Hypothesis Testing

We will begin with a general overview of hypothesis testing—using the third research situation mentioned in "Using Statistics" as an example—and will introduce the more technical considerations and proper terminology throughout the remainder of the chapter. Let us examine this situation in some detail. First, the main question here is "Are people treated in this program more reliable workers than people in the community in general?" In other words, what the researcher would really like to do is compare *all* clients (the *population* of alcoholics treated in the program) with the entire metropolitan area. If she had information for both of these groups, she could answer the question easily, completely, and finally.

The problem is that the researcher does not have the time or money to gather information on the thousands of people who have been treated by the program. Instead, she has drawn a random sample of 127 clients. The absentee rates for the sample and the community are:

Community	Sample of Treated Alcoholics
$\mu = 7.2$ days per year	$\bar{X} = 6.8$ days per year
$\sigma = 1.43$	$N = 127$

We can see that there is a difference: The average rate of absenteeism for the sample is lower than the rate for the community. However, we cannot make any conclusions yet because we are working with a random sample of the population we are interested in, not the population itself (*all* people treated in the program).

Figure 7.1 should clarify these relationships. The community is symbolized by the largest circle because it is the largest group. The population of all treated alcoholics is also symbolized by a large circle because it is a sizable group, although it is only a fraction of the community as a whole. The random sample of 127—the smallest of the three groups—is symbolized by the smallest circle.

The labels on the arrows connecting the circles summarize the major questions and connections in this research situation. As we noted earlier, the main question is "Does the population of *all* treated alcoholics have different absentee rates than the community as a whole?" The population of treated alcoholics—too large to test—is represented by the random sample of 127.

We are interested in what caused the observed difference between the sample mean of 6.8 and the community mean of 7.2. There are two possible explanations for this difference, and we consider them one at a time.

FIGURE 7.1 A Test of Hypothesis for Single-Sample Means

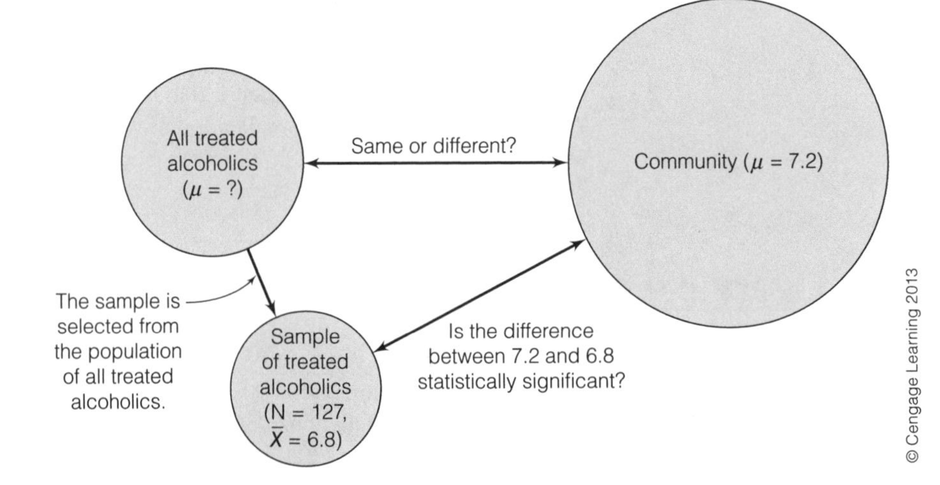

The first explanation is that the difference between the community mean (7.2) and the sample mean (6.8) reflects a real difference in absentee rates between the population of all treated alcoholics and the community. The difference is "statistically significant," which means that it is very unlikely to have occurred by random chance alone. If this explanation is true, the population of all treated alcoholics is different from the community and the sample did *not* come from a population with a mean absentee rate of 7.2 days.

The second explanation is called the **null hypothesis** (symbolized as H_0, or H-sub-zero). It states that the observed difference between sample and community means was caused by mere random chance: There is no important difference between treated alcoholics and the community, and the difference between the sample mean and the community mean is trivial and due to random chance. If the null hypothesis is true, treated alcoholics are just like everyone else and have a mean absentee rate of 7.2 days.

Which explanation is correct? We cannot answer this question with absolute certainty as long as we are working with a sample rather than the entire group. However, we can set up a decision-making procedure so conservative that one of the explanations can be chosen, with the knowledge that the probability of choosing the wrong explanation is very low.

This decision-making process begins with the assumption that the second explanation—the null hypothesis—is correct. Symbolically, the assumption that the mean absentee rate for all treated alcoholics is the same as the rate for the community as a whole can be stated as

$$H_0: \mu = 7.2 \text{ days per year}$$

Remember that μ refers to the mean for *all* treated alcoholics, not just the 127 in the sample. This assumption—$\mu = 7.2$—can be tested statistically.

If the null hypothesis ("The population of treated alcoholics is not different from the community as a whole and has a μ of 7.2") is true, then the probability of getting the observed sample outcome ($\bar{X} = 6.8$) can be found. Let us add an objective decision rule in advance. If the odds of getting the observed difference are less than 0.05 (5 out of 100, or 1 in 20), we will reject the null hypothesis. If this explanation were true, a difference of this size (7.2 days vs. 6.8 days) would be a very rare event, and in hypothesis testing, we always bet *against* rare events.

How can we estimate the probability of the observed sample outcome ($\bar{X} = 6.8$) if the null hypothesis is correct? This value can be determined by using our knowledge of the sampling distribution of all possible sample outcomes. Looking back at the information we have and applying the central limit theorem (see Chapter 6), we can assume that the sampling distribution is normal in shape, and has a mean of 7.2 (because $\mu_{\bar{X}} = \mu$) and a standard deviation of $1.43/\sqrt{127}$ because $\sigma_{\bar{X}} = \sigma/\sqrt{N}$. We also know that the standard normal distribution can be interpreted as a distribution of probabilities (see Chapter 5) and that this sample outcome ($\bar{X} = 6.8$) is one of thousands of possible sample outcomes. The sampling distribution, with the sample outcome noted, is depicted in Figure 7.2.

Using our knowledge of the standardized normal distribution, we can add further useful information to this sampling distribution of sample means. Specifically, we can depict our decision rule (any sample outcome with probability less than 0.05 will cause us to reject the null hypothesis) with Z scores. The probability of

FIGURE 7.2 The Sampling Distribution of All Possible Sample Means

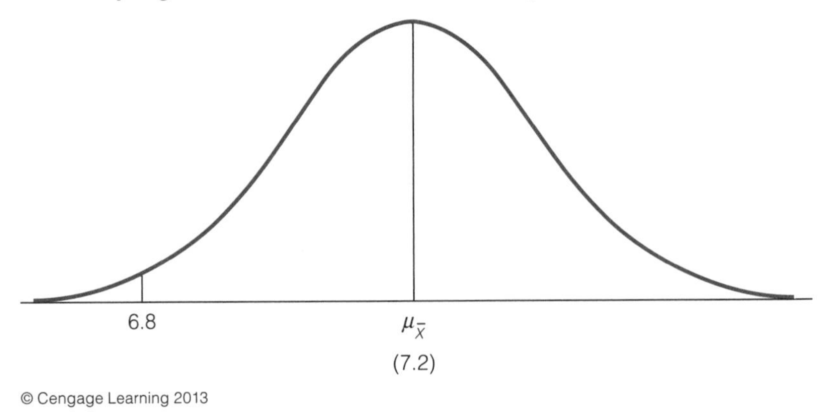

FIGURE 7.3 The Sampling Distribution of All Possible Sample Means

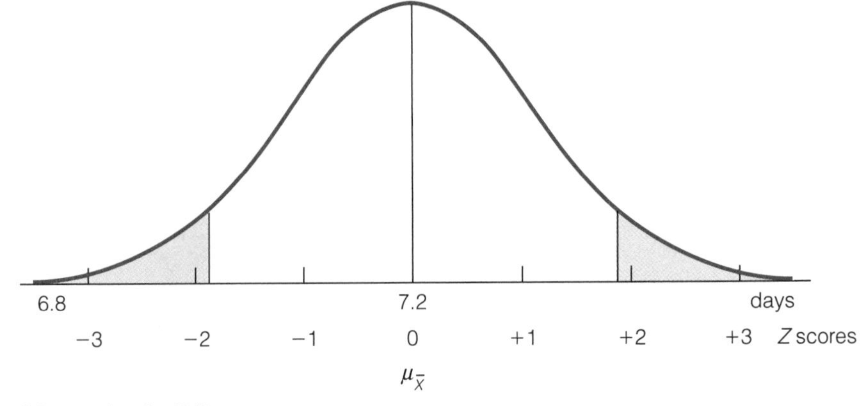

0.05 can be translated into an area and divided equally into the upper and lower tails of the sampling distribution. Using Appendix A, we find that the *Z*-score equivalent of this area is ±1.96. (To review finding *Z* scores from areas or probabilities, see Chapter 6.) The areas and *Z* scores are depicted in Figure 7.3.

The decision rule can now be rephrased. Any sample outcome falling in the shaded areas depicted in Figure 7.3 has a probability of occurrence of less than 0.05. Such an outcome would be a rare event and would cause us to reject the null hypothesis.

All that remains is to translate our sample outcome into a *Z* score to see where it falls on the curve. To do this, we use the standard formula for locating any score under a normal distribution. When we use known or empirical distributions, this formula is expressed as:

$$Z = \frac{X_i - \bar{X}}{s}$$

Or to find the equivalent *Z* score for any raw score, subtract the mean of the distribution from the raw score and divide by the standard deviation of the distribution.

Because we are now concerned with the sampling distribution of all sample means rather than an empirical distribution, the symbols in the formula will change, but the form remains exactly the same:

FORMULA 7.1

$$Z = \frac{\overline{X} - \mu}{\sigma/\sqrt{N}}$$

Or to find the equivalent Z score for any sample mean, subtract the mean of the sampling distribution (which is equal to the population mean, or μ) from the sample mean and then divide by the standard deviation of the sampling distribution.

Recalling the data given for this problem, we can now find the Z-score equivalent of the sample mean:

$$Z = \frac{6.8 - 7.2}{1.43/\sqrt{127}}$$

$$Z = \frac{-0.40}{0.127}$$

$$Z = -3.15$$

In Figure 7.4, this Z score of -3.15 is noted on the distribution of all possible sample means, and we see that the sample outcome does fall in the shaded area. If the null hypothesis is true, this sample outcome has a probability of occurrence of less than 0.05. The sample outcome ($\overline{X} = 6.8$, or $Z = -3.15$) would be rare if the null hypothesis were true, and the researcher may therefore reject the null hypothesis. The sample of 127 treated alcoholics comes from a population that is significantly different from the community on absenteeism. Or to put it another way, the sample does not come from a population that has a mean of 7.2 days of absences.

Keep in mind that our decisions in significance testing are based on random samples. On rare occasions, an EPSEM sample may not be representative of the population from which it was selected. The decision-making process just outlined has a very high probability of resulting in correct decisions, but we always face an element of risk when we must work with samples rather than populations. The decision to reject the null hypothesis might be incorrect if this sample happens to

FIGURE 7.4 The Sampling Distribution of Sample Means With the Sample Outcome ($\overline{X}$ = 6.8) Noted in Z Scores

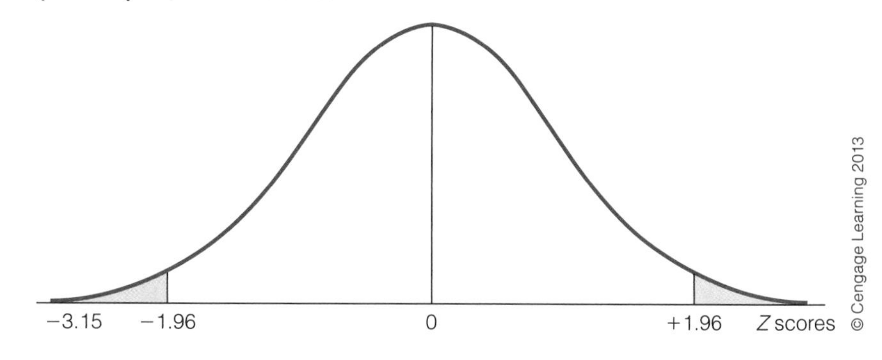

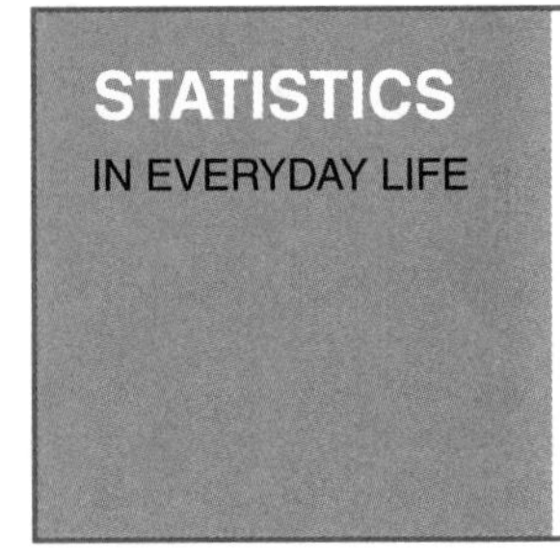

The federal Food and Drug Administration (FDA) estimates that it takes an average of eight years for a new drug to be approved for use in the United States. Before the manufacturer of a drug can apply for FDA approval, the drug must be tested first on animals, then on healthy volunteer humans, and finally on humans who have been diagnosed with the specific disease or condition that the manufacturer claims the drug was designed to treat. The results of these studies must show that the drug is safe and effective and that its benefits outweigh its risks. All the tests are based on the null hypothesis that the drug is ineffective in the treatment of the target disease or condition.

be one of the few that is unrepresentative of the population of alcoholics treated in this program. One important strength of hypothesis testing is that we can estimate the probability of making an incorrect decision. In the example at hand, the null hypothesis was rejected and the probability that this decision was incorrect is less than 0.05—the decision rule established at the beginning of the process. To say that the probability of rejecting the null hypothesis incorrectly is 0.05 means that if we repeated this same test an infinite number of times, we would incorrectly reject the null hypothesis only 5 times out of every 100.

The Five-Step Model for Hypothesis Testing

Now that you have been introduced to hypothesis testing, we will examine a **five-step model** for organizing all hypothesis testing:

Step 1: Making assumptions and meeting test requirements

Step 2: Stating the null hypothesis

Step 3: Selecting the sampling distribution and establishing the critical region

Step 4: Computing the test statistic

Step 5: Making a decision and interpreting the results of the test

We now look at each step individually, using the problem from the previous section as an example throughout.

Step 1. Making Assumptions and Meeting Test Requirements. Any application of statistics requires that we make certain assumptions. Three assumptions about the testing situation and the variables involved have to be satisfied when conducting a test of hypothesis with a mean. First, we must be sure that we are working with a random sample—one that has been selected according to the rules of EPSEM. Second, we must assume that the variable being tested is interval-ratio in level of measurement to justify the computation of a mean. Finally, we must assume that the sampling distribution of all possible sample means is normal in shape so we may use the standardized normal distribution to find areas under the sampling distribution. We can be sure that this assumption is satisfied by using large samples (see the central limit theorem in Chapter 6).

Usually, we will state these assumptions in abbreviated form as a mathematical model for the test. For example:

Model: Random sampling
Level of measurement is interval-ratio
Sampling distribution is normal

Step 2. Stating the Null Hypothesis (H_0). The null hypothesis is always a statement of "no difference," but its exact form will vary depending on the test being conducted. In the single-sample case, the null hypothesis states that the sample comes from a population with a certain characteristic. In our example, the null hypothesis is that the population of treated alcoholics is "no different" from the community as a whole, that their average days of absenteeism is also 7.2, and that the difference between 7.2 and the sample mean of 6.8 is caused by random chance. As we saw previously, the null hypothesis would be stated as

$$H_0: \mu = 7.2$$

where μ refers to the mean of the population of treated alcoholics. The null hypothesis is the central element in any test of hypothesis because the entire process is aimed at rejecting or failing to reject the H_0.

Usually, the researcher believes there is a significant difference and desires to reject the null hypothesis. The researcher's belief is stated in a **research hypothesis (H_1)**—a statement that directly contradicts the null hypothesis. Thus, the researcher's goal in hypothesis testing is often to support the research hypothesis by rejecting the null hypothesis.

The research hypothesis can be stated in several ways. One form would state that the population from which the sample was selected did *not* have a certain characteristic or, in terms of our example, had a mean that was *not* equal to a specific value:

$$(H_1: \mu \neq 7.2)$$

where $\neq$ means "not equal to"

Symbolically, this statement asserts that the population of all treated alcoholics is different from the community as a whole.

The research hypothesis is enclosed in parentheses to emphasize that it has no formal standing in the hypothesis-testing process (except, as we shall see in the next section, in choosing between one-tailed and two-tailed tests). It serves as a reminder of what the researcher believes to be the truth.

Step 3. Selecting the Sampling Distribution and Establishing the Critical Region. The sampling distribution is the probabilistic yardstick against which a particular sample outcome is measured. By assuming that the null hypothesis is true (and *only* by this assumption), we can attach values to the mean and standard deviation of the sampling distribution and thus measure the probability of any specific sample outcome. There are several different sampling distributions, but for now, we will confine our attention to the sampling distribution described by the standard normal curve, as summarized in Appendix A.

The **critical region** consists of the areas under the sampling distribution that include unlikely sample outcomes. We must define what we mean by "unlikely" prior to the test of hypothesis. That is, we must specify in advance those sample outcomes so unlikely that they will lead us to reject the H_0. This decision rule will establish the critical region, or **region of rejection**—the areas under the sampling distribution that contain unlikely sample outcomes. In our earlier example, this area began at a Z score of ± 1.96—called **Z(critical)**—that was graphically displayed in Figures 7.3 and 7.4. The shaded area is the critical region. Any sample outcome for which the Z-score equivalent fell in this area (that is, below -1.96 or above $+1.96$) would have caused us to reject the null hypothesis.

By convention, the size of the critical region is reported as alpha (α)—the proportion of all the area included in the critical region. In our example, our **alpha level** was 0.05. Other commonly used alphas are 0.10, 0.01, 0.001, and 0.0001.

These decisions can be stated in abbreviated form. The critical region is noted by the Z scores that mark its beginnings:

$$\text{Sampling distribution} = Z \text{ distribution}$$
$$\alpha = 0.05$$
$$Z(\text{critical}) = \pm 1.96$$

(For practice in finding Z(critical) scores, see problem 7.1a.)

Step 4. Computing the Test Statistic. To evaluate the probability of the sample outcome, the sample value must be converted into a Z score. Finding this Z score is called computing the **test statistic,** and the resultant value will be referred to as **Z(obtained)** to differentiate the test statistic from the score that marks the beginning of the critical region. In our example, we found a Z(obtained) of 3.15. *(For practice in computing obtained Z scores for means, see problems 7.1c, 7.2 to 7.10, 7.17e and f, and 7.21c–f)*

ONE STEP AT A TIME **Completing Step 4 of the Five-Step Model: Computing Z(obtained)**

Use these procedures if the population standard deviation (σ) is known or the sample size (N) is greater than 100. See the section entitled "The Student's t Distribution" for procedures when σ is unknown and N is less than 100.

Step	***Operation***
Use Formula 7.1 to compute the test statistic.	
1.	Find the square root of N.
2.	Divide the square root of N into the population standard deviation (σ).
3.	Subtract the population mean (μ) from the sample mean ($\bar{X}$).
4.	Divide the quantity you found in step 3 by the quantity you found in step 2. This value is Z(obtained).

TABLE 7.1 Making a Decision in Step 5 and Interpreting the Results of the Test

Situation	Decision	Interpretation
The test statistic is in the critical region.	Reject the null hypothesis (H_0).	The difference is statistically significant.
The test statistic is not in the critical region.	Fail to reject the null hypothesis (H_0).	The difference is not statistically significant.

Step 5. Making a Decision and Interpreting the Results of the Test. Finally, the test statistic is compared with the critical region. If the test statistic falls into the critical region, our decision will be to reject the null hypothesis. If the test statistic does not fall into the critical region, we fail to reject the null hypothesis. In our example, the two values were

$$Z(\text{critical}) = \pm 1.96$$

$$Z(\text{obtained}) = -3.15$$

and we saw that the Z(obtained) fell in the critical region (see Figure 7.4). Our decision was to reject the null hypothesis, which stated: "Treated alcoholics have a mean absentee rate of 7.2 days." When we reject the null hypothesis, we are saying that treated alcoholics do *not* have a mean absentee rate of 7.2 days and that there *is* a difference between them and the community. We can also say that the difference between the sample mean of 6.8 and the community mean of 7.2 is statistically significant or unlikely to be caused by random chance alone.

Note that in order to complete step 5, you have to do two things. First, you reject or fail to reject the null hypothesis (see Table 7.1). Second, you need to say what this decision means. In this case, we rejected the null hypothesis, which means that there is a significant difference between the mean of the sample and the mean for the entire community; therefore, we can conclude that treated alcoholics are different from the community as a whole.

This five-step model will serve us as a framework for decision making throughout the hypothesis-testing chapters. The exact nature and method of expression for our decisions will be different for different situations. However, the five step-model will provide a common frame of reference for all significance testing.

ONE STEP AT A TIME **Completing Step 5 of the Five-Step Model: Making a Decision and Interpreting Results**

Step	***Operation***
1.	Compare the Z(obtained) to the Z(critical). If Z(obtained) is *in* the critical region, *reject* the null hypothesis. If Z(obtained) is *not in* the critical region, *fail to reject* the null hypothesis.
2.	Interpret your decision in terms of the original question. For example, our conclusion for the example problem was "Treated alcoholics miss significantly fewer days of work than the community as a whole."

Think of hypothesis testing as analogous to gambling. Imagine you have been invited to participate in a game of chance involving coin flips: heads you win; tails your opponent wins. You would agree to participate in such a game only if you could **assume** that the coin was honest and that the probability of heads and tails was 0.5. What if your opponent flips 10 tails in a row? Think of these flips as a **test** of your original assumption that the coin is honest. At some point, as the coin shows tail after tail, you must compare the outcomes with your original assumption of honesty and make a **decision**: Either the coin is weighted toward tails or you have just witnessed a very rare series of events. If you conclude that the game is rigged, you have rejected the null hypothesis that there is no difference in the probabilities of heads and tails. While no one could blame you for walking away, note that there is a slight chance that your decision was wrong: It is possible (although very unlikely) that the game is not rigged and the coin is not weighted toward tails. In hypothesis testing, we also **make assumptions** (steps 1 through 3), **test** these assumptions in step 4, and make a **decision** based on probabilities in step 5.

Choosing a One-Tailed or Two-Tailed Test

The five-step model for hypothesis testing is fairly rigid, and the researcher has little room for making choices. Nonetheless, the researcher must still deal with two choices. First, he or she must decide between a one-tailed and a two-tailed test. Second, an alpha level must be selected. In this section, we discuss the former decision; we discuss the latter in the next section.

Choosing a One- or Two-Tailed Test. The choice between a one- and two-tailed test is based on the researcher's expectations about the population from which the sample was selected. These expectations are reflected in the research hypothesis (H_1), which is contradictory to the null hypothesis and usually states what the researcher believes to be "the truth." In most situations, the researcher will wish to support the research hypothesis by rejecting the null hypothesis.

The format for the research hypothesis may take either of two forms depending on the relationship between what the null hypothesis states and what the researcher believes to be the truth. The null hypothesis states that the population has a specific characteristic. In the example that has served us throughout this chapter, the null hypothesis stated, in symbols, "All treated alcoholics have the *same* absentee rate (7.2 days) as the community." The researcher might believe that the population of treated alcoholics actually has *less* absenteeism (their population mean is *lower than* the value stated in the null hypothesis) or *more* absenteeism (their population mean is *greater than* the value stated in the null hypothesis) or he or she might be unsure about the direction of the difference.

If the researcher is unsure about the direction, the research hypothesis would state only that the population mean is "not equal" to the value stated in the null hypothesis. The research hypothesis stated in our example ($\mu \neq 7.2$) was in this format. This is called a **two-tailed test** of significance because it means that the researcher will be equally concerned with the possibility that the true population value is greater than *or* less than the value specified in the null hypothesis.

In other situations, the researcher might be concerned only with differences in a specific direction. If the direction of the difference can be predicted or if the researcher is concerned only with differences in one direction, a **one-tailed test** can be used. A one-tailed test may take one of two forms depending on the researcher's expectations about the direction of the difference. If the researcher believes that the true population value is greater than the value specified in the null hypothesis, the research hypothesis would use the ">" or "greater than" symbol. In our example, if we had predicted that treated alcoholics had *higher* absentee rates than the community (or averaged *more* days of absenteeism), our research hypothesis would have been:

$$(H_1: \mu > 7.2)$$

where $>$ signifies "greater than"

If we predicted that treated alcoholics had lower absentee rates than the community (or averaged *fewer* days of absenteeism than 7.2), our research hypothesis would have been

$$(H_1: \mu < 7.2)$$

where $<$ signifies "less than"

One-tailed tests are often appropriate when programs designed to solve a problem or improve a situation are being evaluated. For example, if the program for treating alcoholics made them *less* reliable workers, the program would be a failure—at least on that criterion. In this situation, the researcher may focus only on outcomes that would indicate that the program is a success (i.e., when treated alcoholics have lower rates) and conduct a one-tailed test with a research hypothesis in the form $H_1{:}\mu < 7.2$. Or consider the evaluation of a program designed to reduce unemployment. The evaluators would be concerned only with outcomes that show a decrease in the unemployment rate. If the rate shows no change or if unemployment increases, the program is a failure and both of these outcomes might be considered equally negative by the researchers. Thus, the researchers could legitimately use a one-tailed test that stated that unemployment rates for graduates of the program would be less than (<) rates in the community.

One- versus Two-Tailed Test. In terms of the five-step model, the choice of a one-tailed or two-tailed test determines what we do with the critical region in step 3. As you recall, in a two-tailed test, we split the critical region equally into the upper and lower tails of the sampling distribution. In a one-tailed test, we place the entire critical area in one tail of the sampling distribution. If we believe that the population characteristic is greater than the value stated in the null hypothesis (if the H_1 includes the $>$ symbol), we place the entire critical region in the upper tail. If we believe that the characteristic is less than the value stated in the null hypothesis (if the H_1 includes the $<$ symbol), the entire critical region goes in the lower tail.

For example, in a two-tailed test with alpha equal to 0.05, the critical region begins at Z(critical) $= \pm 1.96$. In a one-tailed test at the same alpha level, the Z(critical) is $+1.65$ if the upper tail is specified and -1.65 if the lower tail is specified. Table 7.2 summarizes the procedures to follow in terms of

TABLE 7.2 One- Vs. Two-Tailed Tests, $\alpha = 0.05$

If the Research Hypothesis Uses	The Test Is	And Concern Is With	Z(critical) =
$\neq$	Two-tailed	Both tails	± 1.96
$>$	One-tailed	Upper tail	$+1.65$
$<$	One-tailed	Lower tail	-1.65

the nature of the research hypothesis. The difference in placing the critical region is graphically summarized in Figure 7.5, and the critical Z scores for the most common alpha levels are given in Table 7.3 for both one- and two-tailed tests.

Note that the critical Z values for one-tailed tests at all values of alpha are closer to the mean of the sampling distribution. Thus, a one-tailed test is more likely to reject the H_0 without changing the alpha level (assuming that we have specified the correct tail). One-tailed tests are a way of statistically having and eating your cake and should be used whenever (1) the direction of the difference can be confidently predicted or (2) the researcher is concerned only with differences in one tail of the sampling distribution. An example should clarify these procedures.

Using a One-Tailed Test. A sociologist has noted that sociology majors seem more sophisticated, charming, and cosmopolitan than the rest of the student body. A "Sophistication Scale" test has been administered to the entire student body and to a random sample of 100 sociology majors, and these results have been obtained:

Student Body	Sociology Majors
$\mu = 17.3$	$\bar{X} = 19.2$
$\sigma = 7.4$	$N = 100$

We will use the five-step model to test the H_0 of no difference between sociology majors and the general student body.

TABLE 7.3 Finding Critical Z Scores for One-Tailed Tests

		One-Tailed Value	
Alpha	Two-Tailed Value	Upper Tail	Lower Tail
0.10	± 1.65	$+1.29$	-1.29
0.05	± 1.96	$+1.65$	-1.65
0.01	± 2.58	$+2.33$	-2.33
0.001	± 3.32	$+3.10$	-3.10
0.0001	± 3.90	± 3.70	-3.70

FIGURE 7.5 Establishing the Critical Region, One-Tailed Tests versus Two-Tailed Tests (alpha = 0.05)

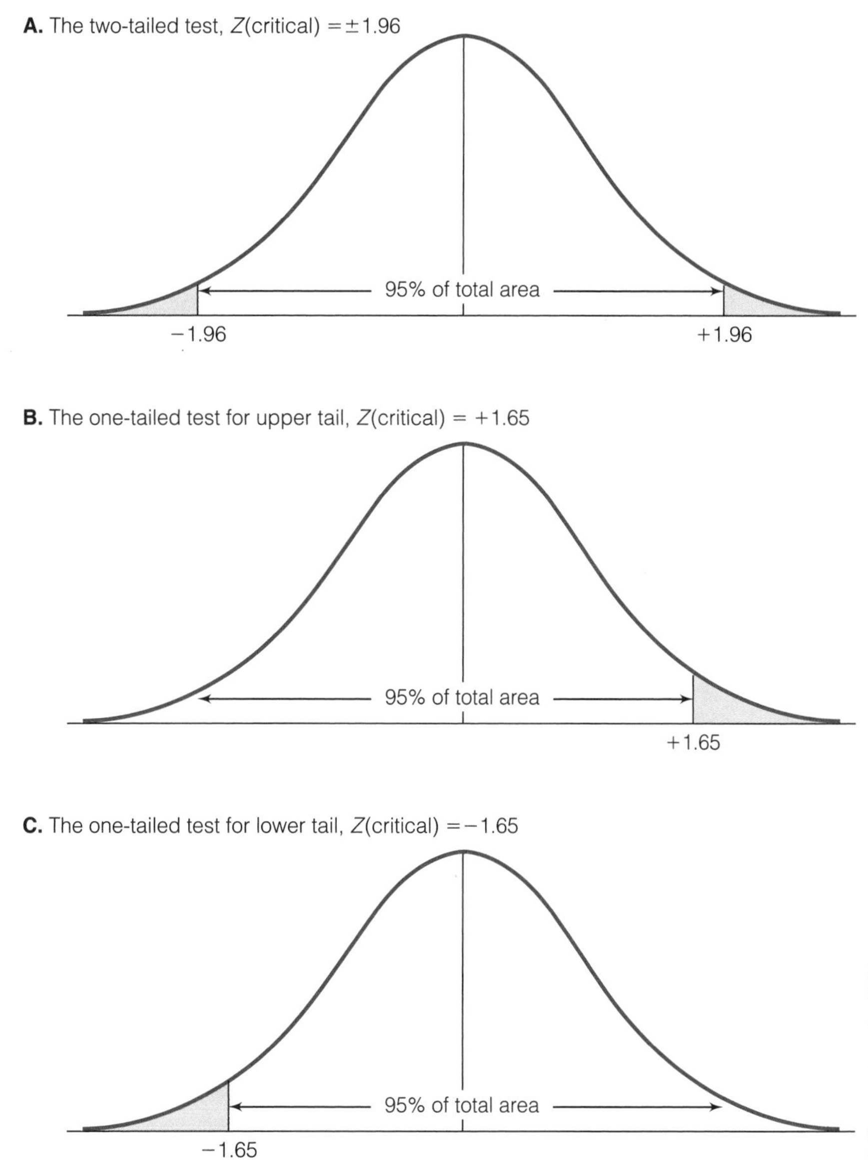

Step 1. Making Assumptions and Meeting Test Requirements. Because we are using a mean to summarize the sample outcome, we must assume that the Sophistication Scale generates interval-ratio-level data. With a sample size of 100, the central limit theorem applies and we can assume that the sampling distribution is normal in shape.

Model: Random sampling
Level of measurement is interval-ratio
Sampling distribution is normal

Step 2. Stating the Null Hypothesis (H_0). The null hypothesis states that there is no difference between sociology majors and the general student body. The research hypothesis (H_1) will also be stated at this point. The researcher has predicted a direction for the difference ("Sociology majors are *more* sophisticated"), so a one-tailed test is justified. The one-tailed research hypothesis asserts that sociology majors have a higher (>) score on the Sophistication Scale. The two hypotheses may be stated as:

$$H_0: \mu = 17.3$$
$$(H_1: \mu > 17.3)$$

Step 3. Selecting the Sampling Distribution and Establishing the Critical Region. We will use Appendix A to find areas under the sampling distribution. If alpha is set at 0.05, the critical region will begin at the Z score $+1.65$. That is, the researcher has predicted that sociology majors are *more* sophisticated and that this sample comes from a population that has a mean *greater than* 17.3, so he or she will be concerned only with sample outcomes in the upper tail of the sampling distribution. If sociology majors are *the same as* other students in terms of sophistication (if the H_0 is true) or if they are *less* sophisticated (and come from a population with a mean less than 17.3), the theory is disproved. These decisions may be summarized as:

$$\text{Sampling distribution} = Z \text{ distribution}$$
$$\alpha = 0.05$$
$$Z(\text{critical}) = +1.65$$

Step 4. Computing the Test Statistic.

$$Z(\text{obtained}) = \frac{\overline{X} - \mu}{\sigma/\sqrt{N}}$$
$$Z(\text{obtained}) = \frac{19.2 - 17.3}{7.4/\sqrt{100}}$$
$$Z(\text{obtained}) = +2.57$$

Step 5 Making a Decision and Interpreting Test Results. In this step, we will compare the Z(obtained) with the Z(critical):

$$Z(\text{critical}) = +1.65$$
$$Z(\text{obtained}) = +2.57$$

The test statistic falls into the critical region, and this outcome is depicted in Figure 7.6. We will reject the null hypothesis because if H_0 were true, a difference of this size would be very unlikely. There is a significant difference between sociology majors and the general student body in terms of sophistication. Because the null hypothesis has been rejected, the research hypothesis (sociology majors are more sophisticated) is supported. *(For practice in dealing with tests of significance for means that may call for one-tailed tests, see problems 7.2 to 7.5, 7.8, 7.10, 7.13, 7.18, and 7.19).*

FIGURE 7.6 Z(Obtained) Versus Z(Critical) (alpha = 0.05, one-tailed test)

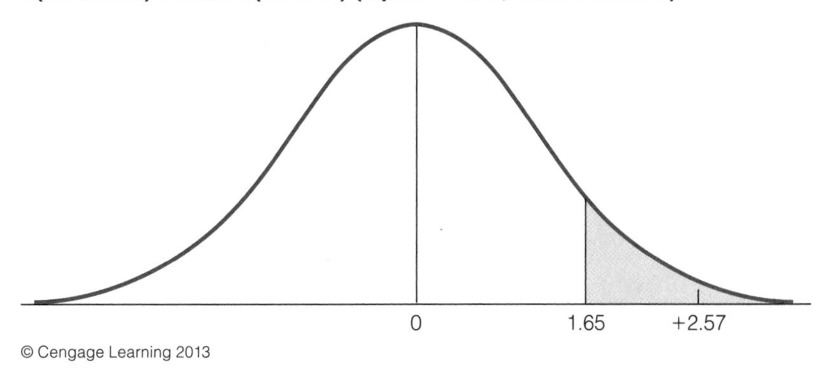

Selecting an Alpha Level

In addition to deciding between one-tailed and two-tailed tests, the researcher must select an alpha level. We have seen that the alpha level plays a crucial role in hypothesis testing. When we assign a value to alpha, we define what we mean by an "unlikely" sample outcome. If the probability of the observed sample outcome is lower than the alpha level (if the test statistic falls into the critical region), then we reject the null hypothesis.

How can we make reasonable decisions about the value of alpha? Recall that in addition to defining what will be meant by *unlikely,* the alpha level is the probability that the decision to reject the null hypothesis if the test statistic falls into the critical region will be incorrect. In hypothesis testing, the error of incorrectly rejecting the null hypothesis—or rejecting a null hypothesis that is actually true—is called **Type I error** (or **alpha error**). To minimize this type of error, use very small values for alpha.

To elaborate: When an alpha level is specified, the sampling distribution is divided into two sets of possible sample outcomes. The critical region includes all unlikely or rare sample outcomes. The remainder of the area consists of all sample outcomes that are not rare. The lower the level of alpha, the smaller the critical region and the greater the distance between the mean of the sampling distribution and the beginnings of the critical region. For the sake of illustration, compare the alpha levels and values for *Z*(critical) for two-tailed tests presented in Table 7.4. As you may recall, Table 6.2 also presented this information.

As alpha goes down, the critical region becomes smaller and moves farther away from the mean of the sampling distribution. The lower the alpha level,

TABLE 7.4 The Relationship Between Alpha and *Z*(critical) for a Two-Tailed Test

If Alpha Equals:	The Two-Tailed Critical Region Will Begin at *Z*(critical) Equal to
0.10	±1.65
0.05	±1.96
0.01	±2.58
0.001	±3.32

the harder it will be to reject the null hypothesis, and because a Type I error can occur only if our decision in step 5 is to reject the null hypothesis, the lower the probability of a Type I error. To minimize the probability of rejecting a null hypothesis that is in fact true, use very low alpha levels.

However, there is a complication. As the critical region decreases in size (as alpha levels decrease), the noncritical region—the area between the two Z(critical) scores in a two-tailed test—becomes larger. All other things being equal, the lower the alpha level, the less likely that the sample outcome will fall into the critical region. This raises the possibility of a second type of incorrect decision called **Type II error** (or **beta error**): failing to reject a null that is, in fact, false. The probability of a Type I error decreases as the alpha level decreases, but the probability of a Type II error increases. Thus, the two types of error are inversely related, and it is not possible to minimize both in the same test. As the probability of one type of error decreases, the probability of the other increases and vice versa.

It may be helpful to clarify in table format the relationships between decision making and errors. Table 7.5 lists the two decisions we can make in step 5 of the five-step model: We either reject or fail to reject the null hypothesis. Table 7.5 also lists the two possible conditions of the null hypothesis: It is either actually true or actually false. The table combines these possibilities into a total of four possible outcomes—two of which are desirable ("OK") and two of which indicate that an error has been made.

Let us consider the two desirable ("OK") outcomes first. We want to reject false null hypotheses and fail to reject true null hypotheses. The goal of any scientific investigation is to verify true statements and reject false statements.

The remaining two combinations are errors or situations we wish to avoid. If we reject a null hypothesis that is actually true, we are saying that a true statement is false. Likewise, if we fail to reject a null hypothesis that is actually false, we are saying that a false statement is true. Obviously, we would always prefer to wind up in one of the boxes labeled "OK"—and always reject false statements and accept the truth when we find it. However, remember that hypothesis testing always carries an element of risk and that it is not possible to minimize the chances of Type I and Type II errors simultaneously.

What all this means, then, is that you must think of selecting an alpha level as an attempt to balance the two types of error. Higher alpha levels will minimize the probability of a Type II error (saying that false statements are true), and lower alpha levels will minimize the probability of a Type I error (saying that true statements are false). Normally, in social science research, we will want to minimize a Type I error, and lower alpha levels (.05, .01, .001, or lower) will be used. The 0.05 level in particular has emerged as a generally recognized indicator

TABLE 7.5 Decision Making and the Five-Step Model

	Decision	
The H_0 Is Actually:	Reject	Fail to Reject
True	Type I, or α, error	OK
False	OK	Type II, or β, error

STATISTICS IN EVERYDAY LIFE

In social science research, the 0.05 alpha level has become the standard indicator of a significant difference. This alpha level means that we will incorrectly reject the null hypothesis only five times out of every 100 tests. These might seem like excellent odds, but research that involves potentially harmful drugs would call for even lower alpha levels (0.001, 0.0001, or even lower) to minimize the possibility of making incorrect decisions and endangering health.

of a significant result. However, the widespread use of the 0.05 level is simply a convention, and there is no reason that alpha cannot be set at virtually any sensible level (such as 0.04, 0.027, 0.083). The researcher has the responsibility of selecting the alpha level that seems most reasonable in terms of the goals of the research project.

The Student's *t* Distribution

To this point, we have considered only situations involving single-sample means in which the value of the population standard deviation (σ) was known. Needless to say, in most research situations, the value of σ will be unknown. However, a value for σ is required in order to compute the standard error of the mean (σ/N), convert our sample outcome into a Z score, and place the Z(obtained) on the sampling distribution (step 4). How can we reasonably obtain a value for the population standard deviation?

It might seem sensible to estimate σ with s, the sample standard deviation. As we noted in Chapter 6, s is a biased estimator of σ, but the degree of bias decreases as the sample size increases. For large samples (that is, samples with 100 or more cases), the sample standard deviation yields an adequate estimate of σ. Thus, for large samples, we simply substitute s for σ in the formula for Z(obtained) in step 4 and continue to use the standard normal curve to find areas under the sampling distribution.[1]

However, for smaller samples, when σ is unknown, an alternative distribution called the **Student's *t* distribution** must be used to find areas under the sampling distribution and establish the critical region. The shape of the t distribution varies as a function of the sample size. The relative shapes of the t and Z distributions are depicted in Figure 7.7. For small samples, the t distribution is much flatter than the Z distribution, but as the sample size increases, the t distribution comes to resemble the Z distribution more and more until the two are essentially identical for sample sizes greater than 120. As N increases, the sample standard deviation (s) becomes a more and more adequate estimator of the population standard deviation (σ) and the t distribution becomes more and more like the Z distribution.

[1]Even though its effect will be minor and will decrease with the sample size, we will always correct for the bias in s by using the term $N - 1$ rather than N in the computation for the standard deviation of the sampling distribution when s is unknown.

FIGURE 7.7 The *t* Distribution and the *Z* Distribution

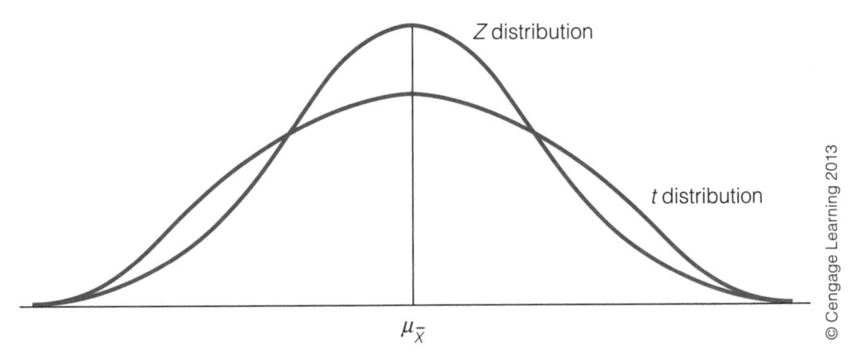

Applying Statistics 7.1 Testing Sample Means for Significance: The One-Sample Case

For a random sample of 152 felony cases tried in a local court, the average prison sentence was 27.3 months. Is this significantly different from the average prison term for felons nationally (28.7 months)? We will use the five-step model to organize the decision-making process.

Step 1. Making Assumptions and Meeting Test Requirements.

Model: Random sampling
Level of measurement is interval-ratio
Sampling distribution is normal

Because this is a large sample ($N > 100$) and the length of a sentence is an interval-ratio variable, we can conclude that the model assumptions are satisfied.

Step 2. Stating the Null Hypothesis (H_0). The null hypothesis would say that the average sentence locally (for *all* felony cases) is equal to the national average. In symbols:

$$H_0: \mu = 28.7$$

The research question does not specify a direction; it only asks if the local sentences are "different from" (not higher or lower than) national averages. This suggests a two-tailed test:

$$(H_1: \mu \neq 28.7)$$

Step 3. Selecting the Sampling Distribution and Establishing the Critical Region. Because this is a large sample, we can use Appendix A to establish the critical region and state the critical scores as *Z* scores (as opposed to *t* scores).

$$\text{Sampling distribution} = Z\text{ distribution}$$
$$\alpha = 0.05$$
$$Z(\text{critical}) = \pm 1.96$$

Step 4. Computing the Test Statistic. The necessary information for conducting a test of the null hypothesis is:

$$\bar{X} = 27.3 \qquad \mu = 28.7$$
$$s = 3.7$$
$$N = 152$$

The test statistic—*Z*(obtained)—would be:

$$Z(\text{obtained}) = \frac{\bar{X} - \mu}{s/\sqrt{N-1}}$$
$$Z(\text{obtained}) = \frac{27.3 - 28.7}{3.7/\sqrt{152-1}}$$
$$Z(\text{obtained}) = \frac{-1.40}{3.7/\sqrt{151}}$$
$$Z(\text{obtained}) = \frac{-1.40}{0.30}$$
$$Z(\text{obtained}) = -4.67$$

Step 5. Making a Decision and Interpreting the Test Results. With alpha set at 0.05, the critical region begins at $Z(\text{critical}) = \pm 1.96$. With an obtained *Z* score of -4.67, the null would be rejected. The difference between the prison sentences of felons convicted in the local court and felons convicted nationally is statistically significant. The difference is so large that we may conclude that it did not occur by random chance. The decision to reject the null hypothesis has a 0.05 probability of being wrong.

The Distribution of *t*: Using Appendix B. The *t* distribution is summarized in Appendix B. The *t* table differs from the *Z* table in several ways. First, there is a column at the left of the table labeled *df* for "degrees of freedom."[2] Because the exact shape of the *t* distribution varies by sample size, the exact location of the critical region also varies. Degrees of freedom, which are equal to $N - 1$ in the case of a single-sample mean, must first be computed before the critical region for any alpha can be located. Second, alpha levels are arrayed across the top of Appendix B in two rows—one row for the one-tailed tests and one for two-tailed tests. To use the table, begin by locating the selected alpha level in the appropriate row.

The third difference is that the entries in the table are the actual scores—called *t*(critical)—that mark the beginnings of the critical regions and not areas under the sampling distribution. To illustrate the use of this table with single-sample means, find the critical region for alpha equal to 0.05, two-tailed test, for $N = 30$. The degrees of freedom will be $N - 1$, or 29; reading down the proper column, you should find a value of 2.045. Thus, the critical region for this test will begin at $t(\text{critical}) = \pm 2.045$.

Notice that this *t*(critical) is larger in value than the comparable *Z*(critical), which for a two-tailed test at an alpha of 0.05 would be ± 1.96. This is because the *t* distribution is flatter than the *Z* distribution (see Figure 7.7). On the *t* distribution, the critical regions will begin farther away from the mean of the sampling distribution; therefore, the null hypothesis will be harder to reject. Furthermore, the smaller the sample size (the lower the degrees of freedom), the larger the value of *t*(obtained) necessary to reject the H_0.

Also note that the values of *t*(critical) decrease as degrees of freedom increase. For one degree of freedom, the *t*(critical) is 12.706 for an alpha of 0.05 with a two-tailed test, but this score grows smaller for larger samples. For degrees of freedom greater than 120, the value of *t*(critical) is the same as the comparable value of *Z*(critical), or ± 1.96. As the sample size increases, the *t* distribution resembles the *Z* distribution more and more until, with sample sizes greater than 120, the two distributions are essentially identical.[3]

Using *t* in a Test. To demonstrate the uses of the *t* distribution in more detail, we will work through an example problem. Note that in terms of the five-step model, the changes occur mostly in steps 3 and 4. In step 3, the sampling distribution will be the *t* distribution and degrees of freedom (*df*) must be computed before locating

[2]Degrees of freedom refers to the number of values in a distribution that are free to vary. For a sample mean, a distribution has $N - 1$ degrees of freedom. This means that for a specific value of the mean and of N, $N - 1$ scores are free to vary. For example, if the mean is 3 and $N = 5$, the distribution of five scores would have $5 - 1 = 4$ degrees of freedom. When the values of four of the scores are known, the value of the fifth is fixed. If four scores are 1, 2, 3, and 4, the fifth must be 5 and no other value.

[3]Appendix B abbreviates the *t* distribution by presenting a limited number of critical *t* scores for degrees of freedom between 31 and 120. If the degrees of freedom for a specific problem equal 77 and alpha equals 0.05, two-tailed, we have a choice between a *t*(critical) of 2.000 ($df = 60$) and a *t*(critical) of 1.980 ($df = 120$). In situations such as these, take the larger table value as *t*(critical). This will make rejection of H_0 less likely and is therefore the more conservative course of action.

the critical region or the *t*(critical) score. In step 4, a slightly different formula for computing the test statistic—*t*(obtained)—will be used. As compared with the formula for *Z*(obtained), s will replace σ and $N - 1$ will replace N.

Specifically:

FORMULA 7.2

$$t\,(\text{obtained}) = \frac{\overline{X} - \mu}{s/\sqrt{N - 1}}$$

A researcher wonders if commuter students are different from the general student body in terms of academic achievement. She has gathered a random sample of 30 commuter students and has learned from the registrar that the mean grade point average for all students is 2.50 ($\mu = 2.50$), but the standard deviation of the population (σ) has never been computed. Sample data are reported here. Is the sample from a population that has a mean of 2.50?

Student Body	Commuter Students
$\mu = 2.50(=\mu_{\overline{X}})$	$\overline{X} = 2.78$
$\sigma = ?$	$s = 1.23$
	$N = 30$

Step 1. Making Assumptions and Meeting Test Requirements.

Model: Random sampling
Level of measurement is interval-ratio
Sampling distribution is normal

Step 2. Stating the Null Hypothesis.

$$H_0: \mu = 2.50$$
$$(H_1: \mu \neq 2.50)$$

You can see from the research hypothesis that the researcher has not predicted a direction for the difference. This will be a two-tailed test.

Step 3. Selecting the Sampling Distribution and Establishing the Critical Regions. Since σ is unknown and the sample size is small, the t distribution will be used to find the critical region. Alpha will be set at 0.01:

$$\text{Sampling distribution} = t \text{ distribution}$$
$$\alpha = 0.01, \text{two-tailed test}$$
$$df = (N - 1) = 29$$
$$t\,(\text{critical}) = \pm 2.756$$

ONE STEP AT A TIME **Completing Step 4 of the Five-Step Model: Computing *t*(obtained)**

Follow these procedures when using Student's t distribution.

To Compute the Test Statistic by Using Formula 7.2

Step	*Operation*
1.	Find the square root of $N - 1$.
2.	Divide the quantity you found in step 1 into the sample standard deviation (s).
3.	Subtract the population mean (μ) from the sample mean ($\bar{X}$).
4.	Divide the quantity you found in step 3 by the quantity you found in step 2. This value is t(obtained).

Step 4. Computing the Test Statistic.

$$t\text{(obtained)} = \frac{\bar{X} - \mu}{s/\sqrt{N-1}}$$

$$t\text{(obtained)} = \frac{2.78 - 2.50}{1.23/\sqrt{29}}$$

$$t\text{(obtained)} = \frac{.28}{.23}$$

$$t\text{(obtained)} = +1.22$$

Step 5. Making a Decision and Interpreting Test Results. The test statistic does not fall into the critical region. Therefore, the researcher fails to reject the H_0. The difference between the sample mean (2.78) and the population mean (2.50) is not statistically significant. The difference is no greater than what would be expected if only random chance were operating. The test statistic and critical regions are displayed in Figure 7.8.

To summarize, when testing single-sample means, we must make a choice regarding the theoretical distribution we will use to establish the critical region. The choice is straightforward. If the population standard deviation (σ) is known or the sample size is large, the *Z* distribution (summarized in Appendix A) will be used. If σ is unknown and the sample is small, the *t* distribution (summarized in Appendix B)

ONE STEP AT A TIME **Completing Step 5 of the Five-Step Model: Making a Decision and Interpreting Results**

Step	*Operation*
1.	Compare the t(obtained) to the t(critical). If t(obtained) is *in* the critical region, *reject* the null hypothesis. If t(obtained) is *not in* the critical region, *fail to reject* the null hypothesis.
2.	Interpret your decision in terms of the original question. For example, our conclusion for the example problem used in this section was "There is no significant difference between the average GPAs of commuter students and the general student body."

FIGURE 7.8 Sampling Distribution Showing *t*(Obtained) versus *t*(Critical) ($\alpha = 0.05$, two-tailed test, *df* = 29)

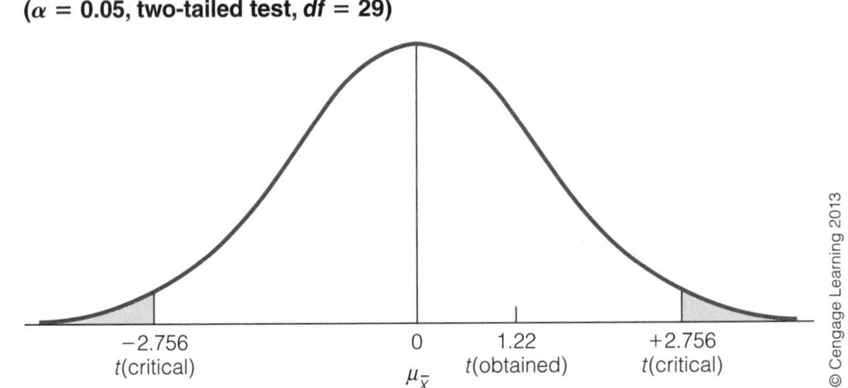

TABLE 7.6 Choosing a Sampling Distribution When Testing Single-Sample Means for Significance

If Population Standard Deviation (σ) Is	Sampling Distribution
Known	*Z* distribution
Unknown and the sample size (*N*) is large	*Z* distribution
Unknown and the sample size (*N*) is small	*t* distribution

will be used. These decisions are summarized in Table 7.6. *(For practice in using the t distribution in a test of hypothesis, see problems 7.8, 7.10 to 7.12, and 7.19 .)*

Tests of Hypotheses for Single-Sample Proportions (Large Samples)

In many cases, we work with variables that are not interval-ratio in level of measurement. One alternative in this situation would be to use a sample proportion (P_s) rather than a sample mean as the test statistic. As we shall see, the overall procedures for testing single-sample proportions are the same as those for testing means. The central question is still "Does the population from which the sample was drawn have a certain characteristic?" We still conduct the test based on the assumption that the null hypothesis is true, and we still evaluate the probability of the obtained sample outcome against a sampling distribution of all possible sample outcomes. Our decision at the end of the test is also the same. If the obtained test statistic falls into the critical region (is unlikely, given the assumption that the H_0 is true), we reject the H_0.

Of course, there are also some important differences in significance tests for sample proportions. These differences are best related in terms of the five-step model. In step 1, we assume that the variable is measured at the nominal level. In step 2, the symbols used to state the null hypothesis are different even though it is still a statement of "no difference."

In step 3, we will use only the standardized normal curve (the Z distribution) to find areas under the sampling distribution and to locate the critical region. This will be appropriate as long as the sample size is large. We will not consider small-sample tests of hypothesis for proportions in this text.

In step 4—computing the test statistic—the form of the formula remains the same. That is, the test statistic—Z(obtained)—equals the sample statistic minus the mean of the sampling distribution divided by the standard deviation of the sampling distribution. However, the symbols will change because we are basing the tests on sample proportions. The formula can be stated as:

FORMULA 7.3

$$Z(\text{obtained}) = \frac{P_s - P_u}{\sqrt{P_u(1 - P_u)/N}}$$

Step 5—making a decision—is exactly the same as before. If the test statistic—Z(obtained)—falls into the critical region, reject the H_0.

Applying Statistics 7.2 Testing Sample Proportions for Significance: The One-sample Case

Seventy-six percent of the respondents in a random sample ($N = 103$) drawn from the most affluent neighborhood in a community voted Republican in the most recent presidential election. For the community as a whole, 66% of the electorate voted Republican. Was the affluent neighborhood significantly more likely to have voted Republican?

Step 1. Making Assumptions and Meeting Test Requirements.

Model: Random sampling
Level of measurement is nominal
Sampling distribution is normal

This is a large sample, so we may assume a normal sampling distribution. The variable "percent Republican" is only nominal in level of measurement.

Step 2. Stating the Null Hypothesis (H_0). The null hypothesis says that the affluent neighborhood is not different from the community as a whole:

$$H_0: P_u = 0.66$$

The original question ("Was the affluent neighborhood *more* likely to vote Republican?") suggests a one-tailed research hypothesis:

$$(H_1: P_u > .66)$$

Step 3. Selecting the Sampling Distribution and Establishing the Critical Region.

Sampling distribution = Z distribution

$$\alpha = 0.05$$

$$Z(\text{critical}) = +1.65$$

The research hypothesis says that we will be concerned only with outcomes in which the neighborhood is *more* likely to vote Republican or with sample outcomes in the upper tail of the sampling distribution.

Step 4. Computing the Test Statistic. The information necessary for a test of the null hypothesis, expressed in the form of proportions, is:

Neighborhood	Community
$P_s = 0.76$ $N = 103$	$P_u = 0.66$

The test statistic—Z(obtained)—would be:

$$Z(\text{obtained}) = \frac{P_s - P_u}{\sqrt{P_u(1 - P_u)/N}}$$

$$Z(\text{obtained}) = \frac{0.76 - 0.66}{\sqrt{(0.66)(1 - 0.66)/103}}$$

$$Z(\text{obtained}) = \frac{0.10}{\sqrt{(0.2244)/103}}$$

(continued next page)

Applying Statistics 7.2 (continued)

$$Z(\text{obtained}) = \frac{0.100}{0.047}$$

$$Z(\text{obtained}) = 2.13$$

Step 5. Making a Decision and Interpreting Test Results. With alpha set at 0.05, one-tailed, the critical region begins at $Z(\text{critical}) = +1.65$. With an obtained Z score of 2.13, the null hypothesis is rejected. The difference between the affluent neighborhood and the community as a whole is statistically significant and in the predicted direction. Residents of the affluent neighborhood were significantly more likely to have voted Republican in the last presidential election.

A Test of Hypothesis by Using Sample Proportions. An example should clarify these procedures. A random sample of 122 households in a low-income neighborhood revealed that 53 (or a proportion of 0.43) of the households were headed by females. In the city as a whole, the proportion of female-headed households is 0.39. Are households in the lower-income neighborhood significantly different from the city as a whole in terms of this characteristic?

Step 1. Making Assumptions and Meeting Test Requirements.

Model: Random sampling
Level of measurement is nominal
Sampling distribution is normal in shape

Step 2. Stating the Null Hypothesis. The research question, as stated earlier, asks only if the sample proportion is *different from* the population proportion. Because we have not predicted a direction for the difference, a two-tailed test will be used.

$$H_0\text{: } P_u = 0.39$$

$$(H_1\text{: } P_u \neq 0.39)$$

Step 3. Selecting the Sampling Distribution and Establishing the Critical Region.

$$\text{Sampling distribution} = Z \text{ distribution}$$

$$\alpha = 0.10, \text{ two-tailed test}$$

$$Z(\text{critical}) = \pm 1.65$$

Step 4. Computing the Test Statistic.

$$Z(\text{obtained}) = \frac{P_s - P_u}{\sqrt{P_u(1 - P_u)/N}}$$

$$Z(\text{obtained}) = \frac{0.43 - 0.39}{\sqrt{(0.39)(0.61)/122}}$$

$$Z(\text{obtained}) = +0.91$$

ONE STEP AT A TIME **Completing Step 4 of the Five-Step model: Computing Z(obtained)**

Use Formula 7.3 to compute the test statistic.

Step	*Operation*
1.	Start with the denominator of Formula 7.3 and then substitute in the value for P_u. This value will be given in the statement of the problem.
2.	Subtract the value of P_u from 1.
3.	Multiply the value you found in step 2 by the value of P_u.
4.	Divide the quantity you found in step 3 by *N*.
5.	Take the square root of the value you found in step 4.
6.	Subtract P_u from P_s.
7.	Divide the quantity you found in step 6 by the quantity you found in step 5. This value is *Z*(obtained).

Step 5. Making a Decision and Interpreting Test Results. The test statistic—*Z*(obtained)—does not fall into the critical region. Therefore, we fail to reject the H_0. There is no statistically significant difference between the low-income community and the city as a whole in terms of the proportion of households headed by females. Figure 7.9 displays the sampling distribution, the critical region, and the *Z*(obtained). *(For practice in tests of significance using sample proportions, see problems 7.1c, 7.13 to 7.16, 7.17a to d, 7.18, and 7.21a and b.)*

ONE STEP AT A TIME **Completing Step 5 of the Five-Step model: Making a Decision and Interpreting Results**

Step	*Operation*
1.	Compare your *Z*(obtained) to your *Z*(critical). If *Z*(obtained) is *in* the critical region, *reject* the null hypothesis. If *Z*(obtained) is *not in* the critical region, *fail to reject* the null hypothesis.
2.	Interpret the decision in terms of the original question. For example, our conclusion for the example problem used in this section was "There is no significant difference between the low-income community and the city as a whole in the proportion of households that are headed by females."

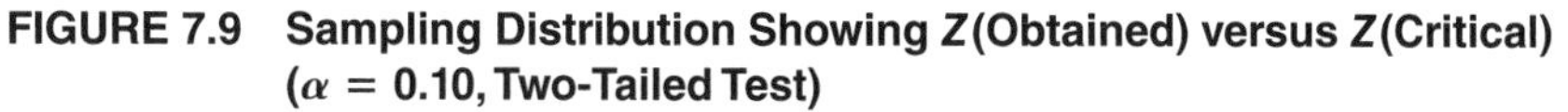

FIGURE 7.9 Sampling Distribution Showing Z(Obtained) versus Z(Critical) (α = 0.10, Two-Tailed Test)

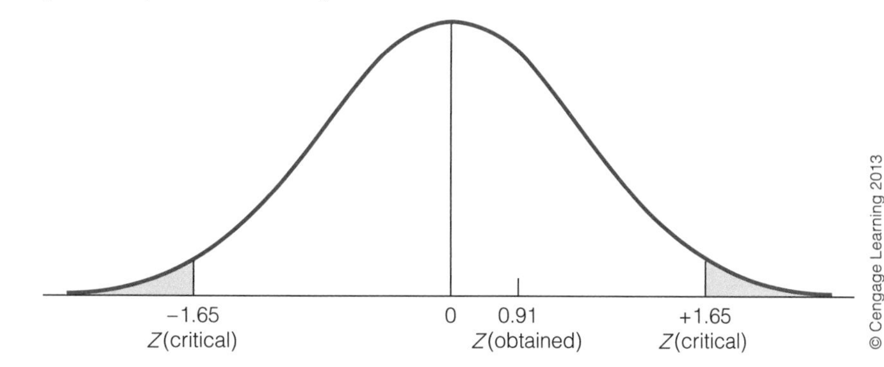

SUMMARY

1. All the basic concepts and techniques for testing hypotheses were presented in this chapter. We saw how to test the null hypothesis of "no difference" for single-sample means and proportions. In both cases, the central question is whether the population represented by the sample has a certain characteristic.
2. All tests of a hypothesis involve finding the probability of the observed sample outcome, given that the null hypothesis is true. If the outcome has a low probability, we reject the null hypothesis. In the usual research situation, we will wish to reject the null hypothesis and thereby support the research hypothesis.
3. The five-step model will be our framework for decision making throughout the hypothesis-testing chapters. However, what we do during each step will vary depending on the specific test being conducted.
4. If we can predict a direction for the difference in stating the research hypothesis, a one-tailed test is called for. If no direction can be predicted, a two-tailed test is appropriate.
5. There are two kinds of errors in hypothesis testing: A Type I, or alpha, error is rejecting a true null; aType II, or beta, error is failing to reject a false null. The probabilities of committing these two types of error are inversely related and cannot be simultaneously minimized in the same test. By selecting an alpha level, we try to balance the probability of these two errors.
6. When testing sample means, the t distribution is used to find the critical region when the population standard deviation is unknown and the sample size is small.
7. Sample proportions can also be tested for significance by using the five-step model. Unlike tests using sample means, tests of sample proportions assume a nominal level of measurement, use different symbols to state the null hypothesis, and use a different formula (7.3) to compute Z(obtained).
8. If you are still confused about the uses of inferential statistics described in this chapter, do not be alarmed or discouraged. A sizable volume of complex material has been presented, and only rarely will someone fully comprehend the unique logic of hypothesis testing on the first exposure. After all, it is not every day that you learn how to test a statement you do not believe (the null hypothesis) against a distribution that does not exist (the sampling distribution)!

SUMMARY OF FORMULAS

FORMULA 7.1 Single-sample means, large samples: $Z\,(\text{obtained}) = \dfrac{\bar{X} - \mu}{\sigma/\sqrt{N}}$

FORMULA 7.2 Single-sample means when samples are small and the population standard deviation is unknown:

$$t\,(\text{obtained}) = \frac{\bar{X} - \mu}{s/\sqrt{N-1}}$$

FORMULA 7.3 Single-sample proportions, large samples: $Z\,(\text{obtained}) = \dfrac{P_s - P_u}{\sqrt{P_u(1 - P_u)/N}}$

GLOSSARY

Alpha level (α). The proportion of area under the sampling distribution that contains unlikely sample outcomes, given that the null hypothesis is true. Also, the probability of a Type I error.

Critical region (region of rejection). The area under the sampling distribution that, in advance of the test itself, is defined as including unlikely sample outcomes, given that the null hypothesis is true.

Five-step model. A step-by-step guideline for conducting tests of hypotheses. A framework that organizes decisions and computations for all tests of significance.

Hypothesis testing. Statistical tests that estimate the probability of sample outcomes if assumptions about the population (the null hypothesis) are true.

Null hypothesis (H_0). A statement of "no difference." In the context of single-sample tests of significance, the population from which the sample was drawn is assumed to have a certain characteristic or value.

One-tailed test. A type of hypothesis test used when (1) the direction of the difference can be predicted or (2) concern focuses on outcomes in only one tail of the sampling distribution.

Research hypothesis (H_1). A statement that contradicts the null hypothesis. In the context of single-sample tests of significance, the research hypothesis says that the population from which the sample was drawn does not have a certain characteristic or value.

Significance testing. See Hypothesis testing.

Student's *t* distribution. A distribution used to find the critical region for tests of sample means when s is unknown and the sample size is small.

***t*(critical).** The t score that marks the beginning of the critical region of a t distribution.

***t*(obtained).** The test statistic computed in step 4 of the five-step model. The sample outcome expressed as a t score.

Test statistic. The value computed in step 4 of the five-step model that converts the sample outcome into either a t score or a Z score.

Two-tailed test. A type of hypothesis test used when (1) the direction of the difference cannot be predicted or (2) concern focuses on outcomes in both tails of the sampling distribution.

Type I error (alpha error). The probability of rejecting a null hypothesis that is, in fact, true.

Type II error (beta error). The probability of failing to reject a null hypothesis that is, in fact, false.

***Z*(critical).** The Z score that marks the beginnings of the critical region on a Z distribution.

***Z*(obtained).** The test statistic computed in step 4 of the five-step model. The sample outcomes expressed as a Z score.

PROBLEMS

7.1 a. For each situation, find Z(critical).

Alpha	Form	Z(Critical)
.05	One-tailed	
.10	Two-tailed	
.06	Two-tailed	
.01	One-tailed	
.02	Two-tailed	

b. For each situation, find the critical t score.

Alpha	Form	N	t(Critical)
.10	Two-tailed	31	
.02	Two-tailed	24	
.01	Two-tailed	121	
.01	One-tailed	31	
.05	One-tailed	61	

c. Compute the appropriate test statistic (Z or t) for each situation:

1. $\mu = 2.40$ $\quad \bar{X} = 2.20$
   $\sigma = 0.75$ $\quad N = 200$
2. $\mu = 17.1$ $\quad \bar{X} = 16.8$
   $s = 0.9$
   $N = 45$
3. $\mu = 10.2$ $\quad \bar{X} = 9.4$
   $s = 1.7$
   $N = 150$
4. $P_u = .57$ $\quad P_s = 0.60$
   $N = 117$
5. $P_u = 0.32$ $\quad P_s = 0.30$
   $N = 322$

7.2 SOC The students at Littlewood Regional High School cut an average of 3.3 classes per week. A random sample of 117 seniors averages 3.8 cuts per week, with a standard deviation of 0.53. Are seniors significantly different from the student body as a whole? *(Hint: The wording of the research question suggests a two-tailed test. This means that the alternative, or research, hypothesis in step 2 will be stated as H_1: $\mu \neq 3.3$ and that the critical region will be split between the upper and lower tails of the sampling distribution. See Table 7.1 for values of Z(critical) for various alpha levels.)*

7.3 What if the research question in problem 7.2 were changed to "Do seniors cut a significantly *greater* number of classes"? How would the test change? *(Hint: This wording implies a one-tailed test of significance. How would the research hypothesis change? For the alpha you used in problem 7.2, what would the value of Z(critical) be?)*

7.4 SW Nationally, social workers average 10.2 years of experience. In a random sample, 203 social workers in greater metropolitan Shinbone average only 8.7 years, with a standard deviation of 0.52. Are social workers in Shinbone significantly less experienced? *(Hint: Note the wording of the research hypotheses. This situation may justify a one-tailed test of significance. If you chose a one-tailed test, what form would the research hypothesis take and where would the critical region begin?)*

7.5 SW The same sample of social workers reports an average annual salary of $25,782, with a standard deviation of $622. Is this figure significantly higher than the national average of $24,509? *(Hint: The wording of the research hypotheses suggests a one-tailed test. What form would the research hypothesis take, and where would the critical region begin?)*

7.6 SOC Nationally, the average score on the college entrance exams (verbal test) is 453, with a standard deviation of 95. A random sample of 137 seniors at Littlewood Regional High School shows a mean score of 502. Is there a significant difference?

7.7 SOC A random sample of 423 Chinese Americans has finished an average of 12.7 years of formal education, with a standard deviation of 1.7. Is this significantly different from the national average of 12.2 years?

7.8 SOC A sample of 105 sanitation workers for the city of Euonymus, Texas, earns an average of $24,375 per year. The average salary for all Euonymus city workers is $24,230, with a standard deviation of $523. Is the sanitation workers' salary significantly different from that of city workers as a whole? Conduct both one- and two-tailed tests.

7.9 GER **a.** Nationally, the population as a whole watches 6.2 hours of TV per day. A random sample of 1,017 senior citizens report watching an average of 5.9 hours per day, with a standard deviation of 0.7. Is the difference significant?

b. The same sample of senior citizens reports that they belong to an average of 2.1 voluntary organizations and clubs, with a standard deviation of 0.5. Nationally, the average is 1.7. Is the difference significant?

7.10 SOC A school system has assigned several hundred "chronic and severe" underachievers to an alternative educational experience. To assess the program, a random sample of 35 students has been selected for comparison with all students in the system.

a. In terms of GPA, did the program work?

Systemwide GPA	Program GPA
$\mu = 2.47$	$\bar{X} = 2.55$
	$s = 0.70$
	$N = 35$

b. In terms of absenteeism (number of days missed per year), what can be said about the success of the program?

Systemwide	Program
$\mu = 6.137$	$\bar{X} = 4.78$
	$s = 1.11$
	$N = 35$

c. In terms of standardized test scores in math and reading, was the program a success?

Math Test—Systemwide	Math Test—Program
$\mu = 103$	$\bar{X} = 106$
	$s = 2.0$
	$N = 35$

Reading Test—Systemwide	Reading Test—Program
$\mu = 110$	$\bar{X} = 113$
	$s = 2.0$
	$N = 35$

(Hint: Note the wording of the research questions. Is a one-tailed test justified? Is the program a success if the students in the program are no different from students systemwide? What if the program students were performing at lower levels? If a one-tailed test is used, what form should the research hypothesis take? Where will the critical region begin?

7.11 SOC A random sample of 26 local sociology graduates scored an average of 458 on the GRE advanced sociology test, with a standard deviation of 20. Is this significantly different from the national average ($\mu = 440$)?

7.12 PA Nationally, the per capita monthly fuel oil bill is $110. A random sample of 36 southeastern cities averages $78, with a standard deviation of $4. Is the difference significant? Summarize your conclusions in a sentence or two.

7.13 GER/CJ A survey shows that 10% of the population is victimized by property crime each year. A random sample of 527 older citizens (65 years or more of age) shows a victimization rate of 14%. Are older people more likely to be victimized? Conduct both one- and two-tailed tests of significance.

7.14 CJ A random sample of 113 felons convicted of nonviolent crimes in a state prison system completed a program designed to improve their education and work skills before being released on parole. Fifty-eight eventually became repeat offenders. Is this recidivism rate significantly different from the rate for all offenders in that state (57%)? Summarize your conclusions in a sentence or two. *(Hint: You must use the information given in the problem to compute a sample proportion. Remember to convert the population percentage to a proportion.)*

7.15 PS In a recent statewide election, 55% of the voters rejected a proposal to institute a state lottery. In a random sample of 150 urban precincts, 49% of the voters rejected the proposal. Is the difference significant? Summarize your conclusions in a sentence or two.

7.16 CJ Statewide, the police clear by arrest 35% of the robberies and 42% of the aggravated assaults reported to them. A researcher takes a random sample of all the robberies ($N = 207$) and aggravated assaults ($N = 178$) reported to a metropolitan police department in one year and finds that 83 of the robberies and 80 of the assaults were cleared by arrest. Are the local arrest rates significantly different from the statewide rate? Write a sentence or two interpreting your decision.

7.17 SOC/SW A researcher has compiled a file of information on a random sample of 317 families that have chronic, long-term patterns of child abuse. Reported here are some of the characteristics of the sample, along with values for the city as a whole. For each trait, test the null hypothesis of "no difference" and summarize your findings.

	Variable	City	Sample
a.	Mothers' educational level (proportion completing high school)	$P_u = 0.65$	$P_s = 0.61$
b.	Family size (proportion of families with four or more children)	$P_u = 0.21$	$P_s = 0.26$
c.	Mothers' work status (proportion of mothers with jobs outside the home)	$P_u = 0.51$	$P_s = 0.27$
d.	Relations with kin (proportion of families that have contact with kin at least once a week)	$P_u = 0.82$	$P_s = 0.43$
e.	Fathers' educational achievement (average years of formal schooling)	$\mu = 12.3$	$\overline{X} = 12.5$ $s = 1.7$
f.	Fathers' occupational stability (average years in present job)	$\mu = 5.2$	$\overline{X} = 3.7$ $s = 0.5$

7.18 SW You are the head of an agency seeking funding for a program to reduce unemployment among teenage males. Nationally, the unemployment rate for this group is 18%. A random sample of 323 teenage males in your area reveals an unemployment rate of 21.7%. Is the difference significant? Can you demonstrate a need for the program? Should you use a one-tailed test in this situation? Why? Explain the result of your test of significance as you would to a funding agency.

7.19 PA The city manager of Shinbone has received a complaint from the local union of firefighters to the effect that they are underpaid. Not having much time, the city manager gathers the records of a random sample of 27 firefighters and finds that their average salary is $38,073, with a standard deviation of $575. If she knows that the average salary nationally is $38,202, how can she respond to the complaint? Should she use a one-tailed test in this situation? Why? What would she say in a memo to the union that would respond to the complaint?

7.20 The following essay questions review the basic principles and concepts of inferential statistics. The order of the questions roughly follows the five-step model.

- **a.** Hypothesis testing or significance testing can be conducted only with a random sample. Why?
- **b.** Under what specific conditions can it be assumed that the sampling distribution is normal in shape?
- **c.** Explain the role of the sampling distribution in a test of hypothesis.
- **d.** The null hypothesis is an assumption about reality that makes it possible to test sample outcomes for their significance. Explain.
- **e.** What is the critical region? How is the size of the critical region determined?
- **f.** Describe a research situation in which a one-tailed test of hypothesis would be appropriate.
- **g.** Thinking about the shape of the sampling distribution, why does use of the *t* distribution (as opposed to the Z distribution) make it more difficult to reject the null hypothesis?
- **h.** What exactly can be concluded in the one-sample case when the test statistic falls into the critical region?

7.21 SOC A researcher is studying changes in the student body at her university and has selected a random sample of 163 from the freshman class. The table below compares their characteristics to those of the student body as a whole. Which differences are significant?

	Variable	Freshmen	All Students
a.	Proportion Republican	$P_s = 0.44$	$P_u = 0.49$
b.	Proportion majoring in math and science	$P_s = 0.17$	$P_u = 0.09$
c.	Average family income	$\overline{X} = 63{,}542$ $s = 1{,}568$	$\mu = 59{,}235$
d.	Average number of siblings	$\overline{X} = 0.9$ $s = 0.2$	$\mu = 1.2$
e.	Average number of telephone calls to parents per week	$\overline{X} = 1.5$ $s = 0.4$	$\mu = 0.4$
f.	Average number of close friends living on campus	$\overline{X} = 4.5$ $s = 0.3$	$\mu = 7.8$

8 Hypothesis Testing II
The Two-Sample Case

LEARNING OBJECTIVES

By the end of this chapter, you will be able to:

1. Identify and cite examples of situations in which the two-sample test of hypothesis is appropriate.
2. Explain the logic of hypothesis testing as applied to the two-sample case.
3. Explain what an independent random sample is.
4. Perform a test of hypothesis for two sample means or two sample proportions following the five-step model and then correctly interpret the results.
5. List and explain each of the factors (especially the sample size) that affect the probability of rejecting the null hypothesis. Explain the differences between statistical significance and importance.

USING STATISTICS

The statistical techniques presented in this chapter are used when we are comparing two random samples. If the samples are significantly different, we conclude that the populations from which they were selected are different. This conclusion has a known probability of error, which is often set at 0.05. Examples of these situations include:

1. A researcher is examining the differences in support for gun control between U.S. men and women. If there are significant differences between samples of men and women, she will conclude that *all* U.S. men and women differ on this issue.
2. Random samples of the freshman and senior classes at a large university are compared for differences in their political views and religious values. If the researcher finds significant differences between the samples, he will conclude that the populations (*all* freshmen and seniors at the university) differ on these dimensions.
3. Is bullying more of a problem in suburban or city schools? Researchers compare random samples of each type of school to see if there are statistically significant differences. If the samples are significantly different, we conclude that there are differences between the populations and that *all* suburban schools are different from *all* city schools.

In Chapter 7, we dealt with hypothesis testing in the one-sample case. In that situation, we were concerned with the significance of the difference between a sample statistic and a population parameter. In this chapter, we consider a new research situation where we will be concerned with the significance of the difference between two separate populations. For example, do men and women in the United States vary in their support for gun control? Obviously, we cannot ask every male and female for their opinions on this issue. Instead, we must draw random samples of both groups and then use the information gathered from these samples to infer population patterns.

The question in hypothesis testing in the two-sample case is: Is the difference between the samples large enough to allow us to conclude (with a known probability of error) that the populations represented by the samples are different? Thus, if we find a large enough difference in support of gun control between random samples of men and women, we can argue that the difference between the samples did not occur by random chance but, rather, represents a real difference between men and women in the population.

In this chapter, we consider tests for the significance of the difference between sample means and sample proportions. In both tests, the five-step model will be the framework for our decision making. The hypothesis-testing process is very similar to that of the one-sample case, but we also need to consider some important differences.

The One-Sample Case versus the Two-Sample Case

There are three important differences between the one-sample case considered in Chapter 7 and the two-sample case covered in this chapter. First, the one-sample case requires that the sample be selected following the principle of EPSEM (each case in the population must have an equal chance of being selected for the sample). The two-sample case requires that the samples be selected independently as well as randomly. This means that the procedure for selecting cases for one of the samples cannot affect the probability that any particular case will be selected for the other sample. In our example concerning gender differences in support of gun control, this would mean that the selection of a specific male for the sample would have no effect on the probability of selecting any particular female. This new requirement will be stated as **independent random sampling** in step 1.

The requirement of independent random sampling can be satisfied by drawing EPSEM samples from separate lists (for example, one for females and one for males). However, it is usually more convenient to draw a single EPSEM sample from a single list of the population and then to subdivide the cases into separate groups (males and females, for example). As long as the original sample is selected randomly, any subsamples created by the researcher will meet the assumption of independent random samples.

The second important difference is in the form of the null hypothesis stated in step 2 of the five-step model. The null is still a statement of "no difference." Now, however, instead of saying that the population from which the sample is drawn has a certain characteristic, it will say that the two populations are the same. ("There is no significant difference between men and women in their support of gun control.") If the test statistic falls in the critical region, the null hypothesis of no difference between the populations can be rejected and the argument that the populations are different will be supported.

A third important new element concerns the sampling distribution or the distribution of all possible sample outcomes. In Chapter 7, the sample outcome was either a mean or a proportion. Now we are dealing with two samples (e.g., samples of men and women), and the sample outcome is the *difference between* the sample statistics. In terms of our example, the sampling distribution would include all possible differences in sample means for support of gun control between men and women. If the null hypothesis is true and men and women do *not* have different views about gun control, the difference between the population means would be zero, the mean of the sampling distribution would be zero, and the huge majority of differences between sample means would be zero or very close to zero. The greater the differences between the sample means, the further the sample outcome (the *difference* between the two sample means) will be from the mean of the sampling distribution (zero) and the more likely that the difference reflects a real difference between the populations represented by the samples.

Hypothesis Testing With Sample Means (Large Samples)

To illustrate the procedure for testing sample means, assume that a researcher has access to a nationally representative random sample and that the individuals in the sample have answered a survey that measures attitudes toward gun control. The sample is divided by sex, and sample statistics are computed for males and females. Assuming that the survey yields interval-ratio-level data, a test for the significance of the difference in sample means can be conducted.

As long as the sample size is large (that is, as long as the combined number of cases in the two samples exceeds 100), the sampling distribution of the differences in sample means will be normal and the normal curve (Appendix A) can be used to establish the critical regions. The test statistic—*Z*(obtained)—will be computed by the usual formula: sample outcome (the difference between the sample means) minus the mean of the sampling distribution divided by the standard deviation of the sampling distribution. The formula is presented as Formula 8.1. Note that numerical subscripts are used to identify the samples and populations. The subscript attached to σ $(\sigma_{\bar{X}-\bar{X}})$ indicates that we are dealing with the sampling distribution of the *differences* in sample means:

FORMULA 8.1

$$Z(\text{obtained}) = \frac{(\bar{X}_1 - \bar{X}_2) - (\mu_1 - \mu_2)}{\sigma_{\bar{X}-\bar{X}}}$$

where: $(\bar{X}_1 - \bar{X}_2)$ = the difference between the sample means
$(\mu_1 - \mu_2)$ = the difference between the population means
$\sigma_{\bar{X}-\bar{X}}$ = the standard deviation of the sampling distribution of sample means

Recall that tests of significance are always based on the assumption that the null hypothesis is true. If the means of the two populations are equal, then the term $(\mu_1 - \mu_2)$ will be zero and can be dropped from the equation. In effect, then, the formula we will actually use to compute the test statistic in step 4 will be:

FORMULA 8.2

$$Z(\text{obtained}) = \frac{\overline{X}_1 - \overline{X}_2}{\sigma_{\overline{X}-\overline{X}}}$$

For large samples, the standard deviation of the sampling distribution of the difference in sample means is defined as:

FORMULA 8.3

$$\sigma_{\overline{X}-\overline{X}} = \sqrt{\frac{\sigma_1^2}{N_1} + \frac{\sigma_2^2}{N_2}}$$

Because we will rarely, if ever, know the values of the population standard deviations (σ_1 and σ_2), we must use the sample standard deviations, corrected for bias, to estimate them. Formula 8.4 displays the equation used to estimate the standard deviation of the sampling distribution in this situation. This is called a **pooled estimate** because it combines information from both samples:

FORMULA 8.4

$$\sigma_{\overline{X}-\overline{X}} = \sqrt{\frac{s_1^2}{N_1 - 1} + \frac{s_2^2}{N_2 - 1}}$$

The sample outcomes for support of gun control are

Sample 1 (Men)	Sample 2 (Women)
$\overline{X}_1 = 6.2$	$\overline{X}_2 = 6.5$
$s_1 = 1.3$	$s_2 = 1.4$
$N_1 = 324$	$N_2 = 317$

We see from the sample statistics that men have a lower average score and are less supportive of gun control. The test of hypothesis will tell us if this difference is large enough to conclude that it did not occur by random chance alone but, rather, reflects an actual difference between the populations of men and women on this issue.

Step 1. Making Assumptions and Meeting Test Requirements. We now assume that the random samples are independent but the rest of the model is the same as in the one-sample case.

Model: Independent random samples
Level of measurement is interval-ratio
Sampling distribution is normal

Step 2. Stating the Null Hypothesis. The null hypothesis states that the *populations* represented by the samples are not different on this variable. No direction for the difference has been predicted, so a two-tailed test is called for:

$$H_0: \mu_1 = \mu_2$$
$$H_1: \mu_1 \neq \mu_2$$

Step 3. Selecting the Sampling Distribution and Establishing the Critical Region. For large samples, the Z distribution can be used to find areas under the sampling distribution and establish the critical region. Alpha will be set at 0.05:

$$\text{Sampling distribution} = Z \text{ distribution}$$
$$\text{Alpha} = 0.05$$
$$Z(\text{critical}) = \pm 1.96$$

Step 4. Computing the Test Statistic. The population standard deviations are unknown, so Formula 8.4 will be used to estimate the standard deviation of the sampling distribution. This value will then be substituted into Formula 8.2 and Z(obtained) will be computed:

$$\sigma_{\bar{X}-\bar{X}} = \sqrt{\frac{s_1^2}{N_1 - 1} + \frac{s_2^2}{N_2 - 1}}$$

$$\sigma_{\bar{X}-\bar{X}} = \sqrt{\frac{(1.3)^2}{324 - 1} + \frac{(1.4)^2}{317 - 1}}$$

$$\sigma_{\bar{X}-\bar{X}} = \sqrt{(0.0052) + (0.0062)}$$

$$\sigma_{\bar{X}-\bar{X}} = \sqrt{0.0114}$$

$$\sigma_{\bar{X}-\bar{X}} = 0.107$$

$$Z(\text{obtained}) = \frac{\bar{X}_1 - \bar{X}_2}{\sigma_{\bar{X}-\bar{X}}}$$

$$Z(\text{obtained}) = \frac{6.2 - 6.5}{0.107}$$

$$Z(\text{obtained}) = \frac{-0.300}{0.107}$$

$$Z(\text{obtained}) = -2.80$$

Step 5. Making a Decision and Interpreting the Results of the Test. Comparing the test statistic with the critical region:

$$Z(\text{obtained}) = -2.80$$
$$Z(\text{critical}) = \pm 1.96$$

We see that the Z score falls into the critical region, which means that a difference as large as -0.30 ($6.2 - 6.5$) between the sample means is unlikely if the null hypothesis is true. The null hypothesis of no difference can be rejected, and the notion that men and women are different in terms of their support of gun control is supported. The decision to reject the null hypothesis has only a 0.05 probability (the alpha level) of being incorrect.

Note that the value for Z(obtained) is negative, indicating that men have significantly lower scores than women for support for gun control. The sign of the test statistics reflects our arbitrary decision to label men sample 1 and women sample 2. If we had reversed the labels and called women sample 1 and men sample 2, the sign of the Z(obtained) would have been positive, but its value (2.80) would have been exactly the same, as would our decision in step 5. *(For practice in testing the significance of the difference between sample means for large samples, see problems 8.1 to 8.6, 8.9, 8.15d to f, and 8.16d to f.)*

ONE STEP AT A TIME Completing Step 4 of the Five-Step Model: Computing Z(obtained)

Use these procedures when samples are large.
Solve Formula 8.4 first and then solve Formula 8.2.

Step	***Operation***
To Solve Formula 8.4	
1.	Subtract 1 from N_1.
2.	Square the value of the standard deviation for the first sample (s_1^2).
3.	Divide the quantity you found in step 2 by the quantity you found in step 1.
4.	Subtract 1 from N_2.
5.	Square the value of the standard deviation for the second sample (s_2^2).
6.	Divide the quantity you found in step 5 by the quantity you found in step 4.
7.	Add the quantity you found in step 6 to the quantity you found in step 3.
8.	Take the square root of the quantity you found in step 7.
To Solve Formula 8.2	
1.	Subtract $\bar{X}_2$ from $\bar{X}_1$.
2.	Divide the value you found in step 1 by the quantity you found in step 8 of "To Solve Formula 8.4." This is Z(obtained).

ONE STEP AT A TIME Completing Step 5 of the Five-Step Model: Making a Decision and Interpreting Results

Step	***Operation***
1.	Compare the Z(obtained) to the Z(critical). If Z(obtained) is *in* the critical region, *reject* the null hypothesis. If Z(obtained) is *not in* the critical region, *fail to reject* the null hypothesis.
2.	Interpret the decision to reject or fail to reject the null hypothesis in terms of the original question. For example, our conclusion for the example problem was "There is a significant difference between men and women in their support for gun control."

Applying Statistics 8.1 A Test of Significance for Sample Means

A scale measuring satisfaction with family life has been administered to a sample of married respondents. On this scale, higher scores indicate greater satisfaction. The sample has been divided into respondents with no children and respondents with at least one child, and means and standard deviations have been computed for both groups. Is there a significant difference in satisfaction with family life between these two groups?

The sample information is:

Sample 1 (No Children)	Sample 2 (At Least One Child)
$\bar{X}_1 = 11.3$	$\bar{X}_2 = 10.8$
$s_1 = 0.6$	$s_2 = 0.5$
$N_1 = 78$	$N_2 = 93$

Applying Statistics 8.1 *(continued)*

We can see from the sample results that respondents with no children are more satisfied. Is this difference significant?

Step 1. Making Assumptions and Meeting Test Requirements.

Model: Independent random samples
Level of measurement is interval-ratio
Sampling distribution is normal

Step 2. Stating the Null Hypothesis.

$$H_0: \mu_1 = \mu_2$$
$$H_1: \mu_1 \neq \mu_2$$

Step 3. Selecting the Sampling Distribution and Establishing the Critical Region.

Sampling distribution = Z distribution
Alpha = 0.05, two-tailed
Z(critical) = ±1.96

Step 4. Computing the Test Statistic.

$$\sigma_{\bar{X}-\bar{X}} = \sqrt{\frac{s_1^2}{N_1 - 1} + \frac{s_2^2}{N_2 - 1}}$$

$$\sigma_{\bar{X}-\bar{X}} = \sqrt{\frac{(0.6)^2}{78 - 1} + \frac{(0.5)^2}{93 - 1}}$$

$$\sigma_{\bar{X}-\bar{X}} = \sqrt{0.008}$$

$$\sigma_{\bar{X}-\bar{X}} = 0.09$$

$$Z(\text{obtained}) = \frac{(\bar{X}_1 - \bar{X}_2)}{\sigma_{\bar{X}-\bar{X}}}$$

$$Z(\text{obtained}) = \frac{11.3 - 10.8}{0.09}$$

$$Z(\text{obtained}) = \frac{0.50}{0.09}$$

$$Z(\text{obtained}) = 5.56$$

Step 5. Making a Decision and Interpreting the Results of the Test. Comparing the test statistic with the critical region,

$$Z(\text{obtained}) = 5.56$$
$$Z(\text{critical}) = \pm 1.96$$

we reject the null hypothesis. Parents and childless couples are significantly different in their satisfaction with family life. Given the direction of the difference, we can also note that childless couples are significantly happier.

STATISTICS IN EVERYDAY LIFE

The reading scores of American children have not improved significantly over recent years. Tests are administered every other year to very large, representative samples of about 180,000 fourth-graders and scores on the test range from 0 to 500. The average score in 2009 was 221, which was significantly higher than the 2005 average (219) but exactly the same as the 2007 average (also 221).* The tests were conducted with alpha set at 0.05. A different report# concludes that girls, on average, have caught up with boys in math skills (that is, there is no significant difference between the average math scores for girls and boys) and remain well ahead of boys in verbal and reading skills. What sociological factors might explain these patterns?

*This report is available at http://nationsreportcard.gov/reading_2009/reading_2009_report.

#This report is available at http://www.cep-dc.org/document/docWindow.cfm?fuseaction=document.viewDocument&documentid=304&documentFormatId=4644.

Hypothesis Testing With Sample Means (Small Samples)

When the population standard deviation is unknown and the sample size is small (combined *N*s of less than 100), the *Z* distribution cannot be used to find areas under the sampling distribution. Instead, we will use the *t* distribution to find the critical region and identify unlikely sample outcomes. To do this, we need to perform one additional calculation and make one additional assumption. The calculation is for degrees of freedom, which we need to use the *t* table (Appendix B). In the two-sample case, degrees of freedom are equal to $N_1 + N_2 - 2$.

The additional assumption is a more complex matter. When samples are small, we must assume that the population variances are equal in order to justify the assumption of a normal sampling distribution and to form a pooled estimate of the standard deviation of the sampling distribution. The assumption of equal variance in the population can be tested by a statistical technique known as the *analysis of variance*, or ANOVA (see Chapter 9). However, for our purposes here, we will simply assume equal population variances without formal testing. This assumption is safe as long as sample sizes are approximately equal.

The Five-Step Model and the *t* Distribution To illustrate this procedure, assume that a researcher believes that center-city families have significantly more children than suburban families. Random samples from both areas are gathered and the following sample statistics computed:

Sample 1 (Suburban)	Sample 2 (Center-City)
$\bar{X}_1 = 2.37$	$\bar{X}_2 = 2.78$
$s_1 = 0.63$	$s_2 = 0.95$
$N_1 = 42$	$N_2 = 37$

The sample data show a difference in the predicted direction. The significance of this observed difference can be tested with the five-step model.

Step 1. Making Assumptions and Meeting Test Requirements. The sample size is small and the population standard deviation is unknown. Hence, we must assume equal population variances in the model.

Model: Independent random samples
Level of measurement is interval-ratio
Population variances are equal
Sampling distribution is normal

Step 2. Stating the Null Hypothesis. Because a direction has been predicted, a one-tailed test will be used. The research hypothesis is stated accordingly:

$$H_0: \mu_1 = \mu_2$$
$$H_1: \mu_1 < \mu_2$$

Step 3. Selecting the Sampling Distribution and Establishing the Critical Region. With small samples, the *t* distribution is used to establish the critical region. Alpha will be set at 0.05 and a one-tailed test will be used:

$$\begin{aligned} \text{Sampling distribution} &= t \text{ distribution} \\ \text{Alpha} &= 0.05, \text{ one-tailed} \\ \text{Degrees of freedom} &= N_1 + N_2 - 2 = 42 + 37 - 2 = 77 \\ t(\text{critical}) &= -1.671 \end{aligned}$$

Note that the critical region is placed in the lower tail of the sampling distribution in accordance with the direction specified in H_1.

Step 4. Computing the Test Statistic. With small samples, a different formula (Formula 8.5) is used for the pooled estimate of the standard deviation of the sampling distribution. This value is then substituted directly into the denominator of the formula for *t*(obtained) in Formula 8.6:

FORMULA 8.5

$$\sigma_{\bar{X}-\bar{X}} = \sqrt{\frac{N_1 s_1^2 + N_2 s_2^2}{N_1 + N_2 - 2}}\sqrt{\frac{N_1 + N_2}{N_1 N_2}}$$

$$\sigma_{\bar{X}-\bar{X}} = \sqrt{\frac{(42)(.63)^2 + (37)(.95)^2}{42 + 37 - 2}}\sqrt{\frac{42 + 37}{(42)(37)}}$$

$$\sigma_{\bar{X}-\bar{X}} = \sqrt{\frac{50.06}{77}}\sqrt{\frac{79}{1{,}554}}$$

$$\sigma_{\bar{X}-\bar{X}} = (.81)(.23)$$

$$\sigma_{\bar{X}-\bar{X}} = 0.19$$

FORMULA 8.6

$$t(\text{obtained}) = \frac{\bar{X}_1 - \bar{X}_2}{\sigma_{\bar{X}-\bar{X}}}$$

$$t(\text{obtained}) = \frac{2.37 - 2.78}{0.19}$$

$$t(\text{obtained}) = \frac{-0.41}{0.19} = -2.16$$

$$t(\text{obtained}) = -2.16$$

Step 5. Making a Decision and Interpreting the Results of the Test. The test statistic falls in the critical region:

$$\begin{aligned} t(\text{obtained}) &= -2.16 \\ t(\text{critical}) &= -1.671 \end{aligned}$$

ONE STEP AT A TIME **Step 4 of the Five-Step Model: Computing *t*(obtained)**

Solve Formula 8.5 first and then solve Formula 8.6 to compute the test statistic.

Step	***Operation***
To Solve Formula 8.5	
1.	Add N_1 and N_2 and then subtract 2 from this total.
2.	Square the standard deviation for the first sample (s_1^2) and then multiply the result by N_1.
3.	Square the standard deviation for the second sample (s_2^2) and then multiply the result by N_2.
4.	Add the quantities you found in steps 2 and 3.
5.	Divide the quantity you found in step 4 by the quantity you found in step 1 and then take the square root of the result.
6.	Multiply N_1 by N_2.
7.	Add N_1 and N_2.
8.	Divide the quantity you found in step 7 by the quantity you found in step 6 and then take the square root of the result.
9.	Multiply the quantity you found in step 8 by the quantity you found in step 5.
To Solve Formula 8.6	
1.	Subtract $\bar{X}_2$ from $\bar{X}_1$.
2.	Divide the difference between the sample means—the quantity you found in step 1—by the quantity you found in step 9 of "To Solve Formula 8.5." This is *t*(obtained).

If the null hypothesis ($\mu_1 = \mu_2$) were true, this would be a very unlikely outcome, so the null hypothesis can be rejected. There is a statistically significant difference (a difference so large that it is unlikely to be due to random chance) in the sizes of center-city and suburban families. Furthermore, center-city families are significantly larger in size. The test statistic and sampling distribution are depicted in Figure 8.1. *(For practice in testing the significance of the difference between sample means for small samples, see problems 8.7 and 8.8.)*

FIGURE 8.1 The Sampling Distribution With Critical Region and Test Statistic Displayed

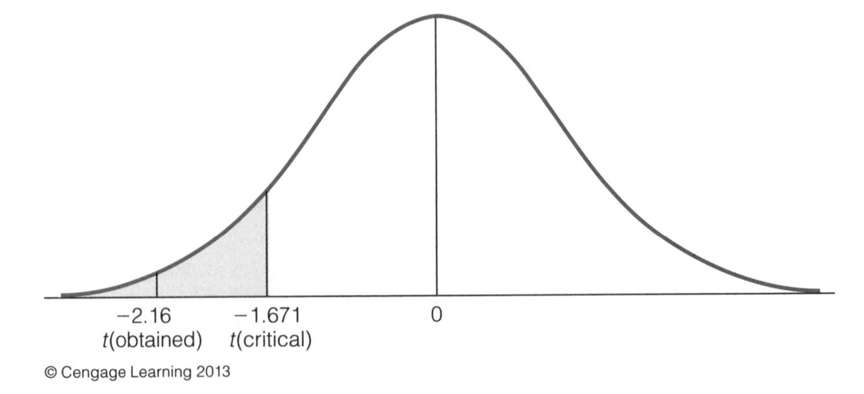

ONE STEP AT A TIME | **Completing Step 5 of the Five-Step Model: Making a Decision and Interpreting Results**

Step	*Operation*
1.	Compare the *t*(obtained) to your *t*(critical). If *t*(obtained) is *in* the critical region, *reject* the null hypothesis. If *t*(obtained) is *not in* the critical region, *fail to reject* the null hypothesis.
2.	Interpret the decision to reject or fail to reject the null hypothesis in terms of the original question. For example, our conclusion for the example problem was "There is a significant difference between the size of center-city and suburban families."

Hypothesis Testing With Sample Proportions (Large Samples)

Testing for the significance of the difference between two sample proportions is analogous to testing sample means. The null hypothesis states that there is no difference between the populations from which the samples are drawn. We compute a test statistic in step 4, which is then compared with the critical region. When sample sizes are large (combined *N*s of more than 100), the *Z* distribution may be used to find the critical region. In this text, we will not consider tests of significance for proportions based on small samples.

In order to find the value of the test statistics, several preliminary equations must be solved. Formula 8.7 uses the values of the two sample proportions (P_s) to give us an estimate of the population proportion (P_u)—the proportion of cases in the population that have the trait under consideration assuming the null hypothesis is true:

FORMULA 8.7

$$P_u = \frac{N_1 P_{s1} + N_2 P_{s2}}{N_1 + N_2}$$

The value of P_u is then used to compute the standard deviation of the sampling distribution of the difference in sample proportions in Formula 8.8:

FORMULA 8.8

$$\sigma_{p-p} = \sqrt{P_u(1 - P_u)}\sqrt{\frac{N_1 + N_2}{N_1 N_2}}$$

This value is then substituted into the formula for computing the test statistic, presented as Formula 8.9:

FORMULA 8.9

$$Z(\text{obtained}) = \frac{(P_{s1} - P_{s2}) - (P_{u1} - P_{u2})}{\sigma_{p-p}}$$

where $(P_{s1} - P_{s2})$ = the difference between the sample proportions
$(P_{u1} - P_{u2})$ = the difference between the population proportions
(σ_{p-p}) = the standard deviation of the sampling distribution of the difference between sample proportions

As was the case with sample means, the second term in the numerator is assumed to be zero by the null hypothesis. Therefore, the formula reduces to:

FORMULA 8.10

$$Z(\text{obtained}) = \frac{(P_{s1} - P_{s2})}{\sigma_{p-p}}$$

Remember to solve these equations in order, starting with Formula 8.7 (and skipping Formula 8.9).

Illustrating a Test of Hypothesis Between Two Sample Proportions (Large Samples) An example will make these procedures clearer. Suppose we are researching social networks among senior citizens and wonder if blacks and whites differ in their number of memberships in clubs and other organizations. Random samples of black and white senior citizens have been selected and classified as high or low in terms of their number of memberships in voluntary associations. Is there a statistically significant difference in the participation patterns of black and white elderly? The proportions of each group classified as "high" in participation and the sample size for both groups are:

Sample 1 (Black Senior Citizens)	Sample 2 (White Senior Citizens)
$P_{s1} = 0.34$	$P_{s2} = 0.25$
$N_1 = 83$	$N_2 = 103$

Step 1. Making Assumptions and Meeting Test Requirements.

Model: Independent random samples
Level of measurement is nominal
Sampling distribution is normal

Step 2. Stating the Null Hypothesis. Because no direction has been predicted, this will be a two-tailed test.

$$H_0: P_{u1} = P_{u2}$$

$$H_1: P_{u1} \neq P_{u2}$$

Step 3. Selecting the Sampling Distribution and Establishing the Critical Region. Because the sample size is large, the Z distribution will be used to establish the critical region. Setting alpha at 0.05, we have:

Sampling distribution = Z distribution
Alpha = 0.05, two-tailed
$Z(\text{critical}) = \pm 1.96$

Step 4. Computing the Test Statistic. Solve Formula 8.7 first, substitute the resultant value into Formula 8.8, and then solve for *Z*(obtained) with Formula 8.10.

$$P_u = \frac{N_1 P_{s1} + N_2 P_{s2}}{N_1 + N_2}$$

$$P_u = \frac{(83)(0.34) + (103)(0.25)}{83 + 103}$$

$$P_u = 0.29$$

$$\sigma_{p-p} = \sqrt{P_u(1 - P_u)}\sqrt{\frac{N_1 + N_2}{N_1 N_2}}$$

$$\sigma_{p-p} = \sqrt{(0.29)(0.71)}\sqrt{\frac{83 + 103}{(83)(103)}}$$

$$\sigma_{p-p} = (0.45)(0.15)$$

$$\sigma_{p-p} = 0.07$$

$$Z(\text{obtained}) = \frac{(P_{s1} - P_{s2})}{\sigma_{p-p}}$$

$$Z(\text{obtained}) = \frac{0.34 - 0.25}{0.07}$$

$$Z(\text{obtained}) = 1.29$$

ONE STEP AT A TIME **Completing Step 4 of the Five-Step Model: Computing Z(obtained)**

Solve Formulas 8.7, 8.8, and 8.10—in that order—to find the test statistic.

Step	***Operation***
	To Solve Formula 8.7
1.	Add N_1 and N_2.
2.	Multiply P_{s1} by N_1.
3.	Multiply P_{s2} by N_2.
4.	Add the quantity you found in step 3 to the quantity you found in step 2.
5.	Divide the quantity you found in step 4 by the quantity you found in step 1. This is P_u.
	To Solve Formula 8.8
1.	Multiply P_u by $(1 - P_u)$.
2.	Take the square root of the quantity you found in step 1.
3.	Multiply N_1 by N_2.
4.	Add N_1 and N_2. (Note: You already found this value when solving Formula 8.7. See step 1.)
5.	Divide the quantity you found in step 4 by the quantity you found in step 3.
6.	Take the square root of the quantity you found in step 5.
7.	Multiply the quantity you found in step 6 by the quantity you found in step 2.

(*continued next page*)

ONE STEP AT A TIME *(continued)*

Step	Operation
To Solve Formula 8.10	
1.	Find the difference between the sample proportions.
2.	Divide the quantity you found in step 1 by the quantity you found in step 7 of "To Solve Formula 8.8." This is *Z*(obtained).

Step 5. Making a Decision and Interpreting the Results of the Test. Because the test statistic—*Z*(obtained) = 1.29—does not fall into the critical region as marked by the *Z*(critical) of ±1.96, we fail to reject the null hypothesis. The difference between the sample proportions is no greater than what would be expected if the null hypothesis were true and only random chance were operating. Black and white senior citizens are not significantly different in terms of participation patterns as measured in this test. *(For practice in testing the significance of the difference between sample proportions, see problems 8.10 to 8.14, 8.15a to c, and 8.16a to c.)*

ONE STEP AT A TIME **Completing Step 5 of the Five-Step Model: Making a Decision and Interpreting Results**

Step	Operation
1.	Compare *Z*(obtained) to *Z*(critical). If *Z*(obtained) is *in* the critical region, *reject* the null hypothesis. If *Z*(obtained) is *not in* the critical region, *fail to reject* the null hypothesis.
2.	Interpret the decision to reject or fail to reject the null hypothesis in terms of the original question. For example, our conclusion for the example problem was "There is no significant difference between the participation patterns of black and white senior citizens."

Applying Statistics 8.2 Testing the Significance of the Difference Between Sample Proportions

Do attitudes toward sex vary by gender? The proportion of each sex that feels that premarital sex is always wrong is:

Females	Males
$P_{s1} = 0.35$	$P_{s2} = 0.32$
$N_1 = 450$	$N_2 = 417$

Females are more likely to say that premarital sex is always wrong. Is the difference significant? The table presents all the information we will need to conduct a test of the null hypothesis following the familiar five-step model with alpha set at .05, two-tailed test.

Step 1. Making Assumptions and Meeting Test Requirements.

Model: Independent random samples
Level of measurement is nominal
Sampling distribution is normal

Step 2. Stating the Null Hypothesis.

$$H_0: P_{u1} = P_{u2}$$
$$H_1: P_{u1} \neq P_{u2}$$

(continued next page)

Applying Statistics 8.2 *(continued)*

Step 3. Selecting the Sampling Distribution and Establishing the Critical Region.

Sampling distribution = Z distribution

Alpha = 0.05, two-tailed

$Z(\text{critical}) = \pm 1.96$

Step 4. Computing the Test Statistic. Remember to start with Formula 8.7, substitute the value for P_u into Formula 8.8, and then substitute that value into Formula 8.10 to solve for Z(obtained):

$$P_u = \frac{N_1 P_{s1} + N_2 P_{s2}}{N_1 + N_2}$$

$$P_u = \frac{(450)(0.35) + (417)(0.32)}{450 + 417}$$

$$P_u = \frac{290.94}{867}$$

$$P_u = 0.34$$

$$\sigma_{p-p} = \sqrt{P_u(1 - P_u)}\sqrt{\frac{N_1 + N_2}{N_1 N_2}}$$

$$\sigma_{p-p} = \sqrt{(0.34)(0.66)}\sqrt{\frac{450 + 417}{(450)(417)}}$$

$$\sigma_{p-p} = \sqrt{0.2244}\sqrt{0.0046}$$

$$\sigma_{p-p} = (0.47)(0.068)$$

$$\sigma_{p-p} = 0.032$$

$$Z(\text{obtained}) = \frac{(P_{s1} - P_{s2})}{\sigma_{p-p}}$$

$$Z(\text{obtained}) = \frac{0.35 - 0.32}{0.032}$$

$$Z(\text{obtained}) = \frac{0.030}{0.032}$$

$$Z(\text{obtained}) = 0.94$$

Step 5. Making a Decision and Interpreting the Results of the Test. With an obtained Z score of 0.94, we would fail to reject the null hypothesis. Females are not significantly more likely to feel that premarital sex is always wrong.

According to the Gallup polls, Republicans and Democrats have become increasingly polarized over the abortion issue. In 1975, two years after the landmark *Roe v. Wade* Supreme Court decision, 18% of Republicans and 19% of Democrats agreed that abortion should be legal "under any circumstances." In 2011, the percentage of Republicans approving abortion under any circumstances had fallen to 13%, while the percentage of Democrats approving had risen to 38%. The one percentage point difference in 1975 might have been statistically significant, but it was clearly not important. The 25% difference in 2011 is statistically significant, important, and extremely consequential for the possibilities of mutual understanding and civil discourse in American politics. These results are based on random samples of about 1,000 adult Americans.

How could we explain these trends? One possibility is that the Republican Party has lost many moderate, pro-choice members since the 1970s, leaving the party smaller and more ideologically homogenous. What information would you need to investigate this possibility?

Source: Saad, Lydia. 2010. "Republicans', Dems' abortion views grow more polarized." Available at http://www.gallup.com/poll/126374/Republicans-Dems-Abortion-Views-Grow-Polarized.aspx?version=print.

Saad, Lydia. 2011. "Americans Still Split Along 'Pro-Choice,' 'Pro-Life' Lines. Available at http://www.gallup.com/poll/147734/Americans-Split-Along-Pro-Choice-Pro-Life-Lines.aspx.

The Limitations of Hypothesis Testing: Significance versus Importance

Given that we are usually interested in rejecting the null hypothesis, we should take a moment to consider systematically the factors that affect our decision in step 5. Generally speaking, the probability of rejecting the null hypothesis is a function of four independent factors:

1. The size of the observed difference
2. The alpha level
3. The use of one- or two-tailed tests
4. The size of the sample

All these factors except the first are under the direct control of the researcher. The relationship between alpha level and the probability of rejection is straightforward. The higher the alpha level, the larger the critical region and the greater the probability of rejection. Thus, it is easier to reject the H_0 at the 0.05 level than at the 0.01 level and easier still at the 0.10 level. The danger here, of course, is that higher alpha levels will lead to more frequent Type I errors, and we might find ourselves declaring small differences to be statistically significant. In similar fashion, using a one-tailed test will increase the probability of rejecting the null hypothesis (assuming that the proper direction has been predicted).

The final factor is the sample size: With all other factors constant, the probability of rejecting H_0 increases with the sample size. In other words, the larger the sample, the more likely we are to reject the null hypothesis, and with very large samples (say, samples with thousands of cases), we may declare small, unimportant differences to be statistically significant.

This relationship may appear to be surprising, but the reasons for it can be appreciated with a brief consideration of the formulas used to compute test statistics in step 4. In all these formulas, for all tests of significance, the sample size (N) is in the "denominator of the denominator." Algebraically, this is equivalent to being in the numerator of the formula and means that the value of the test statistic is directly proportional to N and that the two will increase together.

TABLE 8.1 Test Statistics for Single-Sample Means Computed From Samples of Various Sizes ($\bar{X} = 80$, $\mu = 79$, $s = 5$ throughout)

Sample Size	Test Statistic, Z(Obtained)
100	1.99
200	2.82
500	4.47

To illustrate, consider Table 8.1, which shows the value of the test statistic for single-sample means from samples of various sizes. The value of the test statistic—*Z*(obtained)—increases as *N* increases even though none of the other terms in the formula changes. This pattern of higher probabilities for rejecting H_0 with larger samples holds for all tests of significance.

On one hand, the relationship between the sample size and the probability of rejecting the null hypothesis should not alarm us unduly. Larger samples are, after all, better approximations of the populations they represent. Thus, decisions based on larger samples can be trusted more than decisions based on small samples.

On the other hand, this relationship clearly underlines what is perhaps the most important limitation of hypothesis testing. Simply because a difference is statistically significant does not guarantee that it is important in any other sense. Particularly with very large samples, relatively small differences may be statistically significant. Even with small samples, of course, differences that are otherwise trivial or uninteresting may be statistically significant. The crucial point is that statistical significance and theoretical or practical importance can be two very different things. Statistical significance is a necessary but not sufficient condition for theoretical or practical importance. A difference that is not statistically significant is almost certainly unimportant. However, significance by itself does not guarantee importance. Even when it is clear that the research results were not produced by random chance, the researcher must still assess their importance. Do they firmly support a theory or hypothesis? Are they clearly consistent with a prediction or analysis? Do they strongly indicate a line of action in solving some problem? These are the kinds of questions a researcher must ask when assessing the importance of the results of a statistical test.

Also, we should note that researchers have access to some very powerful ways of analyzing the importance (vs. the statistical significance) of research results. These statistics, including bivariate measures of association and multivariate statistical techniques, are introduced in Parts III and IV of this text.

STATISTICS IN EVERYDAY LIFE

Is the income gap between the genders closing? Comparing samples of full-time workers, the median income for women in 1955 was 64% of the median income for men. In 2007, the gap had closed to 78% but then widened back to 76% in 2009.

Part of what maintains this gap between the genders is the "glass ceiling" on women's chances for promotion to the highest (most lucrative) positions. This means that women are relatively less represented in the very highest income brackets. For example, in 2009, comparing only people with full-time, year-round jobs, 10% of men earned more than $100,000 vs. only 3% of women.

What does the glass ceiling imply about the overall shape of the distribution of men's and women's income? Which distribution will be more positively skewed? What effect will that have on the mean? (See Chapter 4 for a review of the effect of skew on measures of central tendency).

Source: U.S. Bureau of the Census. American FactFinder. Available at www.census.gov.

Applying Statistics 8.3 Testing the Significance of the Difference Between Sample Means

The World Values Survey (WVS) has been administered to random samples in a variety of nations periodically since 1981. Like the General Social Survey used for computer applications in this text, the WVS tests opinions on a variety of issues and concerns. Results of the surveys are available at http://www.worldvaluessurvey.org.

One item on the WVS asks about support for abortion. Are there significant differences in support between males and females in Canada and the United States? Support for abortion was measured with a 10-point scale, with higher scores indicating more support. Results are for 2006.

Canada		United States	
Males	Females	Males	Females
$\bar{X}_1 = 4.6$	$\bar{X}_2 = 4.7$	$\bar{X}_1 = 4.6$	$\bar{X}_2 = 4.4$
$s_1 = 2.8$	$s_2 = 3.0$	$s_1 = 2.9$	$s_2 = 2.8$
$N_1 = 988$	$N_2 = 1{,}074$	$N_1 = 569$	$N_2 = 598$

The differences between the genders seem small. Are they significant? To conserve space, only the most essential steps of the five-step model will be reported. The significance level will be set at 0.05 and the tests will be two-tailed. The null hypothesis will be that there is no difference in support for abortion between *all* males and *all* females in the respective nations (H_0: $\mu_1 = \mu_2$).

We will use Formulas 8.2 and 8.4 to compute the test statistics (step 4):

For Canada:

$$\sigma_{\bar{X}-\bar{X}} = \sqrt{\frac{s_1^2}{N-1} + \frac{s_2^2}{N_2-1}}$$

$$\sigma_{\bar{X}-\bar{X}} = \sqrt{\frac{(2.8)^2}{988-1} + \frac{(3.0)^2}{1{,}074-1}}$$

$$\sigma_{\bar{X}-\bar{X}} = \sqrt{\frac{7.84}{987} + \frac{9.0}{1{,}073}}$$

$$\sigma_{\bar{X}-\bar{X}} = \sqrt{0.008 + 0.008}$$

$$\sigma_{\bar{X}-\bar{X}} = \sqrt{0.016}$$

$$\sigma_{\bar{X}-\bar{X}} = 0.13$$

$$Z(\text{obtained}) = \frac{\bar{X}_1 - \bar{X}_2}{\sigma_{\bar{X}-\bar{X}}} = \frac{4.6 - 4.7}{0.13}$$

$$Z(\text{obtained}) = \frac{0.10}{0.13} = 0.77$$

For the United States:

$$\sigma_{\bar{X}-\bar{X}} = \sqrt{\frac{s_1^2}{N_1-1} + \frac{s_2^2}{N_2-1}}$$

$$\sigma_{\bar{X}-\bar{X}} = \sqrt{\frac{(2.9)^2}{569-1} + \frac{(2.8)^2}{598-1}}$$

$$\sigma_{\bar{X}-\bar{X}} = \sqrt{\frac{8.41}{568} + \frac{7.84}{597}}$$

$$\sigma_{\bar{X}-\bar{X}} = \sqrt{0.015 + 0.013}$$

$$\sigma_{\bar{X}-\bar{X}} = \sqrt{0.028}$$

$$\sigma_{\bar{X}-\bar{X}} = 0.17$$

$$Z(\text{obtained}) = \frac{\bar{X}_1 - \bar{X}_2}{\sigma_{\bar{X}-\bar{X}}} = \frac{4.6 - 4.4}{0.17}$$

$$Z(\text{obtained}) = \frac{0.20}{0.17} = 1.18$$

For both tests, we fail to reject the null hypothesis. For Canada and the United States, there are no significant differences by gender in support for abortion.

BECOMING A CRITICAL CONSUMER: When is a difference a difference?

How big does a difference have to be to be in order to be considered a difference? The question may sound whimsical or even silly, but this is a serious issue because it relates to our ability to identify the truth when we see it. Using the income gap between the genders as an example, how big a difference must there be between average incomes for men and women before the gap becomes evidence of gender inequality? Would we be concerned if U.S. men averaged $55,000 and women averaged $54,500? Across millions of cases, a difference of $500—about 1% of the average incomes—seems small and unimportant. How about if the difference was $1,000? $5,000? $10,000? At what point do we declare the difference to be important evidence of gender income inequality?

Very large or very small differences are easy to deal with. Small differences—relative to the scale of the variable (e.g., a difference of a few dollars in average incomes)—can be dismissed as trivial. At the other extreme, large differences—again thinking in terms of the scale of the variable (e.g., differences of more than $10,000 in average income)—are almost certainly worthy of attention. But what about differences between these two extremes? How big is big?

There are, of course, no absolute rules that would always enable us to identify important differences. However, we can discuss some guidelines that will help us know when a difference is consequential. We will do this in general terms first and then relate this discussion to significance testing.

Differences in General

First, as I suggested earlier, think about the difference in terms of the scale of the variable. A drop of a cent or two in the cost of a gallon of gas when the average is $4.00 will probably not make a difference in the budgets of anyone except very-high-mileage drivers. In very general (and arbitrary) terms, a change of 10% or more (if gas costs $4.00 a gallon, a 10% change would be a rise or fall of 40 cents) might be taken as a signal of an important difference. This rule of thumb is one reasonable guideline for many social indicators—population growth, crime rates, and birthrates, for example.

Secondly, you need to look at the raw frequencies to judge the importance of a change. Most people would be alarmed by a headline that reported a doubling of the number of teen pregnancies in a locality. However, consider two scenarios: one in which the number doubled from 10 to 20 in a town of 100,000 and another when, in a city of the same size, the numbers went from 2,500 to 5,000. The raw frequencies can add a valuable context to the perception of a change (which is one reason they are always reported in the professional research literature).

Third—and another way to add some context to a change—is to look at a broader time period. A report that voter turnout declined by 20% in a locality between 2008 and 2010 would naturally result in a good deal of alarm. However, turnout often declines between years featuring presidential elections (2008) and those that do not (2010). It would be more meaningful to compare 2000 with 2008 and, perhaps, even more revealing to get the data for earlier years.

Differences in Social Research

In social research, the problem of identifying important differences is additionally complicated by the vagaries of random chance when we work with samples rather than populations. That is, the size of a difference between sample statistics may be the result of random chance rather than (or in addition to) actual differences in the population. One of the great strengths of hypothesis testing is that it provides a system for identifying important differences. When we say that a difference is statistically significant, we reject the argument that random chance alone is responsible (with a known probability of error—the alpha level or *p* value) and support the idea that the difference in the sample statistics reflects a difference that also exists in the population. Small differences (e.g., a difference of only a few hundred dollars in average income between the genders) are unlikely to be significant at the 0.05 level. The larger the difference between the sample statistics, the more likely it is to be significant. The larger the difference and the lower the alpha level, the more confidence we can have that the difference reflects actual patterns in the population. Remember to also pay attention to the sample size: Even very small differences can be statistically significant if the sample size is very large. Statistical significance is not the same thing as importance.

(*continued*)

BECOMING A CRITICAL CONSUMER: *(continued)*

Obviously, hypothesis testing is not infallible as a system for identifying important differences. We must remember that there is always a chance of making an incorrect decision: We could declare a trivial difference to be important and, vice versa, we could conclude that an important difference is trivial.

Reading Social Research

When reporting the results of tests of significance, professional researchers use a vocabulary that is much terser than ours. This is partly because of space limitations in scientific journals and partly because professional researchers can assume a certain level of statistical literacy in their audiences. Thus, they omit many of the elements—such as the null hypothesis or the critical region—that we have been so careful to state.

Instead, researchers report only the sample values (for example, means or proportions), the value of the test statistic (for example, a Z or t score), the alpha level, the degrees of freedom (if applicable), and the sample size. The results of the example problem used in the discussion of the t test might be reported in the professional literature as "the difference between the sample means of 2.37 (suburban families) and 2.78 (center-city families) was tested and found to be significant ($t = -2.16$, $df = 77$, $p < 0.05$)." Note that the alpha level is reported as "$p < 0.05$." This is shorthand for "the probability of a difference of this magnitude occurring by chance alone—if the null hypothesis of no difference is true—is less than 0.05" and is a good illustration of how researchers can convey a great deal of information in just a few symbols. In a similar fashion, our somewhat long-winded phrase "the test statistic falls in the critical region and, therefore, the null hypothesis is rejected" is rendered tersely and simply: "the difference . . . was . . . found to be significant."

When researchers need to report the results of many tests of significance, they will often use a summary table to report the sample information and whether the difference is significant at a certain alpha level. If you read the researcher's description and analysis of such tables, you should have little difficulty interpreting and understanding them. These comments about how significance tests are reported in the literature apply to all the tests of hypotheses covered in Part II of this text.

To illustrate the reporting style you would find in the professional research literature, let us look at a research report that analyzed movies aimed at teens and young adults. These films are rich in their portrayals of gender and gender identity issues. Are these portrayals consistent with the behavior and concerns of their intended audiences? What effects—if any—do the portrayals have on their audiences?

Two researchers attempted to address some of these issues by first analyzing the content of the 20 top-grossing "teen" movies released in the United States between 1995 and 2005. Coders counted the number of times movie characters used aggressive tactics—including gossip, backstabbing, and public humiliation—to damage the status or self-esteem of others. They identified 337 such acts in the films and found that female characters were significantly more likely to be the perpetrators ($t = -3.15$, $p < 0.05$) and that aggression was significantly more likely to be rewarded in the film ($t = -2.29$, $p < 0.05$). These findings contrast with studies of real life that find that female friendships are more likely to be supportive and cooperative than friendships among males.

Do these portrayals have any effect on their audiences? Using different statistical techniques, the researchers found evidence that they do, which raises the concern that movies are sending the message that aggression is an effective way of gaining and perpetuating status. Want to learn more? The citation is given below.

Behm-Morawitz, E., and Mastro, D. 2008. "Mean Girls? The Influence of Gender Portrayals in Teen Movies on Emerging Adults' Gender-Based Attitudes and Beliefs." *Journalism and Mass Communications Quarterly* 85: 131–146.

SUMMARY

1. A common research situation is to test for the significance of the difference between two populations. Sample statistics are calculated for random samples of each population and then we test for the significance of the difference between the samples as a way of inferring differences between the specified populations.
2. When sample information is summarized in the form of sample means and N is large, the Z distribution is used to find the critical region. When N is small, the t distribution is used to establish the critical region. In the latter circumstance, we must also assume equal population variances before forming a pooled estimate of the standard deviation of the sampling distribution.
3. Differences in sample proportions may also be tested for significance. For large samples, the Z distribution is used to find the critical region.
4. In all forms of hypothesis testing, a number of factors affect the probability of rejecting the null hypothesis: the size of the difference, the alpha level, the use of one- versus two-tailed tests, and the sample size. Statistical significance is not the same thing as theoretical or practical importance. Even after a difference is found to be statistically significant, the researcher must still demonstrate the relevance or importance of his or her findings. The statistics presented in Parts III and IV of this text will give us the tools we need to deal directly with issues beyond statistical significance.

SUMMARY OF FORMULAS

FORMULA 8.1 Test statistic for two sample means, large samples:

$$Z(\text{obtained}) = \frac{(\bar{X}_1 - \bar{X}_2) - (\mu_1 - \mu_2)}{\sigma_{\bar{X}-\bar{X}}}$$

FORMULA 8.2 Test statistic for two sample means, large samples (simplified): $Z(\text{obtained}) = \dfrac{\bar{X}_1 - \bar{X}_2}{\sigma_{\bar{X}-\bar{X}}}$

FORMULA 8.3 The standard deviation of the sampling distribution of the difference in sample means:

$$\sigma_{\bar{X}-\bar{X}} = \sqrt{\frac{\sigma_1^2}{N_1} + \frac{\sigma_2^2}{N_2}}$$

FORMULA 8.4 Pooled estimate of the standard deviation of the sampling distribution of the difference in sample means: $\sigma_{\bar{X}-\bar{X}} = \sqrt{\dfrac{s_1^2}{N_1 - 1} + \dfrac{s_2^2}{N_2 - 1}}$

FORMULA 8.5 Pooled estimate of the standard deviation of the sampling distribution of the difference in sample means, small samples: $\sigma_{\bar{X}-\bar{X}} = \sqrt{\dfrac{N_1 s_1^2 + N_2 s_2^2}{N_1 + N_2 - 2}}\sqrt{\dfrac{N_1 + N_2}{N_1 N_2}}$

FORMULA 8.6 Test statistic for two sample means, small samples: $t(\text{obtained}) = \dfrac{(\bar{X}_1 - \bar{X}_2)}{\sigma_{\bar{X}-\bar{X}}}$

FORMULA 8.7 Pooled estimate of population proportion, large samples: $P_u = \dfrac{N_1 P_{s1} + N_2 P_{s2}}{N_1 + N_2}$

FORMULA 8.8 Standard deviation of the sampling distribution of the difference in sample proportions, large samples: $\sigma_{p-p} = \sqrt{P_u(1 - P_u)}\sqrt{\dfrac{N_1 + N_2}{N_1 N_2}}$

FORMULA 8.9

Test statistic for two sample proportions, large samples:

$$Z(\text{obtained}) = \frac{(P_{s1} - P_{s2}) - (P_{u1} - P_{u2})}{\sigma_{p-p}}$$

FORMULA 8.10

Test statistic for two sample proportions, large samples (simplified formula):

$$Z(\text{obtained}) = \frac{(P_{s1} - P_{s2})}{\sigma_{p-p}}$$

GLOSSARY

Independent random samples. Random samples gathered in such a way that the selection of a particular case for one sample has no effect on the probability that any other particular case will be selected for the other sample.

Pooled estimate. An estimate of the standard deviation of the sampling distribution of the difference in sample means based on the standard deviations of both samples.

σ_{p-p}. Symbol for the standard deviation of the sampling distribution of the differences in sample proportions.

$\sigma_{\bar{X}-\bar{X}}$. Symbol for the standard deviation of the sampling distribution of the differences in sample means.

PROBLEMS

8.1 For each of the following, test for the significance of the difference in sample statistics by using the five-step model. *(Hint: Remember to solve Formula 8.4 before attempting to solve Formula 8.2. Also, in Formula 8.4, perform the mathematical operations in the proper sequence. First, square each sample standard deviation, divide by the proper N – 1, add the resultant values, and then find the square root of the sum.)*

a.

Sample 1	Sample 2
$\bar{X}_1 = 72.5$	$\bar{X}_2 = 76.0$
$s_1 = 14.3$	$s_2 = 10.2$
$N_1 = 136$	$N_2 = 257$

b.

Sample 1	Sample 2
$\bar{X}_1 = 107$	$\bar{X}_2 = 103$
$s_1 = 14$	$s_2 = 17$
$N_1 = 175$	$N_2 = 200$

8.2 SOC Gessner and Healey administered questionnaires to samples of undergraduates. Among other things, the questionnaires contained a scale that measured attitudes toward interpersonal violence (higher scores indicate greater approval of interpersonal violence). Test the results as reported here for sexual, racial, and social-class differences.

a.

Sample 1 (Males)	Sample 2 (Females)
$\bar{X}_1 = 2.99$	$\bar{X}_2 = 2.29$
$s_1 = 0.88$	$s_2 = 0.91$
$N_1 = 122$	$N_2 = 251$

b.

Sample 1 (Blacks)	Sample 2 (Whites)
$\bar{X}_1 = 2.76$	$\bar{X}_2 = 2.99$
$s_1 = 0.68$	$s_2 = 0.91$
$N_1 = 43$	$N_2 = 304$

c.

Sample 1 (White Collar)	Sample 2 (Blue Collar)
$\bar{X}_1 = 2.46$	$\bar{X}_2 = 2.67$
$s_1 = 0.91$	$s_2 = 0.87$
$N_1 = 249$	$N_2 = 97$

d. Summarize your results in terms of the significance and the direction of the differences. Which of these three factors seems to make the biggest difference in attitudes toward interpersonal violence?

8.3 SOC Do athletes in different sports vary in terms of intelligence? Reported here are College Board scores of random samples of college basketball and football players. Is there a significant difference? Write a sentence or two explaining the difference.

a.

Sample 1 (Basketball Players)	Sample 2 (Football Players)
$\bar{X}_1 = 460$	$\bar{X}_2 = 442$
$s_1 = 92$	$s_2 = 57$
$N_1 = 102$	$N_2 = 117$

b. What about male and female college athletes?

Sample 1 (Males)	Sample 2 (Females)
$\bar{X}_1 = 452$	$\bar{X}_2 = 480$
$s_1 = 88$	$s_2 = 75$
$N_1 = 107$	$N_2 = 105$

8.4 PA A number of years ago, the road and highway maintenance department of Sin City, Nevada, began recruiting minority group members through an affirmative action program. In terms of efficiency ratings as compiled by their superiors, how do the affirmative action employees rate? The ratings of random samples of both groups were collected, and the results are reported here (higher ratings indicate greater efficiency).

Sample 1 (Affirmative Action)	Sample 2 (Regular)
$\bar{X}_1 = 15.2$	$\bar{X}_2 = 15.5$
$s_1 = 3.9$	$s_2 = 2.0$
$N_1 = 97$	$N_2 = 100$

Write a sentence or two of interpretation.

8.5 SOC Are middle-class families more likely than working-class families to maintain contact with kin? Write a paragraph summarizing the results of these tests.

a. A sample of middle-class families reported an average of 7.3 visits per year with close kin, while a sample of working-class families averaged 8.2 visits. Is the difference significant?

Visits	
Sample 1 (Middle Class)	Sample 2 (Working Class)
$\bar{X}_1 = 7.3$	$\bar{X}_2 = 8.2$
$s_1 = 0.3$	$s_2 = 0.5$
$N_1 = 89$	$N_2 = 55$

b. The middle-class families averaged 2.3 phone calls and 8.7 e-mail messages per month with close kin. The working-class families averaged 2.7 calls and 5.7 e-mail messages per month. Are these differences significant?

Phone Calls	
Sample 1 (Middle Class)	Sample 2 (Working Class)
$\bar{X}_1 = 2.3$	$\bar{X}_2 = 2.7$
$s_1 = 0.5$	$s_2 = 0.8$
$N_1 = 89$	$N_2 = 55$

E-Mail Messages	
Sample 1 (Middle Class)	Sample 2 (Working Class)
$\bar{X}_1 = 8.7$	$\bar{X}_2 = 5.7$
$s_1 = 0.3$	$s_2 = 1.1$
$N_1 = 89$	$N_2 = 55$

8.6 SOC Are college students who live in dormitories significantly more involved in campus life than students who commute to campus? The following data report the average number of hours per week students devote to extracurricular activities. Is the difference between these randomly selected samples of commuter and residential students significant?

Sample 1 (Residential)	Sample 2 (Commuter)
$\bar{X}_1 = 10.2$	$\bar{X}_2 = 12.4$
$s_1 = 1.9$	$s_2 = 2.0$
$N_1 = 173$	$N_2 = 158$

8.7 GER Are senior citizens who live in retirement communities more socially active than those who live in age-integrated communities? Write a sentence or two explaining the results of these tests. *(Hint: Remember to use the proper formulas for small sample sizes.)*

a. A random sample of senior citizens living in a retirement village reported that they had an average of 1.42 face-to-face interactions per day with their neighbors. A random sample of those living in age-integrated communities reported 1.58 interactions. Is the difference significant?

Sample 1 (Retirement Community)	Sample 2 (Age-Integrated Neighborhood)
$\bar{X}_1 = 1.42$	$\bar{X}_2 = 1.58$
$s_1 = 0.10$	$s_2 = 0.78$
$N_1 = 43$	$N_2 = 37$

b. Senior citizens living in the retirement village reported that they had an average of 7.43 telephone calls with friends and relatives each week, while those in the age-integrated communities reported an average of 5.50 calls. Is the difference significant?

Sample 1 (Retirement Community)	Sample 2 (Age-integrated Neighborhood)
$\bar{X}_1 = 7.43$	$\bar{X}_2 = 5.50$
$s_1 = 0.75$	$s_2 = 0.25$
$N_1 = 43$	$N_2 = 37$

8.8 SW As the director of the local Boys & Girls Club, you have claimed for years that membership in your club reduces juvenile delinquency. A cynical member of your funding agency has demanded proof of your claim. Fortunately, your local sociology department is on your side and springs to your aid with student assistants, computers, and hand calculators at the ready. Random samples of members and nonmembers are gathered and interviewed with respect to their involvement in delinquent activities. Each respondent is asked to enumerate the number of delinquent acts he has engaged in over the past year. The results are in and reported here (the average number of admitted acts of delinquency). What can you tell the funding agency?

Sample 1 (Members)	Sample 2 (Nonmembers)
$\bar{X}_1 = 10.3$	$\bar{X}_2 = 12.3$
$s_1 = 2.7$	$s_2 = 4.2$
$N_1 = 40$	$N_2 = 55$

8.9 SOC A survey has been administered to random samples of respondents in each of five nations. For each nation, are men and women significantly different in terms of their reported levels of satisfaction? Respondents were asked: "How satisfied are you with your life as a whole?" Responses varied from 1 (very dissatisfied) to 10 (very satisfied). Conduct a test for the significance of the difference in mean scores for each nation.

France	
Males	Females
$\bar{X}_1 = 6.9$	$\bar{X}_2 = 7.0$
$s_1 = 1.8$	$s_2 = 2.0$
$N_1 = 478$	$N_2 = 522$

Egypt	
Males	Females
$\bar{X}_1 = 5.6$	$\bar{X}_2 = 5.9$
$s_1 = 2.7$	$s_2 = 2.7$
$N_1 = 1{,}557$	$N_2 = 1{,}493$

China	
Males	Females
$\bar{X}_1 = 6.8$	$\bar{X}_2 = 6.7$
$s_1 = 2.4$	$s_2 = 2.4$
$N_1 = 899$	$N_2 = 1{,}060$

Mexico	
Males	Females
$\bar{X}_1 = 8.2$	$\bar{X}_2 = 8.3$
$s_1 = 2.0$	$s_2 = 2.0$
$N_1 = 748$	$N_2 = 764$

Japan	
Males	Females
$\bar{X}_1 = 6.9$	$\bar{X}_2 = 7.1$
$s_1 = 1.7$	$s_2 = 1.9$
$N_1 = 474$	$N_2 = 606$

8.10 For each problem, test the sample statistics for the significance of the difference.

a.

Sample 1	Sample 2
$P_{s1} = 0.20$	$P_{s2} = 0.17$
$N_1 = 114$	$N_2 = 101$

b.

Sample 1	Sample 2
$P_{s1} = 0.60$	$P_{s2} = 0.62$
$N_1 = 478$	$N_2 = 532$

8.11 CJ About half of the police officers in Shinbone, Kansas, have completed a special course in investigative procedures. Has the course increased their efficiency in clearing crimes by arrest? The proportions of cases cleared by arrest for samples of trained and untrained officers are reported here.

Sample 1 (Trained)	Sample 2 (Untrained)
$P_{s1} = 0.47$	$P_{s2} = 0.43$
$N_1 = 157$	$N_2 = 113$

8.12 SW A large counseling center needs to evaluate several experimental programs. Write a paragraph summarizing the results of these tests. Did the new programs work?

a. One program is designed for divorce counseling; the key feature of the program is its counselors, who are married couples working in teams. About half of all clients have been randomly assigned to this special program and half to the regular program, and the proportion of cases that eventually ended in divorce was recorded for both. The results for random samples of couples from both programs are reported here. In terms of preventing divorce, did the new program work?

Sample 1 (Special Program)	Sample 2 (Regular Program)
$P_{s1} = 0.53$	$P_{s2} = 0.59$
$N_1 = 78$	$N_2 = 82$

b. The agency is also experimenting with peer counseling for depressed children. About half of all clients were randomly assigned to peer counseling. After the program had run for a year, a random sample of children from the new program was compared with a random sample of children who did not receive peer counseling. In terms of the percentage of children who were judged to be "much improved," did the new program work?

Sample 1 (Peer Counseling)	Sample 2 (No Peer Counseling)
$P_{s1} = 0.10$	$P_{s2} = 0.15$
$N_1 = 52$	$N_2 = 56$

8.13 SOC At St. Algebra College, the sociology and psychology departments have been feuding for years about the respective quality of their programs. In an attempt to resolve the dispute, you have gathered data about the graduate school experience of random samples of both groups of majors. The results are presented here: the proportion of majors who applied to graduate schools, the proportion of majors accepted into their preferred programs, and the proportion of these who completed their programs. As measured by these data, is there a significant difference in program quality?

a. Proportion of majors who applied to graduate school:

Sample 1 (Sociology)	Sample 2 (Psychology)
$P_{s1} = 0.53$	$P_{s2} = 0.40$
$N_1 = 150$	$N_2 = 175$

b. Proportion accepted by program of first choice:

Sample 1 (Sociology)	Sample 2 (Psychology)
$P_{s1} = 0.75$	$P_{s2} = 0.85$
$N_1 = 80$	$N_2 = 70$

c. Proportion completing the programs:

Sample 1 (Sociology)	Sample 2 (Psychology)
$P_{s1} = 0.75$	$P_{s2} = 0.69$
$N_1 = 60$	$N_2 = 60$

8.14 CJ The mayor of Sin City, Nevada, started a "crimeline" program some years ago and wonders if it is really working. The program publicizes unsolved violent crimes in the local media and offers cash rewards for information leading to arrests. Are "featured" crimes more likely to be cleared by arrest than other violent crimes? Results from random samples of both types of crimes are reported as follows:

Sample 1 (Crimeline Crimes Cleared by Arrest)	Sample 2 (Non-Crimeline Crimes Cleared by Arrest)
$P_{s1} = 0.35$	$P_{s2} = 0.25$
$N_1 = 178$	$N_2 = 212$

8.15 SOC Some results from a survey administered to a nationally representative sample are reported here in terms of differences by sex. Which of these differences, if any, are significant? Write a sentence or two of interpretation for each test.

a. Proportion favoring the legalization of marijuana:

Sample 1 (Males)	Sample 2 (Females)
$P_{s1} = 0.37$	$P_{s2} = 0.31$
$N_1 = 202$	$N_2 = 246$

b. Proportion strongly agreeing that "kids are life's greatest joy":

Sample 1 (Males)	Sample 2 (Females)
$P_{s1} = 0.47$	$P_{s2} = 0.58$
$N_1 = 251$	$N_2 = 351$

c. Proportion voting for President Obama in 2008:

Sample 1 (Males)	Sample 2 (Females)
$P_{s1} = 0.59$	$P_{s2} = 0.47$
$N_1 = 399$	$N_2 = 509$

d. Average hours spent with e-mail each week:

Sample 1 (Males)	Sample 2 (Females)
$\bar{X}_1 = 4.18$	$\bar{X}_2 = 3.38$
$s_1 = 7.21$	$s_2 = 5.92$
$N_1 = 431$	$N_2 = 535$

e. Average rate of church attendance: (number of times per year)

Sample 1 (Males)	Sample 2 (Females)
$\bar{X}_1 = 3.19$	$\bar{X}_2 = 3.99$
$s_1 = 2.60$	$s_2 = 2.72$
$N_1 = 641$	$N_2 = 808$

f. Number of children:

Sample 1 (Males)	Sample 2 (Females)
$\bar{X}_1 = 1.49$	$\bar{X}_2 = 1.93$
$s_1 = 1.50$	$s_2 = 1.50$
$N_1 = 635$	$N_2 = 803$

8.16 SOC A researcher is studying the effects of the college experience on attitudes, values, and behaviors and is comparing random samples of freshmen and seniors at the same university. Which of the following differences are significant?

a. Proportion calling parents at least once a day:

Freshmen	Seniors
$P_{s1} = 0.24$	$P_{s2} = 0.08$
$N_1 = 153$	$N_2 = 117$

b. Proportion with a "close friend" from another racial or ethnic group:

Freshmen	Seniors
$P_{s1} = 0.15$	$P_{s2} = 0.26$
$N_1 = 155$	$N_2 = 114$

c. Proportion saying that they are "very interested" in national politics:

Freshmen	Seniors
$P_{s1} = 0.46$	$P_{s2} = 0.48$
$N_1 = 147$	$N_2 = 111$

d. Average score on a scale that measures political ideology. On this scale, 10 means "very conservative" and 1 means "very liberal."

Freshmen	Seniors
$\bar{X}_1 = 5.23$	$\bar{X}_2 = 5.12$
$s_1 = 1.78$	$s_2 = 1.07$
$N_1 = 145$	$N_2 = 105$

e. Average score on a scale that measures support for traditional gender roles. This scale ranges from 7 ("Very traditional") to 1 ("Very nontraditional").

Freshmen	Seniors
$\bar{X}_1 = 2.07$	$\bar{X}_2 = 2.17$
$s_1 = 1.23$	$s_2 = 0.78$
$N_1 = 143$	$N_2 = 111$

YOU ARE THE RESEARCHER: Gender Gaps and Support for Traditional Gender Roles

There are two projects below. The first uses *t* tests to test for significant differences between men and women on four variables of your own choosing. The second uses the **Compute** command to explore attitudes toward abortion or traditional gender roles. You are urged to complete both projects.

PROJECT 1: Exploring the Gender Gap With *t* Tests

In this enlightened age, with its heavy stress on gender equality, how many important differences persist between the sexes? In this section, you will use SPSS to conduct *t* tests with sex as the independent variable. You will select four dependent variables and test to see if there are significant differences between the sexes in the areas measured by your variables.

STEP 1: Choosing Dependent Variables

Select four variables from the 2010 GSS to serve as dependent variables. Choose *only* interval-ratio variables or ordinal variables with three or more scores or categories. As you select variables, you might keep in mind the issues at the forefront of the debate over gender equality: income, education, and other measures of equality. Or you might choose variables that relate to lifestyle choices and patterns of everyday life: religiosity, TV viewing habits, or political ideas.

List your four dependent variables in this table:

Variable	SPSS Name	What Exactly Does This Variable Measure?
1		
2		
3		
4		

STEP 2: Stating Hypotheses

For each dependent variable, state a hypothesis about the difference you expect to find between men and women. For example, you might hypothesize that men will be more liberal or women will be more educated. You can base your hypotheses on your own experiences or on the information about gender differences that you have acquired in your courses or from other sources.

Hypotheses:

1. ______________________________
2. ______________________________
3. ______________________________
4. ______________________________

STEP 3: Getting the Output

SPSS for Windows includes several tests for the significance of the difference between means. In this demonstration, we will use the **Independent-Samples T Test** to test for the significance of the difference between men and women. This test is appropriate when we are working with large samples. If there are statistically significant differences between the sample means for men and women, we can conclude that there are differences between all U.S. men and U.S. women on this variable.

Start *SPSS for Windows* and then load the 2010 GSS database. From the main menu bar, click **Analyze,** then **Compare Means**, and then **Independent-Samples T Test**. The **Independent-Samples T Test** dialog box will open with the usual list of variables on the left. Find and move the cursor over the names of the dependent variables you selected in step 1. Click the top arrow in the middle of the window to move the variable names to the **Test Variable(s)** box.

Next, find and highlight *sex* and then click the bottom arrow in the middle of the window to move *sex* to the **Grouping Variable** box. Two question marks will appear in the **Grouping Variable** box, and the **Define Groups** button will become active. SPSS needs to know which cases go in which groups, and in this case, the instructions we need to supply are straightforward. Males (indicated by a score of 1 on *sex*) go into group 1 and females (a score of 2) will go into group 2.

Click the **Define Groups** button, and the **Define Groups** window will appear. The cursor will be blinking in the box beside Group 1—SPSS is asking for the score that will determine which cases go into this group. Type a 1 in this box (for males) and then click the box next to Group 2 and type a 2 (for females). Click **Continue** to return to the **Independent-Samples T Test** window and then click **OK** to produce your output.

STEP 4: Reading the Output

To illustrate the appearance and interpretation of SPSS *t* test output, I will present a test for the significance of the gender difference in average age (note that age is not a good choice as a dependent variable; it works better as a cause than an effect). Here is what the output looks like. (*NOTE: Several columns of the output have been deleted to conserve space and improve clarity.*)

Group Statistics					
	RESPONDENT'S SEX	N	Mean	Std. Deviation	Std. Error Mean
AGE OF RESPONDENT	MALE	654	48.11	17.069	.667
	FEMALE	838	48.07	18.398	.636

		Levene's Test for Equality of Variances				
		F	Sig.	t	df	Sig. (2-tailed)
AGE OF RESPONDENT	Equal variances assumed	8.227	.004	.038	1490	.970
	Equal variances not assumed			.038	1446.337	.969

The first block of output ("Group Statistics") presents descriptive statistics. There were 654 males in the sample, and their average age was 48.11 years, with a standard deviation of 17.069. The 838 females averaged 48.07 years of age, with a standard deviation of 18.398. We can see from this output that the sample means are different and that, on average, females are slightly younger. Is the difference between the sample means significant?

The results of the test for significance are reported in the next block of output. *SPSS for Windows* does a separate test for each assumption about the population variance [see "Hypothesis Testing With Sample Means (Small Samples)" earlier in this chapter], but we will look only at the "Equal variances assumed" reported in the top row. This is the model used in this chapter.

Skip over the first columns of the output block (which reports the results of a test for equality of the population variances). In the top row, *SPSS for Windows* reports a t value (.038), the degrees of freedom (df = 1,490), and a "Sig. (2-tailed)" of .970. This last piece of information is an alpha level, except that it is the *exact* probability of getting the observed difference in sample means if only chance is operating. Thus, there is no need to look up the test statistic in a t or Z table. This value is much greater than 0.05—our usual indicator of significance. We will fail to reject the null hypothesis and conclude that the difference is not statistically significant. There is no difference in average years of age between men and women in the population.

STEP 5: Recording Your Results

Run ***t*** tests for your dependent variables and gender and then record your results in the table below. Write the SPSS variable name in the first column and then record the descriptive statistics (mean, standard deviation, and N). Next, record the results of the test of significance, using the top row ("Equal variances assumed") of the INDEPENDENT SAMPLES output box. Record the ***t*** score, the degrees of freedom (df), and whether the difference is significant at the 0.05 level. If the value of "Sig (2-tailed)" is *less than* 0.05, reject the null hypothesis and write YES in this column. If the value of "Sig (2-tailed)" is more than 0.05, fail to reject the null hypothesis and write NO in the column.

Dependent Variable		Mean	*s*	*N*	*t* score	*df*	Sig (2-tailed) < 0.05?
	Men						
	Women						
	Men						
	Women						
	Men						
	Women						
	Men						
	Women						

STEP 6: Interpreting Your Results

Summarize your findings. For each dependent variable, write:

1. At least one sentence summarizing the test in which you identify the variables being tested, the sample means for each group, N, the t score, and the significance level. In the professional research literature, you might find the results reported as: "For a sample of 1,492 respondents, there was no significant difference between the average age of men (48.11) and the average age of women (48.07)(t = 0.038, df = 1490, $p > 0.05$)."
2. A sentence relating to your hypotheses. Were they supported? How?

PROJECT 2: Using the Compute Command to Explore Gender Differences

In this project, you will use the **Compute** command, which was introduced in Chapter 4, to construct a summary scale for *either* support for legal abortion *or* support for traditional gender roles. Do these attitudes vary significantly by gender? You will also choose a second independent variable—other than gender—to test for significant differences.

STEP 1: Creating Summary Scales

To refresh your memory, we used the **Compute** command in Chapter 4 to create a summary scale (*abscale*) for attitudes toward abortion by adding the scores on the two constituent items (*abany* and *abpoor*). Remember that once created, a computed variable is added to the active file and can be used like any of the variables actually recorded in the file. If you did not save the data file with ***abscale*** included, you can quickly recreate the variable by following the instructions in Chapter 4.

The GSS data set supplied with this text also includes two variables that measure support for traditional gender roles. One of these (*fefam*) states: "It is much better for everyone involved if the man is the achiever outside the home and the woman takes care of the home and family." There are four possible responses to this item ranging from "Strongly Agree" (1) to "Strongly Disagree" (4). Note that the lowest score (1) is most supportive of traditional gender roles.

The second item (*fechld*) states: "A working mother doesn't hurt the children." This item also has four possible responses, and again, the lowest score—strongly disagree (1)—is the most consistent with support of traditional gender roles. The scores of the two items vary in the same direction (for both, the lowest score indicates support for traditional gender roles), so we can create a summary scale by simply adding the two variables together. Follow the commands in Chapter 4 to create the scale, which we will call *fescale*. The computed variable will have a total of seven possible scores, with lower scores indicating more support for traditional gender roles and higher scores indicating less support. Remember that SPSS will automatically delete any cases that are missing scores on either of the two variables that will make up *fescale*, so any statistical programs run with the scale will be likely to have fewer cases.

STEP 2: Stating a Hypothesis

Choose either scale and then state a hypothesis about the difference you expect to find between men and women. Which gender will be more supportive of legal abortion or traditional gender roles? Why?

STEP 3: Getting and Interpreting the Output

Run the **Independent-Samples T Test** as before, with the computed scale as the **Test Variable** and *sex* as the **Grouping Variable**. See the instructions for Project 1.

STEP 4: Interpreting Your Results

Summarize your results as in Project 1, step 6. Was your hypothesis confirmed? How?

Step 5: Extending the Test by Selecting an Additional Independent Variable

What other independent variable besides gender might be related to the scale you selected? Select another independent variable besides *sex* and conduct an additional *t* test with *abscale* or *fescale* as the dependent variable. Remember that the *t* test requires that the independent variable have *only* two categories. For variables with more than two categories (*relig* or *racecen1*, for example), you can meet this requirement by using the **Define Groups** button in the **Grouping Variables** box to select specific categories of a variable. For example, you could compare Protestants and Catholics on *relig* by choosing scores of 1 (Protestants) and 2 (Catholics).

STEP 6: Stating a Hypothesis

State a hypothesis about what differences you expect to find between the categories of your independent variable. Which category will be more supportive of legal abortion or more supportive of traditional gender roles? Why?

STEP 7: Getting and Interpreting the Output

Run the **Independent-Samples T Test** as before, with the scale you selected as the **Test Variable** and your independent variable as the **Grouping Variable**.

STEP 8: Interpreting Your Results

Summarize your results as in Project 1, step 6. Was your hypothesis confirmed? How?

9 Hypothesis Testing III
The Analysis of Variance

LEARNING OBJECTIVES

By the end of this chapter, you will be able to:

1. Identify and cite examples of situations in which ANOVA is appropriate.
2. Explain the logic of hypothesis testing as applied to ANOVA.
3. Perform the ANOVA test using the five-step model as a guide and then correctly interpret the results.
4. Define and explain the concepts of population variance, total sum of squares, sum of squares between, sum of squares within, and mean square estimates.
5. Explain the difference between the statistical significance and the importance of relationships between variables.

USING STATISTICS

The statistical technique presented in this chapter is used to compare differences between more than two random samples. If the differences are significant, we conclude that the populations from which the samples were selected are different. This conclusion will have a known probability of error, which is often set at the 0.05 level. Examples of situations in which the analysis of variance is appropriate include:

1. A researcher is examining the differences in support for the death penalty between people of various religious affiliations. If there are significant differences between random samples of Protestants, Catholics, and other religions, she will conclude that the respective populations have different views on this issue.
2. Random samples of freshmen, sophomores, juniors, and seniors at a large university are compared for differences in their political views and religious values. If the researcher finds significant differences between the samples, he will conclude that the populations (the classes) differ on these dimensions.
3. Is bullying more of a problem in suburban, city, or rural schools? Researchers compare random samples of each type of school to see if there are statistically significant differences.

In this chapter, we examine a very flexible and widely used test of significance called **the analysis of variance** (often abbreviated as **ANOVA**). This test is designed to be used with interval-ratio-level dependent variables and is a powerful tool for analyzing the most sophisticated and precise measurements you are likely to encounter.

It is perhaps easiest to think of ANOVA as an extension of the test for the significance of the difference between two sample means, which was presented in Chapter 8. Those tests can be used only in situations in which our independent variable has exactly two categories (e.g., Protestants and Catholics). On the other hand, the analysis of variance is appropriate for independent variables with more than two categories (e.g., Protestants, Catholics, Jews, people with no religious affiliation, and so forth).

To illustrate, suppose we were interested in analyzing support for capital punishment. Why does opinion on this issue vary from person to person? Could there be a relationship between religion (the independent variable) and support for capital punishment (the dependent variable)? The death penalty has an obvious moral dimension and may well be affected by a person's religious background.

Suppose we administered a scale that measures support for capital punishment at the interval-ratio level to a randomly selected sample that includes Protestants, Catholics, Jews, people with no religious affiliation ("Nones"), and people from other religions ("Others"). We have five categories of subjects, and we want to see if opinion varies significantly by the category (religion) into which a person is classified. We will also want to raise other issues: Which religion shows the most support for capital punishment? Are Protestants significantly more supportive than Catholics or Jews? How do people with no religious affiliation compare to other people? The analysis of variance provides a very useful statistical context in which the questions can be addressed.

The Logic of the Analysis of Variance

For ANOVA, the null hypothesis is that the populations from which the samples are drawn have the same score on the dependent variable. As applied to our problem, the null hypothesis could be phrased as "People from different religious affiliations do not vary in their support for the death penalty" or, symbolically, as $\mu_1 = \mu_2 = \mu_3 = \ldots \mu_k$. (Note that this is an extended version of the null hypothesis for the two-sample *t* test.) As usual, the researcher will normally want to reject the null hypothesis and, in this case, show that support is related to religion.

If the null hypothesis of "no difference" in the populations is true, then any means calculated from randomly selected samples should be roughly equal in value. The average score for the Protestant sample should be about the same as the average score for the Catholics, the Jews, and so forth. Note that the averages are unlikely to be exactly the same value even if the null hypothesis really is true because we will always encounter some error or chance fluctuations in the measurement process. We are *not* asking: "Are there differences between the samples (or, in our example, the religions)?" Rather, we are asking: "Are the differences between the samples large enough to reject the null hypothesis and justify the conclusion that the populations are different?"

TABLE 9.1 Support for Capital Punishment by Religion (Fictitious Data)

	Protestant	Catholic	Jew	None	Other
Mean =	10.3	11.0	10.1	9.9	10.5
Standard deviation =	2.4	1.9	2.2	1.7	2.0

TABLE 9.2 Support for Capital Punishment by Religion (Fictitious Data)

	Protestant	Catholic	Jew	None	Other
Mean =	14.7	11.3	5.7	8.3	7.1
Standard deviation =	0.9	0.8	1.0	1.1	0.7

Now consider what kinds of outcomes we might encounter if we actually administered a "Support for Capital Punishment Scale" and organized the scores by religion. Of the infinite variety of possibilities, let us focus on the two extreme outcomes presented in Tables 9.1 and 9.2. In the first set of hypothetical results (Table 9.1), we see that the means and standard deviations of the groups are quite similar. The average scores are about the same, and all five groups exhibit about the same dispersion. These results are consistent with the null hypothesis of no difference between the populations on support for capital punishment. Neither the average score nor the dispersion of the scores varies in any important way by religion.

Now consider another set of fictitious results, as displayed in Table 9.2. Here, we see substantial differences in average score, with Jews showing the lowest support and Protestants showing the highest. Also, the standard deviations are low and similar from category to category, indicating that there is not much variation within the religions. Table 9.2 shows marked differences *between* religions combined with homogeneity (or low standard deviations) *within* religions. In other words, the religions are different from each other and there is not much variation within each religion. These results would contradict the null hypothesis and support the notion that support for the death penalty does vary by religion

ANOVA proceeds by making these kinds of comparisons. The test compares the amount of variation between categories (for example, from Protestants to Catholics to Jews to "Nones" to "Others") with the amount of variation within categories (among Protestants, among Catholics, and so forth). The greater the difference *between* categories (as measured by the means) relative to the differences *within* categories (as measured by the standard deviations), the more likely that the null hypothesis of "no difference" is false and can be rejected. If support for capital punishment truly varies by religion, then the sample mean for each religion should be quite different from the others and dispersion within the categories should be relatively low.

The Computation of Anova

Even though we have been thinking of ANOVA as a test for the significance of the difference between sample means, the computational routine actually involves developing two separate estimates of the population variance σ^2

(hence, the name *analysis of variance*). Recall from Chapter 4 that the variance is the standard deviation squared. One estimate of the population variance is based on the amount of variation *within* each of the categories of the independent variable and the other is based on the amount of variation *between* categories.

Before constructing these estimates, we need to introduce some new concepts and statistics. The first new concept is the total variation of the scores, which is measured by a quantity called the **total sum of squares**, or ***SST***:

FORMULA 9.1

$$SST = \Sigma X_i^2 - N\overline{X}^2$$

To solve this formula, first find the sum of the squared scores (in other words, square each score and then add up the squared scores). Next, square the mean of all scores, multiply that value by the total number of cases in the sample (N), and the subtract that quantity from the sum of the squared scores.

Formula 9.1 may seem vaguely familiar. A similar expression—$\Sigma(X_i - \overline{X})^2$—appears in the formula for the standard deviation and variance (see Chapter 4). All three statistics incorporate information about the variation of the scores (or, in the case of *SST*, the squared scores) around the mean (or, in the case of *SST*, the square of the mean multiplied by N). In other words, all three statistics are measures of the variation, or dispersion, of the scores.

To construct the two separate estimates of the population variance, the total variation (*SST*) is divided into two components. One component reflects the pattern of variation *within* each of the categories and is called the **sum of squares within (*SSW*)**. In our example problem, *SSW* would measure the amount of variety in support for the death penalty within each of the religions.

The other component is based on the variation *between* categories and is called the **sum of squares between (*SSB*)**. Again using our example to illustrate, *SSB* measures how different people in each religion are from each other in their support for capital punishment. *SSW* and *SSB* are components of *SST*, as reflected in Formula 9.2:

FORMULA 9.2

$$SST = SSB + SSW$$

Let us with the computation of *SSB*—our measure of the variation between categories. We use the category means as summary statistics to determine the size of the difference from category to category. In other words, we compare the average support for the death penalty for each religion with the average support for all religions combined to determine *SSB*. The formula for the sum of squares between (*SSB*) is:

FORMULA 9.3

$$SSB = \Sigma N_k(\overline{X}_k - \overline{X})^2$$

where:

SSB = the sum of squares between the categories
N_k = the number of cases in a category
$\overline{X}_k$ = the mean of a category

To find *SSB*, subtract the overall mean of all scores $(\overline{X})$ from each category mean $(\overline{X}_k)$, square the difference, multiply by the number of cases in the category, and then add the results across all the categories.

The second estimate of the population variance (*SSW*) is based on the amount of variation within the categories. Formula 9.2 shows that the total sum of squares (*SST*) is equal to the addition of *SSW* and *SSB*. This relationship provides us with an easy method for finding *SSW* by simple subtraction. Formula 9.4 rearranges the symbols in Formula 9.2.

FORMULA 9.4

$$SSW = SST - SSB$$

Let us pause for a second to remember what we are after here. If the null hypothesis is *true*, then there should not be much variation from category to category (see Table 9.1) and *SSW* and *SSB* should be roughly equal. If the null hypothesis is *not true*, there will be large differences between categories (see Table 9.2) relative to the differences within categories and *SSB* should be much larger than *SSW*. *SSB* will increase as the differences *between* category means increases, especially when there is not much variation *within* the categories (*SSW*). The larger *SSB* is compared to *SSW*, the more likely we are to reject the null hypothesis.

The next step in the computational routine is to construct the estimates of the population variance. To do this, we divide each sum of squares by its respective degrees of freedom. To find the degrees of freedom associated with *SSW*, subtract the number of categories (k) from the number of cases (N). The degrees of freedom associated with *SSB* are the number of categories minus 1. In summary:

FORMULA 9.5

$$dfw = N - k$$

where:

dfw = degrees of freedom associated with *SSW*
N = total number of cases
k = number of categories

FORMULA 9.6

$$dfb = k - 1$$

where:

dfb = degrees of freedom associated with *SSB*
k = number of categories

The actual estimates of the population variance—called the **mean square estimates**—are calculated by dividing each sum of squares by its respective degrees of freedom:

FORMULA 9.7

$$\text{Mean square within} = \frac{SSW}{dfw}$$

FORMULA 9.8

$$\text{Mean square between} = \frac{SSB}{dfb}$$

The test statistic calculated in step 4 of the five-step model is called the *F ratio*, and its value is determined by this formula:

FORMULA 9.9

$$F = \frac{\text{Mean square between}}{\text{Mean square within}}$$

ONE STEP AT A TIME **Computing ANOVA**

To compute ANOVA, we will use Formulas 9.1, 9.3, and 9.4 to find *SST*, *SSB*, and *SSW*. Then, we will calculate the degrees of freedom, mean square estimates of the population variance, and the obtained *F* ratio. I strongly recommend you use a computing table like Table 9.3 to organize the computations.

Step	***Operation***
To Find SST by Using Formula 9.1	
1.	Find ΣX_i^2 by squaring each score and adding all the squared scores together.
2.	Find $N\overline{X}_i^2$ by squaring the value of the mean and then multiplying by *N*.
3.	Subtract the quantity you found in step 2 from the quantity you found in step 1.
To Find SSB by Using Formula 9.3	
1.	Subtract the mean of all scores ($\overline{X}$) from the mean of each category ($\overline{X}_k$) and then square each difference.
2.	Multiply each of the squared differences you found in step 1 by the number of cases in the category N_k.
3.	Add the quantities you found in step 2.
To Find SSW by Using Formula 9.4	
1.	Subtract the value of *SSB* from the value of *SST*.
To Calculate Degrees of Freedom	
1.	For *dfw*, subtract the number of categories (*k*) from the number of cases (*N*).
2.	For *dfb*, subtract 1 from the number of categories (*k*).
To Construct the Two Mean Square Estimates of the Population Variance	
1.	To find the *MSW* estimate, divide *SSW* by *dfw*.
2.	To find the *MSB* estimate, divide *SSB* by *dfb*.
To Find the Obtained F Ratio	
1.	Divide the *MSB* estimate by the *MSW* estimate.

As you can see, the value of the *F* ratio will be a function of the amount of variation between categories (based on *SSB*) to the amount of variation within the categories (based on *SSW*). The greater the variation between the categories relative to the variation within, the higher the value of the *F* ratio and the more likely we will reject the null hypothesis.

An Example of Computing the Analysis of Variance Assume we have administered our Support for Capital Punishment Scale to a sample of 20 individuals who are equally divided into the five religions. (Obviously, this sample is much too small for any serious research and is intended solely for purposes of illustration.) All scores are reported in Table 9.3, along with the other quantities needed to complete the computations. The scores (X_i) are listed for each of the five religions, and a column has been added for the squared scores (X_i^2). The sums of X_i and X_i^2 are reported at the bottom of each column. The category means ($\overline{X}_k$) show that the four Protestants averaged 12.5 on the Support for Capital Punishment Scale, the four Catholics averaged 21.0, and so forth. Finally, the overall mean (sometimes called the *grand mean*) is reported in the bottom row of the table. This shows that all 20 respondents averaged 16.6 on the scale.

To organize our computations, we will follow the routine summarized in the "One Step at a Time" box. We begin by finding *SST* by means of Formula 9.1:

$$SST = \Sigma X^2 - N\overline{X}^2$$
$$SST = (666 + 1898 + 1{,}078 + 1794 + 712) - (20)(16.6)^2$$
$$SST = (6{,}148) - (20)(275.56)$$
$$SST = 6{,}148 - 5{,}511.2$$
$$SST = 636.80$$

The sum of squares between (*SSB*) is found by means of Formula 9.3:

$$SSB = \Sigma N_k(\overline{X}_k - \overline{X})^2$$
$$SSB = 4(12.5 - 16.6)^2 + 4(21.0 - 16.6)^2 + 4(16.0 - 16.6)^2 + 4(20.5 - 16.6)^2 + 4(13.0 - 16.6)^2$$
$$SSB = 67.24 + 77.44 + 1.44 + 60.84 + 51.84$$
$$SSB = 258.80$$

Now *SSW* can be found by subtraction (Formula 9.4):

$$SSW = SST - SSB$$
$$SSW = 636.8 - 258.80$$
$$SSW = 378.00$$

To find the degrees of freedom for the two sums of squares, we use Formulas 9.5 and 9.6:

$$dfb = N - k = 20 - 5 = 15$$
$$dfb = k - 1 = 5 - 1 = 4$$

Finally, we are ready to construct the mean square estimates of the population variance. For the estimate based on *SSW*, we use Formula 9.7:

$$\text{Mean square within} = \frac{SSW}{dfw} = \frac{378.00}{15} = 25.20$$

TABLE 9.3 Support for Capital Punishment by Religion (Fictitious Data)

Protestant		Catholic		Jewish		None		Other	
X_i	X_i^2	X_i	X_i^2	X_i	X_i^2	X_i	X_i^2	X_i	X_i^2
8	64	12	144	12	144	15	225	10	100
12	144	20	400	13	169	16	256	18	324
13	169	25	625	18	324	23	529	12	144
17	289	27	729	21	441	28	784	12	144
50	666	84	1,898	64	1,078	82	1,794	52	712
$\overline{X}_k = 12.5$		$\overline{X}_k = 21.0$		$\overline{X}_k = 16.0$		$\overline{X}_k = 20.5$		$\overline{X}_k = 13.0$	
				$\overline{X} = 16.6$					

For the between estimate, we use Formula 9.8:

$$\text{Mean square between} = \frac{SSB}{dfw} = \frac{258.80}{4} = 64.70$$

The test statistic, or F ratio, is found by means of Formula 9.9:

$$F = \frac{\text{Mean square between}}{\text{Mean square within}} = \frac{64.70}{25.20} = 2.57$$

This statistic must still be evaluated for its significance. *(Solve any of the end-of-chapter problems to practice computing these quantities and solving these formulas.)*

A Test of Significance for ANOVA

Now we will see how to test an F ratio for significance. We will also take a look at some of the assumptions underlying the ANOVA test. As usual, we will follow the five-step model as a convenient way of organizing the decision-making process.

Step 1. Making assumptions and meeting test requirements.

Model: Independent random samples
Level of measurement is interval-ratio
Populations are normally distributed
Population variances are equal

The model assumptions are quite strict and underscore the fact that ANOVA should be used only with dependent variables that have been carefully and precisely measured. However, as long as the categories are roughly equal in size, ANOVA can tolerate some violation of the model assumptions. In situations where you are uncertain or have samples of very different size, it is probably advisable to use an alternative test. (Chi square in Chapter 10 is one option.)

Step 2. Stating the null hypothesis. For ANOVA, the null hypothesis always states that the means of the populations from which the samples were drawn are equal. For our example problem, we are concerned with five different populations, or categories, so our null hypothesis would be:

$$H_0: \mu_1 = \mu_2 = \mu_3 = \mu_4 = \mu_5$$

where μ_1 represents the mean for Protestants, μ_2 the mean for Catholics, and so forth.

The alternative hypothesis states simply that at least one of the population means is different. The wording here is important. If we reject the null hypothesis, ANOVA does not identify which mean or means are significantly different.

(H_1: At least one of the population means is different.)

Step 3. Selecting the sampling distribution and establishing the critical region. The sampling distribution for ANOVA is the F distribution, which is summarized in Appendix D. Note that there are separate tables for alphas of .05 and .01.

As with the t table, the value of the critical F score will vary by degrees of freedom. For ANOVA, there are two separate degrees of freedom—one for each estimate of the population variance. The numbers across the top of the table are the degrees of freedom associated with the between estimate (*dfb*), and the numbers down the side of the table are those associated with the within estimate (*dfw*). In our example, *dfb* is $(k - 1)$, or 4, and *dfw* is $(N - k)$, or 15 (see Formulas 9.5 and 9.6). Thus, if we set alpha at .05, our critical F score will be 3.06.

Summarizing these considerations:

$$\begin{aligned} \text{Sampling distribution} &= F \text{ distribution} \\ \text{Alpha} &= 0.05 \\ \text{Degrees of freedom within } (dfw) &= (N - k) = 15 \\ \text{Degrees of freedom between } (dfb) &= (k - 1) = 4 \\ F(\text{critical}) &= 3.06 \end{aligned}$$

Take a moment to inspect the two F tables and you will notice that all the values are greater than 1.00. This is because ANOVA is a one-tailed test and we are concerned only with outcomes in which there is more variance between categories than within categories. F values of less than 1.00 would indicate that the between estimate was lower in value than the within estimate, and because we would always fail to reject the null in such cases, we simply ignore this class of outcomes.

Step 4. Computing the test statistic. This was done in the previous section, where we found an obtained F ratio of 2.57.

Step 5. Making a decision and interpreting the results of the test. Compare the test statistic with the critical value:

$$\begin{aligned} F(\text{critical}) &= 3.06 \\ F(\text{obtained}) &= 2.57 \end{aligned}$$

Because the test statistic does not fall into the critical region, our decision would be to fail to reject the null. Support for capital punishment does not differ significantly by religion, and the variation we observed in the sample means is unimportant.

Applying Statistics 9.1 The Analysis of Variance

An experiment in teaching introductory biology was recently conducted at a large university. One section was taught by the traditional lecture-lab method, a second was taught by an all-lab/demonstration approach with no lectures, and a third was taught entirely by a series of recorded lectures and demonstrations that the students were free to view at any time and as often as they wanted. Students were randomly assigned to each of the three sections, and at the end of the semester, random samples of final exam scores were collected from each section. Is there a significant difference in student performance by teaching method?

Applying Statistics 9.1 *(continued)*

Final Exam Scores by Teaching Method

	Lecture		Demonstration		Recording	
	X	X^2	X	X^2	X	X^2
	55	3,025	56	3,136	50	2,500
	57	3,249	60	3,600	52	2,704
	60	3,600	62	3,844	60	3,600
	63	3,969	67	4,489	61	3,721
	72	5,184	70	4,900	63	3,969
	73	5,329	71	5,041	69	4,761
	79	6,241	82	6,724	71	5,041
	85	7,225	88	7,744	80	6,400
	92	8,464	95	9,025	82	6,724
$\Sigma X =$	636		651		588	
$\Sigma X^2 =$		46,286		48,503		39,420
$\bar{X}_k =$	70.67		72.33		65.33	

$$\bar{X} = 1{,}875/27 = 69.44$$

We can see by inspection that the "Recording" group had the lowest average score and that the "Demonstration" group had the highest average score. The ANOVA test will tell us if these differences are large enough to justify the conclusion that they did not occur by chance alone. We can organize the computations following the steps described in the "One Step at a Time" box about computing ANOVA (see page 229):

$$SST = \Sigma X^2 - N\bar{X}^2$$
$$SST = (46{,}286 + 48{,}503 + 39{,}420) - 27(69.44)^2$$
$$SST = 134{,}209 - 130{,}191.67$$
$$SST = 4017.33$$

$$SSB = \Sigma N_k(\bar{X}_k - \bar{X})^2$$
$$SSB = (9)(70.67 - 69.44)^2 + (9)(72.33 - 69.44)^2 + (9)(65.33 - 69.44)^2$$
$$SSB = 13.62 + 75.17 + 152.03$$
$$SSB = 240.82$$

$$SSW = SST - SSB$$
$$SSW = 4017.33 - 240.82$$
$$SSW = 3776.51$$

$$dfw = N - k = 27 - 3 = 24$$
$$dfb = k - 1 = 3 - 1 = 2$$

$$\text{Mean square within} = \frac{SSW}{dfw} = \frac{3776.51}{24} = 157.36$$

$$\text{Mean square between} = \frac{SSB}{dfb} = \frac{240.82}{2} = 120.41$$

$$F = \frac{\text{Mean square between}}{\text{Mean square within}}$$
$$F = \frac{120.41}{157.36}$$
$$F = 0.77$$

We can now conduct the test of significance.

Step 1. Making Assumptions and Meeting Test Requirements.

Model: Independent random sample
Level of measurement is interval-ratio
Populations are normally distributed
Population variances are equal

Step 2. Stating the Null Hypothesis.

H_0: $\mu_1 = \mu_2 = \mu_3$

H_1: At least one of the population means is different,

Step 3. Selecting the Sampling Distribution and Establishing the Critical Region.

Sampling distribution = F distribution
Alpha = 0.05
Degrees of freedom (within) = $(N - k)$ = $(27 - 3) = 24$
Degrees of freedom (between) = $(k - 1)$ = $(3 - 1) = 2$
F(critical) = 3.40

Step 4. Computing the Test Statistics. We found an obtained F ratio of 0.77.

Step 5. Making a Decision and Interpreting the Results of the Test. Compare the test statistic with the critical value:

$$F(\text{critical}) = 3.40$$
$$F(\text{obtained}) = 0.77$$

We would clearly fail to reject the null hypothesis and conclude that the observed differences among the category means were the results of random chance. Student performance in this course does not vary significantly by teaching method.

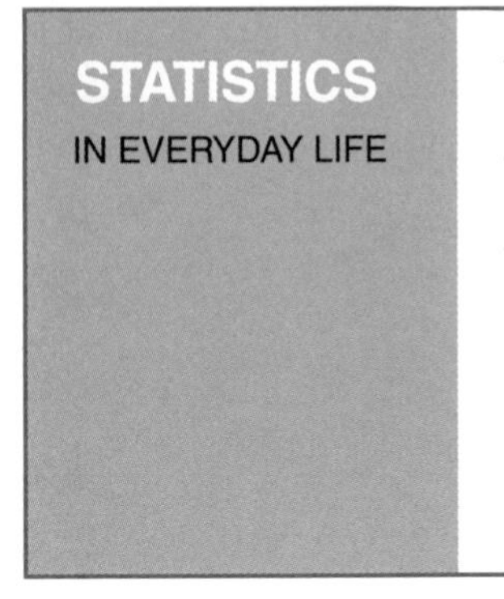

Want to stay slim and trim? A study that followed over 34,000 women for 15 years is not particularly encouraging: On average, the women in the sample put on pounds during the study. However, exercise did help to limit the weight gain. The women were divided into three groups, and those who exercised at the highest levels gained significantly fewer pounds than the women in the moderate and low exercise groups. This study used sophisticated statistics not covered in this text to analyze the data, but the statistical significance of the differences in mean weight gain for the three groups was established by an ANOVA technique.

Source: Lee, I., Djousse, L., Sesso, H., Wang, L., and Buring, J. 2010. *Physical Activity and Weight Gain Prevention. JAMA*: March 24–31, pp. 1173–1179.

An Additional Example of Computing and Testing the Analysis of Variance

In this section, we work through an additional example of the computation and interpretation of the ANOVA test. We first review matters of computation, find the obtained *F* ratio, and then test the statistic for its significance.

A researcher has been asked to evaluate the efficiency with which each of three social service agencies is administering a particular program. One area of concern is the speed of the agencies in processing paperwork and determining the eligibility of potential clients. The researcher has gathered information on the number of days required for processing a random sample of 10 cases in each agency. Is there a significant difference between the agencies? The data are reported in Table 9.4, which also includes some additional information we will need to complete our calculations.

To find *SST* by means of formula 9.1:

$$SST = \Sigma X^2 - N\overline{X}^2$$
$$SST = (524 + 1816 + 2{,}462) - 30(11.67)^2$$
$$SST = 4{,}802 - 30(136.19)$$
$$SST = 4{,}802 - 4{,}085.70$$
$$SST = 716.30$$

To find *SSB* by means of Formula 9.3:

$$SSB = \Sigma N_k(\overline{X}_k - \overline{X})^2$$
$$SSB = (10)(7.0 - 11.67)^2 + (10)(13.0 - 11.67)^2 + (10)(15.0 - 11.67)^2$$
$$SSB = (10)(21.81) + (10)(1.77) + (10)(11.09)$$
$$SSB = 218.10 + 17.70 + 110.90$$
$$SSB = 346.70$$

TABLE 9.4 Number of Days Required to Process Cases for Three Agencies (Fictitious Data)

	Agency A		Agency B		Agency C	
Client	X	X^2	X	X^2	X	X^2
1	5	25	12	144	9	81
2	7	49	10	100	8	64
3	8	64	19	361	12	144
4	10	100	20	400	15	225
5	4	16	12	144	20	400
6	9	81	11	121	21	441
7	6	36	13	169	20	400
8	9	81	14	196	19	361
9	6	36	10	100	15	225
10	6	36	9	81	11	121
$\Sigma X =$	70		130		150	
$\Sigma X^2 =$		524		1,816		2,462
$\bar{X}_k =$	7.0		13.0		15.0	

$$\bar{X} = \frac{350}{30} = 11.67$$

Now we can find *SSW* by using Formula 9.4:

$$SSW = SST - SSB$$
$$SSW = 716.30 - 346.70$$
$$SSW = 369.60$$

The degrees of freedom are found via Formulas 9.5 and 9.6:

$$dfw = N - k = 30 - 3 = 27$$
$$dfb = k - 1 = 3 - 1 = 2$$

The estimates of the population variances are found by means of Formulas 9.7 and 9.8:

$$\text{Mean square within} = \frac{SSW}{dfw} = \frac{369.60}{27} = 13.69$$

$$\text{Mean square between} = \frac{SSW}{dfw} = \frac{346.70}{2} = 173.35$$

The *F* ratio (Formula 9.9) is:

$$F(\text{obtained}) = \frac{\text{Mean square between}}{\text{Mean square within}}$$

$$F(\text{obtained}) = \frac{173.35}{13.69}$$

$$F(\text{obtained}) = 12.66$$

We can now test this value for its significance.

Step 1. Making assumptions and meeting test requirements.

Model: Independent random samples
Level of measurement is interval-ratio
Populations are normally distributed
Population variances are equal

The researcher will always be in a position to judge the adequacy of the first two assumptions in the model. The second two assumptions are more problematical, but remember that ANOVA will tolerate some deviation from its assumptions as long as the sample sizes are roughly equal.

Step 2. Stating the null hypothesis.

$$H_0: \mu_1 = \mu_2 = \mu_3$$

(H_1: At least one of the population means is different.)

Step 3. Selecting the sampling distribution and establishing the critical region.

$$\text{Sampling distribution} = F \text{ distribution}$$
$$\text{Alpha} = 0.05$$
$$\text{Degrees of freedom (within)} = (N - k) = (30 - 3) = 27$$
$$\text{Degrees of freedom (between)} = (k - 1) = (3 - 1) = 2$$
$$F(\text{critical}) = 3.35$$

Step 4. Computing the test statistic. We found an obtained F ratio of 12.66.

Step 5. Making a decision and interpreting the results of the test. Compare the test statistic with the critical value:

$$F(\text{critical}) = 3.35$$
$$F(\text{obtained}) = 12.66$$

The test statistic is in the critical region, and we would reject the null hypothesis of no difference. The differences between the three agencies are very unlikely to have occurred by chance alone. The agencies are significantly different in the speed with which they process paperwork and determine eligibility. *(For practice in conducting the ANOVA test, see problems 9.2 to 9.10. Begin with the lower-numbered problems because they have smaller data sets, fewer categories, and, therefore, the simplest calculations.)*

Applying Statistics 9.2 The Analysis of Variance

A random sample of 15 nations from three levels of development has been selected. "Low Income" nations are largely agricultural and have the lowest quality of life. "High Income" nations are industrial and the most affluent and modern. "Middle Income" nations are between these extremes. Are these general characteristics

Applying Statistics 9.2 *(continued)*

reflected in differences in life expectancy (the number of years the average citizen can expect to live at birth) between the three categories? The data for the 15 nations are for 2009.

Low Income		Middle Income		High Income	
Nation	Life Expectancy	Nation	Life Expectancy	Nation	Life Expectancy
Cambodia	62	China	74	Australia	82
Malawi	54	Indonesia	71	Canada	81
Nepal	68	Pakistan	65	Japan	83
Niger	55	Colombia	73	Poland	76
Ethiopia	56	Turkey	73	United Kingdom	80

Source: Population Reference Bureau. 2011 World Population Data Sheet. Available at http://prb.org/pdf11/2011population-data-sheet_eng.pdf.

To find F(obtained) and conduct the ANOVA test, computations will be organized into table format:

Life Expectancy by Income Level

	X_i	X_i^2	X_i	X_i^2	X_i	X_i^2
	62	3,844	74	5,476	82	6,724
	54	2,916	71	5,041	81	6,561
	68	4,624	65	4,225	83	6,889
	55	3,025	73	5,329	76	5,776
	56	3,136	73	5,329	80	6,400
$\Sigma X =$	295		356		402	
$\Sigma X^2 =$		17,545		25,400		32,350
$\bar{X}_k =$	59.00		71.20		80.40	

$$\bar{X} = 70.20$$

The ANOVA test will tell us if these differences are large enough to justify the conclusion that they did not occur by chance alone. Following the usual computational routine:

$$SST = \Sigma X^2 - N\bar{X}^2$$
$$SST = (17{,}545 + 25{,}400 + 32{,}350) - 15(70.20)^2$$
$$SST = 75{,}295 - 73{,}920.60$$
$$SST = 1{,}374.40$$

$$SSB = \Sigma N_k(\bar{X}_k - \bar{X})^2$$
$$SSB = (5)(59.00 - 70.20)^2 + (5)(71.20 - 70.20)^2 + (5)(80.40 - 70.20)^2$$
$$SSB = 627.20 + 5.00 + 520.20$$
$$SSB = 1{,}152.40$$

$$SSW = SST - SSB$$
$$SSW = 1{,}374.40 - 1{,}152.40$$
$$SSW = 222.00$$

$$dfw = N - k = 15 - 3 = 12$$
$$dfb = k - 1 = 3 - 1 = 2$$

$$\text{Mean square within} = \frac{SSW}{dfw} = \frac{222.00}{12} = 18.50$$

$$\text{Mean square between} = \frac{SSB}{dfb} = \frac{1{,}152.40}{2} = 576.20$$

$$F = \frac{MSB}{MSW} = \frac{576.20}{18.50} = 31.15$$

We can now conduct the test of significance.

Step 1. Making assumptions and meeting test requirements.

Model: Independent random samples
Level of measurement is interval-ratio
Populations are normally distributed
Population variances are equal

Step 2. Stating the null hypothesis.

$$H_0: \mu_1 = \mu_2 = \mu_3$$

(H_1: At least one of the population means is different.)

(continued next page)

Applying Statistics 9.2 *(continued)*

Step 3. Selecting the sampling distribution and establishing the critical region.

$$\text{Sampling distribution} = F \text{ distribution}$$
$$\text{Alpha} = 0.05$$
$$dfw = 12$$
$$dfb = 2$$
$$F(\text{critical}) = 3.88$$

Step 4. Computing the test statistic. We found an obtained F ratio of 31.15.

Step 5. Making a decision and interpreting the results of the test. Compare the test statistic with the critical value:

$$F(\text{critical}) = 3.88$$
$$F(\text{obtained}) = 31.15$$

The null hypothesis ("The population means are equal") can be rejected. The differences in life expectancy between nations at different income levels are statistically significant.

The Limitations of the Test

ANOVA is appropriate whenever you want to test differences between the means of an interval-ratio-level dependent variable across three or more categories of an independent variable. This application is called **one-way analysis of variance** because it involves the effect of a single variable (for example, religion) on another (for example, support for capital punishment). This is the simplest application of ANOVA, and you should be aware that the technique has numerous more advanced and complex forms. For example, you may encounter research projects in which the effects of two separate variables (for example, religion and gender) on some third variable were observed.

One important limitation of ANOVA is that it requires an interval-ratio dependent variable and roughly equal numbers of cases in each category of the independent variable. The former condition may be difficult to meet with complete confidence for many variables of interest to the social sciences. The latter condition may create problems when the research hypothesis calls for comparisons between groups that are, by their nature, unequal in numbers (for example, white versus black Americans) and may call for some unusual sampling schemes in the data-gathering phase of a research project. Neither of these limitations should be particularly crippling because ANOVA can tolerate some deviation from its model assumptions, but you should be aware of these limitations in planning your own research as well as in judging the adequacy of research conducted by others.

A second limitation of ANOVA actually applies to all forms of significance testing and was introduced in Chapter 8. Tests of significance are designed to detect nonrandom differences or differences so large that they are very unlikely to be produced by random chance alone. The problem is that differences that are statistically significant are not necessarily important in any other sense. Parts III and IV of this text provide some statistical techniques that can assess the importance of results directly.

A final limitation of ANOVA relates to the research hypothesis. This hypothesis is not specific; it simply asserts that at least one of the population means is different from the others. Obviously, when we reject the null hypothesis, we would like to know *which* differences between the sample means are significant. We can

BECOMING A CRITICAL CONSUMER: Reading the Professional Literature

It is unlikely that you would encounter a report using ANOVA in everyday life or in the popular media, and I will confine this section to the professional research literature. As I have pointed out previously, reports about tests of significance in social science research journals may be short on detail, but you will still be able to locate all the essential information needed to understand the results of the test.

We can use a recent research article to illustrate how to read ANOVA results in the professional literature. Several researchers were concerned with several competing theories of the relationship between religion and sexual behavior. One theory—the secularization hypothesis—argues that the modern reliance on reason and science has softened the relationship between religious belief and sexual experience. The researchers used a large sample of university students and some of their results (for females only) are presented below. They developed a variety of categories to indicate religious affiliation, including "Spiritual" (which included a variety of modern "New Age" belief systems) and "Monotheist Christian" (which included all mainstream Protestants and Catholics).

The entries in the table are mean scores, and note that they do not vary much across the religious affiliations. This impression is verified by the low *F* ratios in the right-hand column—none of which are statistically significant. These results are broadly consistent with the secularization hypothesis: The participants' professed religious views seem to have little effect on their sexual behaviors. Want to learn more? The citation is given below.

Female Sexual Behavior (means) by Religious Affiliation

	Spiritual	Agnostic	Atheist	None	Monotheist Christian	Fundament-alist	*F* Score
Age at first sexual intercourse	17.33	17.11	17.67	16.38	16.56	16.69	1.66
Number of sex partners in past year	1.50	2.00	2.13	1.89	1.93	1.52	1.48
Times cheated on partner	1.50	1.62	1.44	1.71	1.40	1.31	1.29

Based on Table 2, p. 859.

Farmer, Melissa, Trapnell, Paul, & Meston, Cindy. 2010. "The Relation Between Sexual Behavior and Religiosity Subtypes: A Test of the Secularization Hypothesis." *Archives of Sexual Behavior* 38: 852–865.

sometimes make this determination by simple inspection. For example, in our problem involving social service agencies, it is pretty clear from Table 9.4 that Agency A is the source of most of the differences. However, this informal "eyeball" method can be misleading, and you should exercise caution in drawing conclusions.

SUMMARY

1. One-way analysis of variance is a powerful test of significance that is commonly used when comparisons across more than two categories or samples are of interest. It is perhaps easiest to conceptualize ANOVA as an extension of the test for the difference in sample means.
2. ANOVA compares the amount of variation within categories to the amount of variation between categories. If the null hypothesis of no difference is false, there should be relatively great variation between categories and relatively little variation within categories. The greater the differences from category to category relative to the differences within the categories, the more likely we will be able to reject the null hypothesis.
3. The computational routine for even simple applications of ANOVA can quickly become quite complex. The basic process is to construct separate estimates of the population

variance based on the variation within the categories and the variation between the categories. The test statistic is the F ratio, which is based on a comparison of these two estimates. This is probably an appropriate time to mention the widespread availability of statistical packages, such as SPSS—the purpose of which is to perform complex calculations such as these accurately and quickly. If you have not yet learned how to use such programs, ANOVA may provide you with the necessary incentive.

4. The ANOVA test can be organized into the familiar five-step model for testing the significance of sample outcomes. Although the model assumptions (step 1) require high-quality data, the test can tolerate some deviation as long as the sample sizes are roughly equal. The null hypothesis takes the familiar form of stating that there is no difference of any importance among the population values, while the alternative hypothesis asserts that at least one population mean is different. The sampling distribution is the F distribution, and the test is always one-tailed. The decision to reject or to fail to reject the null is based on a comparison of the obtained F ratio with the critical F ratio as determined for a given alpha level and degrees of freedom. The decision to reject the null hypothesis indicates only that one or more of the population means is different from the others. We can often determine which of the sample means accounts for the difference by inspecting the sample data, but this informal method should be used with caution.

SUMMARY OF FORMULAS

FORMULA 9.1 Total sum of squares: $SST = \Sigma X^2 - N\overline{X}^2$

FORMULA 9.2 The two components of the total sum of squares: $SST = SSB + SSW$

FORMULA 9.3 Sum of squares between: $SSB = \Sigma N_k(\overline{X}_k - \overline{X})^2$

FORMULA 9.4 Sum of squares within: $SSW = SST - SSB$

FORMULA 9.5 Degrees of freedom for SSW: $dfw = N - k$

FORMULA 9.6 Degrees of freedom for SSB: $dfb = k - 1$

FORMULA 9.7 Mean square within: $\text{Mean square within} = \dfrac{SSW}{dfw}$

FORMULA 9.8 Mean square between: $\text{Mean square between} = \dfrac{SSB}{dfb}$

FORMULA 9.9 F ratio: $F = \dfrac{\text{Mean square between}}{\text{Mean square within}}$

GLOSSARY

Analysis of variance. A test of significance appropriate for situations in which we are concerned with the differences among more than two sample means.

ANOVA. See Analysis of variance.

***F* ratio.** The test statistic computed in step 4 of the ANOVA test.

Mean square estimate. An estimate of the variance calculated by dividing the sum of squares within (SSW) or the sum of squares between (SSB) by the proper degrees of freedom.

One-way analysis of variance. Applications of ANOVA in which the effect of a single independent variable on a dependent variable is observed.

Sum of squares between (*SSB*). The sum of the squared deviations of the sample means from the overall mean, weighted by the sample size.

Sum of squares within (*SSW*). The sum of the squared deviations of scores from the category means.

Total sum of squares (*SST*). The sum of the squared deviations of the scores from the overall mean.

PROBLEMS

NOTE: The number of cases in these problems is very low—a fraction of the sample size necessary for any serious research—in order to simplify computations.

9.1 Conduct the ANOVA test for each of the following sets of scores. *(Hint: Follow the computational shortcut outlined in the One Step at a Time Box on p. 229, and keep track of all sums and means by constructing computational tables like Table 9.3 or 9.4.)*

a.

Category		
A	B	C
5	10	12
7	12	16
8	14	18
9	15	20

b.

Category		
A	B	C
1	2	3
10	12	10
9	2	7
20	3	14
8	1	1

c.

Category			
A	B	C	D
13	45	23	10
15	40	78	20
10	47	80	25
11	50	34	27
10	45	30	20

9.2 SOC The Society for Ethical Culture provides nontheistic services in a congregational setting. Until three years ago, the membership of the Persepolis New York Society grew steadily every year, but then it leveled off. The leaders have therefore decided to launch a membership drive, but before they begin, they would like to find out what types of people are most likely to join. A random sample of 15 existing members have been asked how many times they attended services over the past 5 weeks and to provide some other information. Which differences among the members are significant?

a. Attendance by education:

Less Than High School	High School	College
0	1	0
1	3	3
2	3	4
3	4	4
4	5	4

b. Attendance by length of residence in present community:

Less Than 2 Years	2–5 Years	More Than 5 Years
0	0	1
1	2	3
3	3	3
4	4	4
4	5	4

c. Attendance by extent of television watching:

Little or None	Moderate	High
0	3	4
0	3	4
1	3	4
1	3	4
2	4	5

d. Attendance by number of children:

None	One Child	More Than One Child
0	2	0
1	3	3
1	4	4
3	4	4
3	4	5

9.3 SOC In a local community, a random sample of 18 couples has been assessed on a scale that measures the extent to which power and decision making are shared (lower scores) or monopolized by one party (higher scores) and marital happiness (higher scores indicate higher levels of happiness). The couples were also classified by type of relationship: traditional (only the husband works outside the home), dual-career (both parties work), and cohabitational (parties living together but not legally married, regardless of work patterns). Does decision making or happiness vary significantly by type of relationship?

a. Decision Making

Traditional	Dual Career	Cohabitational
7	8	2
8	5	1
2	4	3
5	4	4
7	5	1
6	5	2

b. Happiness

Traditional	Dual Career	Cohabitational
10	12	12
14	12	14
20	12	15
22	14	17
23	15	18
24	20	22

9.4 CJ Two separate afterschool programs for at-risk teenagers have been implemented in the city of Redland. One is a big brother/big sister mentoring program in which adult volunteers are paired with troubled teenagers. The second involves extracurricular team sports competitions. In terms of the percentage reduction in crimes reported to the police over a one-year period, were the programs successful? The results are for random samples of 18 neighborhoods drawn from the entire city.

Mentoring Program	Competition Program	No Program
−10	−21	+30
−20	−15	−10
+10	−80	+14
+20	−10	+80
+70	−50	+50
+10	−10	−20

9.5 Are sexually active teenagers any better informed about AIDS and other potential health problems related to sex than teenagers who are sexually inactive? A 15-item test of general knowledge about sex and health was administered to random samples of teens who are sexually inactive, teens who are sexually active but with only a single partner ("going steady"), and teens who are sexually active with more than one partner. Is there any significant difference in the test scores?

Inactive	Active—One Partner	Active—More Than One Partner
10	11	12
12	11	12
8	6	10
10	5	4
8	15	3
5	10	15

9.6 SOC Does the rate of voter turnout vary significantly by the type of election? A random sample of voting precincts displays the following pattern of voter turnout by election type. Assess the results for significance.

Local Only	State	National
33	35	42
78	56	40
32	35	52
28	40	66
10	45	78
12	42	62
61	65	57
28	62	75
29	25	72
45	47	51
44	52	69
41	55	59

9.7 GER Do older citizens lose interest in politics and current affairs? A brief quiz on recent headline stories was administered to random samples of respondents from each of four different age groups. Is there a significant difference? The following data represent the number of correct responses.

High School (15–18)	Young Adult (21–30)	Middle Aged (30–55)	Retired (65+)
0	0	2	5
1	0	3	6
1	2	3	6
2	2	4	6
2	4	4	7
2	4	5	7
3	4	6	8
5	6	7	10
5	7	7	10
7	7	8	10
7	7	8	10
9	10	10	10

9.8 SOC A small random sample of respondents has been selected from the General Social Survey database. Each respondent has been classified as either a city dweller, a suburbanite, or a rural dweller. Are there statistically significant differences by place of residence for any of the variables listed here?

a. Occupational Prestige

Urban	Suburban	Rural
32	40	30
45	48	40
42	50	40
47	55	45
48	55	45
50	60	50
51	65	52
55	70	55
60	75	55
65	75	60

b. Number of Children

Urban	Suburban	Rural
1	0	1
1	1	4
0	0	2
2	0	3
1	2	3
0	2	2
2	3	5
2	2	0
1	2	4
0	1	6

c. Family Income (See *income06* in Appendix G for coding scheme)

Urban	Suburban	Rural
5	6	5
7	8	5
8	11	11
11	12	10
8	12	9
9	11	6
8	11	10
3	9	7
9	10	9
10	12	8

d. Church Attendance (See *attend* in Appendix G for coding scheme)

Urban	Suburban	Rural
0	0	1
7	0	5
0	2	4
4	5	4
5	8	0
8	5	4
7	8	8
5	7	8
7	2	8
4	6	5

e. Hours of TV Watching per Day

Urban	Suburban	Rural
5	5	3
3	7	7
12	10	5
2	2	0
0	3	1
2	0	8
3	1	5
4	3	10
5	4	3
9	1	1

9.9 SOC Does support for suicide ("death with dignity") vary by social class? Is this relationship different in different nations? Small samples in three nations were asked if it is ever justified for a person with an incurable disease to take his or her own life. Respondents answered in terms of a 10-point scale, on which 10 was "always justified" (the strongest support for "death with dignity") and 1 was "never justified" (the lowest level of support). Results are reported here.

Mexico			
Lower Class	Working Class	Middle Class	Upper Class
5	2	1	2
2	2	1	4
4	1	3	5
5	1	4	7
4	6	1	8
2	5	2	10
3	7	1	10
1	2	5	9
1	3	1	8
3	1	1	8

Canada			
Lower Class	Working Class	Middle Class	Upper Class
7	5	1	5
7	6	3	7
6	7	4	8
4	8	5	9
7	8	7	10
8	9	8	10
9	5	8	8
9	6	9	5
6	7	9	8
5	8	5	9

United States			
Lower Class	Working Class	Middle Class	Upper Class
4	4	4	1
5	5	6	5
6	1	7	8
1	4	5	9
3	3	8	9
3	3	9	9
3	4	9	8
5	2	8	6
3	1	7	9
6	1	2	9

9.10 SOC A researcher is studying the effects of the college experience on attitudes, values, and behaviors and is comparing small random samples from each class at the same university. Which of the following differences are significant?

a. Number of "close friends" from another racial or ethnic group:

Class			
Freshman	Sophomore	Junior	Senior
7	0	10	0
2	3	9	2
5	4	8	3
12	2	5	1
9	7	1	1
1	2	8	4
0	0	7	5
3	3	12	7
4	4	10	10
6	2	11	9

b. Score on a scale that measures political ideology. On this scale, 10 means "very conservative" and 1 means "very liberal."

Class			
Freshman	Sophomore	Junior	Senior
7	7	9	1
7	3	8	1
5	7	5	5
10	8	0	6
9	5	1	5
7	4	1	3
6	1	2	5
5	2	3	6
4	4	5	1
3	2	6	5

c. Number of minutes spent in the library each day studying:

Class			
Freshman	Sophomore	Junior	Senior
0	30	45	60
30	90	45	0
45	120	90	0
60	100	180	120
90	75	145	45
45	60	100	45
30	30	90	30
15	45	110	60
0	20	120	60
30	45	110	180

YOU ARE THE RESEARCHER: Why Are Some People Liberal (or Conservative)?

This project investigates political ideology using *polviews*—a seven-point scale that measures how liberal or conservative a person is—as the dependent variable. This variable is ordinal in level of measurement but will be treated as interval-ratio for this test.

Before you begin the project, read Demonstrations 1 and 2, in which I show you how to conduct an ANOVA test with SPSS and introduce you to a new SPSS command that you will need to complete the project. The new SPSS command—called **Recode**—enables us to collapse scores on a variable. This operation is extremely useful because it allows us to change a variable to fit a particular task.

Demonstration 1: Recoding Variables

Here, I will demonstrate how we can take a variable such as *age*, which has a wide range of scores (respondents in the 2010 GSS ranged in age from 18 to over 89), and transform it into a variable with just a few scores, which we can use as an independent variable in ANOVA. We will create a new version of *age* that has three categories. When we are finished, we will have two versions of the same variable in the data set: the original interval-ratio version with age measured in years and a new ordinal-level version with collapsed categories. If we wish, the new version of age can be added to the permanent data file and used in the future.

The decision to use three categories for *age* is arbitrary, and we could easily have decided on four, five, or even six categories for the new recoded independent variable. If we find that we are unhappy with the three-category version of the variable, we can always recode the original version of *age* into a different set of categories.

We will collapse the values of *age* into three broad categories with a roughly equal number of cases in each category. How can we define these new categories? Begin by running the **Frequencies** command for *age* to inspect the distribution of the scores. I used the cumulative percent column of the frequency distribution to find ages that divided the variable into thirds. I found that 33.4% of the 2010 GSS sample was younger than 37 and that 67.2% were younger than 56. I decided to use these ages as the cutting point for the new categories, as summarized below:

Ages	Percent of Sample
18–37	33.4%
38–56	33.8%
57–89	32.8%

To recode *age* into these categories, follow these steps:

1. In the **SPSS Data Editor** window, click **Transform** from the menu bar and then click **Recode into Different Variables**.
2. The **Recode into Different Variables** window will open. A box containing an alphabetical list of variables will appear on the left. Use the cursor to highlight *age* and then click on the arrow button to move the variable to the **Input Variable → Output Variable** box. The input variable is the old version of age, and the output variable is the new recoded version we will soon create.

3. In the **Output Variable** box on the right, click in the **Name** box and then type a name for the new (output) variable. I suggest *ager* ("**age r**ecoded") for the new variable, but you can assign any name as long as it does not duplicate the name of some other variable in the data set and is no longer than eight characters. Click the **Change** button and the expression *age* → *ager* will appear in the **Input Variable → Output Variable** box.
4. Click on the **Old and New Values** button in the middle of the screen and a new dialog box will open. Read down the left-hand column until you find the **Range** button. Click on the button and then click on the first window below the **Range** command. Type 18 (the youngest age in the sample) in this window and then click in the second window and type 37.
5. In the **New Value** box in the upper right-hand corner of the screen, click the **Value** button if it is not already highlighted. Type 1 in the **Value** dialog box and then click the **Add** button directly below. The expression 18–37 → 1 will appear in the **Old → New** dialog box.
6. Return to the **Range** dialog boxes. Type 38 in the upper box and 56 in the lower box and then click the **Value** button in the **New Values** box. Type 2 in the **Value** dialog box and then click the **Add** button. The expression 38–56 → 2 appears in the **Old → New** dialog box.
7. Finish by returning to the **Range** dialog box and entering the values 57 and 89 in the correct boxes. Click the **Value** button in the **New Values** box. Type 3 in the **Value** dialog box and then click the **Add** button. The expression 56–89 → 3 appears in the **Old → New** dialog box.
8. Click the **Continue** button at the bottom of the screen and you will return to the **Recode into Different Variables** dialog box. Click **OK**, and SPSS will execute the transformation.

You now have a data set with an additional variable named *ager* (or whatever name you gave the recoded variable). You can find the new variable in the last column to the right in the data editor window. You can make the new variable a permanent part of the data set by saving the data file at the end of the session. If you do not wish to save the new expanded data file, click **No** when you are asked if you want to save the data file. If you are using the student version of *SPSS for Windows*, remember that you are limited to a maximum of 50 variables and you will not be able to save the new variable.

Demonstration 2: Using ANOVA to Analyze the Effect of Age on Number of Sex Partners (*partnrs5*)

To demonstrate how to run ANOVA with SPSS, We will conduct a test with recoded age as the independent variable and *partnrs5*—a measure of sexual activity that asks how many different sexual partners the respondent has had over the past five years—as the dependent variable. Note the coding scheme for *partnrs5* (see Appendix G or click **Utilities → Variables** on the SPSS menu bar). The first five scores (zero partners to four partners) are actual numbers, but the higher scores represent broad categories (e.g., "5" means the respondent had between 5 and 10 different partners). This variable is a combination of interval-ratio and ordinal scoring, and we will have to treat the means with some caution.

SPSS provides several different ways of conducting the analysis of variance test. The procedure summarized here is the most accessible of these, but it still incorporates options and capabilities that we have not covered in this chapter. If you wish to explore these possibilities, please use the online **Help** facility.

To use the ANOVA procedure, click **Analyze, Compare Means**, and then **One-way Anova**. The **One-way Anova** window appears. Find *partnrs5* in the variable list on the left and click the arrow to move the variable name into the **Dependent List** box. Note that you can request more than one dependent variable at a time. Next, find the name of the recoded *age* variable (perhaps called *ager?*) and then click the arrow to move the variable name into the **Factor** box.

Click **Options** and then click the box next to **Descriptive** in the **Statistics** box to request means and standard deviations along with the analysis of variance. Click **Continue** and then click **OK** to produce this output:

Descriptives

	N	Mean	Std. Deviation	Std. Error
1.00	455	2.70	2.100	.098
2.00	455	1.74	1.704	.080
3.00	418	1.03	1.405	.069
Total	1328	1.84	1.894	.052

Note: This table has been edited and will not look exactly like the raw SPSS output.

ANOVA
How Many Sex Partners R Had In Last 5 Years

	Sum of Squares	df	Mean Square	F	Sig.
Between Groups	615.452	2	307.726	98.409	.000
Within Groups	4143.282	1325	3.127		
Total	4758.734	1327			

The output box labeled **Descriptives** presents the category means and, not surprisingly, shows that the youngest respondents (category 1 or people 18 to 37 years old) had the highest average number of different partners. The oldest respondents (category 3) had the lowest number of different sex partners and the overall or grand mean was 1.84 (see the Total row).

The output box labeled **ANOVA** includes the various degrees of freedom, all the sums of squares, the "Mean Square" estimates, the *F* ratio (98.409), and, at the far right, the exact probability ("Sig.") of getting these results if the null hypothesis is true. This is reported as .000—much lower than our usual alpha level of .05. The differences in *partnrs5* for the various age groups are statistically significant.

Your turn.

PROJECT: Political Ideology (*polviews*)

In this project, you will investigate political ideology by using scores on *polviews*—a seven-point scale that measures how liberal or conservative a person is—as the dependent variable. This variable is ordinal in level of measurement but will be treated as interval-ratio for this test.

STEP 1: Choosing Independent Variables

Select three variables from the 2010 GSS to serve as independent variables. What factors might help to explain why some people are more liberal and some are more conservative? Choose *only* independent variables with three to six scores or categories. Among other possibilities, you might consider education (use *degree*), religious denomination, age (use the recoded version), or social class. Use the recode command to collapse independent variables with more than six categories into three or four categories. In general, variables that measure characteristics or traits (like gender or race) will work better than those that measure attitude or opinion (like *cappun* or *gunlaw*), which are more likely to be manifestations of political ideology, not causes.

List your independent variables in this table:

Variable	SPSS Name	What Exactly Does This Variable Measure?
1		
2		
3		

STEP 2: Stating Hypotheses

For each independent variable, state a hypothesis about its relationship with *polviews*. For example, you might hypothesize that people with more education will be more liberal or that the more religious will be more conservative. You can base your hypotheses on your own experiences or on information you have acquired in your courses.

1. ____________________
2. ____________________
3. ____________________

STEP 3: Getting and Reading the Output

We discussed these tasks in the previous demonstration.

STEP 4: Recording Results

Record the results of your ANOVA tests in the following table by using as many rows for each independent variable as necessary. Write the SPSS variable name in the first column and then write the names of the categories of that independent variable in the next column. Record the descriptive statistics (mean, standard deviation, and *N*) in the designated columns. Write the value of the *F* ratio, and in the far right-hand column, indicate whether the results are significant at the 0.05 level. If the value in the "Sig." column of the ANOVA output is less than 0.05, write YES in this column. If the value in the "Sig." column of the ANOVA output is more than 0.05, write NO in this column. Finally, for each independent variable, record the grand or overall means, standard deviations, and sample sizes in the row labeled "Totals = ."

Independent Variables:	Categories	Mean	Std. Dev.	*N*	*F* ratio	Sig. at .05 level?
1. ________	1.				_____	_____
	2.					
	3.					
	4.					
	5.					
	6.					
	Totals =					
2. ________	1.				_____	_____
	2.					
	3.					
	4.					
	5.					
	6.					
	Totals =					
3. ________	1.				_____	_____
	2.					
	3.					
	4.					
	5.					
	6.					
	Totals =					

STEP 6: Interpreting your Results

Summarize your findings. For each test, write:

1. At least one sentence summarizing the test in which you identify the variables being tested, the sample means for each group, *N*, the *F* ratio, and the significance level. In the professional research literature, you might find the results reported as: "For a sample of 1,417 respondents, there was no significant difference between the average age of Southerners (46.50), Northerners (44.20), Midwesterners (47.80), and Westerners (47.20) ($F = 1.77$, $p > 0.05$)."
2. A sentence relating to your hypotheses. Were they supported? How?

10 Hypothesis Testing IV
Chi Square

LEARNING OBJECTIVES

By the end of this chapter, you will be able to:

1. Identify and cite examples of situations in which the chi square test is appropriate.
2. Explain the structure of a bivariate table and the concept of independence as applied to expected and observed frequencies in a bivariate table.
3. Explain the logic of hypothesis testing as applied to a bivariate table.
4. Perform the chi square test using the five-step model and correctly interpret the results.
5. Explain the limitations of the chi square test and, especially, the difference between statistical significance and importance.

USING STATISTICS

The statistical techniques presented in this chapter are used to test the statistical significance of the relationship between variables that have been arrayed in table format. If the patterns are significantly different from what would be expected by random chance, we conclude that the variables are related in the population. This conclusion will have a known probability of error, which is often set at the 0.05 level. Examples of situations in which the chi square test is appropriate include:

1. A researcher is studying membership in voluntary associations and hypothesizes that unmarried people will be more involved because they have fewer family obligations and more free time. She gathers a random sample from her community and finds a significant relationship between rates of membership and marital status in the sample. She concludes that the variables are related in the population.

2. Medical researchers wonder if a particular weight loss plan will work equally well with both sexes and people of all ages. A sample of volunteers is recruited from a particular community to try the plan. The sample is divided into six groups by gender and age (e.g., males aged 25–40, females over 60, and so forth). After two months on the program, subjects are scored as "no weight loss," "minor weight loss," and "major weight loss" and the six gender/age groups are compared. Results will be generalized to the community.

3. A researcher wonders about the sources of support for restrictive immigration policies and uses a nationally representative U.S. sample to test the relationships between a variety of variables (including gender, occupation, and education) and support for the idea that immigration should be reduced. If the researcher finds significant relationships in the sample, he will conclude that those variables are related in the population (all adult Americans).

The **chi square (χ^2) test** has probably been the most frequently used test of hypothesis in the social sciences—a popularity that is due largely to the fact that the assumptions and requirements in step 1 of the five-step model are easy to satisfy. Specifically, the test can be conducted with variables measured at the nominal level (the lowest level of measurement), and because it is a **nonparametric**, or "distribution free," test, chi square requires no assumption at all about the shape of the population or sampling distribution.

Why is it an advantage to have easy-to-satisfy assumptions and requirements? The decision to reject the null hypothesis (step 5) is not specific; it means only that one statement in the model (step 1) *or* the null hypothesis (step 2) is wrong. Usually, of course, we single out the null hypothesis for rejection. The more certain we are of the model, the greater our confidence that the null hypothesis is the faulty assumption. A "weak" or easily satisfied model means that our decision to reject the null hypothesis can be made with even greater certainty.

Chi square has also been popular for its flexibility. Not only can it be used with variables at any level of measurement, but it also can be used with variables that have many categories or scores. For example, in Chapter 8, we tested the significance of the difference in the proportions of black and white citizens who were "highly participatory" in voluntary associations. What if the researcher wished to expand the test to include Hispanic and Asian Americans? The two-sample test would no longer be applicable, but chi square handles the more complex variable easily. Also, unlike the ANOVA test presented in Chapter 10, the chi square test can be conducted with variables at any level of measurement.

The Bivariate Table

Chi square is computed from **bivariate tables**—so called because they display the scores of cases on two different variables at the same time. Bivariate tables are used to ascertain if there is a significant relationship between the variables and for other purposes that we will investigate in later chapters. In fact, these tables are very commonly used in research and a detailed examination of them is in order.

First of all, bivariate tables have (of course) two dimensions. We refer to the horizontal dimension (across) in terms of **rows** and the vertical dimension (up and down) in terms of **columns**. Each column or row represents a score on a variable, and the intersections of the rows and columns (**cells**) represent the combined scores on both variables.

Let us use an example to clarify. Suppose a researcher is interested in senior citizens and wants to see if their participation in voluntary groups, community-service organizations, and so forth, is affected by their marital status. To simplify the analysis, the researcher has confined the sample to people who are presently married or not married (including people who are single and divorced) and has measured involvement as a simple dichotomy on which people are classified as either high or low.

By convention, the independent variable (the variable that is taken to be the cause) is placed in the columns and the dependent variable in the rows. In the example at hand, marital status is the causal variable (the question was "Is membership *affected by* marital status?") and each column will represent a score on this variable. On the other hand, each row will represent a score on level of involvement (high or low). Table 10.1 displays the outline of the bivariate table for a sample of 100 senior citizens.

Note some further details of the table. First, subtotals have been added to each column and row. These are called the row or column **marginals**, and in this case, they tell us that 50 members of the sample are married and 50 are not married (the column marginals), while 50 are high in participation and 50 are rated low (the row marginals). Second, the total number of cases in the sample ($N = 100$) is reported at the intersection of the row and column marginals. Finally, take careful note of the labeling of the table. Each row and column is identified, and the table has a descriptive title that includes the names of the variables, with the dependent variable listed first. Clear, complete labels and concise titles must be included in *all* tables, graphs, and charts.

As you have noticed, Table 10.1 lacks one piece of crucial information: the numbers in the body of the table. To finish the table, we need to classify the marital status and level of participation of each member of the sample, keep count of how often each combination of scores occurs, and record these numbers in the appropriate cell of the table. Because each variable has two scores, four combinations of scores are possible—each corresponding to a cell in the table. For example, married people with high levels of participation would be counted in the upper left-hand cell, nonmarried people with low levels of participation would be counted in the lower right-hand cell, and so forth. When we are finished counting, each cell will display the number of times each *combination* of scores occurred.

Finally, note how we could expand the bivariate table to accommodate variables with more scores. If we wished to include people with other marital statuses (widowed, separated, and so forth), we would simply add columns. More elaborate dependent variables could also be easily accommodated. If we had measured participation rates with three categories (e.g., high, moderate, and low) rather than two, we would simply add a row to the table.

TABLE 10.1 Rates of Participation in Voluntary Associations by Marital Status for 100 Senior Citizens

	Marital Status		
Participation Rates	Married	Not Married	
High			50
Low			50
	50	50	100

The Logic of Chi Square

The chi square test has several different uses, but we will cover only the *chi square test for independence*. We have encountered the term *independence* in connection with the requirements for the two-sample case (Chapter 8) and for the ANOVA test (Chapter 9). In the context of chi square, the concept of **independence** refers to the relationship between the variables, not between the samples. Two variables are independent if the classification of a case into a particular category of one variable has no effect on the probability that the case will fall into any particular category of the second variable. For example, marital status and participation in voluntary associations would be independent of each other if the classification of a case as married or not married has no effect on the classification of the case as high or low on participation. In other words, the variables are independent if level of participation and marital status are completely unrelated to each other.

Consider Table 10.1 again. If the variables are independent, the cell frequencies will be determined solely by random chance and we would find that just as an honest coin will show heads about 50% of the time, about half the married respondents will rank high on participation and half will rank low. The same pattern would hold for the 50 nonmarried respondents. In fact, each of the four cells would have 25 cases in it, as illustrated in Table 10.2. This pattern of cell frequencies indicates that marital status has no effect on the probability that a person would be either high or low in participation. The probability of being classified as high or low would be 0.50 for both marital statuses, and the variables would therefore be independent.

The null hypothesis for chi square is that the variables are independent. Under the assumption that the null hypothesis is true, the cell frequencies we would expect to find if only random chance were operating are computed. These frequencies—called **expected frequencies** (symbolized f_e)—are then compared, cell by cell, with the frequencies actually observed in the table (**observed frequencies**, symbolized f_o). If the null hypothesis is true and the variables are independent, then there should be little difference between the expected and observed frequencies. However, if the null hypothesis is false, there should be large differences between the two. The greater the differences between expected (f_e) and observed (f_o) frequencies, the less likely that the variables are independent and the more likely that we will be able to reject the null hypothesis.

TABLE 10.2 The Cell Frequencies That Would Be Expected If Rates of Participation and Marital Status Were Independent

	Marital Status		
Participation Rates	Married	Not Married	
High	25	25	50
Low	25	25	50
	50	50	100

The Computation of Chi Square

As with all tests of hypothesis, with chi square, we compute a test statistic—χ^2 **(obtained)**—from the sample data and then place that value on the sampling distribution of all possible sample outcomes. Specifically, χ^2(obtained) will be compared with the value of χ^2**(critical)** that will be determined by consulting a chi square table (Appendix C) for a particular alpha level and degrees of freedom. Prior to conducting the formal test of hypothesis, let us take a moment to consider the calculation of chi square, as defined by Formula 10.1:

FORMULA 10.1

$$\chi^2\,(\text{obtained}) = \Sigma\frac{(f_o - f_e)^2}{f_e}$$

where: f_o = the cell frequencies observed in the bivariate table
f_e = the cell frequencies that would be expected if the variables were independent

We must work cell by cell to solve this formula. To compute chi square, subtract the expected frequency from the observed frequency for each cell, square the result, divide by the expected frequency for that cell, and then sum the resultant values for all cells.

This formula requires an expected frequency for each cell in the table. In Table 10.2, the marginals are the same value for all rows and columns and the expected frequencies are obvious by intuition: $f_e = 25$ for all four cells. In the more usual case, the expected frequencies will not be obvious, marginals will be unequal, and we must use Formula 10.2 to find the expected frequency for each cell:

FORMULA 10.2

$$f_e = \frac{\text{Row marginal} \times \text{Column marginal}}{N}$$

That is, the expected frequency for any cell is equal to the total number of cases in the row (the row marginal) multiplied by the total number of cases in the column (the column marginal) divided by the total number of cases in the table (N).

A Computational Example

An example using Table 10.3 should clarify these procedures. A random sample of 100 social work majors has been classified in terms of whether the Council on Social Work Education has accredited their undergraduate programs (the column, or independent, variable) and whether they were hired in social work positions within three months of graduation (the row, or dependent, variable).

Beginning with the upper left-hand cell (graduates of accredited programs who are working as social workers), the expected frequency for this cell—using Formula 10.2—is $(40 \times 55)/100$, or 22. For the other cell in this row (graduates of nonaccredited programs who are working as social workers), the expected frequency is $(40 \times 45)/100$, or 18. For the two cells in the bottom row, the expected frequencies are $(60 \times 55)/100$, or 33, and $(60 \times 45)/100$, or 27, respectively. The expected frequencies for all four cells are displayed in Table 10.4.

TABLE 10.3 Employment of 100 Social Work Majors by Accreditation Status of Undergraduate Program

	Accreditation Status		
Employment Status	Accredited	Not Accredited	Totals
Working as a social worker	30	10	40
Not working as a social worker	25	35	60
Totals	55	45	100

TABLE 10.4 Expected Frequencies for Table 10.3

	Accreditation Status		
Employment Status	Accredited	Not Accredited	Totals
Working as a social worker	22	18	40
Not working as a social worker	33	27	60
Totals	55	45	100

TABLE 10.5 Computational Table for Table 10.3

(1)	(2)	(3)	(4)	(5)
f_o	f_e	$f_o - f_e$	$(f_o - f_e)^2$	$(f_o - f_e)^2/f_e$
30	22	8	64	2.91
10	18	–8	64	3.56
25	33	–8	64	1.94
35	27	8	64	2.37
$N = 100$	$N = 100$	0	χ^2(obtained) =	10.78

Note that the row and column marginals as well as the total number of cases in Table 10.4 are exactly the same as those in Table 10.3. The row and column marginals for the expected frequencies must *always* equal those of the observed frequencies—a relationship that provides a convenient way of checking your arithmetic to this point.

The value for chi square for these data can now be found by solving Formula 10.1. It will be helpful to use a computing table, such as Table 10.5, to organize the several steps required to compute chi square. The table lists the observed frequencies (f_o) in column 1 in order from the upper left-hand cell to the lower right-hand cell, moving left to right across the table and top to bottom. The second column lists the expected frequencies (f_e) in exactly the same order. Double-check to make sure you have listed the cell frequencies in the same order for both of these columns.

The next step is to subtract the expected frequency from the observed frequency for each cell and then list these values in the third column. To complete the fourth column, square the value in column 3; then, in column 5, divide the column 4 value by the expected frequency for that cell. Finally, add up column 5. The sum of this column is χ^2(obtained): χ^2(obtained) $= 10.78$.

ONE STEP AT A TIME Computing Chi Square

Begin by preparing a computing table similar to Table 10.5. List the observed frequencies (f_o) in column 1. The total for column 1 is the number of cases (N).

Step	***Operation***
To Find the Expected Frequencies (f_e) by Using Formula 10.2	
1.	Start with the upper left-hand cell and multiply the row marginal by the column marginal for that cell.
2.	Divide the quantity you found in step 1 by N. The result is the expected frequency (f_e) for that cell. Record this value in the second column of the computing table. Make sure you place the value of f_e in the same row as the observed frequency for that cell.
3.	Repeat steps 1 and 2 for each cell in the table. Double-check to make sure you are using the correct row and column marginals. Record each f_e in the second column of the computational table.
4.	Find the total of the expected frequencies column. This total *must* equal the total of the observed frequencies column (which is the same as N). If the two totals do not match (within rounding error), you have made a mistake and need to check your computations.
To Find Chi Square by Using Formula 10.1	
1.	For each cell, subtract the expected frequency (f_e) from the observed frequency (f_o) and then list these values in the third column of the computational table ($f_o - f_e$). Find the total for this column. If this total is not zero, you have made a mistake and need to check your computations.
2.	Square each of the values in the third column of the table and then record the result in the fourth column, labeled $(f_o - f_e)^2$.
3.	Divide each value in column 4 by the expected frequency for that cell and then record the result in the fifth column, labeled $(f_o - f_e)^2/f_e$.
4.	Find the total for the fifth column. This value is χ^2(obtained).

Note that the totals for columns 1 and 2 (f_o and f_e) are exactly the same. This will always be the case, and if the totals do not match, you have made a computational error—probably in the calculation of the expected frequencies. Also note that the sum of column 3 will always be zero—another convenient way to check your math to this point.

This sample value for chi square must still be tested for its significance. *(For practice in computing chi square, see any problem at the end of this chapter.)*

The Chi Square Test for Independence

We are now ready to conduct the chi square test for independence. Recall that if the variables are independent, the score of a case on one variable will have no relationship with its score on the other variable. As always, the five-step model for significance testing will provide the framework for organizing our decision making. The data presented in Table 10.3 will serve as our example.

Step 1. Making Assumptions and Meeting Test Requirements. Note that we make no assumptions at all about the shape of the sampling distribution.

Model: Independent random samples
Level of measurement is nominal

Step 2. Stating the Null Hypothesis. The null hypothesis states that the two variables are independent. If the null hypothesis is true, the differences between the observed and expected frequencies will be small. As usual, the research hypothesis directly contradicts the null hypothesis. Thus, if we reject H_0, the research hypothesis will be supported:

H_0: The two variables are independent.

(H_1: The two variables are dependent.)

Step 3. Selecting the Sampling Distribution and Establishing the Critical Region. The sampling distribution of sample chi squares, unlike the *Z* and *t* distributions, is positively skewed, with higher values of sample chi squares in the upper tail of the distribution (to the right). Thus, with the chi square test, the critical region is established in the upper tail of the sampling distribution.

Values for χ^2(critical) are given in Appendix C. This table is similar to the *t* table, with alpha levels arrayed across the top and degrees of freedom down the side. However, a major difference is that degrees of freedom (*df*) for chi square are found by this formula:

FORMULA 10.3

$$df = (r - 1)(c - 1)$$

A table with two rows and two columns (a 2 × 2 table) has one degree of freedom regardless of the number of cases in the sample.[1] A table with two rows and three columns would have (2 − 1)(3 − 1), or two degrees of freedom. Our sample problem involves a 2 × 2 table with $df = 1$, so if we set alpha at 0.05, the critical chi square score would be 3.841. Summarizing these decisions, we have:

$$\begin{aligned} \text{Sampling distribution} &= \chi^2 \text{ distribution} \\ \text{Alpha} &= 0.05 \\ \text{Degrees of freedom} &= 1 \\ \chi^2(\text{critical}) &= 3.841 \end{aligned}$$

[1]Degrees of freedom are the number of values in a distribution that are free to vary for any particular statistic. A 2 × 2 table has one degree of freedom because for a given set of marginals, once one cell frequency is determined, all other cell frequencies are fixed and no longer free to vary. For example, in Table 10.3, if any cell frequency is known, all others are determined. If the upper left-hand cell is 30, the other cell in that row must be 10 because there are 40 cases in the row and 40 − 30 = 10. Once the cell frequencies in the top row are known, those in the bottom row can be found by subtraction from the column marginal. Incidentally, this relationship can be used to quickly compute expected frequencies. In a 2 × 2 table, only one expected frequency needs to be computed. All others can be found by subtraction.

Step 4. Computing the Test Statistic. The mechanics of these computations were introduced earlier. As you recall, we had:

$$\chi^2(\text{obtained}) = \Sigma \frac{(f_o - f_e)^2}{f_e}$$

$$\chi^2(\text{obtained}) = 10.78$$

Step 5. Making a Decision and Interpreting the Results of the Test. Comparing the test statistic with the critical region,

$$\chi^2(\text{obtained}) = 10.78$$
$$\chi^2(\text{critical}) = 3.841$$

we see that the test statistic falls into the critical region; therefore, we reject the null hypothesis of independence. The pattern of cell frequencies observed in Table 10.3 is unlikely to have occurred by chance alone. The variables are dependent. Specifically, based on these sample data, the probability of securing employment in the field of social work is dependent on the accreditation status of the program. *(For practice in conducting and interpreting the chi square test for independence, see problems 10.2 to 10.16.)*

Let us stress exactly what the chi square test does and does not tell us. A significant chi square means that the variables are (probably) dependent on each other in the population; accreditation status makes a difference in whether a person is working as a social worker. What chi square does *not* tell us is the exact nature of the relationship. In our example, it does not tell us which type of graduate is more likely to be working as a social worker. To make this determination, we must perform some additional calculations. We can figure out how the independent variable (accreditation status) is affecting the dependent variable (employment as a social worker) by computing **column percentages** or by calculating percentages within each column of the bivariate table. This procedure is analogous to calculating percentages for frequency distributions (see Chapter 2).

To calculate column percentages, divide each cell frequency by the total number of cases in the column (the column marginal) and then multiply the result by 100. For Table 10.3, starting in the upper left-hand cell, we see that there are 30 cases in this cell and 55 cases in the column. Thus, 30 of the 55 graduates of accredited programs are working as social workers. The column percentage for this cell is therefore $(30/55) \times 100 = 54.55\%$. For the lower left-hand cell, the column percentage is $(25/55) \times 100 = 45.45\%$. For the two cells in the right-hand column (graduates of nonaccredited programs), the column percentages are $(10/45) \times 100 = 22.22$ and $(35/45) \times 100 = 77.78$. Table 10.6 displays all column percentages for Table 10.3.

Column percentages allow us to examine the bivariate relationship in more detail and show us exactly how the independent variable affects the dependent variable. In this case, they reveal that graduates of accredited programs are more likely to find work as social workers. We explore column percentages more extensively when we discuss bivariate association in Chapters 11 and 12.

Column percentages make the relationship between the variables more obvious, and we can easily see from Table 10.6 that it is the students from accredited programs who are more likely to be working as social workers. Nearly 55% of these students are working as social workers vs. about 22% of the students from

TABLE 10.6 Column Percentages for Table 10.3

Employment Status	Accreditation Status: Accredited	Not Accredited	Totals
Working as a social worker	54.55%	22.22%	40.00%
Not working as a social worker	45.45%	77.78%	60.00%
Totals	100.00%	100.00%	100.00%
	(55)	(45)	

nonaccredited programs. We already know that this relationship is significant (unlikely to be caused by random chance). With the aid of column percentages, we know how the two variables are related. According to these results, graduates from accredited programs have a decided advantage in securing social work jobs.

ONE STEP AT A TIME Computing Column Percentages

Step	*Operation*
1.	Start with the upper left-hand cell. Divide the cell frequency (the number of cases in the cell) by the total number of cases in that column (or the column marginal). Multiply the result by 100 to convert to a percentage.
2.	Move down one cell and then repeat step 1. Continue moving down the column, cell by cell, until you have converted all cell frequencies to percentages.
3.	Move to the next column. Start with the cell in the top row and then repeat step 1 (making sure you use the correct column total in the denominator of the fraction).
4.	Continue moving down the second column until you have converted all cell frequencies to percentages.
5.	Continue these operations, moving from column to column one at a time, until you have converted all cell frequencies to percentages.

Applying Statistics 10.1 The Chi Square Test

Do men and women vary in their opinions about cohabitation? A random sample of 47 males and females have been asked if they approve or disapprove of "living together." The results are:

Support for Cohabitation	Gender: Males	Females	Totals
Approve	15	5	20
Disapprove	10	17	27
Totals	25	22	47

The frequencies we would expect to find if the null hypothesis (H_0: the variables are independent) were true are:

Support for Cohabitation	Group: Males	Females	Totals
Approve	10.64	9.36	20.00
Disapprove	14.36	12.64	27.00
Totals	25.00	22.00	47.00

(*continued next page*)

Applying Statistics 10.1 *(continued)*

Expected frequencies are found on a cell-by-cell basis by means of the formula

$$f_e = \frac{\text{Row marginal} \times \text{Column marginal}}{N}$$

and the calculation of chi square will be organized into a computational table:

(1) f_o	(2) f_e	(3) $f_o - f_e$	(4) $(f_o - f_e)^2$	(5) $(f_o - f_e)^2/f_e$
15	10.64	4.36	19.01	1.79
5	9.36	−4.36	19.01	2.03
10	14.36	−4.36	19.01	1.32
17	12.64	4.36	19.01	1.50
$N = 47$	$N = 47.00$	0.00	χ^2(obtained)	= 6.64

$$\chi^2(\text{obtained}) = 6.64$$

Step 1. Making Assumptions and Meeting Test Requirements.

Model: Independent random samples
Level of measurement is nominal

Step 2. Stating the Null Hypothesis.

H_0: The two variables are independent.
(H_1: The two variables are dependent.)

Step 3. Selecting the Sampling Distribution and Establishing the Critical Region.

Sampling distribution = χ^2 distribution
Alpha = 0.05
Degrees of freedom = 1
$\chi^2(\text{critical}) = 3.841$

Step 4. Computing the Test Statistic.

$$\chi^2\,(\text{obtained}) = \Sigma\frac{(f_o - f_e)^2}{f_e}$$

$$\chi^2\,(\text{obtained}) = 6.64$$

Step 5. Making a Decision and Interpreting the Results of the Test. With an obtained χ^2 of 6.64, we would reject the null hypothesis of independence. For this sample, there is a statistically significant relationship between gender and support for cohabitation.

To complete the analysis, it would be useful to know exactly how the two variables are related. We can determine this by computing and analyzing column percentages:

Support for Cohabitation	Males	Females	Totals
Approve	60.00%	22.73%	42.55%
Disapprove	40.00%	77.27%	57.45%
Totals	100.00%	100.00%	100.00%

The column percentages show that 60% of males in this sample approve of cohabitation vs. only about 23% of females. We have already concluded that the relationship is significant, and now we know the pattern of the relationship: Males are more supportive.

Let us highlight two points in summary:

1. Chi square is a test of statistical significance. It tests the null hypothesis that the variables are independent in the population. If we reject the null hypothesis, we are concluding, with a known probability of error (determined by the alpha level), that the variables are dependent on each other in the population. In the terms of our example, this means that accreditation status makes a difference in the likelihood of finding work as a social worker. However, by itself, chi square does not tell us the nature or pattern of the relationship.

STATISTICS
IN EVERYDAY LIFE

There seems to be a gender gap in American politics: Women tend to be more sympathetic toward Democratic presidential candidates than men. For example, in the 2008 election, Democratic candidate Obama captured 57% of women voters but only 52% of male voters, while Republican McCain attracted 48% of the male voters and only 43% of the female voters. The relationship between gender and presidential vote is statistically significant at well below the 0.05 level. What additional information would you need to confirm or disprove the hypothesis that there is a gender gap in politics?

Presidential Choice by Gender

Candidate	Women	Men
B. Obama (Dem.)	57%	52%
J. McCain (Rep.)	43%	48%
	100%	100%

Source: U.S. Bureau of the Census. http://www.census.gov/prod/2011pubs/11statab/election.pdf.

2. Column percentages allow us to examine the bivariate relationship in more detail. The column percentages show us exactly how the independent variable affects the dependent variable. In this case, they reveal that graduates of accredited programs are more likely to find work as social workers. We explore column percentages more extensively when we discuss bivariate association in Chapters 11 and 12.

The Chi Square Test: An Additional Example

To this point, we have confined our attention to 2 × 2 tables, or tables with two rows and two columns. For purposes of illustration, we will work through the computational routines and decision-making process for a larger table. As you will see, larger tables require more computations (because they have more cells), but in all other essentials, they are dealt with in the same way as the 2 × 2 table.

A researcher is concerned with the possible effects of marital status on the academic progress of college students. Do married students, with their extra burden of family responsibilities, suffer academically as compared to unmarried students? Is academic performance dependent on marital status? A random sample of 453 students is gathered from a local university, and each student is classified as either married or unmarried and as high, moderate, or low in grade point average (GPA). Results are presented in Table 10.7.

For the top left-hand cell (married students with high GPAs), the expected frequency would be (160 × 175)/453, or 61.8. For the other cell in this row, the expected frequency is (160 × 278)/453, or 98.2. In similar fashion, all expected

TABLE 10.7 Grade Point Average (GPA) by Marital Status for 453 College Students

	Marital Status		
GPA	Married	Not Married	Totals
High	70	90	160
Moderate	60	110	170
Low	45	78	123
Totals	175	278	453

TABLE 10.8 Expected Frequencies for Table 10.7

	Marital Status		
GPA	Married	Not Married	Totals
High	61.8	98.2	160
Moderate	65.7	104.3	170
Low	47.5	75.5	123
Totals	175.0	278.0	453

TABLE 10.9 Computational Table for Table 10.7

(1)	(2)	(3)	(4)	(5)
f_o	f_e	$f_o - f_e$	$(f_o - f_e)^2$	$(f_o - f_e)^2/f_e$
70	61.8	8.2	67.24	1.09
90	98.2	−8.2	67.24	0.69
60	65.7	−5.7	32.49	0.49
110	104.3	5.7	32.49	0.31
45	47.5	−2.5	6.25	0.13
78	75.5	2.5	6.25	0.08
$N = 453$	$N = 453.0$	0.0	χ^2 (obtained)	= 2.79

frequencies are computed (being very careful to use the correct row and column marginals) and displayed in Table 10.8.

The next step is to solve the formula for χ^2(obtained)—being very careful to be certain that we are using the proper values for f_o and f_e for each cell. Once again, we will use a computational table (Table 10.9) to organize the calculations and then test the obtained chi square for its statistical significance. Remember that obtained chi square is equal to the total of column 5.

The value of the obtained chi square (2.79) can now be tested for its significance.

Step 1. Making Assumptions and Meeting Test Requirements.

Model: Independent random samples
Level of measurement is nominal

Step 2. Stating the Null Hypothesis.

$$H_0\text{: The two variables are independent.}$$

$$(H_1\text{: The two variables are dependent.})$$

Step 3. Selecting the Sampling Distribution and Establishing the Critical Region.

$$\text{Sampling distribution} = \chi^2 \text{ distribution}$$
$$\text{Alpha} = .05$$
$$\text{Degrees of freedom} = (\text{r} - 1)(\text{c} - 1) = (3 - 1)(2 - 1) = 2$$
$$\chi^2(\text{critical}) = 5.991$$

Step 4. Computing the Test Statistic.

$$\chi^2\,(\text{obtained}) = \Sigma \frac{(f_o - f_e)^2}{f_e}$$

$$\chi^2\,(\text{obtained}) = 2.79$$

Step 5. Making a Decision and Interpreting the Results of the Test. The test statistic—χ^2(obtained) $= 2.79$—does not fall into the critical region, which—for alpha $= 0.05$ and $df = 2$—begins at χ^2(critical) of 5.991. Therefore, we fail to reject the null hypothesis. The observed frequencies are not significantly different from the frequencies we would expect to find if the variables were independent and only random chance were operating. Based on these sample results, we can conclude that the academic performance of college students is not dependent on their marital status. Because we failed to reject the null hypothesis, we will not examine column percentages as we did for Table 10.3.

The Limitations of the Chi Square Test

Like any other test, chi square has limits, and you should be aware of several potential difficulties. First, chi square becomes difficult to interpret when the variables have many categories. For example, two variables with five categories each would generate a 5×5 table with 25 cells—far too many combinations of scores to be easily absorbed or understood. As a very rough rule of thumb, the chi square test is easiest to interpret and understand when both variables have four or fewer scores.

Two further limitations of the test are related to the sample size. When the sample size is small, we can no longer assume that the sampling distribution of all possible sample outcomes is accurately described by the chi square distribution. For chi square, a small sample is defined as one where a high percentage of the cells have expected frequencies (f_e) of 5 or less. Various rules of thumb have been developed to help the researcher decide what constitutes a "high percentage of cells." Probably the safest course is to take corrective action whenever *any* of the cells have expected frequencies of 5 or less.

For 2×2 tables, the value of χ^2(obtained) can be adjusted by applying Yates's correction for continuity, the formula for which is:

FORMULA 10.4

$$\chi_c^2 = \Sigma \frac{(|f_o - f_e| - 0.5)^2}{f_e}$$

where:

χ_c^2 = corrected chi square
$|f_o - f_e|$ = the absolute values of the difference between observed and expected frequency for each cell

The correction factor is applied by reducing the absolute value[2] of the term $(f_o - f_e)$ by 0.5 before squaring the difference and dividing by the expected frequency for the cell.

For tables larger than 2×2, there is no correction formula for computing χ^2 (obtained) for small samples. It may be possible to combine some of the categories of the variables and thereby increase cell sizes. Obviously, however, this course of action should be taken only when it is sensible to do so. In other words, distinctions that have clear theoretical justifications should not be erased merely to conform to the requirements of a statistical test. When you feel that categories cannot be combined to build up cell frequencies and the percentage of cells with expected frequencies of 5 or less is small, it is probably justifiable to continue with the uncorrected chi square test as long as the results are regarded with a suitable amount of caution.

A second potential problem related to the sample size occurs with large samples. I pointed out in Chapter 8 that all tests of hypothesis are sensitive to the sample size. That is, the probability of rejecting the null hypothesis increases as the number of cases increases, regardless of any other factor. Chi square is especially sensitive to the sample size; the value of χ^2 (obtained) will increase at the same rate as the sample size. For example, if N is doubled, the value of χ^2(obtained) will double. Thus, larger samples may lead to the decision to reject the null when the actual relationship is trivial. *(For an illustration of this principle, see problem 10.14.)*

You should be aware of this relationship between the sample size and the value of chi square because it once again raises the distinction between statistical significance and theoretical importance. On one hand, tests of significance play a crucial role in research. When we are working with random samples, we must know if our research results could have been produced by mere random chance.

On the other hand, like any other statistical technique, tests of hypothesis are limited in the range of questions they can answer. Specifically, these tests will tell us whether our results are statistically significant. They will not necessarily tell us if the results are important in any other sense. To deal more directly with questions of importance, we must use an additional set of statistical techniques called *measures of association.* We previewed these techniques in this chapter when we used column percentages, and measures of association are the subject of Part III of this text.

---

[2]Absolute values ignore plus and minus signs.

Applying Statistics 10.2 Using Chi Square to Test for Cross-National Differences in Socialization Values

The World Values Survey, which is administered periodically to randomly selected samples of citizens from around the globe, allows us to test theories in a variety of settings. In this section, we will test the hypothesis that there will be a relationship between gender and socialization goals for children by using randomly selected samples of residents of Canada and Great Britain. Specifically, we will ask if males place more emphasis on obedience than females. Does the relationship between these two variables change from nation to nation?

Respondents to the World Values Survey were given a list of possible socialization goals and asked to select the goals they thought were most important. Those who included obedience in their list were coded as "Yes" in the tables that follow. Both samples were tested in 2006.

Great Britain (frequencies)

	Gender		
Obedience Mentioned?	Male	Female	Totals
Yes	220	260	480
No	278	280	558
Totals	498	540	1,038

And chi square is computed with the following table:

(1) f_0	(2) f_e	(3) $f_0 - f_e$	(4) $(f_0 - f_e)^2$	(5) $(f_0 - f_e)^2/f_e$
220	230.29	−10.29	105.86	0.46
260	249.71	10.29	105.86	0.42
278	267.71	10.29	105.86	0.40
280	290.29	−10.29	105.86	0.36
1,038	1,038.00	0.00		$\chi^2 = 1.64$

This is a 2 × 2 table, so there is one degree of freedom ($df = 1 \times 1 = 1$) and the critical chi square is 3.841. The relationship is not significant, but we will still look at the pattern of the column percentages:

Great Britain (percentages)

	Gender		
Obedience Mentioned?	Male	Female	Totals
Yes	44.2%	48.1%	46.2%
No	55.8%	51.9%	53.8%
Totals	100.0%	100.0%	100.0%

Not only is the relationship not significant, but there is very little difference between British males and females in their support for obedience.

Are the variables related in the Canadian sample?

Canada (frequencies)

	Gender		
Obedience Mentioned?	Male	Female	Totals
Yes	304	340	644
No	729	781	1,510
Totals	1,033	1,121	2,154

We will use a computing table to get the value of chi square:

(1) f_0	(2) f_e	(3) $f_0 - f_e$	(4) $(f_0 - f_e)^2$	(5) $(f_0 - f_e)^2/f_e$
304	308.85	−4.85	23.47	0.08
340	335.15	4.85	23.47	0.07
729	724.15	4.85	23.47	0.03
781	785.85	−4.85	23.47	0.03
2,154	2,154.00	0.00		$\chi^2 = 0.21$

With one degree of freedom, the critical chi square is 3.841. As was the case for the British sample, there is no significant relationship between gender and support for this socialization goal in the Canadian sample.

Looking at the column percentages, there is virtually no difference between males and females in their support for obedience:

Canada (percentages)

	Gender		
Obedience Mentioned?	Male	Female	Totals
Yes	29.4%	30.3%	29.9%
No	70.6%	69.7%	70.1%
Totals	100.0%	100.0%	100.0%

In summary, there is no relationship between gender and support for obedience as a socialization goal in either nation. However, note that there is a difference in support for obedience between the two nations. About 46% of the British mentioned obedience but only 30% of the Canadians did likewise.

BECOMING A CRITICAL CONSUMER: Reading the Professional Literature

As was the case with the analysis of variance, it is extremely unlikely that you will encounter a chi square test in everyday life or in the popular media, so I will again focus our attention on the professional social science research literature. The article I will use as an example uses FBI records to build a statistical portrait of juveniles who attack their parents. The researchers point out that there is not much information about the dynamics and characteristics of child-to-parent violence. They focus on gender differences, and some of their findings for aggravated assault of parents are presented below. To conserve space, only column percentages—not frequencies—are presented. Significant relationships are identified with * ($p < 0.05$) or ** ($p < 0.001$).

Aggravated Parental Assault by Gender (Percentages)

Variable:	Female Offenders	Male Offenders
OFFENDER AGE:**		
13 and under	18.4%	15.4%
14–17	61.9%	53.7%
18–21	19.6%	30.9%
VICTIM SEX:**		
Male	21.2%	45.4%
Female	73.1%	54.6%

Aggravated Parental Assault by Gender (Percentages)

Variable:	Female Offenders	Male Offenders
VICTIM AGE:		
30 and under	2.8%	2.2%
31–40	41.1%	37.4%
41–50	42.8%	44.7%
51 and older	13.3%	15.6%
KNIFE USED?**		
No	57.7%	65.1%
Yes	42.3%	34.9%

The authors conclude that female offenders are significantly younger than males and significantly more likely to attack their mothers or stepmothers. There was no significant difference in the age of the victim, but female offenders were more likely than male offenders to use a knife in the assault.

The authors note that this is only a first step in developing a research knowledge base on this crime. Want to learn more? The citation is given below.

Walsh, Jeffrey, and Krienert, Jessie. 2007. "Child-Parent Violence: An Empirical Analysis of Offender, Victim, and Event Characteristics in a National Sample of Reported Incidents." *Journal of Family Violence* 22: 563–574.

SUMMARY

1. The chi square test for independence is appropriate for situations in which the variables of interest have been organized into table format. The null hypothesis is that the variables are independent or that the classification of a case into a particular category on one variable has no effect on the probability that the case will be classified into any particular category of the second variable.
2. Because chi square is nonparametric and requires only nominally measured variables, its model assumptions are easily satisfied. Furthermore, because it is computed from bivariate tables in which the number of rows and columns can be easily expanded, the chi square test can be used in many situations in which other tests are inapplicable.
3. In the chi square test, we first find the frequencies that would appear in the cells if the variables were independent (f_e) and then compare those frequencies, cell by cell, with the frequencies actually observed in the cells (f_o). If the null is true, expected and observed frequencies should be quite close in value. The greater the difference between the observed and expected frequencies, the more likely we will reject the null hypothesis.
4. The chi square test has several important limitations. It is often difficult to interpret when tables have many (more than four or five) dimensions. Also, as the sample size (N) decreases, the chi square test becomes less trustworthy and corrective action may be required. Finally, with very large samples, we may declare relatively trivial relationships to be statistically significant. As is the case with all tests of hypothesis, statistical significance is not the same thing as "importance" in any other sense. As a general rule, statistical significance is a necessary but not sufficient condition for theoretical or practical importance.

SUMMARY OF FORMULAS

FORMULA 10.1 Chi square (obtained) $\chi^2\,(\text{obtained}) = \Sigma\dfrac{(f_o - f_e)^2}{f_e}$

FORMULA 10.2 Expected frequencies $f_e = \dfrac{(\text{row marginal} \times \text{column marginal})}{N}$

FORMULA 10.3 Degrees of freedom, bivariate tables: $df = (r - 1)(c - 1)$

FORMULA 10.4 Yates's correction for continuity: $\chi_c^2 = \Sigma\dfrac{(|f_o - f_e| - 0.05)^2}{f_e}$

GLOSSARY

Bivariate table. A table that displays the joint frequency distributions of two variables.

Cells. The cross-classification categories of the variables in a bivariate table.

χ^2(critical). The score on the sampling distribution of all possible sample chi squares that marks the beginning of the critical region.

χ^2(obtained). The test statistic as computed from sample results.

Chi square test. A nonparametric test of hypothesis for variables that have been organized into a bivariate table.

Column. The vertical dimension of a bivariate table. By convention, each column represents a score on the independent variable.

Column Percentages. Percentages calculated within each column of a bivariate table.

Expected frequency (f_e). The cell frequencies that would be expected in a bivariate table if the variables were independent.

Independence. The null hypothesis in the chi square test. Two variables are independent if for all cases the classification of a case on one variable has no effect on the probability that the case will be classified in any particular category of the second variable.

Marginals. The row and column subtotals in a bivariate table.

Nonparametric. A "distribution free" test. These tests do not assume a normal sampling distribution.

Observed frequency (f_o). The cell frequencies actually observed in a bivariate table.

Row. The horizontal dimension of a bivariate table, conventionally representing a score on the dependent variable.

PROBLEMS

10.1 For each of the following tables, calculate the obtained chi square. *(Hint: Calculate the expected frequencies for each cell with Formula 10.2. Double-check to make sure you are using the correct row and column marginals for each cell. It may be helpful to record the expected frequencies in table format; see Tables 10.2 10.4, and 10.8. Next, use a computational table to organize the calculation for Formula 10.1; see Tables 10.5 and 10.9. For each cell, subtract expected frequency from observed frequency and then record the result in column 3. Square the value in column 3 and then record the result in column 4. Divide the value in column 4 by the expected frequency for that cell and then record the result in column 5. Remember that the sum of column 5 in the computational table is obtained chi square. As you proceed, double-check to make sure you are using the correct values for each cell.)*

a.

20	25	45
25	20	45
45	45	90

b.

10	15	25
20	30	50
30	45	75

c.

25	15	40
30	30	60
55	45	100

d.

20	45	65
15	20	35
35	65	100

10.2 PS There seems to be a "gender gap" in elections for the U.S. presidency, with women more likely to vote for the Democratic candidate than men. A small sample of university faculty has been asked about their political party preference. Do their responses indicate a significant relationship between gender and party preference?

	Gender		
Party Preference	Male	Female	Totals
Democrat	10	15	25
Republican	15	10	25
Totals	25	25	50

a. Is there a statistically significant relationship between these variables?
b. Compute column percentages for the table to determine the pattern of the relationship. Which gender is more likely to prefer the Democratic Party?

10.3 SW A local politician is concerned that a program for the homeless in her city is discriminating against blacks and other minorities. The following data were taken from a random sample of black and white homeless people.

	Race		
Received Services?	Black	White	Totals
Yes	6	7	13
No	4	9	13
Totals	10	16	26

a. Is there a statistically significant relationship between race and whether the person has received services from the program?
b. Compute column percentages for the table to determine the pattern of the relationship. Which group was more likely to get services?

10.4 SOC A sample of 25 cities have been classified as high or low on their homicide rates and on the number of handguns sold within the metropolitan area. Is there a relationship between these variables? Do cities with higher homicide rates have higher gun sales? Explain your results in a short paragraph.

	Homicide Rate		
Volume of Gun Sales	Low	High	Totals
High	8	5	13
Low	4	8	12
Totals	12	13	25

10.5 PA Is there a relationship between salary levels and unionization for public employees? The following data represent this relationship for fire departments in a random sample of 100 cities of roughly the same size. Salary data have been dichotomized at the median. Summarize your findings.

	Status		
Salary	Union	Nonunion	Totals
High	21	29	50
Low	14	36	50
Totals	35	65	100

a. Is there a statistically significant relationship between these variables?
b. Compute column percentages for the table to determine the pattern of the relationship. Which group was more likely to get high salaries?

10.6 CJ A local judge has been allowing some individuals convicted of "driving under the influence" to work in a hospital emergency room as an alternative to fines and other penalties. A random sample of offenders has been drawn. Do participants in this program have lower rates of recidivism for this offense?

	Status		
Recidivist?	Participants	NonParticipants	Totals
Yes	60	123	183
No	55	108	163
Totals	115	231	346

a. Is there a statistically significant relationship between these variables?
b. Compute column percentages to determine the pattern of the relationship. Which group is more likely to be rearrested for driving under the influence?

10.7 SOC A state's Department of Education has rated a sample of local school systems for compliance with state-mandated guidelines for quality. Is the quality of a school system significantly related to the affluence of the community as measured by per capita income?

	Per Capita Income		
Quality	Low	High	Totals
Low	16	8	24
High	9	17	26
Totals	25	25	50

a. Is there a statistically significant relationship between these variables?
b. Compute column percentages to determine the pattern of the relationship. Are high- or low-income communities more likely to have high-quality schools?

10.8 SOC A program of pet therapy has been running at a local nursing home. Are the participants in the program more alert and responsive than non-participants? The results are drawn from a random sample.

	Status		
Alertness	Participants	Non-Participants	Totals
High	23	15	38
Low	11	18	29
Totals	34	33	67

a. Is there a statistically significant relationship between these variables?
b. Compute column percentages to determine the pattern of the relationship. Which group is more likely to be alert?

10.9 SOC Is there a relationship between length of marriage and satisfaction with marriage? The necessary information has been collected from a random sample of 100 respondents drawn from a local community. Write a sentence or two explaining your decision.

	Length of Marriage (in years)			
Satisfaction	Less Than 5	5–10	More Than 10	Totals
Low	10	20	20	50
High	20	20	10	50
Totals	30	40	30	100

a. Is there a statistically significant relationship between these variables?
b. Compute column percentages to determine the pattern of the relationship. Which group is more likely to be highly satisfied?

10.10 PS Is there a relationship between college class standing and political orientation? Are third- and fourth-year students significantly different from first- and second-year students on this variable? The following table reports the relationship between these two variables for a random sample of 267 college students.

	College Class Standing		
Political Orientation	3rd and 4th Year Students	1st and 2nd year Students	Totals
Liberal	40	43	83
Moderate	50	50	100
Conservative	44	40	84
Totals	134	133	267

a. Is there a statistically significant relationship between these variables?
b. Compute column percentages to determine the pattern of the relationship. Which group is more likely to be conservative?

10.11 SOC At a large urban college, about half the students live off campus in various arrangements and the other half live in dormitories on campus. Is academic performance dependent on living arrangements? The results based on a random sample of 300 students are presented here.

	Residential Status			
GPA	Off Campus With Roommates	Off Campus With Parent	On Campus	Totals
Low	22	20	48	90
Moderate	36	40	54	130
High	32	10	38	80
Totals	90	70	140	300

a. Is there a statistically significant relationship between these variables?
b. Compute column percentages to determine the pattern of the relationship. Which group is more likely to have a high GPA?

10.12 SOC An urban planning commissioner has built a database describing a sample of the neighborhoods in her city and has developed a scale by which each area can be rated for the "quality of life" (this includes measures of pollution, noise, open space, services available, and so on). She has also asked samples of residents of these areas about their level of satisfaction with their neighborhoods. Is there significant agreement between the commissioner's objective ratings of quality and the respondents' self-reports of satisfaction?

	Quality of Life			
Satisfaction	Low	Moderate	High	Totals
Low	21	15	6	42
Moderate	12	25	21	58
High	8	17	32	57
Totals	41	57	59	157

a. Is there a statistically significant relationship between these variables?

b. Compute column percentages to determine the pattern of the relationship. Which group is most likely to have high satisfaction?

10.13 SOC Does support for the legalization of marijuana vary by region of the country? The table displays the relationship between the two variables for a random sample of 1,020 adult citizens. Is the relationship significant?

	Region				
Legalize?	North	Midwest	South	West	Totals
Yes	60	65	42	78	245
No	245	200	180	150	775
Totals	305	265	222	228	1020

a. Is there a statistically significant relationship between these variables?

b. Compute column percentages for the table to determine the pattern of the relationship. Which region is most likely to favor the legalization of marijuana?

10.14 SOC A researcher is concerned with the relationship between attitudes toward violence and violent behavior. If attitudes "cause" behavior (a very debatable proposition), then people who have positive attitudes toward violence should have high rates of violent behavior. A pretest was conducted on 70 respondents, and among other things, the respondents were asked "Have you been involved in a violent incident of any kind over the past six months?" The researcher established the following relationship:

	Attitude Toward Violence		
Involvement	Favorable	Unfavorable	Totals
Yes	16	19	35
No	14	21	35
Totals	30	40	70

The chi square calculated on these data is .23, which is not significant at the .05 level (confirm this conclusion with your own calculations). Undeterred by this result, the researcher proceeded with the project and gathered a random sample of 7,000. In terms of percentage distributions, the results for the full sample were exactly the same as for the pretest:

	Attitude Toward Violence		
Involvement	Favorable	Unfavorable	Totals
Yes	1,600	1,900	3,500
No	1,400	2,100	3,500
Totals	3,000	4,000	7,000

However, the chi square obtained is a very healthy 23.4 (confirm with your own calculations). Why is the full-sample chi square significant when the pretest was not? What happened? Do you think that the second result is important?

10.15 SOC Some results from a survey administered to a nationally representative sample are presented here. For each table, conduct the chi square test of significance and compute column percentages. Write a sentence or two of interpretation for each test.

a. Support for the legal right to an abortion by age:

	Age			
Support?	Younger Than 30	30–49	50 and Older	Totals
Yes	154	360	213	727
No	179	441	429	1,049
Totals	333	801	642	1,776

b. Support for the death penalty by age:

	Age			
Support?	Younger Than 30	30–49	50 and Older	Totals
Favor	361	867	675	1,903
Oppose	144	297	252	693
Totals	505	1,164	927	2,596

c. Fear of walking alone at night by age:

	Age			
Fear?	Younger Than 30	30–49	50 and Older	Totals
Yes	147	325	300	772
No	202	507	368	1,077
Totals	349	832	668	1,849

d. Support for legalizing marijuana by age:

	Age			
Legalize?	Younger Than 30	30–49	50 and Older	Totals
Should	128	254	142	524
Should not	224	534	504	1,262
Totals	352	788	646	1,786

e. Support for suicide when a person has an incurable disease by age:

	Age			
Support?	Younger Than 30	30–49	50 and Older	Totals
Yes	225	537	367	1,129
No	107	270	266	643
Totals	332	807	633	1,772

10.16 PS A random sample of 748 voters in a large city were asked how they voted in the presidential election of 2008. Calculate chi square and the column percentages for each table below and then write a brief report describing the significance of the relationships as well as the patterns you observe.

a. Presidential preference and gender:

	Gender		
Preference	Male	Female	Totals
McCain	165	173	338
Obama	200	210	410
Totals	365	383	748

b. Presidential preference and race/ethnicity:

	Race/Ethnicity			
Preference	White	Black	Latino	Totals
McCain	289	5	44	338
Obama	249	95	66	410
Totals	538	100	110	748

c. Presidential preference by education:

	Education				
Preference	Less Than HS	HS Graduate	College Graduate	Postgrad Degree	Totals
McCain	30	180	118	10	338
Obama	35	120	218	37	410
Totals	65	300	336	47	748

d. Presidential preference by religion:

	Religion					
Preference	Protestant	Catholic	Jewish	None	Other	Totals
McCain	165	110	10	28	25	338
Obama	245	55	20	60	30	410
Totals	410	165	30	88	55	748

YOU ARE THE RESEARCHER: Understanding Political Beliefs

Two projects are presented here, and you are urged to complete both to apply your understanding of the chi square test. In the first, you will examine the sources of people's beliefs about some of the most hotly debated topics in U.S. society: capital punishment, assisted suicide, gay marriage, and immigration. In the second, you will compare various independent variables to see which has the most significant relationship with your chosen dependent variable.

Using Crosstabs

In these projects, you will use a new procedure called **Crosstabs** to produce bivariate tables, chi square, and column percentages. This procedure is very commonly used in social science research at all levels, and you will see many references to **Crosstabs** in chapters to come.

To begin the procedure, click **Analyze** and then click **Descriptive Statistics** and **Crosstabs.** The **Crosstabs** dialog box appears with the variables listed on the left. Highlight the name(s) of your dependent variable(s) and then click the arrow to move the variable name into the **Rows** box. Next, find the name of your independent variable(s) and then move it into the **Columns** box. SPSS will process all combinations of variables in the row and column boxes at one time.

Click the **Statistics** button at the right of the window and then click the box next to Chi-square. Return to the **Crosstabs** window and then click the **Cells** button and select "Column" in the **Percentages** box. This will generate column percentages for the table. Return to the **Crosstabs** window by clicking **Continue** and then click **OK** to produce your output.

Crosstabs Demonstration

I will demonstrate this command by examining the relationship between gender (*sex*) and support for legal abortion "for any reason" (*abany*). Gender is my independent variable and I listed it in the column box and placed *abany* in the row box. Here is the output:

ABORTION IF WOMAN WANTS FOR ANY REASON * RESPONDENT'S SEX Crosstabulation

			RESPONDENT'S SEX		
			MALE	FEMALE	Total
ABORTION IF WOMAN WANTS FOR ANY REASON	YES	Count	185	205	390
		% within RESPONDENT'S SEX	44.7%	41.9%	43.2%
	NO	Count	229	284	513
		% within RESPONDENT'S SEX	55.3%	58.1%	56.8%
Total		Count	414	489	903
		% within RESPONDENT'S SEX	100.0%	100.0%	100.0%

Chi-Square Tests

	Value	98df	Asymp. Sig. (2-sided)	Exact Sig. (2-sided)	Exact Sig. (1-sided)
Pearson Chi-Square	.698[a]	1	.403		
Continuity Correction[b]	.590	1	.442		
Likelihood Ratio	.698	1	.404		
Fisher's Exact Test				.419	.221
Linear-by-Linear Association	.697	1	.404		
N of Valid Cases	903				

a. 0 cells (0.0%) have expected count less than 5. The minimum expected count is 178.80.

b. Computed only for a 2 x 2 table.

Read the crosstab table cell by cell. Each cell displays the number of cases in the cell and the column percentage for that cell. For example, starting with the upper left-hand cell, there were 185 respondents who were male and said "yes" to *abany*, and these were 44.7% of all men in the sample. In contrast, 205 of the females (41.9%) also supported legal abortion "for any reason." The column percentages are similar, which would indicate that *sex* and *abany* will not have a significant relationship.

The results of the chi square test are reported in the output block that follows the table. The value of chi square (obtained) is 0.698, and there was 1 degree of freedom. The exact significance of the chi square, reported in the column labeled "Asymp. Sig (2-sided)," is 0.403. This is far greater than the standard indicator of a significant result (alpha = 0.05), so we may conclude, as we saw with the column percentages, that there is no statistically significant relationship between these variables. Support for legal abortion is not dependent on gender.

PROJECT 1: Explaining Beliefs

In this project, you will analyze beliefs about capital punishment (*cappun*), assisted suicide (*letdie1*), gay marriage (*marhomo*), and immigration (*letin1*). You will select an independent variable, use SPSS to generate chi squares and column percentages, and then analyze and interpret your results.

STEP 1: Choosing an Independent Variable

Select an independent variable that seems likely to be an important cause of people's attitudes about the death penalty, assisted suicide, gay marriage, and immigration. Be sure to select an independent variable that has only 2 to 5 categories. If you select an independent variable with more than 5 scores, use the recode command to reduce the number of categories. You might consider gender, level of education (use *degree*), religion, or age (the recoded version—see Chapter 9) as possible independent variables, but there are many others. Record the variable name and then state exactly what the variable measures in the table below:

SPSS Name	What Exactly Does This Variable Measure?

STEP 2: Stating Hypotheses

State hypotheses about the relationships you expect to find between your independent variable and each of the four dependent variables. State these hypotheses in terms of which category of the independent variable you expect to be associated with which category of the dependent variable (for example, "I expect that men will be more supportive of the legal right to an abortion for any reason").

1. ________________________________

2. ________________________________

3. ________________________________

4. ________________________________

STEP 3: Running Crosstabs

Click **Analyze → Descriptives → Crosstabs** and then place the four dependent variables (*cappun, suicide1, letin1,* and *marhomo*) in the **Rows:** box and the independent variable you selected in the **Columns:** box. Click the **Statistics** button and then click the box next to chi square. Finally, click the **Cells** button and then click the box next to column percentages. Return to the **Crosstabs** window by clicking the **Continue** button and then click **OK**.

STEP 4: Recording Results

Your output will consist of four tables, and it will be helpful to summarize your results in the following table. Remember that the significance of the relationship is found in the column labeled "Asymp. Sig (2-sided)" of the second box in the output. Enter the exact significance of the result (0.403 in our earlier example) in the right-hand column.

Dependent Variable	Chi Square	Degrees of Freedom	Significance
cappun			
letdie1			
letin1			
marhomo			

STEP 5: Analyzing and Interpreting Results

Write a short summary of results of each test in which you:

1. Identify the variables being tested, the value and significance of chi square, *N*, and the pattern (if any) of the column percentages. In the professional research literature, you might find the results reported as: "For a sample of 903 respondents, there was no significant relationship between gender and support for abortion (chi square = 0.698, $df = 1$, $p > 0.05$). About 45% of the men supported the legal right to an abortion "for any reason" vs. about 42% of the women.
2. Explain if your hypotheses were supported.

PROJECT 2: Exploring the Impact of Various Independent Variables

In this project, you will examine the relative ability of a variety of independent variables to explain or account for a single dependent variable. You will again use the **Crosstabs** procedure in SPSS to generate chi squares and column percentages and use the value of alpha to judge which independent variable has the most important relationship with your dependent variable.

STEP 1: Choosing Variables

Select a dependent variable. You may use any of the four from Project 1 in this chapter or select a new dependent variable from the 2010 GSS. Be sure your dependent variable has no more than five values or scores. Use the recode command as necessary to reduce the number of categories. Good choices for dependent variables include any measure of attitudes or opinions. *Do not* select such characteristics as race, sex, or religion as dependent variables.

Select three independent variables that seem likely to be important causes of the dependent variable you selected. Your independent variable should have no more than five or six categories. You might consider gender, level of education (use *degree*), religiosity, or age (the recoded version—see Chapter 9) as possibilities, but there are many others.

Record the variable names and then state exactly what each variable measures in the table below:

SPSS Name	What Exactly Does This Variable Measure?
Dependent Variable	
Independent Variables	

STEP 2: Stating Hypotheses

State hypotheses about the relationships you expect to find between your independent variables and the dependent variable. State these hypotheses in terms of which category of the independent variable you expect to be associated with which category of the dependent variable (for example, "I expect that men will be more supportive of the legal right to an abortion for any reason").

1.

2.

3.

STEP 3: Running Crosstabs

Click **Analyze** → **Descriptives** → **Crosstabs** and then place your dependent variable in the **Rows:** box and all three of your independent variables in the **Columns:** box. Click the **Statistics** button to get chi square and the **Cells** button for column percentages.

STEP 4: Recording Results

Your output will consist of three tables, and it will be helpful to summarize your results in the following table. Remember that the significance of the relationship is found in the column labeled "Asymp. Sig (2-sided)" of the second box in the output.

Independent Variables	Chi Square	Degrees of Freedom	Significance

STEP 5: Analyzing and Interpreting Results

Write a short summary of results of each test in which you:

1. Identify the variables being tested, the value and significance of chi square, *N*, and the pattern (if any) of the column percentages.
2. Explain if your hypotheses were supported.
3. Explain which independent variable had the most significant relationship (lowest value in the "Asymp. Sig 2-tailed" column) with your dependent variable.

Part III Bivariate Measures of Association

The chapters in this part cover the computation and interpretation of a class of statistics known as *measures of association*. These statistics are extremely useful in scientific research and commonly reported in the professional literature. They provide, in a single number, an indication of the strength and—if appropriate—the direction of a bivariate association.

It is important to remember the difference between statistical significance, covered in Part II, and association, the topic of this part. Tests for statistical significance answer certain questions: Were the differences or relationships observed in the sample caused by mere random chance? What is the probability that the sample results reflect patterns in the population(s) from which the sample(s) were selected? Measures of association address a different set of questions: How strong is the relationship between the variables? What is the direction or pattern of the relationship?

Thus, measures of association provide information that complements that supplied by tests of significance. Association and significance are two different things. It is most satisfying to find results that are both statistically significant and strong, but it is common to find mixed or ambiguous results: relationships that are statistically significant but weak, not statistically significant but strong, and so forth.

Chapter 11 introduces the basic ideas behind the analysis of association in terms of bivariate tables and column percentages and presents measures of association for use with nominal-level variables. Chapter 12 covers measures for ordinal-level variables, and Chapter 13 presents Pearson's r—the most important measure of association and the only one designed for interval-ratio-level variables.

11 Association Between Variables Measured at the Nominal Level

LEARNING OBJECTIVES

By the end of this chapter, you will be able to:

1. Explain how we can use measures of association to describe and analyze the importance of relationships (vs. their statistical significance).
2. Define *association* in the context of bivariate tables and in terms of changing conditional distributions.
3. List and explain the three characteristics of a bivariate relationship: existence, strength, and pattern or direction.
4. Investigate a bivariate association by properly calculating percentages for a bivariate table and interpreting the results.
5. Compute and interpret measures of association for variables measured at the nominal level.

Using Statistics

The statistical techniques presented in this chapter are used to measure the association between variables that have been organized in bivariate tables. Also, measures of association for nominal-level variables are presented. Examples of situations in which these techniques are useful include:

1. An industrial sociologist wonders about the relationship between satisfaction and productivity. Are happy workers busy workers? If we know that a worker is very satisfied with her job, can we predict that she will be very productive?
2. Over the years, has there been a relationship between economic hard times and the rate of property crime? Does the crime rate increase when the economy falters? How much?
3. Are there relationships between religiosity and support for traditional gender roles, opposition to gay marriage, and support for the death penalty? If these relationships exist, how strong are they? What patterns do they display?

As we have seen over the past several chapters, tests of statistical significance are extremely important in social science research. When we work with random samples rather than populations, these tests are indispensable for dealing with the possibility that our research results are the products of mere random chance. However, tests of significance are often only the first step in the analysis of research results. These tests do have limitations, and statistical significance is not necessarily the same thing as relevance or importance. Furthermore, all tests of significance are affected by the sample size; tests performed on large samples may result in decisions to reject the null hypothesis when, in fact, the observed differences are quite minor.

Beginning with this chapter, we will be working with **measures of association**. Whereas tests of significance detect nonrandom relationships, measures of association provide information about the strength and direction of relationships—information that is highly relevant for assessing the importance of relationships and testing the power and validity of our theories. The theories that guide scientific research are almost always stated in cause-and-effect terms (for example: "variable *X* causes variable *Y*"). As an example, recall our discussion of the contact hypothesis in Chapter 1. In that theory, the causal (or independent) variable was equal status contacts between groups, and the effect (or dependent) variable was prejudice. The theory asserts that involvement in equal-status-contact situations *causes* prejudice to decline. Measures of association help us trace causal relationships among variables, and they are our most important and powerful statistical tools for documenting, measuring, and analyzing cause-and-effect relationships.

As useful as they are, measures of association—like any class of statistics—do have their limitations. Most importantly, these statistics cannot *prove* that two variables are causally related. Even if there is a strong (and significant) statistical association between two variables, we cannot necessarily conclude that one variable is a cause of the other. We will explore causation in more detail in Part IV, but for now, you should keep in mind that causation and association are two different things. We can use a statistical association between variables as evidence for a causal relationship, but association by itself is not proof that a causal relationship exists.

Another important use for measures of association is prediction. If two variables are associated, we can predict the score of a case on one variable from the score of that case on the other variable. For example, if equal status contacts and prejudice are associated, we can predict that people who have experienced many such contacts will be less prejudiced than those who have had few or no contacts. Note that prediction and causation can be two separate matters. If variables are associated, we can predict from one to the other even if the variables are not causally related.

Bivariate Association: Introduction and Basic Concepts

In this chapter, you will be introduced to the concept of **association** between variables in the context of bivariate tables. You will also learn how to use percentages and several different measures of association to analyze a bivariate relationship. By the end of this chapter, you will have an array of statistical tools you can use to analyze the strength and pattern of an association between two variables.

Association and Bivariate Tables

Most generally, two variables are said to be associated if the distribution of one of them changes under the various categories or scores of the other. For example,

TABLE 11.1 Productivity by Job Satisfaction (Frequencies)

	Job Satisfaction (X)			
Productivity (Y)	Low	Moderate	High	Totals
Low	30	21	7	58
Moderate	20	25	18	63
High	10	15	27	52
Totals	60	61	52	173

suppose an industrial sociologist was concerned with the relationship between job satisfaction and productivity for assembly-line workers. If these two variables are associated, then scores on productivity will change under the different conditions of satisfaction. Highly satisfied workers will have different scores on productivity than workers who are low on satisfaction, and levels of productivity will vary by levels of satisfaction.

We can clarify this relationship with a bivariate table. As you recall (see Chapter 10), bivariate tables display the scores of cases on two different variables. By convention, the **independent**, or ***X*****, variable** (that is, the variable taken as causal) is arrayed in the columns, and the **dependent**, or ***Y*****, variable** in the rows.[1] That is, each column of the table (the vertical dimension) represents a score or category of the independent variable (*X*), and each row (the horizontal dimension) represents a score or category of the dependent variable (*Y*).

Table 11.1 displays a relationship between productivity and job satisfaction for a fictitious sample of 173 factory workers. We focus on the columns to detect the presence of an association between variables displayed in table format. Each column shows the pattern of scores on the dependent variable for each score on the independent variable. For example, the left-hand column indicates that 30 of the 60 workers who were low on job satisfaction were low on productivity, 20 were moderately productive, and 10 were high on productivity. The middle column shows that 21 of the 61 moderately satisfied workers were low on productivity, 25 were moderately productive, and 15 were high on productivity. Of the 52 workers who are highly satisfied (the right-hand column), 7 were low on productivity, 18 were moderate, and 27 were high.

As we read the table from column to column, we can observe the effects of the independent variable on the dependent variable (provided, of course, that the table is constructed with the independent variable in the columns). These "within column" frequency distributions are called the **conditional distributions of *Y*** because they display the distribution of scores on the dependent variable for each condition (or score) of the independent variable.

Table 11.1 indicates that productivity and satisfaction are associated; the distribution of scores on *Y* (productivity) changes across the various conditions of *X* (satisfaction). For example, half the workers who were low on satisfaction were also low on productivity (30 out of 60). On the other hand, over half of the workers who were high on satisfaction were high on productivity (27 out of 52).

[1]In the material that follows, we will often refer to the independent variable as *X* and the dependent variable as *Y*.

Although it is a test of significance, chi square provides another way to detect an association between variables organized into table format. Any nonzero value for obtained chi square indicates that the variables are associated. For example, the obtained chi square for Table 11.1 is 24.2—a value that affirms our previous conclusion, based on the conditional distributions of *Y*, that an association of some sort exists between job satisfaction and productivity.

Often, the researcher will have already computed chi square before considering bivariate association. In such cases, we can use chi square instead of the conditional distributions of *Y* to ascertain whether there is an association. If the obtained chi square is zero, the two variables are independent and not associated. Any value other than zero indicates some association between the variables. However, remember that statistical significance and association are two different things. It is perfectly possible for two variables to be associated (as indicated by a nonzero chi square) but still independent (if we fail to reject the null hypothesis).

In this section, we have defined, in a general way, the concept of association between two variables. We have also shown two different ways to detect the presence of an association: changing conditional distributions of Y and a nonzero value for chi square. In the next section, we extend the analysis and, in a systematic way, show how additional very useful information about the relationship between two variables can be developed.

Three Characteristics of Bivariate Associations

Bivariate associations possess three different characteristics, each of which must be analyzed for a full investigation of the relationship. Investigating these characteristics may be thought of as a process of finding answers to three questions:

1. Does an association exist?

2. If an association does exist, how strong is it?

3. What is the pattern and/or the direction of the association?

We will consider each of these questions separately.

Does an Association Exist? We have already seen that we can detect an association by observing the conditional distributions of *Y* in a table or by using chi square. In Table 11.1, we know that the two variables are associated to some extent because the conditional distributions of productivity (*Y*) are different across the various categories of satisfaction (*X*) and because the chi square statistic is a nonzero value.

Comparisons from column to column in Table 11.1 are relatively easy to make because the column totals are roughly equal. This will not usually be the case, and it is helpful to compute percentages to control for varying column totals. These **column percentages**—introduced in Chapter 10—are computed within each column separately and make the pattern of association more visible.

The general procedure for detecting association with bivariate tables is to compute percentages within the columns (vertically, or down each column) and then compare column to column across the table (horizontally, or across the rows). Table 11.2 presents column percentages calculated from the data in Table 11.1.

TABLE 11.2 Productivity by Job Satisfaction (percentages)

	Job Satisfaction (*X*)			
Productivity (*Y*)	Low	Moderate	High	Totals
Low	50.00%	34.43%	13.46%	33.53% (58)
Moderate	33.33%	40.98%	34.62%	36.42% (63)
High	16.67%	24.59%	51.92%	30.06% (52)
Totals	100.00%.	100.00%	100.00%	100.00%
	(60)	(61)	(52)	(173)

Note that this table reports the row and column marginals but in parentheses. Besides controlling for any differences in column totals, tables in percentage form are usually easier to read because changes in the conditional distributions of *Y* are easier to detect.

In Table 11.2, we can see that the largest cell changes position from column to column. For workers who are low on satisfaction, the single-largest cell is in the top row (low on productivity). For the middle column (moderate on satisfaction), the largest cell is in the middle row (moderate on productivity), and for the right-hand column (high on satisfaction), it is in the bottom row (high on productivity). Even a cursory glance at Table 11.2 reinforces our conclusion that an association does exist between these two variables.

If two variables are not associated, then the conditional distributions of *Y* will not change across the columns. The distribution of *Y* would be the same for each condition of *X*. Table 11.3 illustrates a "perfect nonassociation" between height and productivity. Table 11.3 is only one of many patterns that indicate "no association." The important point is that the conditional distributions of *Y* are the same. Levels of productivity do not change at all for the various heights; therefore, no association exists between these variables. Also, the obtained chi square computed from this table would have a value of zero—again indicating no association.[2]

How Strong Is the Association? Once we know there is an association between two variables, we need to know how strong it is. This is essentially a matter of determining the amount of change in the conditional distributions of *Y*. At one extreme, of course, is the case of "no association," where the conditional distributions of *Y* do not change at all (see Table 11.3). At the other extreme is

TABLE 11.3 Productivity by Height (an illustration of no association)

	Height (*X*)		
Productivity (*Y*)	Short	Medium	Tall
Low	33.33%	33.33%	33.33%
Moderate	33.33%	33.33%	33.33%
High	33.33%	33.33%	33.33%
Totals	100.00%	100.00%	100.00%

[2]See Chapter 10 for instructions on computing column percentages.

TABLE 11.4 Productivity by Height (an illustration of perfect association)

	Height (X)		
Productivity (Y)	Short	Medium	Tall
Low	0%	0%	100%
Moderate	0%	100%	0%
High	100%	0%	0%
Totals	100%	100%	100%

a perfect association—the strongest possible relationship. A perfect association exists between two variables if each value of the dependent variable is associated with one and only one value of the independent variable.[3] In a bivariate table, the variables would have a perfect relationship if all cases in each column are located in a single cell and there is no variation in *Y* for a given value of *X* (see Table 11.4).

A perfect relationship would be taken as very strong evidence of a causal relationship between the variables—at least for the sample at hand. In fact, the results presented in Table 11.4 would indicate that for this sample, height is the sole cause of productivity. Also, in the case of a perfect relationship, predictions from one variable to the other could be made without error. For example, if we know that a particular worker is short, we could be sure that he or she is highly productive.

Of course, the huge majority of relationships will fall somewhere between the two extremes of no association and perfect association, and we need to develop some way of describing these intermediate relationships consistently and meaningfully. For example, Tables 11.1 and 11.2 show an association between productivity and job satisfaction. How could this relationship be described in terms of strength? How close is the relationship to perfect? How far away from no association?

We use measures of association—objective indicators of the strength of a relationship—to answer these questions. Virtually all these statistics are designed so they have a lower limit of 0.00 and an upper limit of 1.00 ($\pm$1.00 for ordinal and interval-ratio measures of association). A measure that equals 0.00 indicates no association between the variables (the conditional distributions of *Y* do not vary), and a measure of 1.00 ($\pm$1.00 in the case of ordinal and interval-ratio measures) indicates a perfect relationship. The exact meaning of values between 0.00 and 1.00 varies from measure to measure, but for all measures, the closer the value is to 1.00, the stronger the relationship (the greater the change in the conditional distributions of *Y*).

We begin to consider the many measures of association later in this chapter. At this point, we consider a more informal way of assessing the strength of a relationship based on comparing column percentages across the rows. The **maximum difference** is a technique that can be used as a "quick and easy" method for assessing the strength of a relationship; it is easy to apply but limited in its usefulness. To find the maximum difference, compute the column percentages as usual and then skim the table across each of the rows to find the largest difference—in any row—between column percentages. For example, the largest difference in column

[3]Each measure of association has its own definition of a "perfect association," and these vary somewhat from statistic to statistic.

percentages in Table 11.2 is in the top row between the "Low" column and the "High" column: 50.00% − 13.46% = 36.54%. The maximum difference in the middle row is between "moderates" and "lows" (40.98% − 33.33% = 7.65%), and in the bottom row, it is between "highs" and "lows" (51.92% − 16.67% = 35.25%). Both of these values are less than the maximum difference in the top row.

Once you have found the maximum difference in the table, the scale presented in Table 11.5 can be used to describe the strength of the relationship. For example, it can help us describe the relationship between productivity and job satisfaction in Table 11.2 as strong.

Be aware that the relationships between the size of the maximum difference and the descriptive terms (weak, moderate, and strong) in Table 11.5 are arbitrary and approximate. We will get more precise and useful information when we compute and analyze a measure of association. Also, maximum differences are easiest to find and most useful for smaller tables. In large tables, with many (say, more than three) columns and rows, it can be cumbersome to find the high and low percentages, and it is advisable to consider only measures of association as indicators of the strength for these tables. Finally, note that the maximum difference is based on only two values (the high and low column percentages within any row). Like the range (see Chapter 4), this statistic can give a misleading impression of the overall strength of the relationship.

As a final caution, do not mistake chi square as an indicator of the strength of a relationship. Even very large values for chi square do not necessarily mean that the relationship is strong. Remember that significance and association are two separate matters and that chi square by itself is not a measure of association. While a nonzero value indicates some association between the variables, the magnitude of chi square bears no particular relationship to the strength of the association. *(For practice in computing percentages and judging the existence and strength of an association, see any of the problems at the end of this chapter.)*

What Is the Pattern and/or the Direction of the Association? To investigate the pattern of the association, we need to ascertain which values or categories of one variable are associated with which values or categories of the other. We have already remarked on the pattern of the relationship between productivity and satisfaction. Table 11.2 indicates that low scores on satisfaction are associated with low scores on productivity, moderate satisfaction with moderate productivity, and high satisfaction with high productivity.

When one or both variables in a bivariate table are measured at the nominal level, we can discuss the pattern of the relationship only. Relationships with nominal-level

TABLE 11.5 The Relationship Between the Maximum Difference and the Strength of the Relationship

Maximum Difference	Strength
If the maximum difference is:	*The strength of the relationship is:*
between 0 and 10 percentage points	weak
between 10 and 30 percentage points	moderate
more than 30 percentage points	strong

TABLE 11.6 Library Use by Education (an illustration of a positive relationship)

	Education		
Library Use	Low	Moderate	High
Low	60%	20%	10%
Moderate	30%	60%	30%
High	10%	20%	60%
Total	100%	100%	100%

TABLE 11.7 Amount of Television Viewing by Education (an illustration of a negative relationship)

	Education		
Television Viewing	Low	Moderate	High
Low	10%	20%	60%
Moderate	30%	60%	30%
High	60%	20%	10%
Total	100%	100%	100%

variables cannot have direction because their scores are nonnumeric. When both variables are at least ordinal in level of measurement, the association between the variables may also be described in terms of direction. The direction of the association can be either positive or negative. An association is positive if the variables vary in the same direction. That is, in a **positive association**, high scores on one variable are associated with high scores on the other variable and low scores on one variable are associated with low scores on the other. In a positive association, as one variable increases in value, the other also increases, and as one variable decreases, the other also decreases. Table 11.6 displays, with fictitious data, a positive relationship between education and the use of public libraries. As education increases (as you move from left to right across the table), library use also increases (the percentage of "high" users increases). The association between job satisfaction and productivity, as displayed in Tables 11.1 and 11.2, is also a positive association.

In a **negative association**, the variables vary in opposite directions. High scores on one variable are associated with low scores on the other, and increases in one variable are accompanied by decreases in the other. Table 11.7 displays a negative relationship—again, with fictitious data—between education and television viewing. The amount of television viewing decreases as education increases. In other words, as you move from left to right across the top of the table (as education increases), the percentage of heavy viewers decreases.

Measures of association for ordinal and interval-ratio variables are designed so they will be positive for positive associations and negative for negative associations. Thus, a measure of association preceded by a plus sign indicates a positive relationship between the two variables, with the value $+1.00$ indicating a perfect positive relationship. A negative sign indicates a negative relationship, with -1.00 indicating a perfect negative relationship. *(For practice in determining the pattern of an association, see any of the end-of-chapter problems. For practice in determining the direction of a relationship, see problems 11.1, 11.5, and 11.8c.)*

STATISTICS
IN EVERYDAY LIFE

Even after professional football was integrated in the middle of the 20th century, racial discrimination continued on a number of levels. Most obviously, perhaps, was the perfect relationship between race and the quarterback position. This central leadership role was reserved for whites—a practice that reflected the widespread stereotype that blacks lacked the intelligence and decision-making ability required to play quarterback. Today, racial discrimination in professional sports is much weaker than in the past, and minority-group athletes face little overt discrimination in their competition for any position, including quarterback. In fact, African Americans are vastly overrepresented in the National Football League (NFL): They are about 13% of the general population and about 67% of all NFL players. This makes the NFL one of the most integrated areas of U.S. society, but African Americans are still only 16% of quarterbacks,* a value that may reflect lingering—if more subtle—prejudices and stereotypes. What data would you need to collect to account for these patterns?

**Source:* Lapchik, R., Kamke, C., and McMechan, D. 2011. *The 2010 Racial and Gender Report Card: National Football League.* Available at http://www.tidesport.org/racialgenderreportcard.html

BECOMING A CRITICAL CONSUMER: Reading Percentages

The first step in analyzing bivariate tables should always be to compute and analyze column percentages. These will give you more detail about the relationship than measures of association, such as phi and lambda, which should be regarded as summary statements about the relationship. Remember that percentages, although among the more humble of statistics, are not necessarily simple and that they can be miscalculated and misunderstood. Errors can occur when there is confusion about which variable is the cause (or independent variable) and which is the effect (or dependent variable). A closely related error can happen when the researcher asks the wrong questions about the relationship.

To illustrate these errors, let us review the process by which we analyze bivariate tables. When we compare column percentages across the table, we are asking "Does Y (the dependent variable) vary by X (the independent variable)?" We conclude that there is evidence for a causal relationship if the values of Y change under the different values of X.

To illustrate further, consider the table below, which shows the relationship between race or ethnic group (limited to black, white, and Hispanic American for the sake of simplicity) and support for affirmative action. The data is taken from the General Social Survey (1972–2010), which is administered to a representative national sample of adult Americans. Race or ethnicity must be the independent or causal variable in this relationship because it may shape people's attitudes and opinions. The reverse cannot be true; a person's opinion cannot cause their race or ethnicity. Group membership is the column variable in the table and the percentages are computed in

TABLE 11.8 Support for Affirmative Action by Racial Group Frequencies and (*Percentages*)

	Racial or Ethnic Group			
Support Affirmative Action?	White	Black	Hispanic	TOTALS
Yes	637 (*11.6%*)	452 (*44.6%*)	87 (*25.5%*)	1,176 (*17.1%*)
No	4,873 (*88.4%*)	561 (*55.4%*)	254 (*74.5%*)	5,688 (*82.9%*)
TOTALS	5,510 (*100.0%*)	1,013 (*100.0%*)	341 (*100.0%*)	6,864 (*100.0%*)

Source: General Social Survey, 1972–2010. Online data analysis at http://sda.berkeley.edu/cgi-bin/hsda?harcsda+gss10.

BECOMING A CRITICAL CONSUMER: *(continued)*

TABLE 11.9 Row Percentages for Table 11.8

	Racial or Ethnic Group			
Support Affirmative Action?	White	Black	Hispanic	TOTALS
Yes	54.2%	38.4%	7.4%	100.0%
No	85.7%	9.9%	4.5%	100.1%

the proper direction. A quick inspection shows that support for affirmative action varies by group, indicating that there may be a causal relationship between these variables. The maximum difference between the columns is about 33 percentage points, indicating that the relationship is moderate to strong.

What if we had misunderstood this causal relationship? For example, if we had computed percentages within each row, race would become the dependent variable. We would be asking "Does race vary by support for affirmative action?" Table 11.9 shows the results of asking this incorrect question.

A glance at the top row of the table seems to indicate a causal relationship because about 54% of the supporters of affirmative action are white, only 38% are black, and a miniscule 7% are Hispanic. If we looked *only* at the top row of the table (as people sometimes do), we would conclude that whites are more supportive of affirmative action than other groups. But the second row shows that whites are also the huge majority (86%) of those who *oppose* the policy. How can this be? The row percentages in this table simply reflect the fact that whites vastly outnumber the other groups; there are 5 times as many whites as blacks in the sample and 16 times as many whites as Hispanics. Computing percentages within the rows would make sense only if racial or ethnic group could vary by attitude or opinion, and Table 11.9 could easily lead to false conclusions about this relationship.

Professional researchers sometimes compute percentages in the wrong direction or ask a question about the relationship incorrectly. You should always check bivariate tables to make sure the analysis agrees with the patterns in the table.

Applying Statistics 11.1 Association and Bivariate Tables

Why are many Americans attracted to movies that emphasize graphic displays of violence? One idea is that the fans of "slasher" movies feel threatened by violence in their daily lives and use these movies as a means of coping with their fears. In the safety of the theater, violence can be vicariously experienced and feelings and fears can be expressed privately. Also, violent movies almost always provide a role model of one character who does deal with violence successfully (usually, of course, with more violence).

Is fear of violence associated with frequent attendance at high-violence movies? The following table reports the joint frequency distributions of "fear" and "attendance" in percentages for a fictitious sample.

	Fear		
Attendance	Low	Moderate	High
Rare	50%	20%	30%
Occasional	30%	60%	30%
Frequent	20%	20%	40%
Totals	100%	100%	100%

The conditional distributions of attendance (Y) change across the values of fear (X), so these variables are associated. The clustering of cases in the diagonal from upper left to lower right suggests a substantial relationship in the predicted direction. People who are low on fear attend

(continued next page)

Applying Statistics 11.1 *(continued)*

violent movies infrequently, and people who are high on fear are frequent attendees. Because the maximum difference in column percentages in the table is 30 (in both the top and middle rows), the relationship can be characterized as moderate to strong.

These results do suggest an important relationship between fear and attendance. However, notice that they also pose an interesting causal problem. The table supports the idea that fearful and threatened people attend violent movies as a coping mechanism (X causes Y) but is also consistent with the reverse causal argument: Attendance at violent movies increases fears for one's personal safety (Y causes X). The results support *both* causal arguments and remind us that association is not the same thing as causation.

Measures of Association

Column percentages provide useful and detailed information about the bivariate associations and should always be computed and analyzed. However, they can be awkward and cumbersome to use, particularly for larger tables. On the other hand, measures of association summarize the overall strength (and direction for ordinal-level variables) of a bivariate association in a single number—a more compact and convenient format for interpretation and discussion.

There are many measures of association, but we will focus on a few of the most widely used. We will cover these statistics by the level of measurement for which they are most appropriate. In this chapter, we consider measures appropriate for nominal variables, and in the next chapter, we will cover measures for ordinal-level variables. Finally, in Chapter 13, we will consider Pearson's r—a measure of association or correlation for interval-ratio-level variables. To analyze relationships for variables that are at different levels (for example, one nominal-level variable and one ordinal-level variable), we generally use the measures of association appropriate for the lower of the two levels of measurement.

Chi Square–Based Measures of Association

Over the years, social science researchers have relied heavily on measures of association based on the value of chi square. When the value of chi square is already known, these measures are easy to calculate. To illustrate, let us reconsider Table 10.3, which displayed, with fictitious data, a relationship between accreditation and employment for social work majors. For the sake of convenience, this table is reproduced here as Table 11.10.

We saw in Chapter 10 that this relationship is statistically significant ($\chi^2 = 10.78$, which is significant at $\alpha = 0.05$), but the question now concerns the *strength* of the association. A brief glance at Table 11.10 shows that the conditional distributions of employment status do change, so the variables are associated. To emphasize this point, it is always helpful to calculate column percentages, as in Table 11.11.

So far, we know that the relationship between these two variables is statistically significant and that there is an association of some kind between accreditation and employment. To assess the strength of the association, we will compute

TABLE 11.10 Employment of 100 Social Work Majors by Accreditation Status of Undergraduate Program (fictitious data)

	Accreditation Status		
Employment Status	Accredited	Not Accredited	Totals
Working as a social worker	30	10	40
Not working as a social worker	25	35	60
Totals	55	45	100

TABLE 11.11 Employment by Accreditation Status (percentages)

	Accreditation Status		
Employment Status	Accredited	Not Accredited	Totals
Working as a social worker	54.6%	22.2%	40.0%
Not working as a social worker	45.5%	77.8%	60.0%
Totals	100.1%	100.0	100.0%

TABLE 11.12 The Relationship Between the Value of Nominal-Level Measures of Association and the Strength of the Relationship

Value	Strength
If the value is:	*The strength of the relationship is:*
between 0.00 and 0.10	weak
between 0.11 and 0.30	moderate
greater than 0.30	strong

a **phi** (ϕ). This statistic is a chi square–based measure of association appropriate for 2 × 2 tables (that is, tables with two rows and two columns).

Before considering the computation of phi, it will be helpful to establish some general guidelines for interpreting the value of measures of association for nominally measured variables, similar to the guidelines we used for interpreting the maximum difference in column percentages. For phi and the other measures introduced in this chapter, the general relationship between the value of the statistic and the strength of the relationship is presented in Table 11.12.

As was the case for Table 11.5, these relationships between the numerical values and the descriptive terms are arbitrary and meant as general guidelines only. Measures of association generally have mathematical definitions that yield interpretations more meaningful and exact than these.

Calculating Phi (ϕ). One of the attractions of phi is that it is easy to calculate. Simply divide the value of the obtained chi square by N and then take the square root of the result. Expressed in symbols, the formula for phi is:

FORMULA 11.1

$$\phi = \sqrt{\frac{\chi^2}{N}}$$

For the data displayed in Table 11.8, the chi square was 10.78. Therefore, phi is:

$$\phi = \sqrt{\frac{\chi^2}{N}}$$

$$\phi = \sqrt{\frac{10.78}{100}}$$

$$\phi = 0.33$$

For a 2 × 2 table, phi ranges in value from 0 (no association) to 1.00 (perfect association). The closer to 1.00, the stronger the relationship; the closer to 0.00, the weaker the relationship. For Table 11.10, we already knew that the relationship was statistically significant at the 0.05 level. As a measure of association, phi adds information about the strength of the relationship; there is a moderate-to-strong relationship between these two variables. As for the pattern of the association, the column percentages in Table 11.11 show that graduates of accredited programs were more often employed as social workers.

Calculating Cramer's *V*. For tables larger than 2 × 2 (or, more exactly, for tables with more than two columns or more than two rows), the upper limit of phi can exceed 1.00. This makes phi difficult to interpret, and a more general form of the statistic called **Cramer's *V*** must be used for larger tables. The formula for Cramer's *V* is:

FORMULA 11.2

$$V = \sqrt{\frac{\chi^2}{(N)(\min r - 1, c - 1)}}$$

where $(\min r - 1, c - 1)$ = the minimum value of $r - 1$ (number of rows minus 1) or $c - 1$ (number of columns minus 1)

In words: To calculate *V*, find the lesser of the number of rows minus 1 $(r - 1)$ or the number of columns minus 1 $(c - 1)$, multiply this value by *N*, divide the result into the value of chi square, and then find the square root. Cramer's *V* has an upper limit of 1.00 for a table of any size and will be the same value as phi if the table has either two rows or two columns. Like phi, Cramer's *V* can be interpreted as an index that measures the strength of the association between two variables.

To illustrate the computation of *V*, suppose you had gathered the data displayed in Table 11.13, which shows the relationship between membership in student organizations and academic achievement for a sample of college students. The obtained chi square for this table is 31.5, which is significant at the 0.05 level. Cramer's *V* is:

$$V = \sqrt{\frac{\chi^2}{(N)(\min r - 1, c - 1)}}$$

$$V = \sqrt{\frac{31.50}{(75)(2)}}$$

$$V = \sqrt{\frac{31.50}{150}}$$

$$V = \sqrt{0.21}$$

$$V = 0.46$$

TABLE 11.13 Academic Achievement by Club Membership

	Membership			
Academic Achievement	Fraternity or Sorority	Other Organization	No Memberships	Totals
Low	4	4	17	25
Moderate	15	6	4	25
High	4	16	5	25
Totals	23	26	26	75

TABLE 11.14 Academic Achievement by Club Membership (percentages)

	Membership			
Academic Achievement	Fraternity or Sorority	Other Organization	No Memberships	Totals
Low	17.4%	15.4%	65.4%	33.3%
Moderate	65.2%	23.1%	15.4%	33.3%
High	17.4%	61.5%	19.2%	33.3%
Totals	100.0%	100.0%	100.0%	99.9%

Because Table 11.13 has the same number of rows and columns, we may use either $(r - 1)$ or $(c - 1)$ in the denominator. In either case, the value of the denominator is N multiplied by $(3 - 1)$, or 2. The computed value of V of 0.46 means there is a strong association between club membership and academic achievement. Column percentages are presented in Table 11.14 to help identify the pattern of this relationship. Fraternity and sorority members tend to be moderate, members of other organizations tend to be high, and nonmembers tend to be low in academic achievement.

ONE STEP AT A TIME Computing and Interpreting Phi and Cramer's *V*

Step ***Operation***

To calculate phi, solve Formula 11.1

1. Divide the value of chi square by *N*.
2. Find the square root of the quantity you found in step 1. The resulting value is phi.
3. Consult Table 11.12 to help interpret the value of phi.

To calculate Cramer's V, solve Formula 11.2

1. Find the number of rows (r) and number of columns (c) in the table. Subtract 1 from the lesser of these two numbers to find min $(r - 1, c - 1)$.
2. Multiply the value you found in step 1 by *N*.
3. Divide the value of chi square by the value you found in step 2.
4. Take the square root of the quantity you found in step 3. The resulting value is *V*.
5. Consult Table 11.12 to help interpret the value of *V*.

The Limitations of Phi and *V*. One limitation of phi and Cramer's *V* is that they are only general indicators of the strength of the relationship. Of course, the closer these measures are to 0.00, the weaker the relationship; the closer to 1.00, the stronger the relationship. Values between 0.00 and 1.00 can be described as weak, moderate, or strong according to the general convention introduced in Table 11.12 but have no direct or meaningful interpretation. On the other hand, phi and *V* are easy to calculate (once the value of chi square has been obtained) and are commonly used indicators of the importance of an association.[4] *(For practice in computing phi and Cramer's V, see any of the problems at the end of this chapter. Remember that for tables that have either two rows or two columns, phi and Cramer's V will have the same value.)*

A Measure of Association Based on Proportional Reduction in Error

In recent years, a group of measures based on a logic known as **proportional reduction in error (PRE)** has been developed to complement the older chi square–based measures of association.

The Logic of Proportional Reduction in Error. The logic of these measures is based on two different predictions about the scores of cases on the dependent variable. In one prediction, we ignore information about the independent variable, and in the second prediction we take the independent variable into account. A PRE measure of association, such as lambda, tells us how much knowledge of the independent variable improves our predictions of the dependent variable. An example will make the logic clearer.

For nominal-level variables, we first predict the category into which each case will fall on the dependent variable (*Y*) while ignoring the independent variable (*X*). Because we would be predicting blindly in this case, we would make many errors of prediction. That is, we would often predict the value of a case on the dependent variable incorrectly.

The second prediction allows us to take the independent variable into account. If the two variables are associated, the additional information supplied by the independent variable will reduce our errors of prediction; that is, we should misclassify fewer cases. The stronger the association between the variables, the greater the reduction in errors. In the case of a perfect association, we would make no errors at all when predicting a score on *Y* from a score on *X*. On the other hand, when there is no association between the variables, knowledge of the independent variable will not improve the accuracy of our predictions. We would make just as many errors of prediction with knowledge

[4]Two other chi square-based measures of association—T^2 and *C* (the contingency coefficient)—are sometimes reported in the literature. Both of these measures have serious limitations. T^2 has an upper limit of 1.00 only for tables with an equal number of rows and columns, and the upper limit of *C* varies depending on the dimensions of the table. These characteristics make these statistics less useful and more difficult to interpret than phi and Cramer's *V*.

of the independent variable as we did without knowledge of the independent variable.

To illustrate further, suppose you were placed in the rather unusual position of having to predict whether each of the next 100 people you meet will be shorter or taller than 5 feet 9 inches under the condition that you would have no knowledge of these people at all. With absolutely no information about these people, your predictions will be wrong quite often (you will frequently misclassify a tall person as short and vice versa).

Now assume you must go through this ordeal twice, but that on the second round, you know the sex of the person whose height you must predict. Because height is associated with sex and females are, on average, shorter than males, the optimal strategy would be to predict that all females are short and all males are tall. Of course, you will still make errors on this second round, but if the variables are associated, the number of errors on the second round will be less than the number of errors on the first. That is, using information about the independent variable will reduce the number of errors (if, of course, the two variables are related). How can these unusual thoughts be translated into a useful statistic?

Lambda. One hundred individuals have been categorized by gender and height, and the data are displayed in Table 11.15. It is clear, even without percentages, that the two variables are associated. To measure the strength of this association, a PRE measure called **lambda** (symbolized by the Greek letter λ) will be calculated. Following the logic introduced in the previous section, we must find two quantities: First, the number of prediction errors made while ignoring the independent variable (gender); and second, the number of prediction errors made while taking gender into account. These two sums will then be compared to derive the statistic.

First, the information given by the independent variable (gender) can be ignored by working only with the row marginals. Two different predictions can be made about height (the dependent variable) by using these marginals. We can predict either that all subjects are tall or that all subjects are short.[5] The first prediction (all subjects are tall) will result in 48 errors. That is, for this prediction, all 100 cases would be placed in the first row. Only 52 of the cases actually belong in this row, so this prediction would result in (100 − 52), or 48, errors. On the other hand, if we had predicted that all subjects were short, we would have made 52 errors (100 − 48 = 52). We will take the *lesser* of these two numbers and refer

TABLE 11.15 Height by Gender

	Gender		
Height	Male	Female	Totals
Tall	44	8	52
Short	6	42	48
Totals	50	50	100

[5]Other predictions are possible, of course, but these are the only two permitted by lambda.

to this quantity as E_1 for the number of errors made while ignoring the independent variable. Thus, $E_1 = 48$.

Next, we predict score on Y (height) again, but this time, we take X (gender) into account by moving from column to column. Because each column is a category of X, we thus take X into account in making our predictions. For each column, predict that all cases will fall in the cell with the largest frequency. For the left-hand column (males), we predict that all 50 cases will be tall and make 6 errors ($50 - 44 = 6$). For the second column (females), our prediction is that all females are short, and 8 errors will be made. By moving from column to column, we have taken X into account and have made a total of 14 errors of prediction—a quantity we will label E_2 ($E_2 = 6 + 8 = 14$).

If the variables are associated, we will make fewer errors under the second procedure than under the first. In other words, E_2 will be smaller than E_1. In this case, we made fewer errors of prediction while taking gender into account ($E_2 = 14$) than while ignoring gender ($E_1 = 48$), so gender and height are clearly associated. Our errors were reduced from 48 to only 14. To find the *proportional* reduction in error, use Formula 11.3:

FORMULA 11.3

$$\lambda = \frac{E_1 - E_2}{E_1}$$

For the sample problem, the value of lambda would be:

$$\lambda = \frac{E_1 - E_2}{E_1}$$

$$\lambda = \frac{48 - 14}{48}$$

$$\lambda = \frac{34}{48}$$

$$\lambda = 0.71$$

The value of lambda ranges from 0.00 to 1.00. Of course, a value of 0.00 means that the variables are not associated at all (E_1 is the same as E_2), and a value of 1.00 means that the association is perfect (E_2 is zero and scores on the dependent variable can be predicted without error from the independent variable). However, unlike phi or V, the numerical value of lambda between the extremes of 0.00 and 1.00 has a precise meaning; it tells us how much the independent variable (X) helps us to predict (or, more loosely, understand) the dependent variable (Y). When multiplied by 100, the value of lambda indicates the strength of the association in terms of the percentage reduction in error. Thus, the lambda calculated for Table 11.15 could be interpreted by saying that knowledge of gender improves our ability to predict height by 71%. That is, we are 71% better off knowing gender when attempting to predict height.

An Additional Example for Calculating and Interpreting Lambda. In this section, we work through another example in order to state the computational routine for lambda in more general terms. Suppose a researcher was concerned with the relationship between religious denomination and attitude toward capital punishment and had collected the data presented in Table 11.16 from a sample of 130 respondents.

TABLE 11.16 Attitude Toward Capital Punishment by Religious Denomination (Fictious Data)

	Religion				
Attitude	Catholic	Protestant	Other	None	Totals
Favors	10	9	5	14	38
Neutral	14	12	10	6	42
Opposes	11	4	25	10	50
Totals	35	25	40	30	130

Find E_1, the number of errors made while ignoring X (religion, in this case), by subtracting the largest row total from N. For Table 11.16 , E_1 will be:

$$E_1 = N - (\text{Largest row total})$$
$$E_1 = 130 - 50$$
$$E_1 = 80$$

To find E_2, begin with the left-hand column (Catholics) and subtract the largest cell frequency from the column total. Repeat these steps for each column and then add the subtotals:

$$\text{For Catholics: } 35 - 14 = 21$$
$$\text{For Protestants: } 25 - 12 = 13$$
$$\text{For Others: } 40 - 25 = 15$$
$$\text{For None: } 30 - 14 = \underline{16}$$
$$E_2 = 65$$

Substitute the values of E_1 and E_2 into Formula 11.3:

$$\lambda = \frac{80 - 65}{80}$$
$$\lambda = \frac{15}{80}$$
$$\lambda = 0.19$$

A lambda of 0.19 means that we are 19% better off using religion to predict attitude toward capital punishment (as opposed to predicting blindly). Or, we could say "Knowledge of a respondent's religious denomination improves the accuracy of our predictions by a factor of 19%." At best, this relationship is moderate in strength (see Table 11.12).

The Limitations of Lambda. Lambda has two characteristics that should be stressed. First, lambda is asymmetric. This means that the value of the statistic will vary depending on which variable is taken as independent. For example, in Table 11.16, lambda would be .14 if attitude toward capital punishment had been taken as the independent variable. Thus, you should exercise some caution in the designation of an independent variable. If you consistently follow the convention of arraying the independent variable in the columns and compute lambda as outlined earlier, the asymmetry of the statistic should not be confusing.

Second, when one of the row totals is much larger than the others, lambda can be misleading. It can be 0.00 even when other measures of association are greater

ONE STEP AT A TIME Computing and Interpreting Lambda

Step	***Operation***
To calculate lambda, solve Formula 11.3	
1.	To find E_1, subtract the largest row subtotal (marginal) from N.
2.	Starting with the far left-hand column, subtract the largest cell frequency in the column from the column total. Repeat this step for all columns in the table.
3.	Add up all the values you found in step 2. The result is E_2.
4.	Subtract E_2 from E_1.
5.	Divide the quantity you found in step 4 by E_1. The result is lambda.
6.	To interpret lambda, multiply the value of lambda by 100. This percentage tells us the extent to which our predictions of the dependent variable are improved by taking the independent variable into account. Also, lambda may be interpreted by using the descriptive terms in Table 11.12.

than 0.00 and the conditional distributions for the table indicate that there is an association between the variables. This anomaly is a function of the way lambda is calculated and suggests that great caution should be exercised in the interpretation of lambda when the row marginals are very unequal. In fact, in the case of very unequal row marginals, a chi square–based measure of association would be the preferred measure of association. *(For practice in computing lambda, see any of the problems at the end of this chapter. As with phi and Cramer's V, it is probably a good idea to start with small samples and 2 × 2 tables.)*

Applying Statistics 11.2 Measures of Association

A random sample of students at a large urban university has been classified as either "traditional" (18–23 years old and unmarried) or "nontraditional" (24 or older or married). Subjects have also been classified as "vocational" if their primary motivation for college attendance is career- or job-oriented or "academic" if their motivation is to pursue knowledge for its own sake. Are these two variables associated?

	Type		
Motivation	Traditional	Nontraditional	Totals
Vocational	25	60	85
Academic	75	15	90
Totals	100	75	175

Because this is a 2 × 2 table, we can compute phi as a measure of association. The chi square for the table is 51.89, so phi is

$$\phi = \sqrt{\frac{\chi^2}{N}}$$

$$\phi = \sqrt{\frac{51.89}{175}}$$

$$\phi = \sqrt{0.30}$$

$$\phi = 0.55$$

which indicates a strong relationship between the two variables.

A lambda can also be computed as an additional measure of association:

$$E_1 = 175 - 90 = 85$$

$$E_2 = (100 - 75) + (75 - 60)$$
$$= 25 + 15 = 40$$

$$\lambda = \frac{E_1 - E_2}{E_1}$$

$$\lambda = \frac{85 - 40}{85}$$

$$\lambda = \frac{45}{85}$$

$$\lambda = 0.53$$

A lambda of 0.53 indicates that we would make 53% fewer errors in predicting motivation *(Y)* from student type *(X)*, as opposed to predicting motivation while ignoring student type. The association is strong, and by inspection of the table, we can see that traditional students are more likely to have academic motivations (75%) and that nontraditional types are more likely to be vocational in motivation (80%).

STATISTICS IN EVERYDAY LIFE

How would you feel if you learned that a Muslim clergyman with strong anti-U.S. views had been invited to speak on your campus? Do you suppose that your gender might affect your feelings? According to the combined 2010 and 2011 General Social Survey, administered to a representative sample of Americans, there is a relationship between these two variables, as illustrated in this table:

Support for Allowing Anti-U.S. Muslim Clergyman to Speak by Gender: Frequencies and (*Percentages*)

	GENDER		
ALLOW?	Males	Females	Total
Yes	583 (*49.5%*)	512 (*36.2%*)	1,095 (*42.2%*)
No	594 (*50.5%*)	903 (*63.8%*)	1,497 (*57.8%*)
	1,177 (*100.0%*)	1,415 (*100.0%*)	2,592 (*100.0%*)

chi square = 46.93 ($p < 0.05$), maximum difference = 13.3, phi = 0.13, lambda = 0.00

The relationship is statistically significant and (disregarding lambda) moderate to weak in strength. Men are split 50-50 on the issue, but a clear majority of women are opposed. What factors might account for this difference between the genders?

SUMMARY

1. Analyzing the association between variables provides information that is complementary to tests of significance. The latter are designed to detect nonrandom relationships, whereas measures of association are designed to quantify the importance or strength of a relationship.
2. Relationships between variables have three characteristics: the existence of an association, the strength of the association, and the direction or pattern of the association. These three characteristics can be investigated by calculating percentages for a bivariate table in the direction of the independent variable (vertically) and then comparing the percentages in the opposite direction (horizontally). It is often useful (as well as quick and easy) to assess the strength of a relationship by finding the maximum difference in column percentages in any row of the table.
3. Tables 11.1 and 11.2 can be analyzed in terms of these three characteristics. Clearly, a relationship does exist between job satisfaction and productivity because the conditional distributions of the dependent variable (productivity) are different for the three different conditions of the independent variable (job satisfaction). Even without a measure of association, we can see that the association is substantial in that the change in Y (productivity) across the three categories of X (satisfaction) is marked. The maximum difference of 36.54% confirms that the relationship is substantial (moderate to strong). Furthermore, the relationship is positive in direction. Productivity increases as job satisfaction rises, and workers who report high job satisfaction also tend to be high on productivity. Workers with little job satisfaction tend to be low on productivity.
4. Given the nature and strength of the relationship, it could be predicted with fair accuracy that highly satisfied workers tend to be highly productive ("Happy workers are busy workers"). These results might be taken as evidence of a causal relationship between these two variables, but they cannot by themselves prove that a causal relationship exists; association is not the same thing as causation. In fact, although we have presumed that job satisfaction is the independent variable ("Happy workers are busy workers"), we could have argued the reverse causal sequence ("Busy workers are happy workers"). The results presented in Tables 11.1 and 11.2 are consistent with both causal arguments.
5. Phi, Cramer's V, and lambda are measures of association, and each is appropriate for a specific situation. Phi is used for nominal-level variables in a 2×2 table and Cramer's V is used for tables larger than 2×2. Lambda is a proportional reduction in error (PRE) measure appropriate for nominal-level variables. These statistics express information about the strength of the relationship *only*. An inspection of the column percentages will reveal patterns in the bivariate relationship as well as other useful information.

GLOSSARY

Association. The relationship between two (or more) variables. Two variables are said to be associated if the distribution of one variable changes for the various categories or scores of the other variable.

Column percentages. Percentages computed with each column of a bivariate table.

Conditional distribution of *Y*. The distribution of scores on the dependent variable for a specific score or category of the independent variable when the variables have been organized into table format.

Cramer's *V*. A chi square–based measure of association for nominally measured variables that have been organized into a bivariate table with any number of rows and columns.

Dependent variable. In a bivariate relationship, the variable that is taken as the effect.

Independent variable. In a bivariate relationship, the variable that is taken as the cause.

Lambda. A proportional reduction in error (PRE) measure for nominally measured variables that have been organized into a bivariate table.

Maximum difference. A way to assess the strength of an association between variables that have been organized into a bivariate table. The maximum difference is the largest difference between column percentages for any row of the table.

Measures of association. Statistics that quantify the strength of the association between variables. Measures of association for ordinal and interval-ratio variables also indicate the direction of the association (see Chapters 12 and 13).

Negative association. A bivariate relationship in which the variables vary in opposite directions. As one variable increases, the other decreases, and high scores on one variable are associated with low scores on the other.

Phi (ϕ). A chi square–based measure of association for nominally measured variables that have been organized into a bivariate table with two rows and two columns (a 2 x 2 table).

Positive association. A bivariate relationship in which the variables vary in the same direction. As one variable increases, the other also increases, and high scores on one variable are associated with high scores on the other.

***X*.** Symbol used for any independent variable.

***Y*.** Symbol used for any dependent variable.

SUMMARY OF FORMULAS

FORMULA 11.1

$$\text{Phi: } \phi = \sqrt{\frac{\chi^2}{N}}$$

FORMULA 11.2

$$\text{Cramer's: } V \quad V = \sqrt{\frac{\chi^2}{(N)(\min r - 1, c - 1)}}$$

FORMULA 11.3

$$\text{Lambda: } \lambda = \frac{E_1 - E_2}{E_1}$$

PROBLEMS

11.1 PA Various supervisors in the city government of Shinbone, Kansas, have been rated on the extent to which they practice authoritarian styles of leadership and decision making. The efficiency of each department has also been rated, and the results are summarized in the following table. Use column percentages, the maximum difference, and measures of association to describe the strength and pattern of this association.

	Authoritarianism		
Efficiency	Low	High	Totals
Low	10	12	22
High	17	5	22
Totals	27	17	44

11.2 SOC The administration of a local college campus has proposed an increase in the mandatory student fee in order to finance an upgrading of the intercollegiate football program. A survey sampling the opinions of the faculty regarding this proposal has been completed. Is there any association between support for raising fees and the gender, discipline, or tenured status of the faculty? Use column percentages, the maximum difference, and measures of association to describe the strength and pattern of these associations.

a. Support for raising fees by gender:

	Gender		
Support	Males	Females	Totals
For	12	7	19
Against	15	13	28
Totals	27	20	47

b. Support for raising fees by discipline:

	Discipline		
Support	Liberal Arts	Science and Business	Totals
For	6	13	19
Against	14	14	28
Totals	20	27	47

c. Support for raising fees by tenured status:

	Status		
Support	Tenured	Nontenured	Totals
For	15	4	19
Against	18	10	28
Totals	33	14	47

11.3 PS How consistent are people in their voting habits? Do people vote for the same party from election to election? Below are the results of a poll in which people were asked if they had voted Democrat or Republican in each of the last two presidential elections. Use column percentages, the maximum difference, and measures of association to describe the strength and pattern of the association.

	2004 Election		
2008 Election	Democrat	Republican	Totals
Democrat	117	23	140
Republican	17	178	195
Totals	134	201	335

11.4 SOC A needs assessment survey has been distributed in a large corporation. Employees have been asked if they thought that a new policy of casual dress on Fridays would boost company morale. Use column percentages, the maximum difference, and measures of association to describe the strength and direction of the association between the type of employee and attitude toward casual Fridays. Write a few sentences describing the relationship.

	Employee Type		
Casual Fridays?	Supervisor	Worker	Totals
Yes	121	426	547
No	175	251	426
Totals	296	677	973

11.5 SW As the state director of mental health programs, you note that some local mental health facilities have very high rates of staff turnover. You believe that part of this problem is a result of the fact that some of the local directors have very little training in administration and poorly developed leadership skills. Before implementing a program to address this problem, you collect some data to make sure your beliefs are supported by the facts. Is there a relationship between staff turnover and the administrative experience of the directors? Use column percentages, the maximum difference, and measures of association to describe the strength and direction of the association. Write a few sentences describing the relationship.

	Director Experienced?		
Turnover	No	Yes	Totals
Low	4	9	13
Moderate	9	8	17
High	15	5	20
Totals	28	22	50

11.6 CJ About half the neighborhoods in a large city have instituted programs to increase citizen involvement in crime prevention. Do these areas experience less crime? Write a few sentences describing the relationship in terms of pattern and strength of the association. Use column percentages, the maximum difference, and measures of association to describe the strength and direction of the association. Write a few sentences describing the relationship.

Crime Rate	Program		Totals
	No	Yes	
Low	28	16	44
Moderate	33	27	60
High	52	45	97
Totals	113	88	201

11.7 GER A survey of senior citizens who live in either a housing development specifically designed for retirees or an age-integrated neighborhood has been conducted. Is the type of living arrangement related to the sense of social isolation?

Sense of Isolation	Living Arrangement		Totals
	Housing Development	Integrated Neighborhood	
Low	80	30	110
High	20	120	140
Totals	100	150	250

11.8 SOC A researcher has given a survey on sexual attitudes to a sample of 317 teenagers. The respondents were asked whether they considered premarital sex to be "always wrong" or "not always wrong." The tables below summarize the relationship between responses to this item and several other variables. For each table, assess the strength and pattern of the relationship and then write a paragraph interpreting these results.

a. Attitudes toward premarital sex by gender:

Premarital Sex	Gender		Totals
	Female	Male	
Always wrong	90	105	195
Not always wrong	65	57	122
Totals	155	162	317

b. Attitudes toward premarital sex by courtship status:

Premarital Sex	Ever "Gone Steady"		Totals
	No	Yes	
Always wrong	148	47	195
Not always wrong	42	80	122
Totals	190	127	317

c. Attitudes toward premarital sex by social class:

Premarital Sex	Social Class		Totals
	Blue Collar	White Collar	
Always wrong	72	123	195
Not always wrong	47	75	122
Totals	119	198	317

11.9 SOC Below are five dependent variables cross-tabulated against gender as an independent variable. Use column percentages, the maximum difference, and an appropriate measure of association to analyze these relationships. Summarize the results of your analysis in a paragraph that describes the strength and pattern of each relationship.

a. Support for the legal right to an abortion by gender:

Right to Abortion?	Gender		Totals
	Male	Female	
Yes	310	418	728
No	432	618	1,050
Totals	742	1,036	1,778

b. Support for capital punishment by gender:

Capital Punishment?	Gender		Totals
	Male	Female	
Favor	908	998	1906
Oppose	246	447	693
Totals	1,154	1,445	2,599

c. Approve of suicide for people with incurable disease by gender:

Right to Suicide?	Gender		Totals
	Male	Female	
Yes	524	608	1132
No	246	398	644
Totals	770	1,006	1,776

d. Support for sex education in public schools by gender:

Sex Education?	Gender		Totals
	Male	Female	
Favor	685	900	1585
Oppose	102	134	236
Totals	787	1,034	1,821

e. Support for traditional gender roles by gender:

Women Should Take Care of Running Their Homes and Leave Running the Country to Men	Gender		
	Male	Female	Totals
Agree	116	164	280
Disagree	669	865	1,534
Totals	785	1,029	1,814

11.10 PS Below are the tables you analyzed with chi square in problem 10.16. As you recall, these are the results of a survey of 748 voters in a large city concerning how they voted in the 2008 presidential election. Use column percentages, the maximum difference, and an appropriate measure of association to analyze these relationships. Summarize the results of your analysis in a paragraph that describes the strength and pattern of each relationship. Based on the chi square test, you already knew if these relationships were statistically significant. What new information do the statistics introduced in this chapter add to your understanding of these relationships?

a. Presidential preference and gender:

	Gender		
Preference	Male	Female	Totals
McCain	165	173	338
Obama	200	210	410
Totals	365	383	748

b. Presidential preference and race/ethnicity:

	Race/Ethnicity			
Preference	White	Black	Latino	Totals
McCain	289	5	44	338
Obama	249	95	66	410
Totals	538	100	110	748

c. Presidential preference by education:

	Education				
Preference	Less Than HS	HS Graduate	College Graduate	Postgrad Degree	Totals
McCain	30	180	118	10	338
Obama	35	120	218	37	410
Totals	65	300	336	47	748

d. Presidential preference by religion:

	Religion					
Preference	Protestant	Catholic	Jewish	None	Other	Totals
McCain	165	110	10	28	25	338
Obama	245	55	20	60	30	410
Totals	410	165	30	88	55	748

YOU ARE THE RESEARCHER: Understanding Political Beliefs, Part II

At the end of Chapter 10, you investigated possible causes of people's beliefs on four controversial issues. Now you will extend your analysis by using **Crosstabs** to produce the statistics presented in this chapter. There will be two projects, and once again, we will begin with a brief demonstration using the relationship between *abany* and *sex*.

Demonstration

With the 2010 GSS loaded, click **Analyze**, **Descriptive Statistics**, and **Crosstabs** and then name *abany* as the row (dependent) variable and *sex* as the column (independent) variable. Click the **Cells** button and then request column percentages by clicking the box next to **Column** in the **Percentages** box. Also, click the **Statistics** button and then request chi square, phi, Cramer's *V*, and lambda by clicking the appropriate boxes. Click **Continue** and **OK** to execute your task.

As you will see, SPSS generates a good deal of output for this procedure, and we have condensed the information into a single table based on how this test might be presented in the professional research literature. Note that the cells of the bivariate table include frequencies and percentages and that all the statistical information is presented in a single short line beneath the table.

ABORTION IF WOMAN WANTS FOR ANY REASON * RESPONDENT'S SEX Crosstabulation

		RESPONDENT'S SEX		
		MALE	FEMALE	Total
Should a woman be able to get a legal abortion if she wants if for any reason?	YES	185	205	390
		44.7%	41.9%	43.2%
	NO	229	284	513
		55.3%	58.1%	56.8%
	Totals	414	489	903
		100.0%	100.0%	100.0%

chi square = 0.70 ($df = 1$, $p > 0.05$); phi = 0.03; lambda = 0.00

The three questions about bivariate association introduced in this chapter provide a useful framework for reading and analyzing these results:

1. *Is there an association?* The column percentages in the bivariate table change so there is an association between support for abortion and gender. This conclusion is verified by the fact that chi square is a nonzero value.
2. *How strong is the association?* We can assess strength in several different ways. First, the maximum difference is 44.7 − 41.9, or 2.8, percentage points. (This calculation is based on the top row, but because this is a 2 × 2 table, using the bottom row would result in exactly the same value). According to Table 11.5, this means the relationship between the variables is weak. Second, we can (and should) use a measure of association to assess the strength of the relationship. We have a choice of two different measures: lambda and (because this is a 2 x 2 table) phi. Looking in the "Directional Measures" output box, we see several different values for lambda. Recall that lambda is an asymmetric measure and that it changes value depending on which variable is taken as dependent. In this case, the dependent variable is "ABORTION IF WOMAN WANTS FOR ANY REASON" and the associated lambda is reported as .000. This indicates that there is no association between the variables, but we have already seen that the variables *are* associated—if only weakly. Remember that lambda can be zero when the variables are associated, but the row totals in the table are very unequal. Lambda might be a little misleading, but it is still telling us that the relationship is weak.

   Turning to phi (under "Symmetric Measures"), we see a value of 0.03. This indicates that the relationship is weak (see Table 11.12), a conclusion that is consistent with the value of the maximum difference and the fact that the relationship is not significant at the 0.05 level.
3. *What is the pattern of the relationship?* Because sex is a nominal-level variable, the relationship cannot have a direction (that is, it cannot be positive or negative). However, we can discuss the pattern of the relationship: how values of the variables seem to go together. Although the maximum difference is small (2.8%), males are more supportive of abortion than females.

In summary, using chi square, the column percentages, and phi, we can say that the relationship between support for legal abortion and gender is weak and not statistically significant. If we were searching for an important cause of attitudes about abortion, we would discard this independent variable and seek another.

Your turn.

PROJECT 1: Explaining Beliefs

In this project, you will once again analyze beliefs about capital punishment (*cappun*), assisted suicide (*letdie1*), gay marriage (*marhomo*), and immigration (*letin1*). You will select an independent variable *other than the one you used in Chapter 10* and then use SPSS to generate chi square, column percentages, phi or Cramer's *V*, and lambda. You will use all these statistics to help analyze and interpret your results.

STEP 1: Choosing an Independent Variable

Select an independent variable that seems likely to be an important cause of people's attitudes about the death penalty, assisted suicide, gay marriage, and immigration. Be sure to select an independent variable that has *only* 2–5 categories, and use the recode command if necessary. You might consider gender, level of education (use *degree*), religion, or age (the recoded version—see Chapter 9) as possibilities, but there are many others. Record the name of your independent variable and then state exactly what it measures in this table:

SPSS Name	What Exactly Does This Variable Measure?

STEP 2: Stating Hypotheses

State hypotheses about the relationships you expect to find between your independent variable and each of the four dependent variables. State these hypotheses in terms of which category of the independent variable you expect to be associated with which category of the dependent variable (for example, "I expect that men will be more supportive of the legal right to an abortion for any reason").

1. ______________________________
2. ______________________________
3. ______________________________
4. ______________________________

STEP 3: Running Crosstabs

Click **Analyze → Descriptives → Crosstabs** and then place the four dependent variables (*cappun*, *letdie1*, *letin1*, and *marhomo*) in the **Rows:** box and the independent variable you selected in the **Columns:** box. Click the **Statistics** button to get chi square, phi or Cramer's *V*, and lambda and then click the **Cells** button for column percentages.

STEP 4: Recording Results

These commands will generate a lot of output, and it will be helpful to summarize your results in the following table.

Dependent Variable	Chi Square Significant at < .0.05?	Maximum Difference	Phi or Cramer's *V*	Lambda
cappun				
letdie1				
letin1				
marhomo				

STEP 5: Analyzing and Interpreting Results

Write a short summary of results for each dependent variable. The summary needs to identify the variables being tested, the results of the chi square test, and the strength and pattern of the relationship. It is probably best to characterize the relationship in general terms and then cite the statistical values in parentheses. For example, we might summarize our test of the relationship between gender and support for abortion by saying: "The relationship between gender and support for abortion was not significant and weak (chi square = 1.853, $df = 1$, $p < .05$. Phi = .05). Men were slightly more supportive of the right to a legal abortion for any reason than women." You should also note whether your hypotheses were supported.

PROJECT 2: Exploring the Impact of Various Independent Variables

In this project, you will examine the relative ability of a variety of independent variables to explain or account for a single dependent variable. You will again use the **Crosstabs** procedure in SPSS to generate statistics and then use the alpha levels and measures of association to judge which independent variable has the most important relationship with your dependent variable.

STEP 1: Choosing Variables

Select a dependent variable. You may use any of the four from Project 1 in this chapter or select a new dependent variable from the 2010 GSS. Be sure your dependent variable has no more than five values or scores. Good choices for dependent variables include any measure of attitudes or opinions. Do not select such characteristics as race, sex, or religion as dependent variables.

Select three independent variables that seem likely to be important causes of the dependent variable you selected. Your independent variable should have no more than five or six categories. You might consider gender, level of education (use *degree*), religion, or age (the recoded version—see Chapter 9) as possibilities, but there are many others.

Record the variable names and then state exactly what each variable measures in this table:

SPSS Name	What Exactly Does This Variable Measure?
Dependent Variable	
Independent Variables	

STEP 2: Stating Hypotheses

State hypotheses about the relationships you expect to find between your independent variables and the dependent variable. State these hypotheses in terms of which category of the independent variable you expect to be associated with which category of the dependent variable (for example, "I expect that men will be more supportive of the legal right to an abortion for any reason").

1. ____________________
2. ____________________
3. ____________________

STEP 3: Running Crosstabs

Click **Analyze** → **Descriptives** → **Crosstabs** and then place your dependent variable in the **Rows:** box and all three of your independent variables in the **Columns:** box. Click the **Statistics** button to get chi square, phi or Cramer's *V*, and lambda. Click the **Cells** button for column percentages.

STEP 4: Recording Results

Your output will consist of three tables, and it will be helpful to summarize your results in the following table. Remember that the significance of the relationship is found in the column labeled "Asymp. Sig (2-sided)" of the second box in the output.

Independent Variables	Chi Square Significant at Less Than 0.05?	Maximum Difference	Phi or Cramer's *V*	Lambda

STEP 5: Analyzing and Interpreting Results

Write a short summary of results for each test by using the same format as in Project 1. Remember to explain whether your hypotheses were supported. Finally, assess which of the independent variables had the most important relationship with your dependent variable. Use the alpha (or the "Asymp. Sig 2-tailed") level and the value of the measures of association to make this judgment.

12 Association Between Variables Measured at the Ordinal Level

LEARNING OBJECTIVES

By the end of this chapter, you will be able to:

1. Calculate and interpret gamma and Spearman's rho.
2. Explain the logic of proportional reduction in error in terms of gamma.
3. Use gamma and Spearman's rho to analyze and describe a bivariate relationship in terms of the three questions introduced in Chapter 11.

USING STATISTICS

The statistical techniques presented in this chapter are used to measure the strength and direction of the association between ordinal-level variables. Examples of situations in which these techniques are useful include:

1. A sociologist believes that burnout or loss of commitment is a huge challenge for public school teachers and investigates the nature of the relationship between burnout and number of years teaching. She finds that the relationship is moderate in strength and positive in direction: the greater the number of years of experience, the higher the level of burnout.
2. A study of crime in the 50 largest U.S. cities finds a moderate-to-strong positive relationship with the percentage of young males in the population: the higher the percentage, the higher the crime rate.
3. A cross-national study of suicide finds a negative correlation with economic growth: the faster the growth, the lower the rate of suicide.

There are two common types of ordinal-level variables. Some have many possible scores and resemble interval-ratio-level variables. We will call these *continuous ordinal variables*. A scale that measured an attitude with many different items would produce this type of variable.

The second type, which we will call a *collapsed ordinal variable*, has only a few (no more than five or six) scores and can be created either by collecting data in collapsed form or by collapsing a continuous scale. For example, we would produce collapsed ordinal variables by measuring social class as upper, middle, or lower or by reducing the scores on an attitude scale into just a few categories: For example, we could collapse the scores on a scale measuring political conservatism into "Very conservative," "Moderately conservative," and "Not conservative."

A number of measures of association have been invented for use with collapsed ordinal-level variables. Rather than attempt to cover them all, we will concentrate on **gamma (*G*)**. Other measures suitable for collapsed ordinal-level data (Somer's *d* and Kendall's tau-*b*) are covered on the website for this text. For "continuous" ordinal variables, a statistic called **Spearman's rho (r_s)** is commonly used, and we cover this measure of association toward the end of this chapter.

This chapter will enable you to expand your understanding of how bivariate associations can be described and analyzed, but it is important to remember that we are still addressing the three questions raised in Chapter 11: Are the variables associated? How strong is the association? What is the direction or pattern of the association?

Proportional Reduction in Error (Pre) for Collapsed Ordinal-Level Variables: Gamma

As we saw in Chapter 11, for nominal-level variables, the logic of PRE is based on two different predictions of the scores of cases on the dependent variable (*Y*): one that ignores the independent variable (*X*) and a second that takes the independent variable into account. The value of lambda shows the extent to which taking the independent variable into account improves accuracy when predicting the score of the dependent variable. The PRE logic for ordinal-level variables is similar, and gamma, like lambda, measures the proportional reduction in error gained by predicting one variable while taking the other into account. The major difference lies in the way predictions are made.

For gamma, we predict the *order* of pairs of cases rather than a specific score on the dependent variable. That is, we predict whether one case will have a higher or lower score than the other. First, we predict the order of a pair of cases on one variable while ignoring their order on the other variable. Second, we repeat the prediction while taking the order on the other variable into account.

To illustrate, assume that a researcher is concerned about the causes of "burnout" (that is, demoralization and loss of commitment) among elementary school teachers and wonders if there is a relationship between burnout and years of service. One way to state the research question is: Do teachers with more years of service have higher levels of burnout? Another way to ask the same question is: Do teachers who *rank higher* on years of service also *rank higher* on burnout? If we knew that teacher A had more years of service than teacher B, could we predict that teacher A is also more "burned out" than teacher B? That is, would knowledge of the order of this pair of cases on one variable help us predict their order on the other?

If the two variables are associated, we will reduce our errors when our prediction about one variable is based on knowledge of the other. Furthermore, the stronger the association, the fewer the errors we will make. When there is no association between the variables, gamma will be 0.00 and knowledge of the order of a pair of cases on one variable will not improve our ability to predict their order on the other. A gamma of ± 1.00 denotes a perfect relationship; the order of all pairs of cases on one variable would be predictable without error from their order on the other variable.

With nominal-level variables, we analyzed the pattern of the relationship between the variables. That is, we looked to see which value on one variable (e.g., "male") was associated with which value on the other variable (e.g., "tall" on height). Recall that ordinally measured variables have scores that can be rank-ordered from high to low or from more to less (see Chapter 1). This means that relationships between ordinal-level variables can have a direction as well as a pattern. In terms of the logic of gamma, variables have a *positive* relationship if cases tend to be ranked in the same order on both variables. If Case A ranks above Case B on one variable, Case A would also be ranked above Case B on the second variable. The relationship suggested earlier between years of service and burnout is positive.

In a *negative* relationship, the order of the cases would be reversed between the two variables. If Case A ranked above Case B on one variable, it would rank below Case B on the second variable. If there were a negative relationship between prejudice and education and Case A was more educated than Case B (that is, ranked above Case B on education), then Case A would be less prejudiced (that is, would rank below Case B on prejudice).

The Computation of Gamma

Table 12.1 summarizes the relationship between "length of service" and "burnout" for a fictitious sample of 100 teachers. We need two sums to compute gamma: the number of pairs of cases that are ranked the same on both variables (labeled N_s) and the number of pairs of cases ranked differently on the variables (N_d). We find these sums by working with the cell frequencies.

To find the number of pairs of cases ranked the same (Ns), begin with the cell containing the cases that were ranked the lowest on both variables. In Table 12.1, this would be the upper left-hand cell. *(Note: Not all tables are constructed with values increasing from left to right and from top to bottom. Check to make sure you have located the proper cell.)* The 20 cases in the upper left-hand cell all rank low on both burnout and length of service, and we will refer to them as low-lows, or LLs.

Now form a pair of cases by selecting one case from this cell and one from any other cell—for example, the middle cell in the table. All 15 cases in this cell are moderate on both variables and can be labeled moderate-moderates, or MMs. Any pair of cases formed between these two cells will be ranked the same on both variables. That is, all LLs are lower than all MMs on both variables (on X, low is less than moderate; on Y, low is less than moderate). The total number of pairs of cases is given by multiplying the cell frequencies. Thus, the contribution of these two cells to the total N_s is (20)(15), or 300.

TABLE 12.1 Burnout by Length of Service (fictitious data)

	Length of Service			
Burnout	Low	Moderate	High	Totals
Low	20	6	4	30
Moderate	10	15	5	30
High	8	11	21	40
Totals	38	32	30	100

Gamma ignores all pairs of cases that are tied on either variable, so any pair of cases formed between the LLs and any other cell in the top row (low on burnout) or the left-hand column (low on length of service) will be tied on one variable. Also, any pair of cases formed within any cell will be tied on *X* and *Y*. Gamma ignores all pairs of cases formed within the same row, column, or cell. This means that in computing *Ns,* we will work with only the pairs of cases that can be formed between each cell and the cells below and to the right of it.

In summary: To find the total number of pairs of cases ranked the same on both variables (N_s), multiply the frequency in each cell by the total of all frequencies below and to the right of that cell. Repeat this procedure for each cell and then add the resultant products. The total of these products is N_s. This procedure is displayed in Figure 12.1 for each cell in Table 12.1. Note that none of the cells in the bottom row or the right-hand column of Table 12.1 can contribute to N_s because they have no cells below and to the right of them. Figure 12.1 shows the direction of multiplication for each of the four cells that in a 3 × 3 table can contribute to N_s. Computing N_s for Table 12.1, we find that a total of 1,831 pairs of cases are ranked the same on both variables.

Our next step is to find the number of pairs of cases ranked differently (N_d) on both variables. To do this, multiply the frequency in each cell by the total of all

FIGURE 12.1 Computing N_s in a 3 × 3 Table

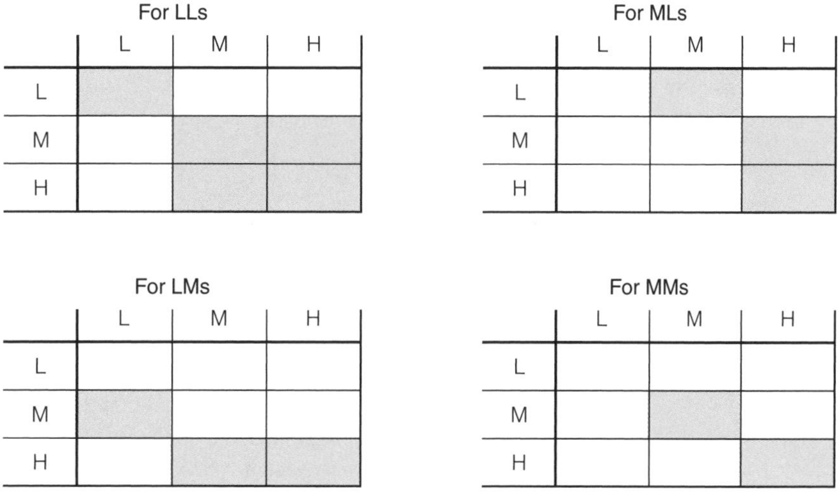

For LLs

	L	M	H
L			
M			
H			

For MLs

	L	M	H
L			
M			
H			

For LMs

	L	M	H
L			
M			
H			

For MMs

	L	M	H
L			
M			
H			

	Contribution to N_s
For LLs, 20(15 + 5 + 11 + 21)	= 1,040
For MLs, 6(21 + 5)	= 156
For HLs, 4(0)	= 0
For LMs, 10(11 + 21)	= 320
For MMs, 15(21)	= 315
For HMs, 5(0)	= 0
For LHs, 8(0)	= 0
For MHs, 11(0)	= 0
For HHs, 21(0)	= 0
	N_s = 1,831

frequencies below and to the left of that cell. Note that the pattern for computing N_d is the reverse of the pattern for N_s. This time, we begin with the upper right-hand cell (high-lows, or HLs) and then multiply the number of cases in the cell by the total frequency of cases below and to the left. The four cases in the upper right-hand cell of Table 12.1 are low on *Y* and high on *X*. If a pair of cases is formed with any case from this cell and any cell below and to the left, the cases will be ranked differently on the two variables. For example, if a pair is formed between any HL case and any case from the middle cell (MMs), the HL case would be less than the MM case on *Y* ("low" is less than "moderate") but more than the MM case on *X* ("high" is greater than "moderate"). The computation of N_d is detailed below and shown graphically in Figure 12.2. In the computations, we have omitted cells that cannot contribute to N_d because they have no cells below and to the left of them.

FIGURE 12.2 Computing N_d in a 3 × 3 Table

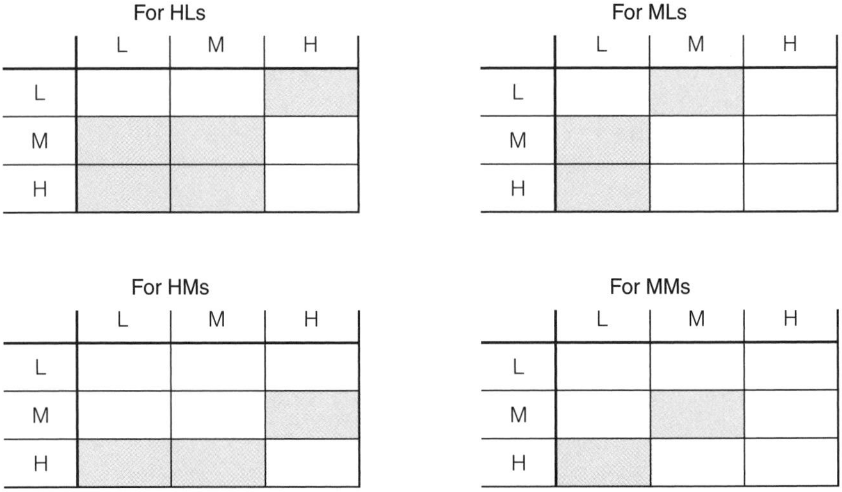

For HLs

	L	M	H
L			
M			
H			

For MLs

	L	M	H
L			
M			
H			

For HMs

	L	M	H
L			
M			
H			

For MMs

	L	M	H
L			
M			
H			

	Contribution to N_d
For LLs, 4(10 + 15 + 8 + 11)	= 176
For MLs, 6(10 + 8)	= 108
For HLs, 5(8 + 11)	= 95
For MMs, 15(8)	= 120
	N_d = 499

Table 12.1 has 499 pairs of cases ranked in different order and 1,831 pairs of cases ranked in the same order. The formula for computing gamma is:

FORMULA 12.1

$$G = \frac{Ns - Nd}{Ns + Nd}$$

where:

N_s = the number of pairs of cases ranked in the same order on both variables

N_d = the number of pairs of cases ranked in different order on the variables

For Table 12.1, the value of gamma would be:

$$G = \frac{N_s - N_d}{N_s + N_d}$$

$$G = \frac{1831 - 499}{1831 + 499}$$

$$G = \frac{1332}{2330}$$

$$G = 0.57$$

The Interpretation of Gamma

A gamma of 0.57 means that we would make 57% fewer errors if we predicted the order of pairs of cases on one variable from the order of pairs of cases on the other (as opposed to predicting order while ignoring the other variable.) Length of service is associated with degree of burnout, and the relationship is positive. Knowing the rankings of two teachers on length of service (Case A is higher on length of service than Case B) will help us predict their ranking on burnout (we would predict that Case A will also be higher than Case B on burnout).

ONE STEP AT A TIME **Computing and Interpreting Gamma**

Step ***Operation***

To Compute Gamma

1. Make sure the table is arranged with the column variable increasing from left to right and the row variable increasing from top to bottom.
2. To compute N_s, start with the upper left-hand cell. Multiply the number of cases in this cell by the total number of cases in all cells below and to the right. Repeat this process for each cell in the table. Add up these subtotals to find N_s.
3. To compute N_d, start with the upper right-hand cell. Multiply the number of cases in this cell by the total number of cases in all cells below and to the left. Repeat this process for each cell in the table. Add up these subtotals to find N_d.
4. Subtract the value of N_d from N_s.
5. Add the value of N_d to N_s.
6. Divide the value you found in step 4 by the value you found in step 5. This value is gamma.

Interpreting the Strength of the Relationship

1. Always begin with the column percentages: the bigger the change in column percentages, the stronger the relationship.
2. There are two ways to use gamma to interpret strength. You can use either or both:
   a. Use Table 12.2 to describe strength in general terms.
   b. Use the logic of proportional reduction in error. Multiply gamma by 100. This value represents the percentage improvement in our predictions of the dependent variable by taking the independent variable into account.

(*continued next page*)

ONE STEP AT A TIME ***(continued)***

Step	***Operation***
Interpreting the Direction of the Relationship	
1.	Always begin by looking at the pattern of the column percentages. If the cases tend to fall in a diagonal from upper left to lower right, the relationship is positive. If the cases tend to fall in a diagonal from lower left to upper right, the relationship is negative.
2.	The sign of the gamma also tells the direction of the relationship. However, be very careful when interpreting direction with ordinal-level variables. Remember that coding schemes for these variables are arbitrary and that a positive gamma may mean that the actual relationship is negative and vice versa.

TABLE 12.2 Relationship Between the Value of Gamma and the Strength of the Relationship

Value	Strength
If the value is:	*The strength of the relationship is:*
between 0.00 and 0.30	weak
between 0.31 and 0.60	moderate
greater than 0.60	strong

Table 12.2 provides some additional assistance for interpreting gamma in a format similar to Tables 11.5 and 11.12. As before, the descriptive terms are arbitrary and intended as general guidelines only. Note, in particular, that strength of a relationship is independent of direction. That is, a gamma of -0.35 is exactly as strong as a gamma of $+0.35$ but is opposite in direction.

To use the computational routine for gamma presented earlier, you must arrange the table in the manner of Table 12.1, with the column variable increasing in value as you move from left to right and the row variable increasing from top to bottom. Be careful to construct your tables according to this format; if you are working with data already in table format, you may have to rearrange the table or rethink the direction of patterns. Gamma is a symmetrical measure of association; that is, the value of gamma will be the same regardless of which variable is taken as independent. *(To practice computing and interpreting gamma, see problems 12.1 to 12.9 and 12.14. Begin with some of the smaller 2 × 2 tables until you are comfortable with these procedures.)*

Applying Statistics 12.1 Gamma

A group of 40 nations have been rated as high or low on religiosity (based on the percentage of a random sample of citizens that described themselves as "very religious") and as high or low in their support for single mothers (based on the percentage of a random sample of citizens who said they would approve of a woman's choosing to be a single parent). Are more religious nations less approving of single mothers?

Applying Statistics 12.1 *(continued)*

	Religiosity		
Approval	Low	High	Totals
Low	4	9	13
High	11	16	27
Totals	15	25	40

The number of pairs of cases ranked in the same order on both variables (N_s) would be:

$$N_s = (4)(16) = 64$$

The number of pairs of cases ranked in different order on both variables (N_d) would be:

$$N_d = (9)(11) = 99$$

Gamma is:

$$G = \frac{N_s - N_d}{N_s + N_d} = \frac{64 - 99}{64 + 99} = \frac{-35}{163} = -0.22$$

A gamma of -0.22 means that we would make 22% fewer errors predicting the order of pairs of cases on the dependent variable (approval of single mothers) when taking the independent variable (religiosity) into account. There is a moderate-to-weak negative association between these two variables. As religiosity increases, approval decreases (that is, more religious nations are less approving of single mothers).

Applying Statistics 12.2 Gamma

For local political elections, do voters turn out at higher rates when the candidates for office spend more money on media advertising? Each of 177 localities has been rated as high or low on voter turnout. Also, the total advertising budgets for all candidates in each locality have been classified as high, moderate, or low. Are these variables associated?

Voter Turnout	Expenditure			Totals
	Low	Moderate	High	
Low	35	32	17	84
High	23	27	43	93
Totals	58	59	60	177

We will use gamma to summarize the strength and direction of the association. The number of pairs of cases ranked in the same order on both variables (N_s) would be:

$$N_s = 35(27 + 43) + 32(43) = 2450 + 1376 = 3826$$

The number of pairs of cases ranked in different order on both variables (N_d) would be:

$$N_d = 17(27 + 23) + 32(23) = 850 + 736 = 1586$$

Gamma is:

$$G = \frac{N_s - N_d}{N_s + N_d}$$

$$G = \frac{3,826 - 1,586}{3,826 + 1,586}$$

$$G = \frac{2,240}{5,412}$$

$$G = 0.41$$

A gamma of 0.41 means that we would make 41% fewer errors by using candidates' advertising expenditures to predict turnout. This is a moderate positive association: Turnout increases as expenditures increase.

Determining the Direction of Relationships

Nominal measures of association like phi and lambda measure only the strength of a bivariate association. Ordinal measures of association, like gamma, are more sophisticated and add information about the overall direction of the relationship (positive or negative). In one way, it is easy to determine direction: If gamma is a plus value, the relationship is positive. A minus sign for gamma indicates a negative relationship. Often, however, direction is confusing when working with ordinal-level variables, so it will be helpful if we focus on the matter specifically.

With gamma, a positive relationship means that the *scores* of cases tend to be ranked in the same order on both variables and that the variables change in the same direction. That is, as scores on one variable increase (or decrease), scores on the other variable also increase (or decrease). Cases tend to have scores in the same range on both variables (i.e., low scores go with low scores, moderate with moderate, and so forth). Table 12.3 illustrates the general shape of a positive relationship. Note that the cases tend to fall along a diagonal from upper left to lower right (assuming, of course, that tables have been constructed with the column variable increasing from left to right and the row variable from top to bottom).

Tables 12.4 and 12.5 present an example of a positive relationship with actual data. The sample consists of 246 preindustrial societies from around the globe. Each has been rated in terms of its degree of stratification or inequality and the type of political institution it has. In the societies that are the lowest in stratification, people share resources equally and the degree of inequality increases from left to right across the columns of the table. In "stateless" societies, there is no formal political institution and the political institution becomes more elaborate

TABLE 12.3 A Generalized Positive Relationship

	Variable *X*		
Variable *Y*	Low	Moderate	High
Low	X		
Moderate		X	
High			X

TABLE 12.4 State Structure by Degree of Stratification* (frequencies)

	Degree of Stratification		
Type of State	Low	Medium	High
Stateless	77	5	0
Semi-state	28	15	4
State	12	19	26
Totals	117	39	30

*Data are from the Human Relation Area File, standard cross-cultural sample.

TABLE 12.5 State Structure by Degree of Stratification (percentages)

	Degree of Stratification		
Type of State	Low	Medium	High
Stateless	66%	13%	0%
Semi state	24%	38%	13%
State	10%	49%	87%
Totals	100%	100%	100%

and stronger as you move down the rows from top to bottom. The gamma for this table is 0.86, so the relationship is strong and positive. By inspection, we can see that most cases fall in the diagonal from upper left to lower right.

The percentages in Table 12.5 make it even clearer that societies with little inequality tend to be stateless and that the political institution becomes more elaborate as inequality increases. The great majority of the least stratified societies had no political institution, and none of the highly stratified societies were stateless.

In negative relationships, low scores on one variable are associated with high scores on the other and high scores with low scores. This means that the cases will tend to fall along a diagonal from lower left to upper right. Table 12.6 illustrates a generalized negative relationship. The cases with higher scores on variable *X* tend to have lower scores on variable *Y*, and score on *Y* decreases as score on *X* increases.

Tables 12.7 and 12.8 present an example of a negative relationship, with data taken from a public opinion poll. The independent variable is church attendance, and the dependent variable is approval of cohabitation ("Is it all right for a couple to live together without intending to get married?"). Note that rates of attendance increase from left to right, and approval of cohabitation increases from top to bottom of the table.

TABLE 12.6 A Generalized Negative Relationship

	Variable X		
Variable *Y*	Low	Moderate	High
Low			X
Moderate		X	
High	X		

TABLE 12.7 Approval of Cohabitation by Church Attendance (frequencies)

	Attendance		
Approval	Never	Monthly or Yearly	Weekly
Low	37	186	195
Moderate	25	126	46
High	156	324	52
Totals	218	636	293

TABLE 12.8 Approval of Cohabitation by Church Attendance (percentages)

	Attendance		
Approval	Never	Monthly or Yearly	Weekly
Low	17.0	29.2	66.6
Moderate	11.5	19.8	15.7
High	71.6	50.9	17.7
Totals	100.1%	99.9%	100.0 %

Once again, the percentages in Table 12.8 make the pattern more obvious. The great majority of people who were low on attendance ("never") were high on approval of cohabitation, and most people who were high on attendance ("weekly") were low on approval. As attendance increases, approval of cohabitation tends to decrease. The gamma for this table is -0.57, indicating a moderate to strong negative relationship between these variables.

You should be aware of an additional complication. The coding for ordinal-level variables is arbitrary, and a higher score may mean "more" or "less" of the variable being measured. For example, if we measured social class as upper, middle, or lower, we could assign scores to the categories in either of two ways:

A	B
(1) Upper	(3) Upper
(2) Middle	(2) Middle
(3) Lower	(1) Lower

While coding scheme B might seem preferable (because higher scores go with higher class position), *both* schemes are perfectly legitimate and the direction of a relationship will change depending on which scheme is selected. Using scheme B, we would find a positive gamma between social class and (for example) education: As education increased, so would class.

However, with scheme A, the same relationship would appear to be negative because the numerical scores (1, 2, 3) are coded in reverse order: The highest

STATISTICS IN EVERYDAY LIFE

It is not uncommon for teenagers and young adults to have disagreements with their parents and grandparents about issues of morality, especially as these relate to dating and sex. One recent poll found large generation gaps on some—but not all—moral issues. For example, there were big differences on pornography and premarital sex, but the differences on opinions about stem cell research, divorce, and doctor-assisted suicide were quite small. To illustrate, the table below shows the relationship between several items and age group.

Perceived Moral Acceptability by Age Group (percent "yes")

	Age Group		
Issue	*18 to 34*	*35 to 54*	*55 and older*
Pornography	42%	29%	19%
Out-of-wedlock births	62%	56%	46%
Cloning humans	18%	10%	9%
Doctor-assisted suicide	46%	45%	43%
Extramarital affairs	8%	7%	7%

The difference in perception of pornography is statistically significant, moderate in strength, and positive in direction (disapproval increases as age increases). On other issues, such as cloning and suicide, the relationship with age group is insignificant and weak.

Source: Saad, Lydia. 2011. Accessed at http://www.gallup.com/poll/147842/Doctor-Assisted-Suicide-Moral-Issue-Dividing-Americans.aspx?version=print.

social class is assigned the lowest score and so forth. If you did not check the coding scheme, you might conclude that class decreases as education increases, when, actually, the opposite is true.

Unfortunately, this source of confusion cannot be avoided when working with ordinal-level variables. Coding schemes will always be arbitrary for these variables, so you need to exercise additional caution when interpreting the direction of ordinal-level variables.

Spearman's Rho (R_s)

In the previous section, we considered ordinal variables that have a limited number of categories (possible values) and are presented in tables. However, many ordinal-level variables have a broad range and many distinct scores. These variables may be collapsed into a few broad categories (such as high, moderate, and low), organized into a bivariate table, and analyzed with gamma. Collapsing scores in this manner may be desirable in many instances but may also obscure or lose some important distinctions between cases. For example, suppose a researcher wished to test the claim that jogging is beneficial, not only physically but also psychologically. Do joggers have an enhanced sense of self-esteem? To deal with this issue, 10 female joggers are evaluated on two scales—the first measuring involvement in jogging and the other measuring self-esteem. Scores are reported in Table 12.9.

These variables could be collapsed by splitting the scores into two values (high and low). There are two potential difficulties with collapsing scores in this way. First, the scores seem continuous and there are no obvious or natural division points in the distribution that would allow us to distinguish between high scores and low ones. Second—and more importantly—grouping these cases into broader categories will lose information. That is, if Wendy and Debbie are placed in the category "high" on involvement, the fact that they had different scores would be lost. If differences like this are important and meaningful, then we should choose a measure of association that permits the retention of as much detail and precision in the scores as possible.

TABLE 12.9 The Scores of 10 Subjects on Involvement in Jogging and a Measure of Self-Esteem

Jogger	Involvement in Jogging (*X*)	Self-Esteem (*Y*)
Wendy	18	15
Debbie	17	18
Phyllis	15	12
Stacey	12	16
Evelyn	10	6
Tricia	9	10
Christy	8	8
Patsy	8	7
Marsha	5	5
Lynn	1	2

The Computation of Spearman's Rho

Spearman's rho (r_s) is a measure of association for ordinal-level variables that have a broad range of many different scores and few ties between cases on either variable. Scores on ordinal-level variables cannot, of course, be manipulated mathematically except for judgments of "greater than" or "less than." To compute Spearman's rho, the cases are first ranked from high to low on each variable and then the ranks (not the scores) are manipulated to produce the final measure. Table 12.10 displays the original scores and the rankings of the cases on both variables.

To rank the cases, first find the highest score on each variable and assign it rank 1. Wendy has the high score on X (18) and is thus ranked number 1. On the other hand, Debbie is highest on Y and is ranked first on that variable. All other cases are then ranked in descending order of scores. If any cases have the same score on a variable, assign them the average of the ranks they would have used up had they not been tied. Christy and Patsy have identical scores of 8 on involvement. Had they not been tied, they would have used up ranks 7 and 8. The average of these two ranks is 7.5, and this average of used ranks is assigned to all tied cases. (For example, if Marsha had also had a score of 8, three ranks—7, 8, and 9—would have been used and all three tied cases would have been ranked eighth.)

The formula for Spearman's rho is:

FORMULA 12.2

$$r_s = 1 - \frac{6\Sigma D^2}{N(N^2 - 1)}$$

where:

ΣD^2 = the sum of the differences in ranks, the quantity squared

To compute ΣD^2, the rank of each case on Y is first subtracted from its rank on X (D is the difference between rank on Y and rank on X). A column has been

TABLE 12.10 Computing Spearman's Rho

	Involvement(X)	Rank	Self-Image(Y)	Rank	D	D^2
Wendy	18	1	15	3	−2	4
Debbie	17	2	18	1	1	1
Phyllis	15	3	12	4	−1	1
Stacey	12	4	16	2	2	4
Evelyn	10	5	6	8	−3	9
Tricia	9	6	10	5	1	1
Christy	8	7.5	8	6	1.5	2.25
Patsy	8	7.5	7	7	0.5	0.25
Marsha	5	9	5	9	0	0
Lynn	1	10	2	10	0	0
					$\Sigma D = 0$	$\Sigma D^2 = 22.50$

provided in Table 12.10 to record the differences case by case. Note that the sum of this column (ΣD) is 0. That is, the negative differences in rank are equal to the positive differences, as will always be the case, and you should find the total of this column as a check on your computations to this point. If ΣD is not equal to 0, you have made a mistake either in ranking the cases or in subtracting the differences.

In the column headed D^2, each difference is squared to eliminate negative signs. The sum of this column is ΣD^2, and this quantity is entered directly into the formula. For our sample problem:

$$r_s = 1 - \frac{6\Sigma D^2}{N(N^2 - 1)}$$

$$r_s = 1 - \frac{6(22.5)}{10(100 - 1)}$$

$$r_s = 1 - \frac{135}{990}$$

$$r_s = 1 - 0.14$$

$$r_s = 0.86$$

The Interpretation of Rho

Spearman's rho is an index of the strength of association between the variables; it ranges from 0 (no association) to ± 1.00 (perfect association). A perfect positive association ($r_s = +1.00$) would exist if there were no disagreements in ranks between the two variables (if cases were ranked in exactly the same order on both variables). A perfect negative relationship ($r_s = -1.00$) would exist if the ranks were in perfect disagreement (if the case ranked highest on one variable were lowest on the other and so forth). A Spearman's rho of 0.86 indicates a strong positive relationship between these two variables. The respondents who were highly involved in jogging also ranked high on self-image. These results are supportive of claims regarding the psychological benefits of jogging.

Spearman's rho is an index of the relative strength of a relationship, and values between 0 and ± 1.00 have no direct interpretation. However, if the value of rho is squared, a PRE interpretation is possible. Rho squared (r_s^2) represents the proportional reduction in errors of prediction when predicting rank on one variable from rank on the other variable, as compared to predicting rank while ignoring the other variable. In our example, r_s was 0.86 and r_s^2 would be 0.74. Thus, our errors of prediction would be reduced by 74% if we used the rank on jogging to predict rank on self-image. *(For practice in computing and interpreting Spearman's rho, see problems 12.10 to 12.13. Problem 12.10 has the fewest number of cases and is probably a good choice for a first attempt at these procedures.)*

ONE STEP AT A TIME Computing and Interpreting Spearmans Rho

Step	**Operation**
To Compute Spearman's Rho	
1.	Set up a computing table like Table 12.10 to organize the computations. In the far left-hand column, list the cases in order, with the case with the highest score on the independent variable (*X*) stated first.
2.	In the next column, list the scores on *X*.
3.	In the third column, list the rank of each case on *X*, beginning with rank 1 for the highest score. If any cases have the same score, assign them the average of the ranks they would have used up had they not been tied.
4.	In the fourth column, list the score of each case on *Y*; in the fifth column, rank the cases on *Y* from high to low. Assign the rank of 1 to the case with the highest score on *Y*, and assign any tied cases the average of the ranks they would have used up had they not been tied.
5.	For each case, subtract the rank on *Y* from the rank on *X* and then write the difference (*D*) in the sixth column. Add up this column. If the sum is not zero, you have made a mistake and need to recompute.
6.	Square the value of each *D* and then record the result in the seventh column. Add this column up to find ΣD^2 and then substitute this value into the numerator of Formula 12.2.
7.	Multiply ΣD^2 by 6.
8.	Square *N* and subtract 1 from the result.
9.	Multiply the quantity you found in step 8 by *N*.
10.	Divide the quantity you found in step 7 by the quantity you found in step 9.
11.	Subtract the quantity you found in step 10 from 1. The result is r_s.
Interpreting the Strength of the Relationship	
1.	Use either or both of the following to interpret strength: a. Use Table 12.2 to describe strength in general terms. b. Square the value of rho and then multiply by 100. This value represents the percentage improvement in our predictions of the dependent variable by taking the independent variable into account.
Interpreting the Direction of the Relationship	
1.	Use the sign of r_s. Be careful when interpreting direction with ordinal-level variables. Remember that coding schemes for these variables are arbitrary and that a positive r_s may mean that the actual relationship is negative and vice versa.

STATISTICS IN EVERYDAY LIFE

Rankings have become a staple of the media and the Internet, and there seem to be infinite lists of everything imaginable: best places to live, cities with the worst traffic, most important news stories, most popular celebrities, and so forth. These lists are bound to be somewhat arbitrary and subjective but can also provide useful insights and perspectives. A search of the Internet for "top ten" lists provides a

STATISTICS IN EVERYDAY LIFE

(continued)

multitude of examples—one of the most interesting of which was a list of the most haunted cities in the United States:

Rank	City	Comment
1	New Orleans, LA	A city full of haunted mansions, taverns, and graveyards
2	Savannah, GA	Fits the bill of a haunted city as well as any in the United States
3	Gettysburg, PA	Almost 50,000 died during the battle
4	Chicago, IL	The great fire, gangsters, and "Resurrection Mary"
5	Salem, MA	Witches!
6	Charleston, SC	An older city with plenty of time to accumulate ghost stories
7	Portland, OR	The ghostliest city in the Northwest
8	Athens, OH	One-time home of a "Lunatic Asylum"
9	Key West, FL	Pirates, rumrunners, Ernest Hemingway, and Robert the Doll
10	San Francisco, CA	A rich cultural legacy, disasters, and Alcatraz

More information is available at the website cited below.

Source: http://www.toptenz.net/top-10-most-haunted-cities-in-the-u-s.php.

Applying Statistics 12.3 Spearman's Rho

Five cities have been rated on an index that measures the quality of life. Also, the percentage of the population that has moved into each city over the past year has been determined. Have cities with higher quality-of-life scores attracted more new residents? The table below summarizes the scores, ranks, and differences in ranks for each of the five cities.

Spearman's rho for these variables is:

$$r_s = 1 - \frac{6\Sigma D^2}{N(N^2 - 1)}$$

$$r_s = 1 - \frac{(6)(4)}{5(25 - 1)}$$

$$r_s = 1 - \left(\frac{24}{120}\right)$$

$$r_s = 1 - 0.20$$

$$r_s = 0.80$$

These variables have a strong positive association. The higher the quality-of-life score, the greater the percentage of new residents. The value of r_s^2 is 0.64($[0.80]^2 = 0.64$), which indicates that we will make 64% fewer errors when predicting rank on one variable from rank on the other, as opposed to ignoring rank on the other variable.

City	Quality of Life	Rank	% New Residents	Rank	D	D^2
A	30	1	17	1	0	0
B	25	2	14	3	–1	1
C	20	3	15	2	1	1
D	10	4	3	5	–1	1
E	2	5	5	4	1	1
					$\Sigma D^2 = 0$	$\Sigma D^2 = 4$

STATISTICS IN EVERYDAY LIFE

Is it possible to buy success? Can you spend your way to the top? The answer in many areas of life is probably yes—but what about professional sports? Some teams are commonly accused of trying to buy championships by stockpiling the best—and most expensive—players. Does this strategy work? To find out, I looked at Major League Baseball and ranked all 30 teams on their opening day payroll for the 2010 season and on the number of wins they had that year. We do not have space to display all 30 teams, but here are the top 10 teams ranked by payroll and number of wins:

Teams	Rank on Opening Day Payroll, 2010	Rank on Number of Wins in 2010
NY Yankees	1 (highest)	3
Boston Red Sox	2	10
Chicago Cubs	3	23
Philadelphia Phillies	4	1
NY Mets	5	20
Detroit Tigers	6	15
Chicago White Sox	7	11
Los Angeles Angels	8	17
Seattle Mariners	9	29
San Francisco Giants	10	5

The Spearman's rho for this relationship is 0.27—positive in direction but only weak to moderate in strength. Most of the teams with the 10 highest payrolls were not in the top 10 in wins. The bottom line is that it is hard to buy a championship, and even teams with relatively few stars—and a small payroll—have a shot at success.

SUMMARY

1. A measure of association for variables with collapsed ordinal scales (gamma) was covered along with a measure (Spearman's rho) appropriate for "continuous" ordinal variables. Both measures summarize the overall strength and direction of the association between the variables.
2. Gamma is a PRE-based measure that shows the improvement in our ability to predict the order of pairs of cases on one variable from the order of pairs of cases on the other variable, as opposed to ignoring the order of the pairs of cases on the other variable.
3. Spearman's rho is computed from the ranks of the scores of the cases on two "continuous" ordinal variables and, when squared, can be interpreted by the logic of PRE.

SUMMARY OF FORMULAS

FORMULA 12.1 Gamma: $$G = \frac{N_s - N_d}{N_s + N_d}$$

FORMULA 12.2 Spearman's rho: $$r_s = 1 - \frac{6\Sigma D^2}{N(N^2 - 1)}$$

GLOSSARY

Gamma (*G*). A measure of association appropriate for variables measured with "collapsed" ordinal scales that have been organized into table format.

N_d. The number of pairs of cases ranked in different order on two variables.

N_s. The number of pairs of cases ranked in the same order on two variables.

Spearman's rho (r_s). A measure of association appropriate for ordinally measured variables that are "continuous" in form.

PROBLEMS

For problems 12.1 to 12.9 and 12.14, calculate percentages for the bivariate tables. Use the percentages to help analyze the strength and direction of the association.

12.1 SOC A small sample of non-English-speaking immigrants to the United States has been interviewed about their level of assimilation. Is the pattern of adjustment affected by length of residence in the United States? For each table, compute gamma and summarize the relationship in terms of strength and direction. *(HINT: In 2 × 2 tables, only two cells can contribute to N_s or N_d. To compute N_s, multiply the number of cases in the upper left-hand cell by the number of cases in the lower right-hand cell. For N_d, multiply the number of cases in the upper right-hand cell by the number of cases in the lower left-hand cell.)*

a. Facility in English:

	Length of Residence		
English Facility	Less Than Five Years (Low)	More Than Five Years (High)	Totals
Low	20	10	30
High	5	15	20
Totals	25	25	50

b. Total family income:

	Length of Residence		
Income	Less Than Five Years (Low)	More Than Five Years (High)	Totals
Below national average (1)	18	8	26
Above national average (2)	7	17	24
Totals	25	25	50

c. Extent of contact with country of origin:

	Length of Residence		
Contact	Less Than Five Years (Low)	More Than Five Years (High)	Totals
Rare (1)	5	20	25
Frequent (2)	20	5	25
Totals	25	25	50

12.2 Compute gamma for the table presented in problem 11.1 and reproduced below. Compare gamma with the nominal measure of association you computed in Chapter 11. Are they similar in value? Do they characterize the strength for the association in the same way? What information about the relationship does gamma provide that is not available from nominal measures of association?

	Authoritarianism		
Efficiency	Low	High	Totals
Low	10	12	22
High	17	5	22
Totals	27	17	44

12.3 CJ A sample of 150 cities has been classified as small, medium, or large by population and as high or low on crime rate. Is there a relationship between city size and crime rate? Describe the strength and direction of the relationship.

	City Size			
Crime Rate	Small	Medium	Large	Totals
Low	21	17	8	46
High	29	33	42	104
Totals	50	50	50	150

12.4 SOC Some researchers have argued that families vary by how they socialize their children to sports, games, and other leisure-time activities. In middle-class families, such activities are carefully monitored by parents and are, in general, dominated by adults (for example, Little League baseball). In working-class families, children more often organize and initiate such activities themselves, and parents are much less involved (for example, sandlot or playground baseball games). Are the following data consistent with the ideas presented by the researchers? Summarize your conclusions in a few sentences.

As a Child, Did You Play Mostly Organized or Sandlot Sports?	Social Class		
	Middle Class	Working Class	Totals
Organized	155	123	278
Sandlot	101	138	239
Totals	256	261	517

12.5 Is there a relationship between age and support for traditional gender roles? Is the relationship between the variables different for different nations? The World Values Survey has been administered to random samples drawn from Canada, the United States, and Mexico. Respondents were asked if they agree or disagree that a university education is more important for boys than for girls. Compute column percentages and gamma for each table. Is there a relationship? Describe the strength and direction of the relationship. Which age group is most supportive of traditional gender roles (agrees with the statement). How does the relationship change from nation to nation?

a. Canada

"University Is More Important for Boys"	Age Group		
	15–29	30–49	50 and older
Agree	17	24	64
Disagree	384	804	841
	401	828	905

b. United States

"University Is More Important for Boys"	Age Group		
	15–29	30–49	50 and older
Agree	29	29	29
Disagree	238	418	478
	267	447	507

c. Mexico

"University Is More Important for Boys"	Age Group		
	15–29	30–49	50 and older
Agree	109	154	114
Disagree	385	489	259
	494	643	373

12.6 PA All applicants to the local police academy are given an aptitude test, but the test has never been evaluated to see if test scores are in any way related to job performance. The following table reports aptitude test scores and job performance ratings for a sample of 78 policemen.

Efficiency Ratings	Test Scores			
	Low	Moderate	High	Totals
Low	12	6	7	25
Moderate	9	11	9	29
High	5	9	10	24
Totals	26	26	26	78

a. Are these two variables associated? Describe the strength and direction of the relationship in a sentence or two.

b. Should the aptitude test continue to be administered? Why or why not?

12.7 SW A sample of children has been observed and rated for symptoms of depression. Their parents have been rated for authoritarianism. Is there any relationship between these variables? Write a few sentences stating your conclusions.

Symptoms of Depression	Authoritarianism			
	Low	Moderate	High	Totals
Few	7	8	9	24
Some	15	10	18	43
Many	8	12	3	23
Totals	30	30	30	90

12.8 SOC Are prejudice and level of education related? State your conclusion in a few sentences.

Prejudice	Level of Education				
	Elementary School	High School	Some College	College Graduate	Totals
Low	48	50	61	42	201
High	45	43	33	27	148
Totals	93	93	94	69	349

12.9 SOC In a recent survey, a random sample of respondents was asked to indicate how happy they were with their situations in life. Are their responses related to income level? Describe the strength and direction of the relationship.

	Income			
Happiness	Low	Moderate	High	Totals
Not happy	101	82	36	219
Pretty happy	40	227	100	367
Very happy	216	198	203	617
Totals	357	507	339	1,203

12.10 SOC A random sample of 11 neighborhoods in Leonia, Illinois, has been rated by an urban sociologist on a "quality-of-life" scale (which includes measures of affluence, availability of medical care, and recreational facilities) and a social cohesion scale. The results are presented here in scores. Higher scores indicate higher "quality of life" and greater social cohesion. Are the variables associated? What is the strength and direction of the relationship? (*Hint: Do not forget to square the value of Spearman's rho for a PRE interpretation.*)

Neighborhood	Quality of Life	Social Cohesion
Beaconsdale	17	8.8
Brentwood	40	3.9
Chesapeake Shores	47	4.0
College Park	90	3.1
Denbigh Plantation	35	7.5
Kingswood	52	3.5
North End	23	6.3
Phoebus	67	1.7
Riverview	65	9.2
Queens Lake	63	3.0
Windsor Forest	100	5.3

12.11 SW Several years ago, a job-training program began, and a team of social workers screened the candidates for suitability for employment. Now the screening process is being evaluated, and the actual work performance of a sample of hired candidates has been rated. Did the screening process work? Is there a relationship between the original scores and performance evaluation on the job?

Case	Original Score	Performance Evaluation
A	17	78
B	17	85
C	15	82
D	13	92
E	13	75
F	13	72
G	11	70
H	10	75
I	10	92
J	10	70
K	9	32
L	8	55
M	7	21
N	5	45
O	2	25

12.12 SOC Following are the scores of a sample of 15 nations on a measure of ethnic diversity (the higher the number, the greater the diversity) and a measure of economic inequality (the higher the score, the greater the inequality). Are these variables related? Are ethnically diverse nations more economically unequal?

Nation	Diversity	Inequality
India	91	29.7
South Africa	87	58.4
Kenya	83	57.5
Canada	75	31.5
Malaysia	72	48.4
Kazakhstan	69	32.7
Egypt	65	32.0
United States	63	41.0
Sri Lanka	57	30.1
Mexico	50	50.3
Spain	44	32.5
Australia	31	33.7
Finland	16	25.6
Ireland	4	35.9
Poland	3	27.2

12.13 Twenty ethnic, racial, or national groups were rated by a random sample of white and black students on a Social Distance Scale. Lower scores represent less social distance and less prejudice. How similar are these rankings?

Group	Average Social Distance Scale Score	
	White Students	Black Students
1 White Americans	1.2	2.6
2 English	1.4	2.9
3 Canadians	1.5	3.6
4 Irish	1.6	3.6
5 Germans	1.8	3.9
6 Italians	1.9	3.3
7 Norwegians	2.0	3.8
8 American Indians	2.1	2.7
9 Spanish	2.2	3.0
10 Jews	2.3	3.3
11 Poles	2.4	4.2
12 Black Americans	2.4	1.3
13 Japanese	2.8	3.5
14 Mexicans	2.9	3.4
15 Koreans	3.4	3.7
16 Russians	3.7	5.1
17 Arabs	3.9	3.9
18 Vietnamese	3.9	4.1
19 Turks	4.2	4.4
20 Iranians	5.3	5.4

12.14 In problem 11.9, we looked at the relationships between five dependent variables and sex. Now we will use income as an independent variable and assess its relationship with this same set of variables. For each table, calculate percentages and gamma. Describe the strength and direction of each relationship in a few sentences. *Be careful in interpreting direction.*

a. Support for the legal right to an abortion by income:

Right to an Abortion?	Income			Totals
	Low	Moderate	High	
Yes	220	218	226	664
No	366	299	250	915
Totals	586	517	476	1,579

b. Support for capital punishment by income:

Capital Punishment?	Income			Totals
	Low	Moderate	High	
Favor	567	574	552	1,693
Oppose	270	183	160	613
Totals	837	757	712	2,306

c. Approval of suicide for people with an incurable disease by income:

Right to Suicide?	Income			Totals
	Low	Moderate	High	
Approve	343	341	338	1,022
Oppose	227	194	147	568
Totals	570	535	485	1,590

d. Support for sex education in public schools by income:

Sex Education?	Income			Totals
	Low	Moderate	High	
For	492	478	451	1,421
Against	85	68	53	206
Totals	577	546	504	1,627

e. Support for traditional gender roles by income:

Women Should Take Care of Running Their Homes and Leave Running the Country to Men	Income			Totals
	Low	Moderate	High	
Agree	130	71	39	240
Disagree	448	479	461	1,388
Totals	578	550	500	1,628

YOU ARE THE RESEARCHER: Exploring Sexual Attitudes and Behavior

Two projects are presented to help you apply the skills developed in this chapter. Both focus on sex—a subject of great fascination to many people. The first project uses bivariate tables and gamma to explore the possible causes of attitudes and opinions about premarital sex. The second uses Spearman's rho to investigate sexual behavior—specifically the number of different sexual partners people have had over the past five years.

PROJECT 1: Who Approves of Premarital Sex?

What type of person is most likely to approve of sex before marriage? In this exercise, you will take attitudes toward premarital sex (*premarsx*) as the dependent variable. The wording and coding scheme for the variable are presented in the table below. Remember that the coding scheme for ordinal level variables is arbitrary. In this case, higher scores indicate greater approval of premarital sex and lower scores mean greater disapproval. Keep the coding scheme in mind as you analyze the direction of the relationships you find.

> *There has been a lot of discussion about the way morals and attitudes about sex are changing in this country. If a man and a woman have sex relations before marriage, do you think this is:*

1	Always wrong
2	Almost always wrong
3	Sometimes wrong
4	Not wrong at all

STEP 1: Choosing Independent Variables

Select four variables from the 2010 GSS that you think might be important causes of attitudes toward premarital sex and list the variables in the table below. Your independent variables *cannot* be nominal in then level of measurement and *cannot* have more than four to five categories or scores. Some of the ordinal level variables in which you might be interested (such as *attend*, a measure of church attendance) have more than four to five categories and should be recoded, as should such interval-ratio variables as *age* or *income06*. See Chapter 9 for instructions on recoding. Select independent variables that seem likely to be an important cause of people's attitudes about sex. Be sure to note the coding scheme for each variable.

SPSS Name	What Exactly Does This Variable Measure?

STEP 2: Stating Hypotheses

State hypotheses about the relationships you expect to find between your independent variables and *premarsx.* State these hypotheses in terms of the direction of the relationship you expect to find. For example, you might hypothesize that approval of premarital sex will decline as age increases (a negative relationship).

1. ______________________________
2. ______________________________
3. ______________________________
4. ______________________________

STEP 3: Running Crosstabs

Click **Analyze** → **Descriptives** → **Crosstabs** and then place the dependent variable (*premarsx*) in the **Rows:** box and all the independent variables you selected in the **Columns:** box. Click the **Statistics** button to get chi square and gamma and the **Cells** button to get column percentages.

STEP 4: Recording Results

These commands will generate a lot of output, and it will be helpful to summarize your results in the following table.

Independent Variable	Chi Square Significant at < 0.05? (*Yes* or *No*)	Gamma

STEP 5: Analyzing and Interpreting Results

For each independent variable, analyze and interpret the significance, strength, and direction of the relationship. For example, you might say the following: "There was a moderate negative relationship between age and approval of premarital sex (chi square = 4.26; $df = 2$; $p < 0.05$; gamma = 0.35). Older respondents tended to be more opposed." Be sure to *explain* the direction of the relationship, and do not just characterize it as negative or positive. *Be careful when interpreting direction with ordinal variables.*

Independent Variable	Interpretation
1	
2	
3	
4	

STEP 6: Testing Hypotheses

Write a few sentences of overall summary for this test. Were your hypotheses supported? Which independent variable had the strongest relationship with approval of premarital sex? Why do you think this is so?

PROJECT 2: Sexual Behavior

In this exercise, your dependent variable will be *partnrs5,* which measures how many different sex partners a person has had over the past five years. The wording and coding scheme for the variable are presented in the table below. For this variable, higher scores mean a greater number of partners, so interpreting the direction of relationships should be relatively straightforward.

How Many Sex Partners Have You Had Over the Past Five Years?	
0	No partners
1	1 partner
2	2 partners
3	3 partners
4	4 partners
5	5–10 partners
6	11–20 partners
7	21–100 partners
8	More than 100 partners

Note that *partnrs5* is interval-ratio for the first five categories and then becomes ordinal for higher scores. In other words, the variable has a true zero point (a score of 0 means no sex partners at all) and increases by equal, defined units from one partner through four partners. However, higher scores represent broad categories, not exact numbers. This mixture of levels of measurement makes Spearman's rho an appropriate statistic to use to measure the strength and direction of relationships

To access Spearman's rho, click **Analyze → Correlate → Bivariate**, and the **Bivariate Correlations** window will open. The variables are listed in the window on the left. Below that window is a box labeled **Correlation Coefficients**. Click **Spearman**—and unclick Pearson—from the options to get Spearman's rho. Next, select your variables from the list on the left; click the arrow to move them to the **Variables:** window on the right. SPSS will compute Spearman's rho for all pairs of variables included in the **Variables:** window.

Let us take a brief look at the output produced by this procedure. I looked at some relationships that are *not* particularly sensible so as not to usurp any of the more interesting ideas you might have for this (or other) projects. I took *attend* (frequency of church attendance) as the dependent variable and *papres 80* (the relative prestige of the respondent's father's occupation) as the independent variable.

The output from the **Correlate** procedure is a table showing the bivariate correlations of all possible combinations of variables, including the relationship of the variable with itself. The table is called a *correlation matrix* and will look like the table below. We will deal with the correlation matrix in more detail in Chapter 13.

Correlations				
			HOW OFTEN r ATTENDS RELIGIOUS SERVICES	FATHER'S OCCUPATIONAL PRESTIGE SCORE (1980)
Spearman's rho	HOW OFTEN r ATTENDS RELIGIOUS SERVICES	Correlation Coefficient	1.000	−.023
		Sig. (2-tailed)	.	.437
		N	1,489	1,160
	FATHER'S OCCUPATIONAL PRESTIGE SCORE (1980)	Correlation Coefficient	−.023	1.000
		Sig. (2-tailed)	.437	.
		N	1,160	1,162

The table has four cells, and there are three pieces of information in each cell: the value of Spearman's rho, the statistical significance of r_s, and the number of cases. The cell on the upper left shows the relationship between *attend* and itself; the next cell to the right shows the relationship between *attend* and *papres 80.* The output shows a weak negative relationship between the variables. Now it is your turn.

STEP 1: Choosing Independent Variables for *Partnrs5*

Select four variables from the 2010 GSS you think might be important causes of this dimension of sexual behavior. Your independent variables *cannot* be nominal in level of measurement and should have more than three categories or scores. List the variables and then describe exactly what they measure in the table below.

SPSS Name	What Exactly Does This Variable Measure?

STEP 2: Stating Hypotheses

State hypotheses about the relationships you expect to find between your independent variables and *partnrs 5*. State these hypotheses in terms of the direction of the relationship you expect to find. For example, you might hypothesize that the number of sexual partners will decline as age increases.

1. ______
2. ______
3. ______
4. ______

STEP 3: Running Bivariate Correlations

Click **Analyze** → **Bivariate** → **Correlate** and then place all variables in the **Variables:** box. Click **OK** to get your results.

STEP 4: Recording Results

Use the table below to summarize your results. Enter the r_s for each independent variable in each cell. Ignore correlations of variables with themselves and redundant information.

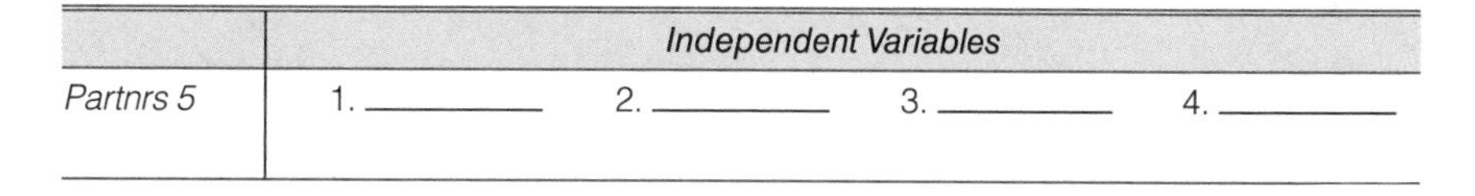

	Independent Variables			
Partnrs 5	1. ________	2. ________	3. ________	4. ________

STEP 5: Analyzing And Interpreting Results

Write a short summary of results for each independent variable. Your summary needs to identify the variables being tested and the strength and direction of the relationship. It is probably best to characterize the relationship in general terms and then cite the statistical values in parentheses. Be sure to note whether your hypotheses were supported. Be careful when interpreting direction, and refer back to the coding scheme to make sure you understand the relationship.

13 Association Between Variables Measured at the Interval-Ratio Level

LEARNING OBJECTIVES

By the end of this chapter, you will be able to:

1. Interpret a scattergram.
2. Calculate and interpret slope (b), Y intercept (a), and Pearson's r and r^2.
3. Find and explain the least-squares regression line and then use it to predict values of Y.
4. Explain the concepts of total, explained, and unexplained variance.
5. Use regression and correlation techniques to analyze and describe a bivariate relationship in terms of the three questions introduced in Chapter 11.

USING STATISTICS

The statistical techniques presented in this chapter are used to measure the strength and direction of the association between variables measured at the interval-ratio level. Examples of research situations in which these techniques are useful include:

1. A family sociologist is studying the division of labor between dual-wage-earner married couples and wonders if husbands increase their contribution to housework as the number of children in the family increases.
2. A criminologist is studying the relationship between poverty and crime and has gathered information on both variables for 542 counties across the United States. Does the crime rate increase as poverty increases? How strong is the relationship?
3. A demographer is interested in the relationship between fertility (average lifetime number of births) and education for females. She has information from 76 nations and wonders if fertility decreases as the average number of years of schooling for women increases. How strong is the association?

In this chapter, we will consider statistical techniques for analyzing the association, or correlation, between variables measured at the interval-ratio level.[1] The techniques presented here are also commonly used on ordinal-level variables, especially when they are "continuous" or have many values (see Chapter 12).

[1]The term *correlation* is commonly used instead of *association* when discussing the relationship between interval-ratio variables. We will use the two terms interchangeably.

As we shall see, these statistics are rather different in their logic and computation from the measures of association covered in Chapters 11 and 12. Therefore, let me stress at the outset that we are still asking the same three questions: Is there a relationship between the variables? How strong is the relationship? What is the direction of the relationship? I introduce some new and complex computational routines, so remind yourself occasionally that our ultimate goals are unchanged: We are trying to understand bivariate relationships, explore possible causal ties between variables, and improve our ability to predict scores.

Scattergrams and Regression Analysis

The usual first step in analyzing a relationship between interval-ratio variables is to construct and examine a **scattergram**. Like bivariate tables, these graphs allow us to identify several important features of the relationship. An example will illustrate how to construct and interpret scattergrams.

Suppose a researcher is interested in analyzing how dual-wage-earner families (that is, families where husband and wife have jobs outside the home) cope with housework. Specifically, the researcher wonders if the number of children in the family is related to the amount of time the husband contributes to housekeeping chores. The relevant data for a sample of 12 families are displayed in Table 13.1.

TABLE 13.1 Number of Children and Husband's Contribution to Housework (fictitious data)

Family	Number of Children	Hours per Week Husband Spends on Housework
A	1	1
B	1	2
C	1	3
D	1	5
E	2	3
F	2	1
G	3	5
H	3	0
I	4	6
J	4	3
K	5	7
L	5	4

FIGURE 13.1 Husband's Housework by Number of Children

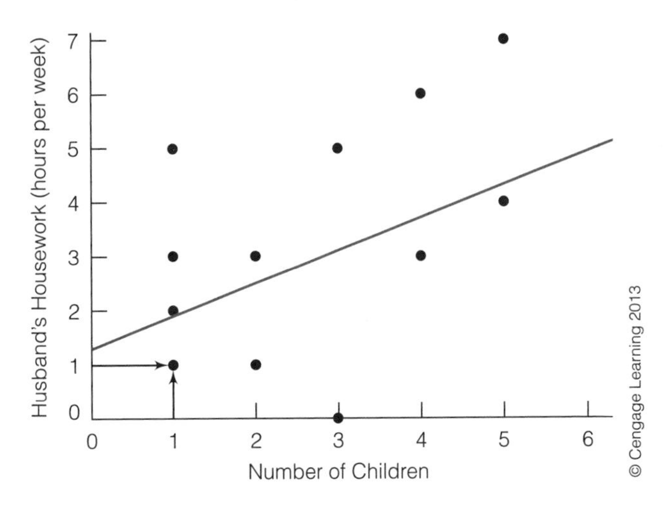

Constructing Scattergrams

A scattergram, like a bivariate table, has two dimensions. The scores of the independent (*X*) variable are arrayed along the horizontal axis, and the scores of the dependent (*Y*) variable along the vertical axis. Each dot on the scattergram represents a case in the sample, and the dot's location represents the scores of the case on both variables. Figure 13.1 shows a scattergram displaying the relationship between "number of children" and "husband's housework" for the sample of 12 families presented in Table 13.1. Family A has a score of 1 on the *X* variable (number of children) and 1 on the *Y* variable (husband's housework) and is represented by the dot above the score of 1 on the *X* axis and directly to the right of the score of 1 on the *Y* axis. All 12 cases are similarly represented by dots on Figure 13.1. Also note that, as always, the scattergram is clearly titled and both axes are labeled.

Interpreting Scattergrams

The overall pattern of the dots or cases summarizes the nature of the relationship between the two variables. The clarity of the pattern can be enhanced by drawing a straight line through the cluster of dots such that the line touches every dot or comes as close to doing so as possible. We will soon consider a precise technique for fitting this line to the pattern of the dots, but for now, an "eyeball" approximation will suffice. This summarizing line—called the **regression line**—has already been added to the scattergram.

Even when they are crudely drawn, scattergrams can be used for a variety of purposes. They provide at least impressionistic information about the existence, strength, and direction of the relationship and can also be used to check the relationship for linearity (that is, how well the pattern of dots can be approximated with a straight line). Finally, the scattergram can be used to predict the score of a case on one variable from the score of that case on the other variable. We now briefly examine each of these uses in terms of the three questions we first asked in Chapter 11.

The data used in our example are fictitious. Studies of the actual division of labor in American households commonly find large differences between the contributions of husbands and wives. One study* found that the differences depended on who was doing the reporting. Based on a nationally representative sample, the researchers found no significant difference between men and women in their estimates of the number of hours that women contribute to housework; both sexes estimated about 18 hours per week.

Both genders agreed that men spend much less time on housework, but there were large and statistically significant differences in the estimates of men's hours. Men estimated that they contributed an average of 8.6 hours (less than half of their estimate for women) and women estimated about 6 hours (about a third of the estimate for women). Do you think that men are exaggerating their contribution? Or are women underestimating it? How would you explore this difference? What other questions would you ask?

*Achen, Alexandra, and Stafford, Frank. 2005. *Data Quality of Housework Hours in the Panel Study of Income Dynamics: Who Really Does the Dishes?* Available at http://psidonline.isr.umich.edu/Publications/Papers/achenproxyreports04.pdf.

Does a Relationship Exist? The basic definition of an association was stated in Chapter 11: Two variables are associated if the distributions of Y (the dependent variable) change for the various conditions of X (the independent variable). In Figure 13.1, scores on X (number of children) are arrayed along the horizontal axis. The dots above each score on X are the scores (or conditional distributions) of Y. That is, the dots represent scores on Y for each value of X. There is a relationship between these variables because the conditional distributions of Y (the dots above each score on X) change as X changes. The existence of an association is further reinforced by the fact that the regression line lies at an angle to the X axis. If these two variables had not been associated, the conditional distributions of Y would not have changed and the regression line would have been parallel to the horizontal axis.

How Strong Is the Relationship? The strength of the bivariate association can be judged by observing the spread of the dots around the regression line. In a perfect association, all dots would lie on the regression line. The more the dots are clustered around the regression line, the stronger the association.

What Is the Direction of the Relationship? The direction of the relationship can be detected by observing the angle of the regression line. Figure 13.1 shows a positive relationship: As X (number of children) increases, the husband's housework (Y) also increases. Husbands in families with more children tend to do more housework. If the relationship had been negative, the regression line would have sloped in the opposite direction to indicate that high scores on one variable were associated with low scores on the other.

To summarize these points, Figure 13.2 shows a perfect positive and a perfect negative relationship and a "zero relationship," or "nonrelationship," between two variables.

Linearity

One key assumption underlying the statistical techniques to be introduced later in this chapter is that the two variables have an essentially **linear relationship**.

FIGURE 13.2 Perfect Positive, Perfect Negative, and Zero Relationships

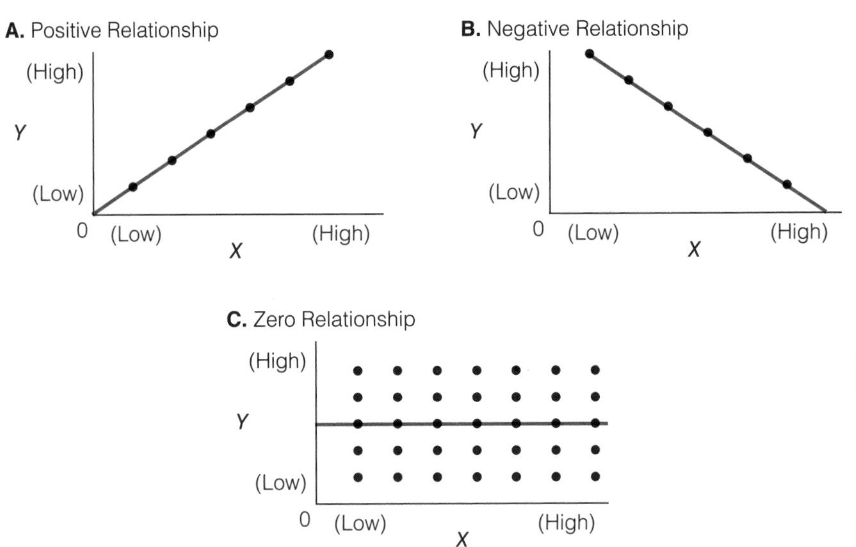

FIGURE 13.3 Some Nonlinear Relationships

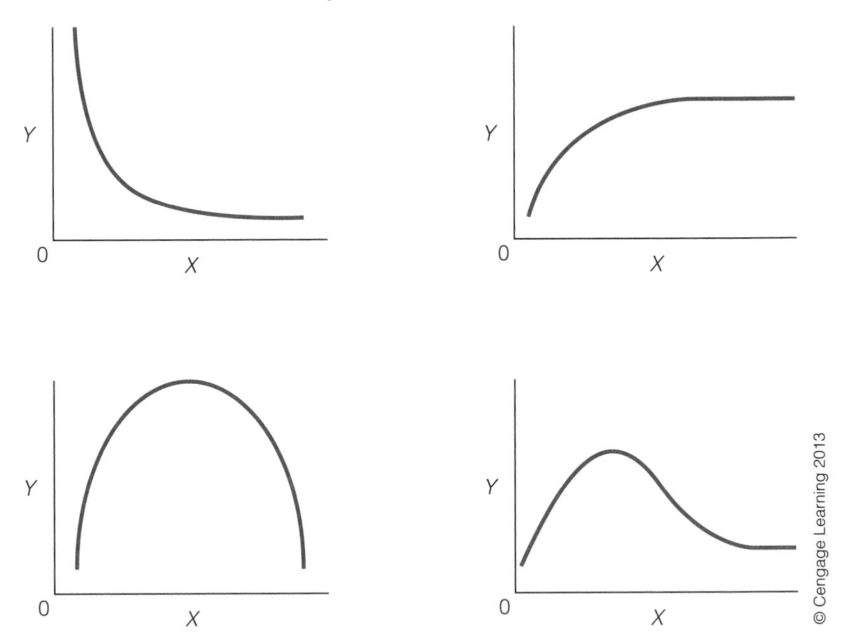

In other words, the observation points or dots in the scattergram must form a pattern that can be approximated with a straight line. Significant departures from linearity would require the use of statistical techniques beyond the scope of this text. Examples of some common curvilinear relationships are presented in Figure 13.3. If the scattergram shows that the variables have a nonlinear relationship, the techniques described in this chapter should be used with great caution or not at all. Checking for the linearity of the relationship is perhaps the most important

FIGURE 13.4 Predicting Husband's Housework

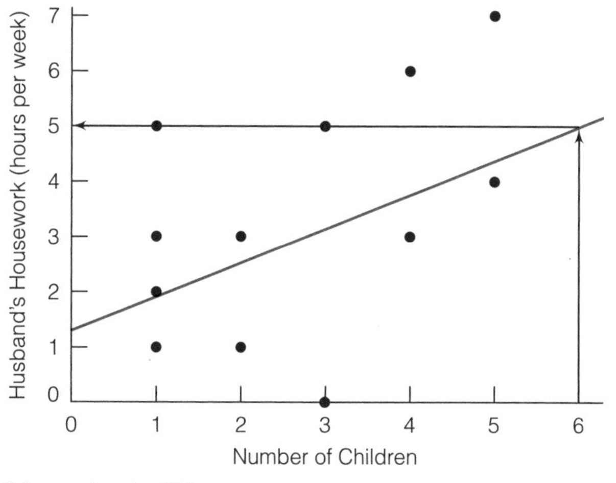

reason for examining the scattergram before proceeding with the statistical analysis. If the relationship is nonlinear, you might need to treat the variables as if they were ordinal rather than interval-ratio in level of measurement. *(For practice in constructing and interpreting scattergrams, see problems 13.1 to 13.5.)*

Using Scattergrams for Prediction

A final use of the scattergram is to predict the scores of cases on one variable from their score on the other. To illustrate, suppose that based on the relationship between number of children and husband's housework displayed in Figure 13.1, we wish to predict the number of hours of housework a husband with a family of six children would do each week. The sample has no families with six children, but if we extend the axes and regression line in Figure 13.1 to incorporate this score, a prediction is possible. Figure 13.4 reproduces the scattergram and illustrates how the prediction would be made.

The predicted score on Y—which is symbolized as Y' or "Y-prime" to distinguish predictions of Y from actual Y scores—is found by first locating the relevant score on X ($X = 6$ in this case) and then drawing a straight line from that point to the regression line. From the regression line, another straight line parallel to the X axis is drawn across to the Y axis. The predicted Y score (Y') is found at the point where the line crosses the Y axis. In our example, we would predict that in a dual-wage-earner family with six children, the husband would devote about five hours per week to housework.

The Regression Line

The prediction technique introduced in the previous section is crude, and the value of Y' can change depending on how accurately the freehand regression line is drawn. One way to eliminate this source of error would be to find the straight line that most accurately describes the relationship between the two variables. How can we find this line?

Recall that our criterion for the freehand regression line was that it touch all the dots or come as close to doing so as possible. Also, recall that the dots above each value of X can be thought of as conditional distributions of Y, the dependent variable. Within each conditional distribution of Y, the mean is the point around which the variation of the scores is at a minimum. In Chapter 3, we noted that the mean of any distribution of scores is the point around which the variation of the scores, as measured by squared deviations, is minimized:

$$\Sigma(X_i - \overline{X})^2 = \text{minimum}$$

Thus, if the regression line is drawn so it touches each **conditional mean** of Y, it would be the straight line that comes as close as possible to all the scores.

Conditional means are found by summing all Y values for each value of X and then dividing by the number of cases. For example, four families had one child ($X = 1$) and the husbands of these four families devoted 1, 2, 3, and 5 hours of housework. The conditional mean of Y for $X = 1$ is 2.75 ($11/4 = 2.75$). Husbands in families with one child worked an average of 2.75 hours per week doing housekeeping chores. Conditional means of Y are computed in the same way for each value of X. They are displayed in Table 13.2 and plotted in Figure 13.5.

Let us quickly remind ourselves of the reason for these calculations. We are seeking the single best-fitting regression line for summarizing the relationship between X and Y, and we have seen that a line drawn through the conditional means of Y will minimize the spread of the observation points. It will come as close to all the scores as possible and will therefore be the single best-fitting regression line.

A line drawn through the points on Figure 13.5 (the conditional means of Y) will be the best-fitting line we are seeking, but you can see from the scattergram that the line will not be straight. In fact, only rarely (when there is a perfect relationship between X and Y) will conditional means fall in a perfectly straight line. Because we still must meet the condition of linearity, let us revise our criterion to define the regression line as the unique straight line that touches all conditional means of Y or comes as close to doing so as possible. Formula 13.1 defines the

TABLE 13.2 Conditional Means of Y (husband's housework) for Various Values of *X* (number of children)

Number of Children (*X*)	Husband's Housework (*Y*)	Conditional Mean of *Y*
1	1,2,3,5	2.75
2	3,1	2.00
3	5,0	2.50
4	6,3	4.50
5	7,4	5.50

FIGURE 13.5 Conditional Means of *Y*

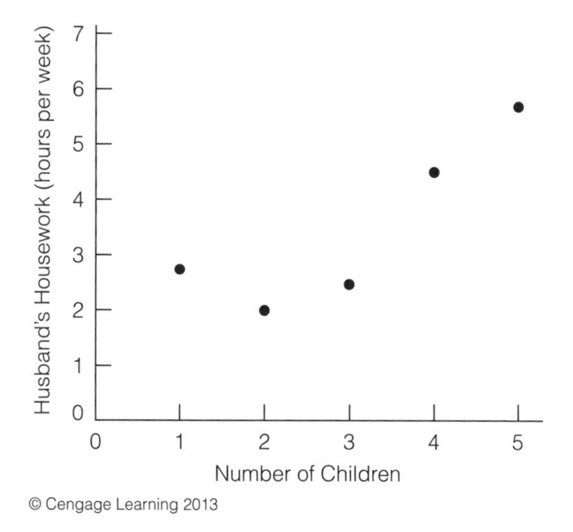

"least squares" regression line or the single straight regression line that best fits the pattern of the data points:

FORMULA 13.1

$$Y = a + bX$$

where: Y = score on the dependent variable
a = the Y intercept, or the point where the regression line crosses the Y axis
b = the slope of the regression line, or the amount of change produced in Y by a unit change in X
X = score on the independent variable

The formula introduces two new concepts. First, the ***Y* intercept (*a*)** is the point at which the regression line crosses the vertical, or *Y*, axis. Second, the ***slope* (*b*)** of the least-squares regression line is the amount of change produced in the dependent variable (*Y*) by a unit change in the independent variable (*X*). Think of the slope of the regression line as a measure of the effect of the *X* variable on the *Y* variable. If the variables have a strong association, then changes in the value of *X* will be accompanied by substantial changes in the value of *Y* and the slope (*b*) will have a high value. The weaker the effect of *X* on *Y* (the weaker the association between the variables), the lower the value of the slope (*b*). If the two variables are unrelated, the least-squares regression line would be parallel to the *X* axis and *b* would be 0.00 (the line would have no slope).

With the least-squares formula (Formula 13.1), we can predict values of *Y* in a much less arbitrary and impressionistic way than through mere eyeballing. This is because the least-squares regression line is the single straight line that best fits the data and comes as close as possible to all the conditional means of *Y*. However, before seeing how predictions of *Y* can be made, we must first calculate *a* and *b*. *(For practice in using the regression line to predict scores on Y from scores on X, see problems 13.1 to 13.3 and 13.5.)*

Regression analysis has many practical applications and is used routinely in many professions—city planning, criminal justice, business, demography, and even military science—to model and predict trends. For example, medical researchers are using regression analysis to study the relationship between weather conditions at mass events (e.g., the Indianapolis 500 or rock concerts) and the need for emergency services.* This knowledge may be extremely useful for planners, who might be able to accurately forecast the need for medical services.

Of course, these applications use techniques that are much more sophisticated than the bivariate relationships explored in this chapter. However, the basic regression equation ($Y = a + bX$) is the foundation for even the most advanced and powerful of these models.

* Baird, M., O'Conner, R., Williamson, A., Sojka, B., and Brady, W. 2010. "The Impact of Warm Weather on Mass Event Medical Need: A Review of the Literature." *American Journal of Emergency Medicine* 28: 224–229.

Computing and Interpreting the Regression Coefficients (*a* and *b*)

In this section, we cover how to compute and interpret the coefficients in the equation for the regression line: the slope (b) and the Y intercept (a). Because the value of b is needed to compute a, we begin with the computation of the slope.

Computing the Slope (*b*). The formula for the slope is:

FORMULA 13.2

$$b = \frac{\Sigma(X - \overline{X})(Y - \overline{Y})}{\Sigma(X - \overline{X})^2}$$

The numerator of this formula is called the *covariation* of X and Y. It is a measure of how X and Y vary together, and its value will reflect the direction and the strength of the relationship. The denominator is the sum of the squared deviations around the mean of X.

The calculations necessary for computing the slope should be organized into a computational table, as in Table 13.3, which has a column for each of the quantities needed to solve the formula. The data are from the dual-wage-earner family sample (see Table 13.1).

The first column of the table lists the original X scores for each case, and the second column shows the deviations of these scores around their mean. The third and fourth columns repeat this information for the Y scores and the deviations of the Y scores. Column 5 shows the covariation of the X and Y scores. The entries in this column are found by multiplying the deviation of the X score (column 2) by the deviation of the Y score (column 4) for each case. Finally, the entries in column 6 are found by squaring the value in column 2 for each case.

Table 13.3 gives us all the quantities we need to solve Formula 13.2. Substitute the total of column 5 in Table 13.3 in the numerator and the total of column 6 in the denominator:

$$b = \frac{\Sigma(X - \overline{X})(Y - \overline{Y})}{\Sigma(X - \overline{X})^2}$$

$$b = \frac{18.33}{26.67}$$

$$b = 0.69$$

TABLE 13.3 Computation of the Slope (*b*)

1	2	3	4	5	6
X	$X - \bar{X}$	Y	$Y - \bar{Y}$	$(X - \bar{X})(Y - \bar{Y})$	$(X - \bar{X})^2$
1	−1.67	1	−2.33	3.89	2.79
1	−1.67	2	−1.33	2.22	2.79
1	−1.67	3	−0.33	0.55	2.79
1	−1.67	5	1.67	−2.79	2.79
2	−0.67	3	−0.33	0.22	0.45
2	−0.67	1	−2.33	1.56	0.45
3	0.33	5	1.67	0.55	0.11
3	0.33	0	−3.33	−1.10	0.11
4	1.33	6	2.67	3.55	1.77
4	1.33	3	−0.33	−0.44	1.77
5	2.33	7	3.67	8.55	5.43
5	2.33	4	0.67	1.56	5.43
32	−0.04	40	0.04	18.33	26.67

$$\bar{X} = \frac{32}{12} = 2.67$$

$$\bar{Y} = \frac{40}{12} = 3.33$$

A slope of 0.69 indicates that for each unit change in X, there is an increase of 0.69 units in Y. For our example, the addition of each child (an increase of one unit in X) results in an increase of 0.69 hours of housework being done by the husband (an increase of 0.69 units— or hours—in Y).

Computing the *Y* intercept (*a*). Once the slope has been calculated, finding the intercept (a) is relatively easy. We computed the means of X and Y while calculating the slope, and we enter these figures into Formula 13.3:

FORMULA 13.3

$$a = \bar{Y} - b\bar{X}$$

For our sample problem, the value of a would be:

$$a = \bar{Y} - b\bar{X}$$
$$a = 3.33 - (0.69)(2.67)$$
$$a = 3.33 - 1.84$$
$$a = 1.49$$

Thus, the least-squares regression line will cross the Y axis at the point where Y equals 1.49.

Now that we have values for the slope and the Y intercept, we can state the full least-squares regression line for our sample data:

$$Y = a + bX$$
$$Y = 1.49 + (0.69)X$$

Predicting Scores on *Y* With the Least-Squares Regression Line

The regression formula can be used to estimate, or predict, scores on Y for any value of X. Previously, we used the freehand regression line to predict a score on Y (husband's housework) for a family with six children ($X = 6$). Our prediction was that in families of six children, husbands would contribute about five hours per week to housekeeping chores. By using the least-squares regression line, we can see how close our impressionistic eyeball prediction was:

$$Y = a + bX$$

$$Y = 1.49 + (0.69)(6)$$

$$Y = 1.49 + 4.14$$

$$Y = 5.63$$

Based on the least-squares regression line, we would predict that in a dual-wage-earner family with six children, the husband would devote 5.63 hours a week to housework. What would our prediction of the husband's housework be for a family of seven children ($X = 7$)?

Note that our predictions of Y scores are "educated guesses." We will be unlikely to predict values of Y exactly except in the (relatively rare) case where the bivariate relationship is perfect and perfectly linear. However, also note that the accuracy of our predictions will increase as relationships become stronger. This is because the dots are more clustered around the least-squares regression line in stronger relationships. *(The slope and Y intercept may be computed for any problem at the end of this chapter.)*

ONE STEP AT A TIME Computing the Slope (*b*)

Step	***Operation***
	To compute the slope (b), solve Formula 13.2:
1.	Use a computing table like Table 13.3 to help organize the computations. List the scores of the cases on the independent variable (X) in column 1.
2.	Compute the mean of X by dividing the total of column 1 by the number of cases.
3.	Subtract the mean of X from each X score and then list the results in column 2. *Note: The sum of column 2 must be zero (except for rounding error). If this sum is not zero, you have made a mistake in computations.*
4.	List the score of each case on Y in column 3. Compute the mean of Y by dividing the total of column 3 by the number of cases.
5.	Subtract the mean of Y from each Y score and then list the results in column 4. *Note: The sum of column 4 must be zero (except for rounding error). If this sum is not zero, you have made a mistake in computations.*
6.	For each case, multiply the value in column 2 by the value in column 4. Place the result in column 5. Find the sum of this column.
7.	Square each value in column 2 and then place the result in column 6. Find the sum of this column.
8.	Divide the sum of column 5 by the sum of column 6. The result is the slope.

ONE STEP AT A TIME **Computing the Y Intercept (a)**

Step	***Operation***
	To compute the Y intercept (a), solve Formula 13.3:
1.	Multiply the slope (b) by the mean of X.
2.	Subtract the value you found in step 1 from the mean of Y. This value is a, or the Y intercept.

ONE STEP AT A TIME **Using the Regression Line to Predict Scores on Y**

Step	***Operation***
1.	Choose a value for X. Multiply this value by the value of the slope (b).
2.	Add the value you found in step 1 to the value of a, the Y intercept. The resulting value is the predicted score on Y.

The Correlation Coefficient (Pearson's *R*)

The slope of the least-squares regression line (b) is a measure of the effect of X on Y, and it will increase in value as the relationship increases in strength. However, b does not vary between 0 and 1 and is therefore awkward to use as a measure of association. Instead, researchers rely heavily (almost exclusively) on a statistic called **Pearson's *r***—or the correlation coefficient—to measure the association between interval-ratio variables. Like the ordinal measures of association discussed in Chapter 12, Pearson's r varies from 0.00 to ± 1.00, with 0.00 indicating no association and $+1.00$ and -1.00 indicating perfect positive and perfect negative relationships, respectively. The formula for Pearson's r is:

FORMULA 13.4

$$r = \frac{\Sigma(X - \overline{X})(Y - \overline{Y})}{\sqrt{[\Sigma(X - \overline{X})^2][\Sigma(Y - \overline{Y})^2]}}$$

Note that the numerator of this formula is the covariation of X and Y, as was the case with Formula 13.2. A computing table like Table 13.3, with a column added for the sum of $(Y - \overline{Y})^2$, is strongly recommended as a way of organizing the quantities needed to solve this equation (see Table 13.4).

For our sample problem involving dual-wage-earner families, the quantities displayed in Table 13.4 can be substituted directly into Formula 13.4:

$$r = \frac{\Sigma(X - \overline{X})(Y - \overline{Y})}{\sqrt{[\Sigma(X - \overline{X})^2][\Sigma(Y - \overline{Y})^2]}}$$

$$r = \frac{18.33}{\sqrt{(26.67)(50.67)}}$$

$$r = \frac{18.33}{\sqrt{1{,}351.37}}$$

$$r = \frac{18.33}{36.76}$$

$$r = 0.50$$

TABLE 13.4 Computation of Pearson's *r*

1	2	3	4	5	6	7
X	$X - \overline{X}$	Y	$Y - \overline{Y}$	$(X - \overline{X})(Y - \overline{Y})$	$(X - \overline{X})^2$	$(Y - \overline{Y})^2$
1	−1.67	1	−2.33	3.89	2.79	5.43
1	−1.67	2	−1.33	2.22	2.79	1.77
1	−1.67	3	−0.33	0.55	2.79	0.11
1	−1.67	5	1.67	−2.79	2.79	2.79
2	−0.67	3	−0.33	0.22	0.45	0.11
2	−0.67	1	−2.33	1.56	0.45	5.43
3	0.33	5	1.67	0.55	0.11	2.79
3	0.33	0	−3.33	−1.10	0.11	11.09
4	1.33	6	2.67	3.55	1.77	7.13
4	1.33	3	−0.33	−0.44	1.77	0.11
5	2.33	7	3.67	8.55	5.43	13.47
5	2.33	4	0.67	1.56	5.43	0.45
32	−0.04	40	0.04	18.33	26.67	50.67

An *r* value of 0.50 indicates a moderately strong positive linear relationship between the variables. As the number of children in the family increases, the hourly contribution of husbands to housekeeping duties also increases. *(Every problem at the end of this chapter requires the computation of Pearson's r. It is probably a good idea to practice with smaller data sets and easier computations first; see problem 13.1 in particular.)*

ONE STEP AT A TIME Computing Pearson's *r*

Step ***Operation***

1. Add a column to the computing table you used to compute the slope (b). Square the value of $(Y - \overline{Y})$ and then record the result in this column (column 7).
2. Find the sum of column 7.

Find the value of Pearson's r by solving Formula 13.4:

3. Multiply the sum of column 6 by the sum of column 7.
4. Take the square root of the value you found in step 3.
5. Divide the quantity you found in step 4 into the sum of column 5 or the sum of the cross-products. The result is Pearson's *r*.

To interpret the strength of Pearson's r, you can either:

1. Use table 12.2 to describe strength in general terms.
2. Square the value of *r* and then multiply by 100. This value represents the percentage of variation in *Y* that is explained by *X*.

ONE STEP AT A TIME ***(continued)***

Step	***Operation***
To interpret the direction of the relationship:	
1.	Look at the sign of *r*. If *r* has a plus sign (or if there is no sign), the relationship is positive and the variables change in the same direction. If *r* has a minus sign, the relationship is negative and the variables change in opposite directions.

Interpreting the Correlation Coefficient: r^2

Pearson's r is an index of the strength of the linear relationship between two variables. While a value of 0.00 indicates no linear relationship and a value of ±1.00 indicates a perfect linear relationship, values between these extremes have no direct interpretation. We can, of course, describe relationships in terms of how closely they approach the extremes (for example, coefficients approaching 0.00 can be described as "weak" and those approaching ±1.00 as "strong"), but this description is somewhat subjective. Also, we can use the guidelines stated in Table 12.2 for gamma to attach descriptive words to the specific values of Pearson's r. In other words, values between 0.00 and 0.30 would be described as weak, values between 0.30 and 0.60 would be moderate, and values greater than 0.60 would be strong. Remember, of course, that these labels are arbitrary guidelines and will not be appropriate or useful in all possible research situations.

The Coefficient of Determination

Fortunately, there is a less arbitrary, more direct interpretation of r by calculating an additional statistic called the **coefficient of determination**. This statistic, which is simply the square of Pearson's r (r^2), can be interpreted with a type of logic similar to proportional reduction in error (PRE). As you recall, the logic of PRE measures of association is to predict the value of the dependent variable under two different conditions. First, Y is predicted while ignoring the information supplied by X; second, the independent variable is taken into account. With r^2, the method of prediction and the construction of the final statistic require the introduction of some new concepts.

Predicting *Y* Without *X*

When working with variables measured at the interval-ratio level, the predictions of the Y scores under the first condition (while ignoring X) will be the mean of the Y. Given no information on X, this prediction strategy will be optimal because we know that the mean of any distribution is closer to all the scores than any other point in the distribution. Recall the principle of minimized variation introduced in Chapter 3 and expressed as:

$$\Sigma(Y - \bar{Y})^2 = \text{minimum}$$

FIGURE 13.6 Predicting *Y* Without *X* (dual-career families)

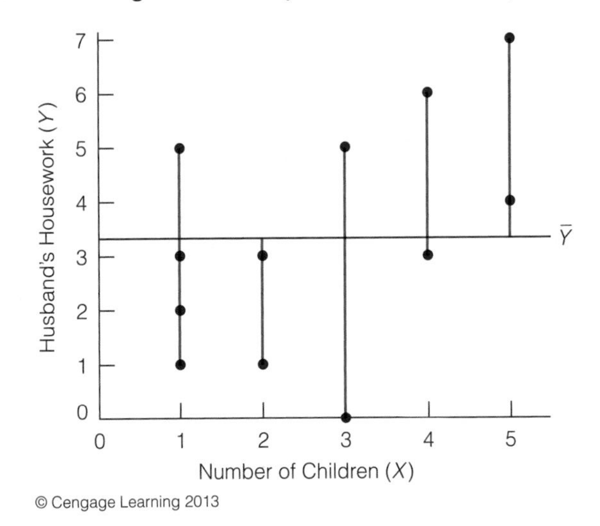

The scores of any variable vary less around the mean than around any other point. If we predict the mean of Y for every case, we will make fewer errors of prediction than if we predict any other value for Y.

Of course, we will still make errors in predicting Y even if we faithfully follow this strategy. The amount of error is represented in Figure 13.6, which displays the relationship between number of children and husband's housework, with the mean of Y noted. The vertical lines from the actual scores to the predicted score represent the amount of error we would make when predicting Y while ignoring X.

We can define the extent of our prediction error under the first condition (while ignoring X) by subtracting the mean of Y from each actual Y score and squaring and summing these deviations. The resultant figure, which can be noted as $\Sigma(Y - \overline{Y})^2$, is called the **total variation** in Y. We now have a visual representation (Figure 13.6) and a method for calculating the error we incur by predicting Y without knowledge of X. As we shall see, we do not actually need to calculate the total variation to find the value of the coefficient of determination: r^2.

Predicting *Y* With *X*

Our next step will be to determine the extent to which knowledge of X improves our ability to predict Y. If the two variables have a linear relationship, then predicting scores on Y from the least-squares regression equation will incorporate knowledge of X and reduce our errors of prediction. So thus, under the second condition, our predicted Y score for each value of X will be:

$$Y' = a + bX$$

Figure 13.7 displays the data from the dual-career families with the regression line drawn in. The vertical lines from each actual score on Y to the regression line represent the amount of error in predicting Y that remains even after X has been taken into account.

FIGURE 13.7 Predicting *Y* With *X* (dual-career families)

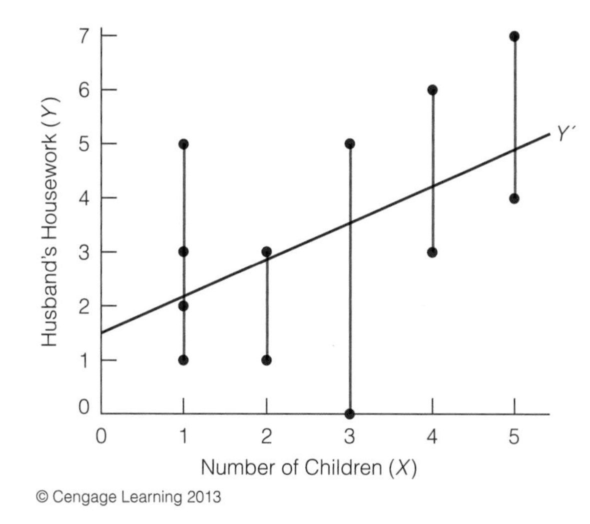

Explained, Unexplained, and Total Variation

We can precisely define the reduction in error that results from taking X into account. Specifically, two different sums can be found and then compared with the total variation of Y to construct a statistic that will indicate the improvement in prediction.

The first sum—called the **explained variation**—represents the improvement in our ability to predict Y when taking X into account. This sum is found by subtracting $\overline{Y}$ (our predicted Y score without X) from the score predicted by the regression equation (Y', or the Y score predicted with knowledge of X) for each case and then squaring and summing these differences. These operations can be summarized as $\Sigma(Y' - \overline{Y})^2$, and the resultant figure could then be compared with the total variation in Y to ascertain the extent to which our knowledge of X improves our ability to predict Y. Specifically, it can be shown mathematically that:

FORMULA 13.5

$$r^2 = \frac{\Sigma\,(Y' - \overline{Y})^2}{\Sigma\,(Y - \overline{Y})^2} = \frac{\text{Explained variation}}{\text{Total variation}}$$

Thus, the coefficient of determination, or r^2, is the proportion of the total variation in Y attributable to or explained by X. Like other PRE measures, r^2 indicates precisely the extent to which X helps us predict, understand, or explain Y.

Earlier, we referred to the improvement in predicting Y with X as the explained variation. The use of this term suggests that some of the variation in Y will be "unexplained," or not attributable to the influence of X. In fact, the vertical lines in Figure 13.7 represent the **unexplained variation**—or the difference between our best prediction of Y with X and the actual scores. The unexplained variation is thus the scattering of the actual scores around the regression line and can be found by subtracting the predicted Y scores from the actual Y scores for each case and then squaring and summing the differences. These operations can be summarized as $\Sigma(Y - Y')^2$, and the resultant sum would measure the amount of error in predicting Y that remains even after X has been taken into account. The proportion of the total variation in Y unexplained by X can be found by subtracting the value of r^2 from 1.00. Unexplained variation is usually attributed to the influence of some combination of other variables, measurement error, and random chance.

As you may have recognized by this time, the explained and unexplained variations bear a reciprocal relationship with one another. As one increases in value, the other decreases. Also, the stronger the linear relationship between X and Y, the greater the value of the explained variation and the lower the unexplained variation. In the case of a perfect relationship ($r = \pm 1.00$), the unexplained variation would be 0 and r^2 would be 1.00. This would indicate that X explains or accounts for all the variation in Y and that we could predict Y from X without error. On the other hand, when X and Y are not linearly related ($r = 0.00$), the explained variation would be 0 and r^2 would be 0.00. In such a case, we would conclude that X explains none of the variation in Y and does not improve our ability to predict Y.

Relationships intermediate between these two extremes can be interpreted in terms of how much X increases our ability to predict or explain Y. For the dual-career families, we calculated an r of 0.50. Squaring this value yields a coefficient of determination of 0.25 ($r^2 = 0.25$), which indicates that the number of children (X) explains 25% of the total variation in husband's housework (Y). When predicting the number of hours per week that husbands in such families would devote to housework, we will make 25% fewer errors by basing the predictions on number of children and predicting from the regression line, as opposed to ignoring the X variable and predicting the mean of Y for every case. Also, 75% of the variation in Y is unexplained by X and presumably due to some combination of the influence of other variables, measurement error, and random chance. *(For practice in the interpretation of r^2, see any of the problems at the end of this chapter.)*

STATISTICS
IN EVERYDAY LIFE

Does popularity pay? According to a recent study*, having many friends in high school is correlated with economic success later in life. Using a regression-based statistical model, researchers found that each additional friend in high school correlates with a 2% gain in wages 35 years later.# Statistically speaking, popularity seemed to have an effect that was roughly equal to about half the gain from an extra year of schooling; in other words, two additional high school friends would statistically equal the benefit of one additional year of college.

Remembering that correlation is not the same thing as causation, how else could this relationship be explained? What additional information would you need to demonstrate that the relationship between popularity and income was causal?

*Conti, G., Galeotti, A., Mueller, G., and Pudney, S. 2009. "Popularity." Available at http://www.iser.essex.ac.uk/publications/working-papers/iser/2009-03.pdf.

#Although it is based on more advanced statistics, this statement could be interpreted as a characterization of the slope of the bivariate relationship between popularity and economic success.

Applying Statistics 13.1 Computing the Regression Coefficients and Pearson's *r*.

For five cities, information has been collected on the number of civil disturbances (riots, strikes, and so forth) over the past year and on unemployment rate. Are these variables associated? The data are presented in the following table. Columns have been added for all sums necessary for the computation of the slope (b) and Pearson's r.

City	Unemployment Rate (X)	$(X - \bar{X})$	Civil Disturbances (Y)	$(Y - \bar{Y})$	$(X - \bar{X})(Y - \bar{Y})$	$(X - \bar{X})^2$	$(Y - \bar{Y})^2$
A	22	6.8	25	14.4	97.92	46.24	207.36
B	20	4.8	13	2.4	11.52	23.04	5.76
C	10	−5.2	10	−0.6	3.12	27.04	0.36
D	15	−0.2	5	−5.6	1.12	0.04	31.36
E	9	−6.2	0	−10.6	65.72	38.44	112.36
	76	0.0	53	0.0	179.40	134.80	357.20

$$\bar{X} = 15.2$$

$$\bar{Y} = 10.6$$

The slope (b) is:

$$b = \frac{\Sigma(X - \bar{X})(Y - \bar{Y})}{\Sigma(X - \bar{X})^2}$$

$$b = \frac{179.4}{134.8}$$

$$b = 1.33$$

A slope of 1.33 means that for every unit change in X (for every increase of 1 in the unemployment rate), there was a change of 1.33 units in Y (the number of civil disturbances increased by 1.33). The Y intercept (a) is:

$$a = \bar{Y} - b\bar{X}$$

$$a = \frac{53}{5} - (1.33)\left(\frac{76}{5}\right)$$

$$a = 10.6 - (1.33)(15.20)$$

$$a = 10.6 - 20.2$$

$$a = -9.6$$

The least-squares regression equation is:

$$Y = a + bX = -9.6 + (1.33)X$$

The correlation coefficient is:

$$r = \frac{\Sigma(X - \bar{X})(Y - \bar{Y})}{\sqrt{[\Sigma(X - \bar{X})^2][\Sigma(Y - \bar{Y})^2]}}$$

$$r = \frac{179.40}{\sqrt{(134.80)(357.20)}}$$

$$r = \frac{179.40}{\sqrt{48{,}150.56}}$$

$$r = \frac{179.40}{219.43}$$

$$r = 0.82$$

These variables have a strong positive association. The number of civil disturbances increases as the unemployment rate increases. The coefficient of determination (r^2), is $(0.82)^2$, or 0.67. This indicates that 67% of the variance in civil disturbances is explained by the unemployment rate.

The Correlation Matrix

Social science research projects usually include many variables, and the data analysis phase of a project often begins with the examination of a **correlation matrix**—a table that shows the relationships between all possible pairs of variables. The correlation matrix gives a quick, easy-to-read overview of the interrelationships in the data set and may suggest strategies or "leads" for further analysis. These tables are commonly included in the professional research literature, and it will be useful to have some experience reading them.

An example of a correlation matrix, which displays some demographic variables for 115 nations as of 2011, is presented in Table 13.5. The matrix uses variable names as rows and columns, and the cells in the table show the bivariate correlation (Pearson's r) for each combination of variables. Note that the row headings duplicate the column headings. To read the table, begin with birthrate, the variable in the far left-hand column (column 1) and top row (row 1). Read down column 1 or across row 1 to see the correlations of this variable with all other variables, including the correlation of birthrate with itself (1.00) in the top cell. To see the relationships between other variables, move from column to column or row to row.

Note that the diagonal from upper left to lower right of the matrix contains the correlation of each variable with itself. Values along this diagonal will always be 1.00, and because this information is not useful, it could easily be deleted from the table. Also, note that the cells below and to the left of the diagonal are redundant with the cells above and to the right of the diagonal. For example, look at the second cell down (row 2) in column 1. This cell displays the correlation between birthrate and infant mortality rate, as does the cell in the top row (row 1) of column 2. In other words, the cells below and to the left of the diagonal are mirror images of the cells above and to the right of the diagonal. Commonly, research articles in the professional literature will delete the redundant cells in order to make the table more readable.

What does this matrix tell us? Starting at the upper left of the table (column 1), we can see that birthrate has a strong positive relationship with infant mortality rate. This means that infant mortality increases as birthrate increases; nations with higher rates of birth also have higher rates of infant deaths. Birthrate has strong negative relationships with life expectancy (the higher the birthrate, the lower the life expectancy) and a moderate negative relationship with migration rate (nations with higher birthrates tend to be lower on migration rate).

To assess the other relationships, move from column to column and row to row—one variable at a time. For each subsequent variable, there will be one fewer cell of new information. For example, consider infant mortality rate, the variable

TABLE 13.5 A Correlation Matrix Showing the Interrelationships of Four Variables for 115 Nations

		1 Birthrate	2 Infant Mortality Rate	3 Life Expectancy	4 Migration Rate
1	Birthrate	1.00	0.90	−0.86	−0.36
2	Infant Mortality Rate	0.90	1.00	−0.91	−0.29
3	Life Expectancy	−0.86	−0.91	1.00	0.32
4	Migration Rate	−0.36	−0.29	0.32	1.00

Source: Population Reference Bureau. 2011. 2011 World Population Data Sheet. Available at http://prb.org/Publications/Datasheets/2011/world-population-data-sheet/data-sheet.aspx.

in column 2 and row 2. We have already noted its strong positive relationship with birthrate, and, of course, we can ignore the correlation of the variable with itself. This leaves only two new relationships, which can be read by moving down column 2 or across row 2. Infant mortality rate has a strong negative relationship with life expectancy (the higher the infant mortality rate, the lower the life expectancy) and a moderate to weak negative relationship with migration rate (nations with higher infant mortality rates tend to have lower migration rates). For life expectancy, the variable in column 3, there is only one new relationship: a moderate positive correlation with migration rate (the higher the life expectancy, the higher the migration rate).

In closing, we should make two points. First, these correlations are unusually strong for social science research. This is largely because the variables all measure a single underlying dimension: level of development or modernization. For example, low-income nations generally have higher birthrates, higher infant mortality rates, and lower life expectancy than high-income nations. The correlations are strong because they compare apples with apples and oranges with oranges.

Second, the cells in a correlation matrix will often include other information in addition to the bivariate correlations. For example, it is common to include the number of cases on which the correlation is based and, if relevant, an indication of the statistical significance of the relationship.

Applying Statistics 13.2 Regression and Correlation

Is there a relationship between poverty and crime? Data on homicide rate and poverty have been gathered for 10 states. The independent (X) variable is the percent of individuals in the state living below the federally defined poverty line, and the dependent variable (Y) is the rate of homicides per 100,000 population. The data are presented in the following table, and columns have been added for all computations and sums.

State	X	$X - \bar{X}$	Y	$Y - \bar{Y}$	$(X - \bar{X})(Y - \bar{Y})$	$(X - \bar{X})^2$	$(Y - \bar{Y})^2$
Connecticut	9.3	−4.05	3.0	−2.72	11.02	16.40	7.40
New York	13.6	0.25	4.0	−1.72	−0.43	0.06	2.96
Illinois	12.2	−1.15	6.0	0.28	−0.32	1.32	0.08
Michigan	14.4	1.05	6.3	0.58	0.60	1.10	0.34
Nebraska	10.8	−2.55	2.2	−3.52	8.97	6.50	12.39
Tennessee	15.5	2.15	7.3	1.58	3.39	4.62	2.50
Louisiana	17.3	3.95	11.8	6.08	24.01	15.60	36.97
Texas	15.8	2.45	5.4	−0.32	−0.78	6.00	0.10
Nevada	11.3	−2.05	5.9	0.18	−0.36	4.20	0.03
California	13.3	−0.05	5.3	−0.42	0.02	0.00	0.18
TOTALS	133.5	0.00	57.2	0.00	46.13	55.83	62.94

$$\bar{X} = 13.35$$

$$\bar{Y} = 5.72$$

The slope (b) is: $$b = \frac{\Sigma(X - \bar{X})(Y - \bar{Y})}{\Sigma(X - \bar{X})^2} = \frac{46.13}{55.83} = 0.83$$

A slope of 0.83 means that for every increase of 1% in the poverty rate (a unit change in X), there is an increase of 0.83 points in the rate of homicide in these 10 states.

(*continued next page*)

Applying Statistics 13.2 *(continued)*

The Y intercept (a) is:

$$a = \bar{Y} - b\bar{X}$$

$$a = 5.72 - (0.83)(13.35)$$

$$a = 5.72 - 11.08$$

$$a = -5.36$$

The least-squares regression equation is:

$$Y = a + bX = -5.36 + (0.83)(X)$$

The correlation coefficient is:

$$r = \frac{\Sigma(X - \bar{X})(Y - \bar{Y})}{\sqrt{[\Sigma(X - \bar{X})^2 \Sigma(Y - \bar{Y})^2]}}$$

$$r = \frac{46.13}{\sqrt{(55.83)(62.94)}}$$

$$r = \frac{46.13}{\sqrt{3{,}513.94}}$$

$$r = \frac{46.13}{59.28}$$

$$r = 0.78$$

For these 10 states, poverty and homicide have a strong, positive relationship. Homicide rate increases as poverty increases. The coefficient of determination (r^2), is $(0.78)^2$, or 0.61. This indicates that 61% of the variance in homicide rate is explained by poverty for this sample of 10 states.

This is a very strong relationship for social science research, and it is tempting to conclude that poverty *causes* violence in general and murder in particular, even though this correlation is based on only 10 states and 1 year. This conclusion is consistent with much of the research on homicide. However, remember that correlation is not the same thing as causation. Violence is a complex and challenging topic, and this simple exercise does not justify any hard and fast conclusions.

BECOMING A CRITICAL CONSUMER: Correlation, Causation, and Cancer

Causation—how variables affect each other—is a central concern of the scientific enterprise. Virtually every social science theory argues that some variable(s) cause some other variable(s), and the central goal of social science research is to ascertain the strength and direction of these causal relationships.

Causation is not just a concern of science; we encounter claims about causal relationships between variables in the popular media and in everyday conversation. For example, you might hear a news commentator say that a downturn in the economy will lead to higher crime rates or that higher gasoline prices will cause people to change their driving habits and result in fewer highway deaths or that the attractions of cable TV and the Internet have led to lower rates of community involvement. How can we know when such causal claims are true? How can we judge the credibility of arguments that one variable causes another?

Probably the most obvious evidence for a causal relationship between two variables comes from measures of association, the statistics covered in this part of the text. Any of the measures introduced in this or previous chapters—phi, gamma, Pearson's *r*, etc.—can be used as evidence for the existence of a causal association. The larger the value of the measure, the stronger the evidence for causation, whereas measures of zero—or close to zero—make it extremely difficult to argue for a causal relationship.

However, even very strong correlations are not proof of causation. A common adage in social science research is that "correlation is not the same thing as causation" and, in fact, I made this point when we first took up the topic of bivariate association at the beginning of Chapter 11.

If correlation by itself does not prove causation, what other evidence is required? To build a case for causation beyond the strength of the association, we generally need to satisfy two more tests: First, we should be able to show that the independent variable occurred before the dependent variable in time; second, no other variable can explain the bivariate relationship. Let us explore these criteria by seeing how they have been applied to one of the most serious public health problems that has affected (and continues to affect) U.S. society: smoking and cancer.

Today, virtually everyone knows that smoking tobacco causes cancer. However, this information was not part of the common wisdom just a few generations ago. As recently as the 1950s, about half of all men smoked, and smoking was equated with sophistication and mature adulthood, not illness and disease. Since that time, medical research has established links between smoking, cancer, and a number of other health risks, and just as importantly, these connections have been widely broadcast by public and private agencies. The effect has been dramatic; only about 20% of adults smoke.

Statistics, especially measures of association like Pearson's *r*, played an important role in establishing the links that led to the public campaign against smoking. The most convincing studies followed large samples of individuals over long periods of time. Researchers collected a variety of medical information for each respondent, including their smoking habits and the incidence of cancer and other health problems. For example, one study conducted by the office of the U.S. Surgeon General studied women from 1976 to 1988, and in the graph below, the connection between smoking and cancer is clear. The graph plots number of cigarettes per day for the smokers against the relative risk of contracting cancer. (The relative risk is the actual cancer death rate for the smokers compared to the cancer death rate for non smokers, controlling for age and a number of medical conditions).

Even though this relationship is not perfectly linear, it is clear that there is a strong correlation between the number of cigarettes smoked and cancer risk. Women who smoked more than 35 cigarettes a day had almost twice the relative risk of women who smoked less than 14 cigarettes a day. It was graphs like this—along with strong measures of association and tons of other medical evidence, of course—that persuaded the medical profession and the government that smoking was a very significant public health risk.

Some argued against the growing evidence of a causal link between smoking and cancer by pointing out, quite legitimately, that correlation is not the same thing as causation. Researchers developed a convincing response to this objection.

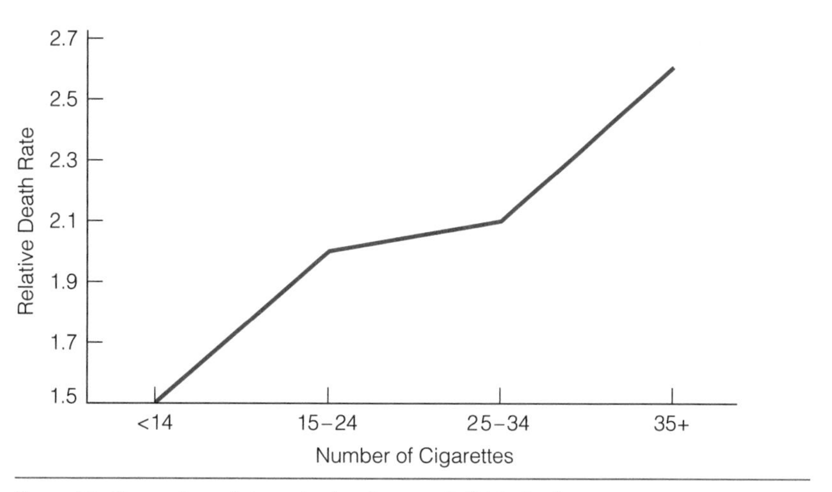

Source: http://www.cdc.gov/tobacco/sgr/sgr_forwomen/pdfs/chp3.pdf.

(*continued next page*)

BECOMING A CRITICAL CONSUMER: ***(continued)***

First, because the best studies followed people across time, researchers were able to demonstrate a *time order* between the variables. These studies started with a large number of smokers in good health. If the smokers developed cancer later in the study, cancer must be the dependent variable. That is, in this design, smoking can be the cause of cancer, but the reverse cannot be true: Given the time order, getting cancer could not have caused people to start smoking.

By itself, time order does not prove that the relationship is causal. Other variables might have been involved and might explain why people smoke and why they get cancer. For example, maybe very nervous or anxious people smoke to "calm their nerves" and are more likely to contract cancer because of their nervousness, not because of their smoking. If this were the case, the relationship between smoking and cancer would be spurious or false: What looks like a bivariate, causal relationship between the two variables is actually caused by a third variable (nervousness or anxiety).

How did the researchers respond to this possibility? The best, most convincing studies use various means to control for and examine the effect of third variables—such as anxiety, family history, gender, race, and so forth—and were able to statistically eliminate the possibility that these factors affected the bivariate relationship. We will examine some of the multivariate statistics used by these studies in the next chapter.

In addition, the case against smoking was further reinforced by several other statistical relationships, including the fact that people who smoke more are more at risk (e.g., see the graph earlier) and the fact that people who quit smoking get a health dividend—they are less likely to contract cancer. Finally, tests show that the chemicals in tobacco smoke cause cancer in rats and other animals. Taken together, these various studies provided overwhelming evidence for a causal relationship.

Whenever you encounter a claim that one variable causes another, think through these questions: How strong is the measure of association? Can it be shown that the independent variable occurred first? What other variables might affect the relationship? Have they been controlled for? The greater the extent to which a relationship satisfies these criteria, the stronger the case for a causal relationship.

Correlation, Regression, Level of Measurement, and Dummy Variables

Correlation and regression are very powerful and useful techniques—so much so that they are often used to analyze relationships between ordinal-level variables. This practice is generally not a problem, especially with ordinal variables that are "continuous" (see Chapter 12). However, this flexibility does not extend to nominal-level variables, such as religious denomination or gender. The scores of nominal variables are not numbers and have no mathematical quality. We might represent a Protestant with a score of "2" and a Catholic with a score of "1," but the former score is not "twice as much" as the latter. The scores of nominal-level variables are labels, not numbers, and it makes no sense to compute a slope or to discuss positive or negative relationships for these variables.

This is an unfortunate situation. Many of the variables that are most important in everyday social life—gender, marital status, race, and ethnicity—are nominal in level of measurement and cannot be included in a regression equation or a correlational analysis—two of the most powerful and sophisticated tools available for social science research.

Fortunately, researchers have developed a way to change the scoring on nominal-level variables by creating **dummy variables**. Dummy variables can be any level of measurement, including nominal, and have exactly two categories: one coded as 0 and the other coded as 1. Treated this way, nominal-level variables, such as gender (for example, with males coded as 0 and females coded as 1), race (with whites coded as 0 and non whites as 1), and religious denomination (Catholics coded as 1 and non-Catholics coded as 0), are commonly included in regression equations.

To illustrate, imagine we are concerned with the relationship between racial or ethnic group and education as measured by the number of years of schooling completed. If we coded whites as 0 and non whites as 1, we could compute a slope and Y intercept, write a regression equation by using race as an independent variable, and examine the correlation between the two variables. Suppose we measured the education of a sample of non white (coded as 1) and white (coded as 0) Americans and found the following regression equation:

$$Y = a + bX$$
$$Y = 12.0 + (-0.5)X$$

Education is the dependent (or Y) variable and racial or ethnic group (X) is the independent variable. The regression line crosses the vertical axis of the scattergram at the point where $Y = 12.0$. The value for the slope ($b = -0.5$) indicates a negative relationship: As racial or ethnic group "increases" (or moves toward the higher score associated with non whites), education tends to decrease. In other words, the non white respondents in this sample averaged fewer years of schooling than the white respondents. Note that the *sign* of the slope (b) would have been positive had we reversed the coding scheme and labeled whites as 1 and non whites as 0, but the *value* of b would have stayed exactly the same. The coding scheme for dummy variables is arbitrary, and as with ordinal-level variables, the researcher needs to be clear about what the values of a dummy variable indicate.

We can also use Pearson's r to assess the strength and direction of relationships with dummy variables. If we found an r of -0.23 between ethnic or racial group and education, we would conclude that there was a weak-to-moderate negative relationship between these variables for this sample. Consistent with the sign of the slope, we could also say that education decreased as ethnic or racial group "increased," or moved from white to non white. Also, using the coefficient of determination, we can say that racial or ethnic group explains, or accounts for, 5% ($r^2 = 0.23^2 = .05$) of the variance in education. (*For experience in working with dummy variables, see problem 13.9b.*)

SUMMARY

This summary is based on the example used throughout this chapter.

1. We began with a question: Is the number of children in dual-wage-earner families related to the number of hours per week husbands devote to housework? We presented the observations in a scattergram (Figure 13.1), and our visual impression was that the variables were associated in a positive direction. The pattern formed by the observation points in the scattergram could be approximated with a straight line; thus, the relationship was roughly linear.
2. Values of Y can be predicted with the freehand regression line, but predictions are more accurate if the least-squares regression line is used. The least-squares regression line is the line that best fits the data by minimizing the variation in Y. Using the formula that defines the least-squares

regression line ($Y = a + bX$), we found a slope (b) of 0.69, which indicates that each additional child (a unit change in X) is accompanied by an increase of 0.69 hours of housework per week for the husbands. We also predicted based on this formula that in a dual-wage-earner family with six children ($X = 6$), husbands would contribute 5.63 hours of housework a week ($Y = 5.63$ for $X = 6$).

3. Pearson's r is a statistic that measures the overall linear association between X and Y. Our impression from the scattergram of a substantial positive relationship was confirmed by the computed r of 0.50. We also saw that this relationship yields an r^2 of .25, which indicates that 25% of the total variation in Y (husband's housework) is accounted for, or explained, by X (number of children).
4. We acquired a great deal of information about this bivariate relationship. We know the strength and direction of the relationship and have also identified the regression line that best summarizes the effect of X on Y. We know the amount of change we can expect in Y for a unit change in X. In short, we have a greater volume of more precise information about this association between interval-ratio variables than we ever did about associations between ordinal or nominal variables. This is possible, of course, because the data generated by interval-ratio measurement are more precise and flexible than those produced by ordinal or nominal measurement techniques.

SUMMARY OF FORMULAS

FORMULA 13.1 Least-squares regression line: $Y = a + bX$

FORMULA 13.2 Slope (b): $$b = \frac{\Sigma(X - \bar{X})(Y - \bar{Y})}{\Sigma(X - \bar{X})^2}$$

FORMULA 13.3 Y intercept (a): $$a = \bar{Y} - b\bar{X}$$

FORMULA 13.4 Pearson's r: $$r = \frac{\Sigma(X - \bar{X})(Y - \bar{Y})}{\sqrt{[\Sigma(X - \bar{X})^2][\Sigma(Y - \bar{Y})^2]}}.$$

FORMULA 13.5 Coefficient of determination: $$r^2 = \frac{\Sigma(Y' - \bar{Y})^2}{\Sigma(Y - \bar{Y})^2}$$

GLOSSARY

Coefficient of determination (r^2). The proportion of all variation in Y that is explained by X. Found by squaring the value of Pearson's r.

Conditional means of Y. The mean of all scores on Y for each value of X.

Dummy variable. A nominal-level variable dichotomized so it can be used in regression analysis. A dummy variable has two scores: one coded as 0 and the other as 1.

Explained variation. The proportion of all variation in Y that is attributed to the effect of X. Equal to $\Sigma(Y' - \bar{Y})^2$.

Linear relationship. A relationship between two variables in which the observation points (dots) in the scattergram can be approximated with a straight line.

Pearson's r (r). A measure of association for variables that have been measured at the interval-ratio level.

Regression line. The single best-fitting straight line that summarizes the relationship between two variables. Regression lines are fitted to the data points by the least-squares criterion, whereby the line touches all conditional means of Y or comes as close to doing so as possible.

Scattergram. Graphic display device that depicts the relationship between two variables.

Slope (b). The amount of change in one variable per unit change in the other; b is the symbol for the slope of a regression line.

Total variation. The spread of the Y scores around the mean of Y. Equal to $\Sigma(Y' - \bar{Y})^2$.

Unexplained variation. The proportion of the total variation in Y that is not accounted for by X. Equal to $\Sigma(Y - Y')^2\,\Sigma(Y - Y')^2$.

Y intercept (a). The point where the regression line crosses the Y axis.

Y'. Symbol for predicted score on Y.

PROBLEMS

13.1 PS Why does voter turnout vary from election to election? For municipal elections in five different cities, information has been gathered on the percentage of registered voters who actually voted, unemployment rate, average years of education for the city, and the percentage of all political ads that used "negative campaigning" (personal attacks, negative portrayals of the opponent's record, etc.). For each relationship:

a. Draw a scattergram and a freehand regression line.

b. Compute the slope (b) and then find the Y intercept (a). *(Hint: Remember to compute b before computing a. A computing table like Table 13.3 is highly recommended.)*

c. State the least-squares regression line and then predict the voter turnout for a city in which the unemployment rate was 12%—a city in which the average years of schooling was 11, and a city that had an election in which 90% of the ads were negative.

d. Compute r and r^2. *(Hint: A computing table like Table 13.4 is highly recommended. If you constructed one for computing b, you already have most of the quantities you will need to solve for r.)*

e. Describe the strength and direction of each relationship in a sentence or two. Which factor had the strongest effect on turnout?

Turnout and Unemployment

City	Turnout	Unemployment Rate
A	55	5%
B	60	8%
C	65	9%
D	68	9%
E	70	10%

Turnout and Level of Education

City	Turnout	Average Years of School
A	55	11.9
B	60	12.1
C	65	12.7
D	68	12.8
E	70	13.0

Turnout and Negative Campaigning

City	Turnout	% of Negative Ads
A	55	60
B	60	63
C	65	55
D	68	53
E	70	48

13.2 SOC Occupational prestige scores for a sample of fathers and their oldest son and oldest daughter are presented below.

Family	Father's Prestige	Son's Prestige	Daughter's Prestige
A	80	85	82
B	78	80	77
C	75	70	68
D	70	75	77
E	69	72	60
F	66	60	52
G	64	48	48
H	52	55	57

Analyze the relationship between father's and son's prestige and the relationship between father's and daughter's prestige. For each relationship:

a. Draw a scattergram and a freehand regression line.

b. Compute the slope (b) and then find the Y intercept (a).

c. State the least-squares regression line. What prestige score would you predict for a son whose father had a prestige score of 72? What prestige score would you predict for a daughter whose father had a prestige score of 72?

d. Compute r and r^2.

e. Describe the strength and direction of the relationships in a sentence or two. Does the occupational prestige of the father have an impact on his children? Does it have the same impact for daughters as it does for sons?

13.3 GER The residents of a housing development for senior citizens have completed a survey in which they indicated how physically active they are and how many visitors they receive each week. Are these two variables related for the 10 cases reported here? Draw a scattergram and then compute r and r^2. Find the least-squares regression line. What would be the predicted number of visitors for a person whose level of activity was a 5? How about a person who scored 18 on level of activity?

Case	Level of Activity	Number of Visitors
A	10	14
B	11	12
C	12	10
D	10	9
E	15	8
F	9	7
G	7	10
H	3	15
I	10	12
J	9	2

13.4 PS The following variables were collected for a random sample of 10 precincts during the last national election. Draw scattergrams and then compute r and r^2 for each combination of variables. Write a paragraph interpreting the relationship between these variables.

Precinct	Percent Democrat	Percent Minority	Voter Turnout
A	49	9	36
B	78	15	60
C	62	18	64
D	50	10	56
E	45	12	55
F	56	8	52
G	85	20	25
H	25	0	42
I	13	5	89
J	33	9	88

13.5 SOC/CJ The table below presents the scores of 10 states on each of six variables, which are three measures of criminal activity as dependent variables and three possible independent variables: unemployment rate, education, and poverty. Crime rates are the number of incidents per 100,000 population.

For each combination of crime rate and the three independent variables:

a. Draw a scattergram and a freehand regression line.

b. Compute the slope (b) and then find the Y intercept (a).

c. State the least-squares regression line. What homicide rate would you predict for a state with a poverty rate of 15? What robbery rate would you predict for a state where the percentage of high school graduates was 75? What auto theft rate would you predict for a state in which 12% of the population was unemployed?

d. Compute r and r^2.

e. Describe the strength and direction of each of these relationships in a sentence or two.

	Crime Rates, 2009			Independent Variables		
State	Homicide	Robbery	Car Theft	% Families in Poverty, 2010	% HS Grads, 2009	% Unemployed, 2009
Maine	2	30	77	9	90	8
New York	4	144	112	10	84	8
Ohio	5	154	198	10	88	10
Iowa	1	210	129	7	90	6
Alabama	7	132	235	12	82	11
Kentucky	4	135	141	13	81	11
Texas	5	153	309	12	80	8
Louisiana	12	136	260	13	81	7
Washington	3	100	355	8	90	9
California	5	173	443	10	80	11

Sources: U.S. Bureau of the Census, *Statistical Abstract of the United States, 2011*. Available at http://www.census.gov/compendia/statab/2011edition.html. FBI, *Uniform Crime Report*. Available at http://www2.fbi.gov/ucr/cius2009/index.html.

13.6 Data on three variables have been collected for 15 nations. The variables are fertility rate (average number of children born to each woman), education for females (expressed as a percentage of all students at the secondary level who are female), and percent of married women ages 15–49 using all forms of contraception.

Nation	Fertility Rate[1]	Education of Females[1]	Percent of women using contraception[2]
Niger	7.7	38.4	11
Cambodia	3.9	35.6	51
Guatemala	4.3	46.9	54
Ghana	4.1	44.7	24
Bolivia	3.7	48.2	61
Kenya	5.0	47.6	46
Dominican Republic	2.7	54.8	73
Mexico	2.1	50.5	73
Vietnam	1.8	47.1	78
Turkey	2.0	37.5	73
United States	2.0	49.0	79
China	1.8	45.3	85
United Kingdom	2.4	52.7	84
Japan	1.3	49.0	54
Italy	1.3	48.1	60

Sources: Nationmaster.com. World Population Data Sheet, 2011. Population Reference Bureau (www.prb.org).

a. Compute r and r^2 for each combination of variables.

b. Summarize these relationships in terms of strength and direction.

13.7 The basketball coach at a small local college believes that his team plays better and scores more points in front of larger crowds. The number of points scored and attendance for all home games last season are reported here. Do these data support the coach's argument?

Game	Points Scored	Attendance
1	54	378
2	57	350
3	59	320
4	80	478
5	82	451
6	75	250
7	73	489
8	53	451
9	67	410
10	78	215
11	67	113
12	56	250
13	85	450
14	101	489
15	99	472

13.8 The following table presents the scores of 15 states on three variables. Compute r and r^2 for each combination of variables. Write a paragraph interpreting the relationship among these three variables.

State	Per Capita Expenditures on Education, 2007	Percent High School Graduates, 2010	Rank in per Capita Income
Arkansas	1,705	83.8	45
Colorado	1,863	88.9	15
Connecticut	2,475	88.6	1
Florida	1,728	85.2	24
Illinois	1,955	85.9	14
Kansas	1,995	89.5	23
Louisiana	1,835	81.2	46
Maryland	1,930	88.0	4
Michigan	1,965	88.1	37
Mississippi	1,345	79.9	50
Nebraska	1,637	90.1	22
New Hampshire	1,919	90.9	8
North Carolina	1,522	83.6	35
Pennsylvania	1,990	87.5	18
Wyoming	2,900	91.7	6

Source: U.S. Bureau of the Census, 2011. *Statistical Abstract of the United States, 2011.* Available at http://www.census.gov/compendia/statab.

13.9 Fifteen individuals were randomly selected from the respondents to a public opinion survey and their scores are reproduced here.

a. Calculate Pearson's r for age (X) and four dependent variables (Y's): prestige, number of children, support for abortion, and hours of TV watching.

b. Calculate Pearson's r for gender (X) and four dependent variables (Y's): support for abortion, prestige, hours of TV watching, and number of children.

c.

Occupational Prestige	Number of Children	Age	Support for Legal Abortion	Hours of TV per Day	Gender
32	3	34	3	1	1
50	0	41	1	3	1
17	0	52	5	2	1
69	3	67	1	5	0
17	0	40	1	5	0
52	0	22	2	3	1
32	3	31	3	4	0
50	0	23	4	4	0
19	9	64	1	6	1
37	4	55	4	2	0
14	3	66	5	5	1
51	0	22	2	0	0
45	0	19	3	7	0
44	0	21	4	1	1
46	4	58	2	0	1

KEY TO VARIABLES:

"Number of Children," "Age," and "Hours of TV" are actual values. "Occupational Prestige" is a continuous ordinal scale that indicates the amount of respect or esteem associated with occupation; the higher the score, the greater the prestige. "Support for Legal Abortion" is a collapsed ordinal scale that measures the respondent's strength of agreement or disagreement with the statement "A women should be able to get a legal abortion for any reason," where 1 = strongly agree, 2 = agree, 3 = neither agree nor disagree, 4 = disagree, and 5 = strongly disagree. "Gender" is a nominal-level variable coded as a dummy variable, with 0 = female and 1 = male.

YOU ARE THE RESEARCHER: Who Watches TV? Who Succeeds in Life?

Two projects are presented to help you apply the statistical skills developed in this chapter. In the first, you will analyze the correlates of time spent watching TV. You will select four independent variables and then assess their impact on *tvhours* (number of hours per day spent watching TV). In the second project, you will choose either *income 06* or *prestg 80* as your dependent variable. Both variables measure social class standing, and you will select independent variables you believe might be associated with an individual's level of success.

In both projects, you will have the option to analyze males and females separately to see if the pattern of correlation varies by sex. You will use a SPSS command called **Split File**, which allows us to examine subgroups within the sample independently.

There is no need for a detailed demonstration of how to generate Pearson's *r* with SPSS. We will use essentially the same command used in Chapter 12 to generate Spearman's rho. Briefly: Click **Analyze → Correlate → Bivariate**. The **Bivariate Correlations** dialog box will open and "Pearson" will already be checked. Find your variables in the list on the left and then transfer them to the **Variables:** window, click **OK**, and then SPSS will produce a correlation matrix showing the relationship of all the variables with each other. Recall that the cells in the output present three pieces of information: the value of the measure of association (Pearson's *r* in this case), its statistical significance, and the number of cases.

Optional Analysis. The **Split File** command is used when we want to see if variables are related in the same way for various subgroups in the sample. To demonstrate this command, let us look at the relationship between religiosity (measured by *attend*) and support for gay marriage (*marhomo*). The correlation between these variables for the entire sample is 0.32. This is a moderate positive relationship: As frequency of church attendance rises (as score on *attend* goes up), opposition to gay marriage also rises (score on *marhomo* increases). Would the same relationship hold for males and females?

To observe the effect of sex on the bivariate relationship, we split the GSS sample into two subfiles. Men and women will then be processed separately, and SPSS will produce one correlation matrix for men and another for women.

Click **Data** from the main menu and then click **Split File**. In the **Split File** window, click the button next to "organize output by groups." This will generate separate outputs for men and women. Select *sex* from the variable list and then click the arrow to move the variable name into the **Groups Based On** window. Click **OK**, and all procedures requested will be done separately for men and women. To restore the full sample, call up the **Split File** window again and then click the **Reset** button. For now, click **Statistics → Correlate → Bivariate** to rerun the correlation *attend* and *marhomo*.

The output shows a correlation of 0.26 for males and 0.39 for females. This result suggests that religiosity has a slightly more powerful impact for females than for males. For both groups, there is a weak-to-moderate positive relationship between religiosity and opposition to gay marriage. Be sure to return to the **Split File** window and then click **Reset** to restore the full sample before starting the projects.

PROJECT 1: Who Watches TV?

In this exercise, your dependent variable will be *tvhours*. Scores on this variable range from 0 to a high of 24 hours (all day!), with most cases clustered around one to three hours of TV watching per day.

STEP 1: Choosing Independent Variables

Select four variables from the 2010 GSS you think might be important causes of *tvhours*. Your independent variables *cannot* be nominal in level of measurement unless you recode the variable into a dummy variable (see below). You may use any interval-ratio or ordinal level variable.

Dummy variables. To include a dummy variable in your analysis, recode the variable so it has only two values: 0 and 1. Some possibilities include recoding *sex* so males = 0 and females = 1, *racecen 1* so whites = 0 and non whites = 1, or *relig* so Protestants = 0 and non-Protestants = 1. See Chapter 10 for instructions on recoding.

Once you have selected your variables, list them in the table below and describe exactly what they measure.

SPSS variable Name	What Exactly Does This Variable Measure?

STEP 2: Stating Hypotheses

State hypotheses about the relationships you expect to find between your independent variables and *tvhours*. State these hypotheses in terms of the direction of the relationship you expect to find. For example, you might hypothesize that TV time will increase as age increases.

1. ______________________________
2. ______________________________
3. ______________________________
4. ______________________________

STEP 3: Running Bivariate Correlations

Click **Analyze** → **Bivariate** → **Correlate** and then place all variables in the **Variables:** box. Click **OK** to get your results.

STEP 4: Recording Results

Use the table below to summarize your results. Enter the *r* for each independent variable in each cell. As you read the correlation matrix, ignore correlations of variables with themselves and any redundant information.

	Independent Variables			
tvhours	1. ________	2. ________	3. ________	4. ________

STEP 5: Analyzing and Interpreting Results

Write a short summary of results for each independent variable. Your summary needs to identify the variables being tested and the strength and direction of the relationship. It is probably best to characterize the relationship in general terms and then cite the statistical values in parentheses. Be sure to note whether your hypotheses were supported. Be careful when interpreting direction, and refer back to the coding scheme to make sure you understand the relationship.

STEP 6: Optional Analysis by Using Split Files Command

You can analyze the correlations for different subgroups by splitting the sample. *Do not* use any of your independent variables to split the sample, and follow the instructions earlier to use the command. Disregard the results for very small groups. Are the variables related in essentially the same way for the subgroups as for the entire sample? If not, what experiences might account for the differences?

PROJECT 2: Who Succeeds?

For this exercise, choose either income (*income 06*) or occupational prestige (*prestg 80*) as your dependent variable. Most Americans would regard these variables as measures of success in life. What are the correlates and antecedents of affluence and prestige?

STEP 1: Choosing Independent Variables

Select four variables that you think might be important causes of the dependent variable you selected. An obvious choice is education (use *educ*, the interval-ratio version, not *degree*, the ordinal version). Remember that independent variables *cannot* be nominal in level of measurement unless you recode the variable into a dummy variable (see below). You may use any interval-ratio or ordinal level variable.

Dummy variables. To include a dummy variable in your analysis, recode the variable so it has only two values: 0 and 1. Some possibilities include recoding *sex* so males = 0 and females = 1, *racecen 1* so whites = 0 and non whites = 1, or *relig* so Protestants = 0 and non-Protestants = 1. See Chapter 10 for instructions on recoding.

Once you have selected your variables, list them in the table below and then describe exactly what they measure.

SPSS variable Name	What Exactly Does This Variable Measure?

STEP 2: Stating Hypotheses

State hypotheses about the relationships you expect to find between your independent variables and the dependent variable you selected. State these hypotheses in terms of the direction of the relationship you expect to find. For example, you might hypothesize that income will increase as age increases.

1. ______
2. ______
3. ______
4. ______

STEP 3: Running Bivariate Correlations

Click **Analyze** → **Bivariate** → **Correlate** and then place all variables in the **Variables:** box. Click **OK** to get your results.

STEP 4: Recording Results

Use the table below to summarize your results. Enter the *r* for each independent variable in each cell. As you read the correlation matrix, ignore correlations of variables with themselves and any redundant information.

	Independent Variables			
Name of Dependent Variable:	1. ______	2. ______	3. ______	4. ______

STEP 5: Analyzing and Interpreting Results

Write a short summary of results for each independent variable. Your summary needs to identify the variables being tested and the strength and direction of the relationship. It is probably best to characterize the relationship in general terms and then cite the statistical values in parentheses. Be sure to note whether your hypotheses were supported. Be careful when interpreting direction, and refer back to the coding scheme to make sure you understand the relationship.

STEP 6: Optional Analysis by Using Split Files Command

You can analyze the correlations for different subgroups by splitting the sample. *Do not* use any of your independent variables to split the sample, and follow the instructions earlier to use the command. Disregard the results for very small groups. Are the variables related in essentially the same way for the subgroups as for the entire sample? If not, what experiences might account for the differences?

Part IV Multivariate Techniques

Chapter 14 introduces multivariate analytical techniques: statistics that allow us to analyze the relationships between more than two variables at a time. These statistics are extremely useful for probing possible causal relationships between variables and are commonly reported in the professional literature. In particular, this chapter introduces regression analysis, which is the basis for many of the most popular and powerful statistical techniques in use today. These techniques are designed to be used with variables measured at the interval-ratio level of measurement, and the mathematics underlying these statistics can become very complicated, so this chapter focuses on the simplest possible applications and stresses interpretation.

14 Partial Correlation and Multiple Regression and Correlation

LEARNING OBJECTIVES

By the end of this chapter, you will be able to:

1. Compute and interpret partial correlation coefficients.
2. Find and interpret the least-squares multiple regression equation with partial slopes.
3. Find and interpret standardized partial slopes or beta-weights (b^*).
4. Calculate and interpret the coefficient of multiple determination (R^2).
5. Explain the limitations of partial and multiple regression analysis.

USING STATISTICS

The statistical techniques presented in this chapter are used to measure the strength and direction of the association between more than two interval-ratio level variables. Examples of research situations in which these techniques are useful include:

1. A family sociologist is studying the division of household labor in dual-wage earner couples and wonders if husbands at all levels of education increase their contribution to housework as the number of children in the family increases.
2. A criminologist has data that shows a moderate and positive correlation between poverty and crime rate for a sample of 542 U.S. counties and wonders if the relationship retains its strength in every region of the country. Is the relatiosnhip the same in densely populated urban counties and less populated rural counties? Is the relatiosnhip the same in counties with high and low levels of education?
3. A demographer is investigating the relationship between fertility rates and the educational levels of females in 76 nations. Her data shows a moderate negative relationship between the variables; as education increases, fertility rates tend to decrease. Is the relationship the same in Christian and non-Christian nations? Does the relationship retain its strength for nations at all levels of affluence?

Few (if any) worthwhile research questions can be answered through a statistical analysis of only two variables. Social science research is, by nature, multivariate and involves the simultaneous analysis of scores of variables. Some of the most powerful and widely used statistical tools for multivariate analysis are introduced in this chapter. We cover techniques that are used to analyze causal relationships and to make predictions—both crucial endeavors in any science.

These techniques are based on Pearson's r (see Chapter 13) and are most appropriately used with high-quality, precisely measured interval-ratio variables. As we have noted on many occasions, such data are relatively rare in social science research, and the techniques presented in this chapter are commonly used on variables measured at the ordinal level and with nominal-level variables in the form of dummy variables (see Chapter 13).

We first consider partial correlation analysis—a technique that allows us to examine bivariate relationships while controlling for a third variable. The second technique—multiple regression and correlation—allows researchers to assess the effects—separately and in combination—of more than one independent variable on the dependent variable.

Throughout this chapter, we focus on research situations involving three variables. This is the simplest application, but extensions to situations involving four or more variables are relatively straightforward. However, computations become very complex when dealing with more than two independent variables, and you should use computerized statistical packages (such as SPSS) for these situations.

Partial Correlation

In Chapter 13, we used Pearson's r to measure the strength and direction of bivariate relationships. To provide an example, we looked at the relationship between husband's contribution to housework (the dependent, or Y, variable) and number of children (the independent, or X, variable) for a sample of 12 families. We found a positive relationship of moderate strength $(r = 0.50)$ and concluded that husband's contribution to housework tends to increase as the number of children increases.

You might wonder, as researchers commonly do, if this relationship holds true for *all* types of families. For example, might husbands in strongly religious families respond differently than those in less religious families? Would politically conservative husbands behave differently than husbands who are politically liberal? How about more educated husbands? Would they respond differently than less educated husbands? We can address these kinds of issues by means of a technique called **partial correlation**, in which we observe how the bivariate relationship changes when a third variable, such as religiosity, education, or ethnicity, is introduced. Third variables are often referred to as Z variables or **control variables**.

Partial correlation proceeds by first computing Pearson's r for the bivariate (or **zero-order**) relationship and then computing the partial (or **first-order**) correlation coefficient. If the partial correlation coefficient differs from the zero-order correlation coefficient, we conclude that the third variable has an effect on the bivariate relationship. For example, if well-educated husbands respond differently to an additional child than less-well-educated husbands, the partial correlation coefficient will differ in strength (and perhaps in direction) from the bivariate correlation coefficient.

Before considering matters of computation, we will consider the possible relationships between the partial and bivariate correlation coefficients and what they might mean. There are three possible patterns, and we will consider each in turn.

Types of Relationships

Direct Relationship. One possible outcome is that the partial correlation coefficient is essentially the same value as the bivariate coefficient. For example, imagine that after we controlled for husband's education, we found a partial correlation coefficient of +0.49, compared to the zero-order Pearson's *r* of +0.50. This would mean that the third variable (husband's education) has no effect on the relationship between number of children and husband's hours of housework. In other words, regardless of their education, husbands respond in a similar way to additional children. This outcome is consistent with the conclusion that there is a direct or causal relationship (see Figure 14.1) between *X* and *Y* and that the third variable (*Z*) is irrelevant and should be discarded from further consideration, although the researcher would probably run additional tests with other likely control variables (e.g., the researcher might control for the religion or ethnicity of the family).

FIGURE 14.1 A Direct Relationship Between *X* and *Y*

$X \longrightarrow Y$

Spurious and Intervening Relationships. A second possible outcome occurs when the partial correlation coefficient is much weaker than the bivariate correlation—perhaps even dropping to zero. This outcome is consistent with two different relationships. The first is called a *spurious relationship*: The control variable (*Z*) is a cause of the independent (*X*) and the dependent (*Y*) variable (see Figure 14.2). This outcome would mean that *X* and *Y* are not actually related. They appear to be related only because both are dependent on a common cause (*Z*). Once *Z* is taken into account, the apparent relationship between *X* and *Y* disappears.

What would a spurious relationship look like? Imagine that we controlled for the political ideology of parents and found that the partial correlation coefficient was much weaker than the bivariate Pearson's *r*. This might indicate that the number of children does not actually change the husband's contribution to housework (that is, the relationship between *X* and *Y* is not direct). Rather, political ideology is the mutual cause of both of the other variables: More conservative families are more likely to follow traditional gender role patterns (in which husbands contribute less to housework) *and* have more children.

This pattern (partial correlation much weaker than the bivariate correlation) is also consistent with an **intervening** relationship between the variables (see Figure 14.3). In this situation, *X* and *Y* are not linked directly but are connected through the control variable. Again, once *Z* is controlled, the apparent relationship between *X* and *Y* disappears.

How can we tell the difference between spurious and intervening relationships? This distinction cannot be made on statistical grounds: Spurious and intervening relationships look exactly the same in terms of statistics. The researcher may be able to distinguish between these two relationships in terms of

FIGURE 14.2 A Spurious Relationship Between *X* and *Y*

FIGURE 14.3 An Intervening Relationship Between *X* and *Y*

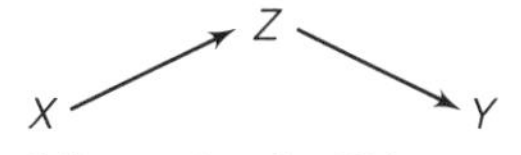

Figure 14.4 An Interactive Relationship Between *X*, *Y*, and *Z*

Z_1 + X Y Z_2 −

the time order of the variables (i.e., which came first) or on theoretical grounds, but not on statistical grounds.

Interaction. A third possible relationship between variables should be mentioned, even though it *cannot* be detected by partial correlation analysis. This relationship—called *interaction*—occurs when the relationship between *X* and *Y* changes markedly under the various values of *Z*. For example, if we controlled for social class and found that husbands in middle-class families increased their contribution to housework as the number of children increased while husbands in working-class families did just the reverse, we would conclude that there was interaction between these three variables. In other words, there would be a positive relationship between *X* and *Y* for one category of *Z* and a negative relationship for the other category, as illustrated in Figure 14.4.

Computing and Interpreting the Partial Correlation Coefficient

The logic and computation of the partial correlation coefficient requires some new concepts and terminology. We will introduce the terminology first and then move on to the formula.

Terminology and Formula. Partial correlation requires that we deal with more than one bivariate relationship, and we need to differentiate between them with subscripts. Thus, the symbol r_{yx} will refer to the correlation coefficient between variable *Y* and variable *X*; r_{yz} will refer to the correlation coefficient between *Y* and *Z*; and r_{xz} will refer to the correlation coefficient between *X* and *Z*. Recall that correlation coefficients calculated for bivariate relationships are often referred to as zero-order correlations.

Partial correlation coefficients, or first-order partials, are symbolized as $r_{yx.z}$. The variable to the right of the dot is the control variable. Thus, $r_{yx.z}$ refers to the partial correlation coefficient that measures the relationship between variables *X* and *Y* while controlling for variable *Z*. The formula for the first-order partial is:

FORMULA 14.1

$$r_{yx.z} = \frac{r_{yx} - (r_{yz})(r_{xz})}{\sqrt{1 - r_{yz}^2}\sqrt{1 - r_{xz}^2}}$$

Note that you must first calculate the zero-order coefficients between all possible pairs of variables (variables *X* and *Y*, *X* and *Z*, and *Y* and *X*) before solving this formula.

Computation. To illustrate the computation of a first-order partial, we will return to the relationship between number of children (*X*) and husband's contribution to housework (*Y*) for 12 dual-career families. The zero-order *r* between these two variables ($r_{yx} = 0.50$) indicate a moderate positive relationship (as number of children increased, husbands tended to contribute more to housework). Suppose the researcher wished to investigate the possible effects of husband's education on the bivariate relationship. The original data (from Table 13.1) and the scores of the 12 families on the new variable are presented in Table 14.1.

The zero-order correlations, as presented in the correlation matrix in Table 14.2, indicate that the husband's contribution to housework is positively related to number of children ($r_{yx} = 0.50$), that better-educated husbands tend to do less housework ($r_{yz} = -0.30$), and that families with better-educated husbands have fewer children ($r_{xz} = -0.47$).

TABLE 14.1 Scores on Three Variables for 12 Dual-Wage-Earner Families and Zero-Order Correlations

Family	Husband's Housework (Y)	Number of Children (X)	Husband's Years of Education (Z)
A	1	1	12
B	2	1	14
C	3	1	16
D	5	1	16
E	3	2	18
F	1	2	16
G	5	3	12
H	0	3	12
I	6	4	10
J	3	4	12
K	7	5	10
L	4	5	16

TABLE 14.2 Zero-Order Correlations

	Husband's Housework (Y)	Number of Children (X)	Husband's Years of Education (Z)
Husband's Housework (Y)	1.00	0.50	−0.30
Number of Children (X)		1.00	−0.47
Husband's Years of Education (Z)			1.00

Is the relationship between husband's housework (*X*) and number of children (*Y*) affected by husband's years of education? Substituting the zero-order correlations into Formula 14.1, we would have:

$$r_{yx.z} = \frac{r_{yx} - (r_{yz})(r_{xz})}{\sqrt{1 - r_{yz}^2}\sqrt{1 - r_{xz}^2}}$$

$$r_{yx.z} = \frac{(0.50) - (-0.30)(-0.47)}{\sqrt{1 - (-0.30)^2}\sqrt{1 - (0.47)^2}}$$

$$r_{yx.z} = \frac{(0.50) - (0.14)}{\sqrt{1 - 0.09}\sqrt{1 - 0.22}}$$

$$r_{yx.z} = \frac{0.36}{\sqrt{0.91}\sqrt{0.78}}$$

$$r_{yx.z} = \frac{0.36}{(0.95)(0.88)}$$

$$r_{yx.z} = \frac{0.36}{0.84}$$

$$r_{yx.z} = 0.43$$

Interpretation. The first-order partial ($r_{yx.z} = 0.43$), which measures the strength of the relationship between husband's housework (*Y*) and number of children (*X*) while controlling for husband's education (*Z*), is lower in value than the zero-order coefficient (r_{yx} = 0.50), but the difference in the two values is not great. This result suggests a direct relationship between variables X and Y. That is, when controlling for husband's education, the statistical relationship between husband's housework and number of children is essentially unchanged. Regardless of education, husband's hours of housework increase with the number of children.

Our next step in statistical analysis would probably be to select another control variable. The more the bivariate relationship retains its strength across a series of controls for third variables (Zs), the stronger the evidence for a direct relationship between X and Y.

Figure 14.5 A Possible Causal relationship Among three Variables

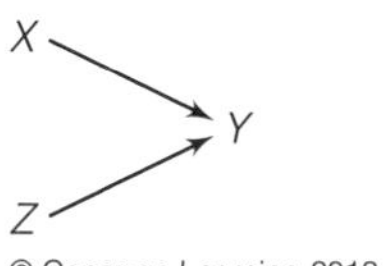

In closing, I should mention an additional possible outcome, in which the partial correlation coefficient is greater in value than the zero-order coefficient ($r_{yx.z} > r_{yx}$). This outcome would be consistent with a causal model in which the variable taken as independent and the control variable each had a separate effect on the dependent variable and were uncorrelated with each other. This relationship is depicted in Figure 14.5. The absence of an arrow between X and Z indicates that they have no mutual relationship.

This pattern means that X and Z should be treated as independent variables, and the next step in the statistical analysis would probably involve multiple correlation and regression—techniques that will be presented in the remainder of this chapter. (*For practice in computing and interpreting partial correlation coefficients, see problems 14.1 to 14.3.*)

ONE STEP AT A TIME Computing and Interpreting the Partial Correlation Coefficient

To begin, compute Pearson's *r* for all pairs of variables. Be clear about which variable is independent (*X*), which is dependent (*Y*), and which is the control (*Z*).

Step ***Operation***

To find the partial correlation coefficient, solve Formula 14.1:

1. Multiply r_{yz} by r_{xz}.
2. Subtract the value you found in step 1 from r_{yx}. *This value is the numerator of Formula 14.1.*
3. Square the value of r_{YZ}.
4. Subtract the quantity you found in step 3 from 1.
5. Take the square root of the quantity you found in step 4.
6. Square the value of r_{XZ}.
7. Subtract the quantity you found in step 6 from 1.
8. Take the square root of the quantity you found in step 7.
9. Multiply the quantity you found in step 8 by the value you found in step 5. *This value is the denominator of Formula 14.1.*
10. Divide the quantity you found in step 2 by the quantity you found in step 9. *This is the partial correlation coefficient.*

To interpret the partial correlation coefficient, compare its value to the zero-order correlation:

Choose the description below that comes closest to matching the relationship between the two values:

1. The partial correlation coefficient is roughly the same value as the zero-order or bivariate correlation. A good rule of thumb for "roughly the same" is a difference of less than 0.10.

(*continued next page*)

ONE STEP AT A TIME ***(continued)***

This outcome is evidence that the control variable (*Z*) has little or no effect and that the relationship between *X* and *Y* is direct.

2. The partial correlation coefficient is much less (say, more than 0.10 less) than the bivariate correlation. This is evidence that the control variable (*Z*) changes the relationship between *X* and *Y*. The relationship between *X* and *Y* is either spurious (*Z* causes *X* and *Y*) or intervening (*X* and *Y* are linked by *Z*).

Be aware that *X*, *Y*, and *Z* may have an interactive relationship in which the relationship between *X* and *Y* changes for each category of *Z*. Partial correlation analysis cannot detect interactive relationships.

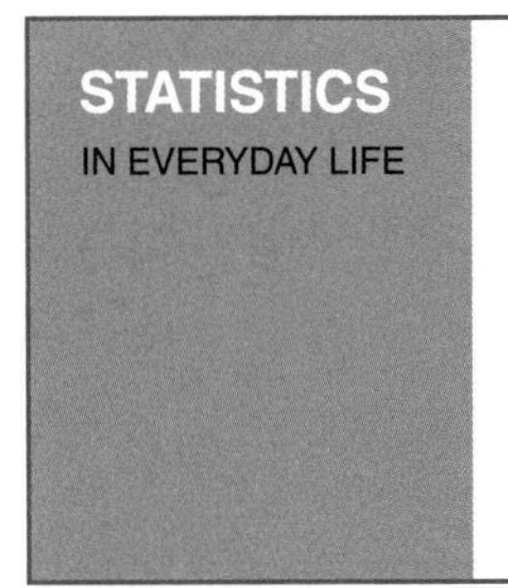

An analysis of a representative sample of Americans aged 25 to 65 found a correlation of 0.43 between years of schooling and income—a moderate positive relationship. Income increases as education increases and education explains almost 19% of the variation in income.

Is this relationship affected by age? The partial correlation between education and income while controlling for age is also 0.43, the same value as the zero-order correlation. This indicates that the relationship between education and income is direct: Regardless of age, income increases as education increases.

People with higher levels of education tend to have higher income regardless of their age—a finding that you may find reassuring as you move to the completion of your degree.

Multiple Regression: Predicting the Dependent Variable

In Chapter 13, the least-squares regression line was introduced as a way of describing the overall linear relationship between two interval-ratio variables and of predicting scores on *Y* from scores on *X*. This line was the best-fitting line to summarize the bivariate relationship and was defined by this formula:

FORMULA 14.2

$$Y = a + bX$$

where: a = the Y intercept
b = the slope

The least-squares regression line can be modified to include (theoretically) any number of independent variables. This technique is called **multiple regression**. For ease of explication, we will confine our attention to the case involving two independent variables. The least-squares multiple regression equation is:

FORMULA 14.3

$$Y = a + b_1X_1 + b_2X_2$$

where: b_1 = the partial slope of the linear relationship between the first independent variable and Y
b_2 = the partial slope of the linear relationship between the second independent variable and Y

Some new notation and some new concepts are introduced in this formula. First, while the dependent variable is still symbolized as Y, the independent variables are differentiated by subscripts. Thus, X_1 identifies the first independent variable and X_2 the second. The symbol for the slope (b) is also subscripted to identify the independent variable with which it is associated.

Partial Slopes

A major difference between the multiple and bivariate regression equations concerns the slopes (b's). In the case of multiple regression, the b's are called **partial slopes**, and they show the amount of change in Y for a unit change in the independent variable while controlling for the effects of the other independent variables in the equation. The partial slopes are thus analogous to partial correlation coefficients and represent the direct effect of the associated independent variable on Y.

Computing Partial Slopes. The partial slopes for the independent variables are determined by Formulas 14.4 and 14.5:

FORMULA 14.4

$$b_1 = \left(\frac{s_y}{s_1}\right)\left(\frac{r_{y1} - r_{y2}r_{12}}{1 - r_{12}^2}\right)$$

FORMULA 14.5

$$b_2 = \left(\frac{s_y}{s_2}\right)\left(\frac{r_{y2} - r_{y1}r_{12}}{1 - r_{12}^2}\right)$$

where: b_1 = the partial slope of X_1 on Y
b_2 = the partial slope of X_2 on Y
s_y = the standard deviation of Y
s_1 = the standard deviation of the first independent variable (X_1)
s_2 = the standard deviation of the second independent variable (X_2)
r_{y1} = the bivariate correlation between Y and X_1
r_{y2} = the bivariate correlation between Y and X_2
r_{12} = the bivariate correlation between X_1 and X_2

To illustrate the computation of the partial slopes, we will assess the combined effects of number of children (X_1) and husband's education (X_2) on husband's contribution to housework. All the relevant information can be calculated from Table 14.1 and is reproduced here:

Husband's Housework	Number of Children	Husband's Education
$\bar{Y} = 3.3$	$\bar{X}_1 = 2.7$	$\bar{X}_2 = 13.7$
$s_y = 2.1$	$s_1 = 1.5$	$s_2 = 2.6$
	Zero-Order Correlations	
	$r_{y1} = 0.50$	
	$r_{y2} = -0.30$	
	$r_{12} = -0.47$	

The partial slope for the first independent variable (X_1) is:

$$b_1 = \left(\frac{s_y}{s_1}\right)\left(\frac{r_{y1} - r_{y2}r_{12}}{1 - r_{12}^2}\right)$$

$$b_1 = \left(\frac{2.1}{1.5}\right)\left(\frac{0.50 - (-0.30)(-0.47)}{1 - (-0.47)^2}\right)$$

$$b_1 = (1.4)\left(\frac{0.50 - 0.14}{1 - 0.22}\right)$$

$$b_1 = (1.4)\left(\frac{0.36}{0.78}\right)$$

$$b_1 = (1.4)(0.46)$$

$$b_1 = 0.65$$

For the second independent variable (SES or X_2), the partial slope is:

$$b_2 = \left(\frac{s_y}{s_2}\right)\left(\frac{r_{y2} - r_{y1}r_{12}}{1 - r_{12}^2}\right)$$

$$b_2 = \left(\frac{2.1}{2.6}\right)\left(\frac{-0.30 - (+.50)(-0.47)}{1 - (-0.47)^2}\right)$$

$$b_2 = (0.81)\left(\frac{-0.30 - (-0.24)}{1 - 0.22}\right)$$

$$b_2 = (0.81)\left(\frac{-0.30 + 0.24}{0.78}\right)$$

$$b_2 = (0.81)\left(\frac{-0.06}{0.78}\right)$$

$$b_2 = (0.81)(-0.08)$$

$$b_2 = -0.07$$

ONE STEP AT A TIME Computing and Interpreting Partial Slopes

These procedures apply when there are two independent variables and one dependent variable. For more complex situations, use a computerized statistical package, such as SPSS, to do the calculations:

Step	***Operation***
	To compute the partial slope associated with the first independent variable by using Formula 14.4:
1.	Divide s_y by s_1.
2.	Multiply r_{y2} by r_{12}.
3.	Subtract the value you computed in step 2 from r_{y1}.
4.	Square the value of r_{12}.
5.	Subtract the value you computed in step 4 (r_{12}^2) from 1.
6.	Divide the value you computed in step 3 by the value you computed in step 5.
7.	Multiply the value you computed in step 6 by the value you computed in step 1. *This value is the partial slope associated with the first independent variable.*

ONE STEP AT A TIME ***(continued)***

To compute the partial slope associated with the second independent variable by using Formula 14.5:

1. Divide s_y by s_2.
2. Multiply r_{y1} by r_{12}.
3. Subtract the value you computed in step 2 from r_{y2}.
4. Square the value of r_{12}.
5. Subtract the value you computed in step 4 from 1.
6. Divide the value you computed in step 3 by the value you computed in step 5.
7. Multiply the value you computed in step 6 by the value you computed in step 1. *This value is the partial slope associated with the second independent variable.*

To Interpret Partial Slopes

The value of a partial slope is the increase in the value of Y for a unit increase in the value of the associated independent variable while controlling for the effects of the other independent variable.

Finding the Y Intercept. Now that partial slopes have been determined for both independent variables, the Y intercept (a) can be found. Note that a is calculated from the mean of the dependent variable (symbolized as $\bar{Y}$) and the means of the two independent variables ($\bar{X}_1$ and $\bar{X}_2$):

FORMULA 14.6

$$a = \bar{Y} - b_1\bar{X}_1 - b_2\bar{X}_2$$

Substituting the proper values for the example problem at hand, we would have:

$$a = \bar{Y} - b_1\bar{X}_1 - b_2\bar{X}_2$$
$$a = 3.3 - (0.65)(2.7) - (-0.07)(13.7)$$
$$a = 3.3 - (1.8) - (-1.0)$$
$$a = 3.3 - 1.8 + 1.0$$
$$a = 2.5$$

The Least-Squares Multiple Regression Line and Predicting Y'. For our example problem, the full least-squares multiple regression equation would be:

$$Y = a + b_1X_1 + b_2X_2$$
$$Y = 2.5 + (0.65)X_1 + (-0.07)X_2$$

ONE STEP AT A TIME **Computing the Y Intercept**

Step ***Operation***

To find the Y intercept by using Formula 14.6:

1. Multiply the mean of X_2 by b_2.
2. Multiply the mean of X_1 by b_1.
3. Subtract the quantity you found in step 1 from the quantity you found in step 2.
4. Subtract the quantity you found in step 3 from the mean of Y. *The result is the value of a: the Y intercept.*

ONE STEP AT A TIME	Using the Multiple Regression Line to Predict Scores on Y
Step	***Operation***
1.	Choose a value for X_1. Multiply this value by the value of b_1.
2.	Choose a value for X_2. Multiply this value by the value of b_2.
3.	Add the values you found in steps 1 and 2 to the value of *a*: the *Y* intercept. *The resulting value is the predicted score on Y.*

As was the case with the bivariate regression line, this formula can be used to predict scores on the dependent variable from scores on the independent variables. For example, what would be our best prediction of husband's housework (Y') for a family of four children ($X_1 = 4$), where the husband had completed 11 years of schooling ($X_2 = 11$)? Substituting these values into the least-squares formula, we would have:

$$Y' = 2.5 + (0.65)(4) + (-.07)(11)$$
$$Y' = 2.5 + 2.6 - 0.8$$
$$Y' = 4.3$$

Our prediction would be that this husband would contribute 4.3 hours per week to housework. This prediction is, of course, a kind of "educated guess," which is unlikely to be perfectly accurate. However, we will make fewer errors of prediction using the least-squares line (and, thus, incorporating information from the independent variables) than we would using any other method of prediction (assuming, of course, that there is a linear association between the independent and the dependent variables). (*For practice in predicting Y scores and in computing slopes and the Y intercept, see problems 14.1 to 14.6.*)

Multiple Regression: Assessing the Effects of the Independent Variables

The least-squares multiple regression equation (Formula 14.3) is used to isolate the separate effects of the independents and to predict scores on the dependent variable. However, in many situations, using this formula to determine the relative importance of the various independent variables will be awkward, especially when the independent variables differ in terms of units of measurement (e.g., number of children vs. years of education). When the units of measurement differ, a comparison of the partial slopes will not necessarily tell us which independent variable has the strongest effect and is thus the most important. Comparing the partial slopes of variables that differ in units of measurement is a little like comparing apples and oranges.

We can make it easier to compare the effects of the independent variables by converting all variables in the equation to a common scale and thereby eliminating variations in the values of the partial slopes that are solely a function of differences in units of measurement. For example, we can standardize all distributions by changing the scores of all variables to *Z* scores. Each distribution of scores would then have a mean of 0 and a standard deviation of 1 (see Chapter 5), and comparisons between the independent variables would be much more meaningful.

Computing the Standardized Regression Coefficients

To standardize the variables to the normal curve, we could actually convert all scores into the equivalent *Z* scores and then recompute the slopes and the *Y* intercept. This would require a good deal of work; fortunately, a shortcut is available for computing the slopes of the standardized scores directly. These **standardized partial slopes**—called **beta-weights**—are symbolized b^*.

Beta-Weights. The beta-weights show the amount of change in the standardized scores of *Y* for a one-unit change in the standardized scores of each independent variable while controlling for the effects of all other independent variables.

Formulas and Computation for Beta-Weights. When we have two independent variables, the beta-weight for each is found by using Formulas 14.7 and 14.8:

FORMULA 14.7

$$b_1^* = b_1\left(\frac{s_1}{s_y}\right)$$

FORMULA 14.8

$$b_2^* = b_2\left(\frac{s_2}{s_y}\right)$$

We can now compute the beta-weights for our sample problem to see which of the two independent variables has the stronger effect on the dependent. For the first independent variable, number of children (X_1):

$$b_1^* = b_1\left(\frac{s_1}{s_y}\right)$$

$$b_1^* = (0.65)\left(\frac{1.5}{2.1}\right)$$

$$b_1^* = (0.65)(0.71)$$

$$b_1^* = 0.46$$

For the second independent variable, SES (X_2):

$$b_2^* = b_2\left(\frac{s_2}{s_y}\right)$$

$$b_2^* = (-0.07)\left(\frac{2.6}{2.1}\right)$$

$$b_2^* = (-0.07)(1.24)$$

$$b_2^* = -0.09$$

ONE STEP AT A TIME **Computing and Interpreting Beta-Weights (b^*)**

These procedures apply when there are two independent variables and one dependent variable. For more complex situations, use a computerized statistical package, such as SPSS, to do the calculations.

Step	***Operation***
To compute the beta-weight associated with the first independent variable using Formula 14.7:	
1.	Divide s_1 by s_y.
2.	Multiply the value you found in step 1 by the partial slope of the first independent variable (b_1). *This value is the beta-weight associated with the first independent variable.*
To compute the beta-weight associated with the second independent variable using Formula 14.8:	
1.	Divide s_2 by s_y.
2.	Multiply the value you found in step 1 by the partial slope of the second independent variable (b_2). *This value is the beta-weight associated with the second independent variable.*

To Interpret Beta Weights

A beta-weight (or standardized partial slope) shows the increase in the value of Y for a unit increase in the value of the associated independent variable while controlling for the effects of the other independent variable after all variables have been standardized (or transformed to Z scores).

Comparing the value of the beta-weights, we see that number of children has a stronger effect than SES on husband's housework. Furthermore, the net effect (after controlling for the effect of SES) of the first independent variable is positive, while the net effect of the second independent variable is negative.

The Standardized Least-Squares Regression Line. Using standardized scores, the least-squares regression equation can be written as:

FORMULA 14.9

$$Z_y = a_z + b_1^* Z_1 + b_2^* Z_2$$

where: Z indicates that all scores have been standardized to the normal curve

The standardized regression equation can be further simplified by dropping the term for the Y intercept because this term will always be zero when scores have been standardized. This value is the point where the regression line crosses the Y axis and is equal to the mean of Y when all independent variables equal 0. This relationship can be seen by substituting 0 for all independent variables in Formula 14.6:

$$a = \bar{Y} - b_1\bar{X}_1 - b_2\bar{X}_2$$
$$a = \bar{Y} - b_1(0) - b_2(0)$$
$$a = \bar{Y}$$

Because the mean of any standardized distribution of scores is zero, the mean of the standardized Y scores will be zero and the Y intercept will also be zero $(a = \bar{Y} = 0)$. Thus, Formula 14.9 simplifies to:

FORMULA 14.10

$$Z_y = b_1^* Z_1 + b_2^* Z_2$$

The standardized regression equation, with beta-weights noted, would be

$$Z_y = (0.46)Z_1 + (-0.09)Z_2$$

and it is immediately obvious that the first independent variable has a much stronger direct effect on Y than does the second independent variable.

As we have seen, multiple regression analysis permits the researcher to summarize the linear relationship among two or more independents and a dependent variable. The unstandardized regression equation (Formula 14.2) permits values of Y to be predicted from the independent variables in the original units of the variables. The standardized regression equation (Formula 14.10) allows the researcher to easily assess the relative importance of the various independent variables by comparing the beta-weights. (*For practice in computing and interpreting beta-weights, see any of the problems at the end of this chapter. It is probably a good idea to start with problem 14.1 because it has the smallest data set and the least complex computations.*)

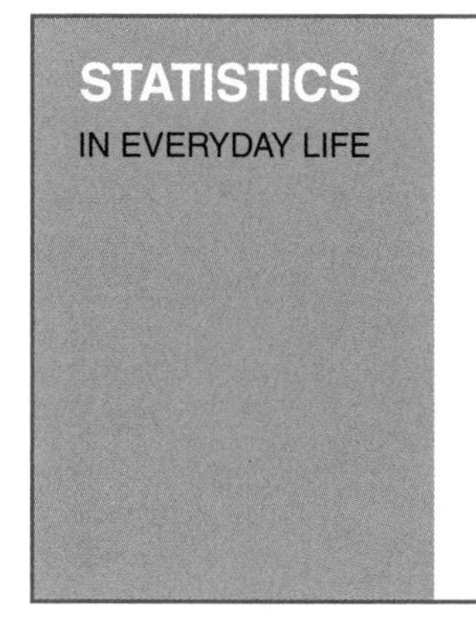

The research literature on the crime of rape generally finds that the underlying motivation is about power and dominance. However, an analysis by economist Todd Kendall suggests that sexual desire may be the important force behind these attacks. He finds a correlation between the availability of pornography on the Internet and the rate of rape. In fact, his regression analysis "... implies that an increase in home internet access of 10 percentage points is associated with a 7.3% decline in rape" (p. 23) net of the effect of other variables in the regression equation (poverty, unemployment, alcohol consumption, and so forth). These results suggest that pornography is a substitute for rape—a finding that is in sharp contrast to most previous literature.

Source: Kendall, Todd. 2006. *Pornography, Rape, and the Internet.* Available from http://www.toddkendall.net/internetcrime.pdf.

Applying Statistics 14.1 Multiple Regression and Correlation

Five recently divorced men have been asked to rate subjectively the success of their adjustment to single life on a scale ranging from 5 (very successful adjustment) to 1 (very poor adjustment). Is adjustment related to the length of time married? Is adjustment related to socioeconomic status as measured by yearly income?

Case	Adjustment (Y)	Years Married (X_1)	Income (dollars) (X_2)
A	5	5	30,000
B	4	7	45,000
C	4	10	25,000
D	3	2	27,000
E	1	15	17,000
	$\bar{Y} = 3.4$	$\bar{X}_1 = 7.8$	$\bar{X}_2 = 28{,}800$
	$s = 1.4$	$s = 4.5$	$s = 9{,}173.88$

(continued next page)

Applying Statistics 14.1 *(continued)*

The zero-order correlations among these three variables are:

	Adjustment (Y)	Years Married (X_1)	Income (X_2)
Adjustment (Y)	1.00	–0.62	0.62
Years married (X_1)		1.00	−0.49
Income (X_2)			1.00

These results suggest a strong but opposite relationship between each independent variable and adjustment. Adjustment decreases as years married increases, and it increases as income increases.

To find the multiple regression equation, we must find the partial slopes.

For years married (X_1):

$$b_1 = \left(\frac{s_y}{s_1}\right)\left(\frac{r_{y1} - r_{y2}r_{12}}{1 - r_{12}^2}\right)$$

$$b_1 = \left(\frac{1.4}{4.5}\right)\left(\frac{(-0.62) - (0.62)(-0.49)}{1 - (-0.49)^2}\right)$$

$$b_1 = (0.31)\left(\frac{(-0.62) - (-0.30)}{1 - 0.24}\right)$$

$$b_1 = (0.31)\left(\frac{-0.32}{0.76}\right)$$

$$b_1 = (0.31)(-0.42)$$

$$b_1 = -0.13$$

For income (X_2):

$$b_2 = \left(\frac{s_y}{s_2}\right)\left(\frac{r_{y2} - r_{y1}r_{12}}{1 - r_{12}^2}\right)$$

$$b_2 = \left(\frac{1.4}{9{,}173.88}\right)\left(\frac{(0.62) - (-0.62)(-0.49)}{1 - (-0.49)^2}\right)$$

$$b_2 = (0.00015)\left(\frac{0.62 - 0.30}{1 - 0.24}\right)$$

$$b_2 = (0.00015)\left(\frac{0.32}{0.76}\right)$$

$$b_2 = (0.00015)(0.42)$$

$$b_2 = 0.000063$$

The Y intercept would be:

$$a = \bar{Y} - b_1\bar{X}_1 - b_2\bar{X}_2$$

$$a = 3.4 - (-0.13)(7.8) - (0.000063)(28{,}800)$$

$$a = 3.4 - (-1.01) - (1.81)$$

$$a = 3.4 + 1.01 - 1.81$$

$$a = 2.60$$

The multiple regression equation is:

$$Y = a + b_1X_1 + b_2X_2$$

$$Y = 2.60 + (-0.13)X_1 + (0.000063)X_2$$

What adjustment score could we predict for a male who had been married 30 years ($X_1 = 30$) and had an income of \$50,000 ($X_2 = 50{,}000$)?

$$Y' = 2.60 + (-0.13)(30) + (0.000063)(50{,}000)$$

$$Y' = 2.60 + (-3.90) + (3.15)$$

$$Y' = 1.85$$

To assess which of the two independents has the stronger effect on adjustment, the standardized partial slopes must be computed.

For years married (X_1):

$$b_1^* = b_1\left(\frac{s_1}{s_y}\right)$$

$$b_1^* = (-0.13)\left(\frac{4.5}{1.4}\right)$$

$$b_1^* = -0.42$$

For income (X_2):

$$b_2^* = b_2\left(\frac{s_2}{s_y}\right)$$

$$b_2^* = (0.000063)\left(\frac{9173.88}{1.4}\right)$$

$$b_2^* = 0.41$$

The standardized regression equation is:

$$Z_y = b_1^*Z_1 + b_2^*Z_2$$

$$Z_y = (-0.42)Z_1 + (0.41)Z_2$$

and the independent variables have nearly equal but opposite effects on adjustment. To assess the combined effects of the two independents on adjustment, the coefficient of multiple determination must be computed:

$$R^2 = r_{y1}^2 + r_{y2.1}^2(1 - r_{y1}^2)$$

$$R^2 = (-0.62)^2 + (0.46)^2(1 - (-0.62)^2)$$

(continued next page)

Applying Statistics 14.1 *(continued)*

$$R^2 = 0.38 + (0.21)(1 - 0.38)$$
$$R^2 = 0.38 + (0.21)(0.62)$$
$$R^2 = 0.38 + 0.13$$
$$R^2 = 0.51$$

The first independent variable—years married—explains 38% of the variation in adjustment by itself. To this quantity, income explains an additional 13% of the variation in adjustment. Taken together, the two independent variables explain a total of 51% of the variation in adjustment.

Applying Statistics 14.2 R^2 and Beta-Weights

The following table presents information on three variables for a small sample of 10 nations. The dependent variable is the percentage of respondents who said they are "very happy" on a survey administered to random samples from each nation. The independent variables measure health and physical well-being (life expectancy, or the number of years the average citizen can expect to live at birth) and income inequality (the amount of total income that goes to the richest 20% of the population). Our expectation is that happiness will have a positive correlation with life expectancy (the greater the health, the happier the population) and a negative relationship with inequality (the greater the inequality, the greater the discontent and the lower the level of happiness). In this analysis, we focus on R^2 and the beta-weights only.

Nation	Percent "Very Happy" (Y)	Life Expectancy (X_1)	Income Inequality (X_2)
Brazil	34.1	73	56.99
Canada	46.4	81	32.56
Germany	19.9	80	29.31
Ghana	50.1	59	40.79
India	29.0	64	36.80
Japan	29.2	83	24.85
Mexico	58.5	75	46.05
Russia	21.2	68	39.93
Rwanda	11.9	48	46.79
United States	34.4	78	40.81
Mean =	33.47	70.90	39.48
Standard deviation =	13.82	10.61	8.83

A correlation matrix presents the zero-order correlations:

	Happiness	Life Expectancy	Inequality
Happiness	1.00	0.29	0.13
Life expectancy		1.00	–0.46
Inequality			1.00

Consistent with our expectations, there is a weak-to-moderate positive relationship between life expectancy and happiness. Unexpectedly, however, the relationship between inequality and happiness is positive, although weak in strength. The relationship between the two independent variables is moderate and negative, indicating that nations with more income inequality have lower life expectancy.

(continued next page)

Applying Statistics 14.2 *(continued)*

The combined effect of life expectancy and inequality on happiness is found by computing R^2:

$$R^2 = r^2y_1 + r^2y_{2.1}(1 - r^2y_1)$$
$$R^2 = (0.29)^2 + (0.31)^2(1 - 0.29^2)$$
$$R^2 = 0.08 + (0.10)(0.92)$$
$$R^2 = 0.08 + 0.09$$
$$R^2 = 0.17$$

By itself, life expectancy explains 8% of the variance in happiness. To this, income inequality adds another 9%, for a total of 17%. This leaves about 83% of the variance unexplained—a sizable proportion but not unusually large in social science research.

We must calculate the beta-weights to assess the separate effects of the two independent variables. The unstandardized partial slopes needed for this computation are 0.58 for X_1 (life expectancy) and 0.53 for X_2 (income inequality).

For the first independent variable (life expectancy):

$$b_1^* = b1\left(\frac{s_1}{sy}\right)$$
$$b_1^* = (0.58)\left(\frac{10.61}{13.82}\right)$$
$$b_1^* = (0.58)(0.77)$$
$$b_1^* = 0.45$$

For the second independent variable (income inequality):

$$b_2^* = b2\left(\frac{s_2}{sy}\right)$$
$$b_2^* = (0.53)\left(\frac{8.83}{13.82}\right)$$
$$b_2^* = (0.53)(0.64)$$
$$b_2^* = 0.34$$

Recall that the beta-weights show the effect of each independent variable on the dependent variable while controlling for the other independent variables in the equation. In this case, life expectancy has the stronger effect and the relationship is positive. The effect of income inequality is also positive.

In summary, for these nations, level of happiness has a moderate positive relationship with life expectancy and a weaker positive relationship with income inequality. Taken together, the independent variables explain 17% of the variation in happiness.

Multiple Correlation

We use the multiple regression equations to disentangle the separate direct effects of each independent variable on the dependent variable. Using **multiple correlation** techniques, we can also ascertain the combined effects of all independent variables on the dependent variable. We do so by computing the **multiple correlation coefficient (R)** and the **coefficient of multiple determination (R)^2**. The value of the latter statistic represents the proportion of the variance in Y that is explained by all the independent variables combined.

We have seen that number of children (X_1) explains 25% of the variance in Y ($r^2_{y1} = (.50)^2 = .25 \times 100 = 25\%$) by itself and that husband's education explains 9% of the variance in Y ($r^2_{y2} = (-.30)^2 = .09 \times 100 = 9\%$). The two zero-order correlations cannot simply be added together to ascertain their combined effect on Y because the two independent variables are also correlated with each other; therefore, they will "overlap" in their effects on Y and explain some of the same variance. This overlap is eliminated in Formula 14.11:

FORMULA 14.11

$$R^2 = r^2_{y1} + r^2_{y2.1}(1 - r^2_{y1})$$

where: R^2 = the coefficient of multiple determination
r^2_{y1} = the zero-order correlation between Y and X_1, the quantity squared

$r^2_{y2.1}$ = the partial correlation of Y and X_2, while controlling for X_1, the quantity squared

The first term in this formula (r^2_{y1}) is the coefficient of determination for Y and X_1 or the amount of variation in Y explained by X_1 by itself. To this quantity, we add the amount of the variation remaining in Y (given by $1 - r^2_{y1}$) that can be explained by X_2 after the effect of X_1 is controlled ($r^2_{y2.1}$). Basically, Formula 14.11 allows X_1 to explain as much of Y as it can and then adds in the effect of X_2 after X_1 is controlled (thus eliminating the "overlap" in the variance of Y that X_1 and X_2 have in common).

Computing and Interpreting *R* and *R*². To observe the combined effects of number of children (X_1) and husband's education (X_2) on husband's housework (Y), we need two quantities. The correlation between X_1 and Y ($r_{y1} = .50$) has already been found. Before we can solve Formula 14.11, we must first calculate the partial correlation of Y and X_2 while controlling for X_1 ($r_{y2.1}$):

$$r_{y2.1} = \frac{r_{y2} - (r_{y1})(r_{12})}{\sqrt{1 - r^2_{y1}}\sqrt{1 - r^2_{y2}}}$$

$$r_{y2.1} = \frac{(-0.30) - (0.50)(-0.47)}{\sqrt{1 - (0.50)^2}\sqrt{1 - (-0.47)^2}}$$

$$r_{y2.1} = \frac{(-0.30) - (-0.24)}{\sqrt{0.75}\sqrt{0.78}}$$

$$r_{y2.1} = \frac{-0.06}{0.77}$$

$$r_{y2.1} = -0.08$$

Formula 14.11 can now be solved for our sample problem:

$$R^2 = r^2_{y1} + r^2_{y2.1}(1 - r^2_{y1})$$

$$R^2 = (0.50)^2 + (-0.08)^2(1 - 0.50^2)$$

$$R^2 = 0.25 + (0.006)(1 - 0.25)$$

$$R^2 = 0.25 + 0.005$$

$$R^2 = 0.255$$

The first independent variable (X_1)—number of children—explains 25% of the variance in Y by itself. To this total, the second independent (X_2)—husband's education—adds only a half a percent, for a total explained variance of 25.5%. In combination, the two independent variables explain a total of 25.5% of the variation in the dependent variable. (*For practice in computing and interpreting R and R², see any of the problems at the end of this chapter. It is probably a good idea to start with problem 14.1 because it has the smallest data set and the least complex computations.*)

ONE STEP AT A TIME **Computing and Interpreting the Multiple Correlation Coefficient (R^2)**

These procedures apply when there are two independent variables and one dependent variable. For more complex situations, use a computerized statistical package, such as SPSS, to do the calculations.

Step ***Operation***

To compute the coefficient of multiple determination by using Formula 14.11:

1. Find the value of the partial correlation coefficient for $r_{y2.1}$.
2. Square the value you found in step 1.
3. Square the value of r_{y1}.
4. Subtract the value you found in step 3 from 1.00.
5. Multiply the value you found in step 4 by the value you found in step 2.
6. Add the value you found in step 5 to r^2_{y1}. *The result is the coefficient of multiple determination* (R^2).

To Interpret the Coefficient of Multiple Determination (R^2)

The coefficient of multiple determination is the total amount of the variation in Y explained by all independent variables combined.

The Limitations of Multiple Regression and Correlation

Partial correlation and multiple regression and correlation are very powerful tools for analyzing the interrelationships among three or more variables. The techniques presented in this chapter permit the researcher to predict scores on one variable from two or more other variables, to distinguish between independent variables in terms of the importance of their direct effects on a dependent variable, and to ascertain the total effect of a set of independent variables on a dependent variable. In terms of the flexibility of the techniques and the volume of information they can supply, multiple regression and correlation represent some of the most powerful statistical techniques available to social science researchers.

Powerful tools are not cheap. They demand high-quality data, and measurement at the interval-ratio level is difficult to accomplish in social science research. Furthermore, these techniques assume that the interrelationships among the variables follow a particular form. First, they assume that each independent variable has a linear relationship with the dependent variable. How well a given set of variables meets this assumption can be quickly checked with scattergrams.

Second, these techniques assume there is no interaction among the variables in the equation. If there is interaction among the variables, it will not be possible to estimate or predict the dependent variable accurately simply by adding the effects of the independents. There are techniques for handling interaction among the variables in the set, but these techniques are beyond the scope of this text.

Third, multiple regression and correlation assume that the independent variables are uncorrelated with each other. Strictly speaking, this condition means that the zero-order correlation among all pairs of independents should be zero; but practically, we act as if this assumption has been met if the intercorrelations among the independents are low.

To the extent that these assumptions are violated, the regression coefficients (especially partial and standardized slopes) and the coefficient of multiple determination (R^2) become less and less trustworthy and the technique less and less useful. A careful inspection of the bivariate scattergrams will help assess the reasonableness of the assumptions.

Finally, we should note that we have covered only the simplest applications of partial correlation and multiple regression and correlation. In terms of logic and interpretation, the extensions to situations involving more variables are relatively straightforward. However, the computations for these situations are extremely complex. If you are faced with a situation involving more than three variables, turn to one of the computerized statistical packages commonly available on college campuses (e.g., SPSS or SAS). These programs require minimal computer literacy and can handle complex calculations in literally the blink of an eye. Efficient use of these packages will enable you to avoid drudgery and will free you to do what social scientists everywhere enjoy doing most: pondering the meaning of your results and, by extension, the nature of social life.

STATISTICS
IN EVERYDAY LIFE

Baseball is a paradise for statistics junkies. Virtually every aspect of what happens on the field is recorded, and there are decades of data to analyze. Can this mountain of information be used for purposes other than settling arguments over trivia among baseball fans?

The answer is a resounding yes, according to Michael Lewis, author of *Moneyball.** Lewis relates the story of Billy Beane, who, as general manager of the Oakland A's, was faced with the challenge of fielding a competitive team with one of the smallest budgets in the sport. Beane accomplished this in part by hiring statisticians who conducted a series of regression analyses with won/lost records as the dependent variable and every conceivable indicator of performance as independent variables. Beane and his statisticians were able to identify a number of measures of player effectiveness that had strong relationships with success (i.e., had high beta-weights) but were not widely valued by other baseball executives. He used these criteria to identify and recruit highly effective players for relatively low wages. Under Beane's management, the A's were remarkably successful for many years, although they did not reach the World Series.

Source: Lewis, Michael. 2004. *Moneyball.* New York: Norton.

*A movie based on this book was released in fall 2011.

BECOMING A CRITICAL CONSUMER: Reading the Professional Literature

It is unlikely that you will encounter multiple regression in the popular press or in your everyday conversations. On the other hand, multiple regression analysis has become an extremely important and widely used tool in social science research, and it is very likely that you will have to deal with articles using this technique—or one of its many variants—in the professional research literature. These articles might appear to be hopelessly complex at first glance—well beyond the understanding of an undergraduate student. Indeed, they are written for other professionals in the field and assume a high level of statistical sophistication on the part of the reader.

Nevertheless, without denying the challenge, it is quite possible for "ordinary people" to distill the essence of these articles by following a few guidelines and looking for some central elements. The key is to focus on the words, not the numbers. That is, read the text to see what the authors have to say about their results, not the (perhaps impenetrable) array of numbers and symbols. The details of the statistical analysis may be beyond your understanding, but you can almost always decipher the words.

Let us take a look at some actual research and see what we can make of it. We will focus on a project that takes a serious look at a topic we have used as an example in this and the previous chapter: the division of domestic labor between husbands and wives. In this case, the researchers used samples of Mexican American and white (Anglo) families, and they were particularly concerned with the role of both cultural factors—such as gender role attitudes—and structural factors—such as time constraints and the quantity of resources available—in shaping the division of labor.

The researchers conducted several different tests and models by using regression analysis to test the predictors of how much time Mexican and Anglo wives spent on household chores. The table to the right summarizes some of their findings.

Regression Coefficients (Beta-Weights) of Mother's Time on Household Chores

Independent Variables	Mexican	Anglo
Hours of Paid Work		
Mother	−0.19 **	−0.14*
Father	0.08	−0.07
Household Size		
# Adults	−1.29	−1.85
# Children	2.08*	1.71*
Resources		
Mother's % of Income	−20.74**	−10.59
Father's Education	1.08**	0.56
Attitudes		
Traditional Gender Role Attitudes		
Father	−0.01	0.45**
Mother	0.48	0.41
Familism		
Father	0.20	−0.02
Mother	0.48	0.41
R^2	0.32	0.24

*$p < 0.01$ **$p < 0.05$

Going down the table, we can see that structural factors, such as the number of hours worked by the mother, affect Mexican and Anglo women in similar ways: the greater the hours of paid work, the lower the number of hours devoted to household chores. In the same way, such factors as education and household size seem to have similar effects on both groups. The variables measuring attitudes—contrary to stereotypes about machismo—had significant effects only for Anglo women. Altogether, the independent variables explained about 32% of the variation for Mexican women and 24% of the variation for Anglo women. Want to learn more? The citation is given below.

Source: Pinto, K., & Coltrane, S. 2009. "Divisions of Labor in Mexican Origin and Anglo Families: Structure and Culture." *Sex Roles* 60: 482–495.

SUMMARY

1. Partial correlation involves controlling for third variables and permits the detection of direct and spurious or intervening relationships between X and Y.
2. Multiple regression includes statistical techniques by which predictions of the dependent variable from more than one independent variable can be made (by partial

slopes and the multiple regression equation) and by which we can disentangle the relative importance of the independent variables (by standardized partial slopes).

3. The coefficient of multiple determination (R^2) summarizes the combined effects of all independent variables on the dependent variable in terms of the proportion of the total variation in Y that is explained by all the independent variables combined.
4. Partial correlation and multiple regression and correlation are among the most powerful tools available to researchers and demand high-quality measurement and relationships among the variables that are linear and noninteractive. Furthermore, correlations among the independent variables must be low (preferably zero). Although the price is high, these techniques pay considerable dividends in the volume of precise and detailed information they generate about the interrelationships among the variables.

SUMMARY OF FORMULAS

FORMULA 14.1 Partial correlation coefficient: $r_{yx.z} = \dfrac{r_{yx} - (r_{yz})(r_{xz})}{\sqrt{1 - r_{yz}^2}\sqrt{1 - r_{xz}^2}}$

FORMULA 14.2 Least-squares regression line (bivariate): $Y = a + bX$

FORMULA 14.3 Least-squares multiple regression line: $Y = a + b_1X_1 + b_2X_2$

FORMULA 14.4 Partial slope for X_1: $b_1 = \left(\dfrac{s_y}{s_1}\right)\left(\dfrac{r_{y1} - r_{y2}r_{12}}{1 - r_{12}^2}\right)$

FORMULA 14.5 Partial slope for X_2: $b_2 = \left(\dfrac{s_y}{s_2}\right)\left(\dfrac{r_{y2} - r_{y1}r_{12}}{1 - r_{12}^2}\right)$

FORMULA 14.6 Y intercept: $a = \bar{Y} - b_1\bar{X}_1 - b_2\bar{X}_2$

FORMULA 14.7 Standardized partial slope (beta-weight) for X_1: $b_1^* = b_1\left(\dfrac{s_1}{s_y}\right)$

FORMULA 14.8 Standardized partial slope (beta-weight) for X_2: $b_2^* = b_2\left(\dfrac{s_2}{s_y}\right)$

FORMULA 14.9 Standardized least-squares regression line: $Z_y = a_z + b_1^*Z_1 + b_2^*Z_2$

FORMULA 14.10 Standardized least-squares regression line (simplified): $Z_y = b_1^*Z_1 + b_2^*Z_2$

FORMULA 14.11 Coefficient of multiple determination: $R^2 = r^2_{y1} + r^2_{y2.1}(1 - r^2_{y1})$

GLOSSARY

Beta-weight (b^*). See Standardized partial slope.

Coefficient of multiple determination (R^2). A statistic that equals the total variation explained in the dependent variable by all independent variables combined.

Multiple correlation. A multivariate technique for examining the combined effects of more than one independent variable on a dependent variable.

Multiple correlation coefficient (R). A statistic that indicates the strength of the correlation between a dependent variable and two or more independent variables.

Multiple regression. A multivariate technique that breaks down the separate effects of the independent variables on the dependent variable; used to make predictions of the dependent variable.

Partial correlation. A multivariate technique for examining a bivariate relationship while controlling for other variables.

Partial correlation coefficient. A statistic that shows the relationship between two variables while controlling for other variables; $r_{yx.z}$ is the symbol for the partial correlation coefficient when controlling for one variable.

Partial slope. In a multiple regression equation, the slope of the relationship between a particular independent variable and the dependent variable while controlling for all other independent variables in the equation.

Standardized partial slope (beta-weight). The slope of the relationship between a particular independent variable and the dependent variable when all scores have been normalized.

Zero-order correlation. Correlation coefficient for bivariate relationships.

PROBLEMS

14.1 PS In problem 13.1, data regarding voter turnout in five cities was presented. For the sake of convenience, the data for three of the variables are presented again here along with descriptive statistics and zero-order correlations.

City	Turnout	Unemployment Rate	% Negative Ads
A	55	5	60
B	60	8	63
C	65	9	55
D	68	9	53
E	70	10	48
Mean =	63.6	8.2	55.8
s =	5.5	1.7	5.3

	Unemployment Rate	Negative Ads
Turnout	0.95	−0.87
Unemployment Rate		−0.70

a. Compute the partial correlation coefficient for the relationship between turnout (Y) and unemployment (X) while controlling for the effect of negative advertising (Z). What effect does this control variable have on the bivariate relationship? Is the relationship between turnout and unemployment direct? (*HINT: Use Formula 14.1 and see "Types of Relationships" in the section on partial correlation.*)

b. Compute the partial correlation coefficient for the relationship between turnout (Y) and negative advertising (X) while controlling for the effect of unemployment (Z). What effect does this have on the bivariate relationship? Is the relationship between turnout and negative advertising direct? (*Hint: Use Formula 14.1 and see "Types of Relationships" in the section on partial correlation. You will need this partial correlation to compute the multiple correlation coefficient.*)

c. Find the unstandardized multiple regression equation with unemployment (X_1) and negative ads (X_2) as the independent variables. What turnout would be expected in a city in which the unemployment rate was 10% and 75% of the campaign ads were negative? (*Hint: Use Formulas 14.4 and 14.5 to compute the partial slopes and then use Formula 14.6 to find a, the Y intercept. The regression line is stated in Formula 14.3. Substitute 10 for X_1 and 75 for X_2 to compute predicted Y.*)

d. Compute beta-weights for each independent variable. Which has the stronger impact on turnout? (*Hint: Use Formulas 14.7 and 14.8 to calculate the beta-weights.*)

e. Compute the multiple correlation coefficient (R) and the coefficient of multiple determination (R^2). How much of the variance in voter turnout is explained by the two independent variables? (*Hint: Use Formula 14.11. You calculated $r^2_{y2.1}$ in part b of this problem.*)

f. Write a paragraph summarizing your conclusions about the relationships among these three variables.

14.2 SOC A scale measuring support for increases in the national defense budget has been administered to a sample. The respondents have also been asked to indicate how many years of school they have completed and how many years, if any, they served in the military. Take "support" as the dependent variable. Compute the zero-order correlations among the three variables.

Case	Support	Years of School	Years of Service
A	20	12	2
B	15	12	4
C	20	16	20
D	10	10	10
E	10	16	20
F	5	8	0
G	8	14	2
H	20	12	20
I	10	10	4
J	20	16	0

a. Compute the partial correlation coefficient for the relationship between support (Y) and years of school (X) while controlling for the effect of years of service (Z). What effect does this have on the bivariate relationship? Is the relationship between support and years of school direct?

b. Compute the partial correlation coefficient for the relationship between support (Y) and years of service (X) while controlling for the effect of years of school (Z). What effect does this have on the bivariate relationship? Is the relationship between support and years of service direct? (*Hint: You will need this partial correlation to compute the multiple correlation coefficient.*)

c. Find the unstandardized multiple regression equation with school (X_1) and service (X_2) as the independent variables. What level of support would be expected in a person with 13 years of school and 15 years of service?

d. Compute beta-weights for each independent variable. Which has the stronger impact on support?

e. Compute the multiple correlation coefficient (R) and the coefficient of multiple determination (R^2). How much of the variance in support is explained by the two independent variables? (*Hint: You calculated* $r^2_{y2.1}$ *in part b of this problem.*)

f. Write a paragraph summarizing your conclusions about the relationships among these three variables.

14.3 SOC Data on civil strife (number of incidents), unemployment, and urbanization have been gathered for 10 nations. Take civil strife as the dependent variable. Compute the zero-order correlations among the three variables.

Number of Incidents of Civil Strife	Unemployment Rate	Percentage of Population Living in Urban Areas
0	5.3	60
1	1.0	65
5	2.7	55
7	2.8	68
10	3.0	69
23	2.5	70
25	6.0	45
26	5.2	40
30	7.8	75
53	9.2	80

a. Compute the partial correlation coefficient for the relationship between strife (Y) and unemployment (X) while controlling for the effect of urbanization (Z). What effect does this have on the bivariate relationship? Is the relationship between strife and unemployment direct?

b. Compute the partial correlation coefficient for the relationship between strife (Y) and urbanization (X) while controlling for the effect of unemployment (Z). What effect does this have on the bivariate relationship? Is the relationship between strife and urbanization direct? (*Hint: You will need this partial correlation to compute the multiple correlation coefficient.*)

c. Find the unstandardized multiple regression equation with unemployment (X_1) and urbanization (X_2) as the independent variables. What level of strife would be expected in a nation in which the unemployment rate was 10% and where 90% of the population lived in urban areas?

d. Compute beta-weights for each independent variable. Which has the stronger impact on civil strife?

e. Compute the multiple correlation coefficient (R) and the coefficient of multiple determination (R^2). How much of the variance in strife is explained by the two independent variables?

f. Write a paragraph summarizing your conclusions about the relationships among these three variables.

14.4 SOC/CJ In problem 13.5, three measures of crime and three other variables were presented for each of 10 states. The data are reproduced on the next page.

	Crime Rates, 2009			Independent Variables		
State	Homicide	Robbery	Car Theft	% Families in Poverty, 2010	% HS Grads, 2009	% Unemployed, 2009
Maine	2	30	77	9	90	8
New York	4	144	112	10	84	8
Ohio	5	154	198	10	88	10
Iowa	1	210	129	7	90	6
Alabama	7	132	235	12	82	11
Kentucky	4	135	141	13	81	11
Texas	5	153	309	12	80	8
Louisiana	12	136	260	13	81	7
Washington	3	100	355	8	90	9
California	5	173	443	10	80	11

Sources: U.S. Bureau of the Census, *Statistical Abstracts of the United States, 2011*. Available at http://www.census.gov/compendia/statab/2011edition.html.

Take the three crime variables as the dependent variables (one at a time) and then do the following:

a. Find the multiple regression equations (unstandardized) with each crime variable (one at a time) and poverty and education as independent variables.
b. Make a prediction for each crime variable for a state with a 5% poverty rate and a score of 75 for the percent of high school graduates.
c. Compute beta-weights for each of the independent variables (poverty and education) in each equation and then compare their relative effect on each dependent variable.
d. Compute R and R^2 for each crime variable. Write a paragraph summarizing your findings.

14.5 PS Problem 13.4 presented data on 10 precincts. The information is reproduced here.

Percent	Percent Democrat	Percent Minority	Voter Turnout
A	50	10	56
B	45	12	55
C	56	8	52
D	78	15	60
E	13	5	89
F	85	20	25
G	62	18	64
H	33	9	88
I	25	0	42
J	49	9	36

Take voter turnout as the dependent variable and do the following:

a. Find the multiple regression equations (unstandardized).
b. What turnout would you expect for a precinct in which 0% of the voters were Democrats and 5% were minorities?
c. Compute beta-weights for each independent variable and then compare their relative effect on turnout. Which was the more important factor?
d. Compute R and R^2.
e. Write a paragraph summarizing your findings.

14.6 Twelve families have been referred to a counselor, and she has rated each of them on a cohesiveness scale. Also, she has information on family income and the number of children currently living at home. Take family cohesion as the dependent variable.

Family	Cohesion Score	Income	Number of Children
A	10	30,000	5
B	10	70,000	4
C	9	35,000	4
D	5	25,000	0
E	1	55,000	3
F	7	40,000	0
G	2	60,000	2
H	5	30,000	3
I	8	50,000	5
J	3	25,000	4
K	2	45,000	3
L	4	50,000	0

a. Find the multiple regression equations (unstandardized).
b. What level of cohesion would be expected in a family with an income of $20,000 and 6 children?
c. Compute beta-weights for each independent variable and then compare their relative effect on cohesion. Which was the more important factor?
d. Compute R and R^2.
e. Write a paragraph summarizing your findings.

14.7 Problem 13.8 presented per capita expenditures on education for 15 states, along with rank on income per capita and the percentage of the population that has graduated from high school. The data are reproduced here.

State	Per Capita Expenditures on Education, 2007	Percent High School Graduates, 2010	Rank in per Capita Income
Arkansas	1,705	83.8	45
Colorado	1,863	88.9	15
Connecticut	2,475	88.6	1
Florida	1,728	85.2	24
Illinois	1,955	85.9	14
Kansas	1,995	89.5	23
Louisiana	1,835	81.2	46
Maryland	1,930	88.0	4
Michigan	1,965	88.1	37
Mississippi	1,345	79.9	50
Nebraska	1,637	90.1	22
New Hampshire	1,919	90.9	8
North Carolina	1,522	83.6	35
Pennsylvania	1,990	87.5	18
Wyoming	2,900	91.7	6

Source: U.S. Bureau of the Census, 2011. *Statistical Abstract of the United States, 2011.* Available at http://www.census.gov/compendia/statab.

Take per capita expenditures as the dependent variable:

a. Compute beta-weights for each independent variable and then compare their relative effect on expenditures. Which was the more important factor?
b. Compute R and R^2.
c. Write a paragraph summarizing your findings.

14.8 SOC The scores on four variables for 20 individuals are reported here: hours of TV (average number of hours of TV viewing each day), occupational prestige (higher scores indicate greater prestige), number of children, and age. Take TV viewing as the dependent variable and then select two of the remaining variables as independent variables.

Hours of TV	Occupational Prestige	Number of Children	Age
4	50	2	43
3	36	3	58
3	36	1	34
4	50	2	42
2	45	2	27

Hours of TV	Occupational Prestige	Number of Children	Age
3	50	5	60
4	50	0	28
7	40	3	55
1	57	2	46
3	33	2	65
1	46	3	56
3	31	1	29
1	19	2	41
0	52	0	50
2	48	1	62
4	36	1	24
3	48	0	25
1	62	1	87
5	50	0	45
1	27	3	62

a. Compute beta-weights for each of the independent variables you selected and then compare their relative effect on the hours of television watching. Which was the more important factor?

b. Compute R and R^2.

c. Write a paragraph summarizing your findings.

14.9 Problem 13.6 presented data on three variables for 15 nations. The scores are reproduced below.

NATION	Fertility Rate	Education of Females	Percent of Women Using Contraception
Niger	7.7	38.4	11
Cambodia	3.9	35.6	51
Guatemala	4.3	46.9	54
Ghana	4.1	44.7	24
Bolivia	3.7	48.2	61
Kenya	5.0	47.6	46
Dominican Republic	2.7	54.8	73
Mexico	2.1	50.5	73
Vietnam	1.8	47.1	78
Turkey	2.0	37.5	73
United States	2.0	49.0	79
China	1.8	45.3	85
United Kingdom	2.4	52.7	84
Japan	1.3	49.0	54
Italy	1.3	48.1	60

Sources: Nationmaster.com. World Population Data Sheet, 2011. Population Reference Bureau (www.prb.org).

Take fertility as the dependent variable:

a. Compute beta-weights for each independent variable and then compare their relative effect on fertility. Which was the more important factor? Interpret the direction of each relationship with fertility.

b. Compute R and R^2.

c. Write a paragraph summarizing your findings.

YOU ARE THE RESEARCHER: A Multivariate Analysis of TV Viewing and Success

The two projects presented here continue the investigations begun in Chapter 13. You will use a SPSS procedure called **Regression** to analyze the combined effects of your independent variables (including dummy variables if you used any) on the dependent variables from Chapter 13. The **Regression** procedure produces a much greater volume of output than the **Correlate** procedure used in Chapter 13. Among other things, **Regression** displays the slope (*b*) and the *Y* intercept (*a*), so we can use this procedure to find least-squares regression lines. In these projects, we will use very little of the power of the **Regression** command, and we will be extremely economical in our choice of all the options available. I urge you to explore some of the variations and capabilities of this powerful data-analysis procedure on your own.

Let us begin with a demonstration using *marhomo* (approval of gay marriage) as the dependent variable and *attend* (church attendance) and *educ* as independent variables.

Click **Analyze**, **Regression**, and **Linear**, and the **Linear Regression** window will appear. Move *marhomo* into the **Dependent:** box and *attend* and *educ* into the **Independent(s):** box. If you wish, you may click the **Statistics** button and then click **Descriptives** to get zero-order correlations, means, and standard deviations for the variables. Click **OK** in the **Linear Regression Window**: and the following output will appear (descriptive information about the variables and the zero-order correlations are omitted to conserve space).

Model Summary

Model	R	R Square	Adjusted R Square	Std. Error of the Estimate
1	.367[a]	.135	.133	1.404

a. Predictors: (Constant), HIGHEST YEAR OF SCHOOL COMPLETED, HOW OFTEN R ATTENDS RELIGIOUS SERVICES

Coefficients[a]

Model		Unstandardized Coefficients		Standardized Coefficients		
		B	Std. Error	Beta	t	Sig.
1	(Constant)	3.571	.209		17.067	.000
	HOW OFTEN R ATTENDS RELIGIOUS SERVICES	.173	.017	.318	10.380	.000
	HIGHEST YEAR OF SCHOOL COMPLETED	−.087	.014	−.185	−6.041	.000

a. Dependent Variable: HOMOSEXUALS SHOULD HAVE RIGHT TO MARRY

The "Model Summary" block reports the multiple R (.367) and R^2 (.135). The ANOVA output block shows the significance of the relationship (Sig. = .000). In the last output block, under "Unstandardized Coefficients," we see the Y intercept—reported as a constant of 3.571—and the slopes (B) of the independent variables on *marhomo*. From this information, we can build a regression equation (see Formula 14.3) to predict scores on *marhomo*:

$$Y = 3.571 + (0.173)(attend) + (-0.087)(educ)$$

Under "Standardized Coefficients," we find the standardized partial slopes (Beta). The beta for *attend* (.318) is greater in value than the beta for *educ* (−.185). This tells us that religiosity (*attend*) is the more important of the two independent variables. It has a positive relationship with *marhomo* after the effects of education have been controlled. As religiosity increases, with education controlled, disapproval of gay marriage increases. Education has a negative relationship with the dependent variable after religiosity has been controlled: As education increases, with church attendance controlled, disapproval of gay marriage decreases.

Your turn.

PROJECT 1: Who Watches TV?

This project follows up on Project 1 from Chapter 13. The dependent variable is *tvhours*, which, as you recall, measures the number of hours per day the respondent watches televsion.

STEP 1: Choosing Independent Variables

Choose two of the four independent variables you selected in Project 1 from Chapter 13. All other things being equal, you might choose the variables that had the strongest relationship with *tvhours*. Remember that independent variables *cannot* be nominal in level of measurement unless you recode the variable into a dummy variable. You may use any interval-ratio or ordinal-level variables with more than three categories or scores. Once you have selected your variables, list them in the table below and then describe exactly what they measure.

SPSS Variable Name	What Exactly Does This Variable Measure?
1.	
2.	

STEP 2: Stating Hypotheses

Restate your hypotheses from Chapter 13 and then add some ideas about the relative importance of your two independent variables. Which do you expect to have the stronger effect? Why?

STEP 3: Running the Regression Procedure

Click **Analyze → Regression → Linear** and then place *tvhours* in the **Dependent:** box and your selected independent variables in the **Independent(s):** box. Click **Statistics** and **Descriptives** and then click **OK** in the **Linear Regression** Window to get your results.

STEP 4: Recording Results

Summarize your results by filling in the blanks below with information from the **Coefficients** box. The *Y* intercept (*a*) is reported in the top row of the box as a "constant." The slopes (*b*) are reported in the "Unstandardized Coefficients" column under "B," and the beta-weights (b^*) are reported in the "Standardized Coefficients" column under "Beta."

1. a = ________
2. a. Slope for your first independent variable (b_1) = ____.
   b. Slope for your second independent variable (b_2) = ____.
3. State the least squares multiple regression equation (see Formula 14.3).

$$Y = \_\_\_ + (\_)\,X_1 + (\_)\,X_2$$

4. a. Beta-weight of your first independent variable (b^*_1) = _____.
   b. Beta-weight of your second independent variable (b^*_2) = _____.
5. State the standardized least-square regression line (see Formula 15.10).

$$Z_y = (\ \_\ )\,Z_1 + (\ \_\_\ )\,Z_2$$

6. R^2 = ____

STEP 5: Analyzing and Interpreting Results

Write a short summary of your results. Your summary needs to identify your variables and distinguish between independent and dependent variables. You also need to report the strength and direction of relationships as indicated by the slopes and beta-weights and the total explained variance (R^2). How much of the variance does your first independent variable explain? How much more is explained by your second independent variable?

PROJECT 2: What Are the Correlates of Success?

In this second project, we will continue the investigation into the correlates of success, which we began in Chapter 13.

STEP 1: Choosing Independent Variables

Use two of the four independent variables you selected in Project 2 from Chapter 13. All other things being equal, you might choose the variables that had the strongest relationship with the variable you selected to be the dependent variable, which will be either *income06* or *prestg80*. Remember that independent variables *cannot* be nominal in level of measurement unless you recode the variable into a dummy variable. You may use any interval-ratio or ordinal-level variables with more than three categories or scores. Once you have selected your variables, list them in the table below and then describe exactly what they measure.

SPSS Variable Name	What Exactly Does This Variable Measure?
1.	
2.	

STEP 2: Stating Hypotheses

Restate your hypotheses from Chapter 13 and then add some ideas about the relative importance of your two independent variables. Which do you expect to have the stronger effect? Why?

STEP 3: Running the Regression Procedure

Click **Analyze** → **Regression** → **Linear** and then place your dependent variable in the **Dependent:** box and your selected independent variables in the **Independent(s):** box. Click **Statistics** and **Descriptives** and then click **OK** to get your results.

STEP 4: Recording Results

Summarize your results by filling in the blanks below with information from the **Coefficients** box. The *Y* intercept (*a*) is reported in the top row of the box as a "constant." The slopes are reported in the "Unstandardized Coefficients" column under "B," and the beta-weights are reported in the "Standardized Coefficients" column under "Beta."

1. $a =$ ________
2. a. Slope for your first independent variable (b_1) = ____.
   b. Slope for your second independent variable (b_2) = ____.
3. State the least-squares multiple regression equation (see Formula 15.3).

$$Y = \_\_ + (\_)\,X_1 + (\_)\,X_2$$

4. a. Beta-weight of your first independent variable (b_1^*) = _____.
   b. Beta-weight of your second independent variable (b_2^*) = _____.
5. State the standardized least-squares regression line (see Formula 15.10).

$$Z_y = (\_\_)\,Z_1 + (\_\_\_)\,Z_2$$

6. $R^2 =$ ____

STEP 5: Analyzing and Interpreting Results

Write a short summary of your results. Your summary needs to identify your variables and distinguish between independent and dependent variables. You also need to report the strength and direction of relationships as indicated by the slopes and beta-weights and the total explained variance (R^2). How much of the variance does your first independent variable explain? How much more is explained by your second independent variable?

Appendix A Area Under the Normal Curve

Column (a) lists Z scores from 0.00 to 4.00. Only positive scores are displayed, but because the normal curve is symmetrical, the areas for negative scores will be exactly the same as areas for positive scores. Column (b) lists the proportion of the total area between the Z score and the mean. Figure A.1 displays areas of this type. Column (c) lists the proportion of the area beyond the Z score, and Figure A.2 displays this type of area.

FIGURE A.1 Area Between Mean and *Z*

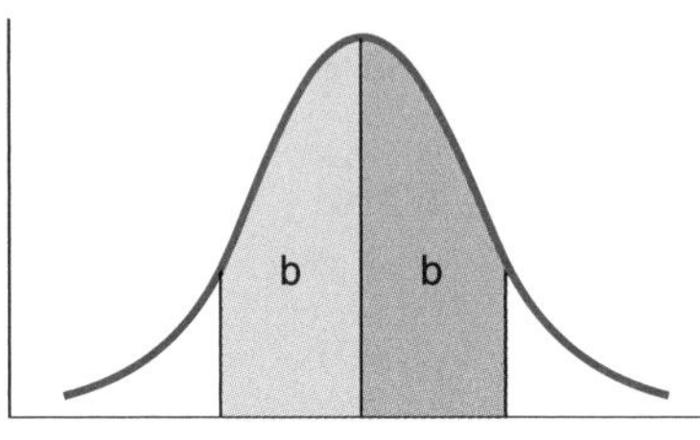

FIGURE A.2 Area Beyond *Z*

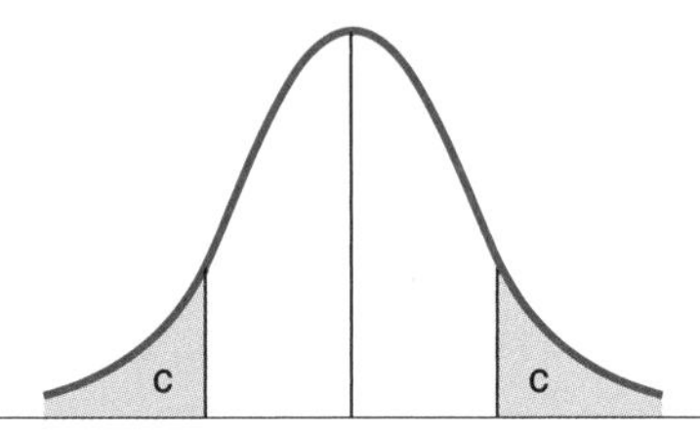

(a) *Z*	(b) Area Between Mean and *Z*	(c) Area Beyond *Z*
0.00	0.0000	0.5000
0.01	0.0040	0.4960
0.02	0.0080	0.4920
0.03	0.0120	0.4880
0.04	0.0160	0.4840
0.05	0.0199	0.4801
0.06	0.0239	0.4761
0.07	0.0279	0.4721
0.08	0.0319	0.4681
0.09	0.0359	0.4641
0.10	0.0398	0.4602
0.11	0.0438	0.4562
0.12	0.0478	0.4522
0.13	0.0517	0.4483
0.14	0.0557	0.4443
0.15	0.0596	0.4404
0.16	0.0636	0.4364
0.17	0.0675	0.4325
0.18	0.0714	0.4286
0.19	0.0753	0.4247
0.20	0.0793	0.4207
0.21	0.0832	0.4168
0.22	0.0871	0.4129
0.23	0.0910	0.4090
0.24	0.0948	0.4052
0.25	0.0987	0.4013
0.26	0.1026	0.3974
0.27	0.1064	0.3936
0.28	0.1103	0.3897
0.29	0.1141	0.3859
0.30	0.1179	0.3821
0.31	0.1217	0.3783
0.32	0.1255	0.3745
0.33	0.1293	0.3707
0.34	0.1331	0.3669
0.35	0.1368	0.3632
0.36	0.1406	0.3594
0.37	0.1443	0.3557
0.38	0.1480	0.3520
0.39	0.1517	0.3483
0.40	0.1554	0.3446
0.41	0.1591	0.3409
0.42	0.1628	0.3372
0.43	0.1664	0.3336
0.44	0.1700	0.3300
0.45	0.1736	0.3264

(a) Z	(b) Area Between Mean and Z	(c) Area Beyond Z	(a) Z	(b) Area Between Mean and Z	(c) Area Beyond Z
0.46	0.1772	0.3228	1.01	0.3438	0.1562
0.47	0.1808	0.3192	1.02	0.3461	0.1539
0.48	0.1844	0.3156	1.03	0.3485	0.1515
0.49	0.1879	0.3121	1.04	0.3508	0.1492
0.50	0.1915	0.3085	1.05	0.3531	0.1469
0.51	0.1950	0.3050	1.06	0.3554	0.1446
0.52	0.1985	0.3015	1.07	0.3577	0.1423
0.53	0.2019	0.2981	1.08	0.3599	0.1401
0.54	0.2054	0.2946	1.09	0.3621	0.1379
0.55	0.2088	0.2912	1.10	0.3643	0.1357
0.56	0.2123	0.2877	1.11	0.3665	0.1335
0.57	0.2157	0.2843	1.12	0.3686	0.1314
0.58	0.2190	0.2810	1.13	0.3708	0.1292
0.59	0.2224	0.2776	1.14	0.3729	0.1271
0.60	0.2257	0.2743	1.15	0.3749	0.1251
0.61	0.2291	0.2709	1.16	0.3770	0.1230
0.62	0.2324	0.2676	1.17	0.3790	0.1210
0.63	0.2357	0.2643	1.18	0.3810	0.1190
0.64	0.2389	0.2611	1.19	0.3830	0.1170
0.65	0.2422	0.2578	1.20	0.3849	0.1151
0.66	0.2454	0.2546	1.21	0.3869	0.1131
0.67	0.2486	0.2514	1.22	0.3888	0.1112
0.68	0.2517	0.2483	1.23	0.3907	0.1093
0.69	0.2549	0.2451	1.24	0.3925	0.1075
0.70	0.2580	0.2420	1.25	0.3944	0.1056
0.71	0.2611	0.2389	1.26	0.3962	0.1038
0.72	0.2642	0.2358	1.27	0.3980	0.1020
0.73	0.2673	0.2327	1.28	0.3997	0.1003
0.74	0.2703	0.2297	1.29	0.4015	0.0985
0.75	0.2734	0.2266	1.30	0.4032	0.0968
0.76	0.2764	0.2236	1.31	0.4049	0.0951
0.77	0.2794	0.2206	1.32	0.4066	0.0934
0.78	0.2823	0.2177	1.33	0.4082	0.0918
0.79	0.2852	0.2148	1.34	0.4099	0.0901
0.80	0.2881	0.2119	1.35	0.4115	0.0885
0.81	0.2910	0.2090	1.36	0.4131	0.0869
0.82	0.2939	0.2061	1.37	0.4147	0.0853
0.83	0.2967	0.2033	1.38	0.4162	0.0838
0.84	0.2995	0.2005	1.39	0.4177	0.0823
0.85	0.3023	0.1977	1.40	0.4192	0.0808
0.86	0.3051	0.1949	1.41	0.4207	0.0793
0.87	0.3078	0.1922	1.42	0.4222	0.0778
0.88	0.3106	0.1894	1.43	0.4236	0.0764
0.89	0.3133	0.1867	1.44	0.4251	0.0749
0.90	0.3159	0.1841	1.45	0.4265	0.0735
0.91	0.3186	0.1814	1.46	0.4279	0.0721
0.92	0.3212	0.1788	1.47	0.4292	0.0708
0.93	0.3238	0.1762	1.48	0.4306	0.0694
0.94	0.3264	0.1736	1.49	0.4319	0.0681
0.95	0.3289	0.1711	1.50	0.4332	0.0668
0.96	0.3315	0.1685	1.51	0.4345	0.0655
0.97	0.3340	0.1660	1.52	0.4357	0.0643
0.98	0.3365	0.1635	1.53	0.4370	0.0630
0.99	0.3389	0.1611	1.54	0.4382	0.0618
1.00	0.3413	0.1587	1.55	0.4394	0.0606

(a) Z	(b) Area Between Mean and Z	(c) Area Beyond Z
1.56	0.4406	0.0594
1.57	0.4418	0.0582
1.58	0.4429	0.0571
1.59	0.4441	0.0559
1.60	0.4452	0.0548
1.61	0.4463	0.0537
1.62	0.4474	0.0526
1.63	0.4484	0.0516
1.64	0.4495	0.0505
1.65	0.4505	0.0495
1.66	0.4515	0.0485
1.67	0.4525	0.0475
1.68	0.4535	0.0465
1.69	0.4545	0.0455
1.70	0.4554	0.0446
1.71	0.4564	0.0436
1.72	0.4573	0.0427
1.73	0.4582	0.0418
1.74	0.4591	0.0409
1.75	0.4599	0.0401
1.76	0.4608	0.0392
1.77	0.4616	0.0384
1.78	0.4625	0.0375
1.79	0.4633	0.0367
1.80	0.4641	0.0359
1.81	0.4649	0.0351
1.82	0.4656	0.0344
1.83	0.4664	0.0336
1.84	0.4671	0.0329
1.85	0.4678	0.0322
1.86	0.4686	0.0314
1.87	0.4693	0.0307
1.88	0.4699	0.0301
1.89	0.4706	0.0294
1.90	0.4713	0.0287
1.91	0.4719	0.0281
1.92	0.4726	0.0274
1.93	0.4732	0.0268
1.94	0.4738	0.0262
1.95	0.4744	0.0256
1.96	0.4750	0.0250
1.97	0.4756	0.0244
1.98	0.4761	0.0239
1.99	0.4767	0.0233
2.00	0.4772	0.0228
2.01	0.4778	0.0222
2.02	0.4783	0.0217
2.03	0.4788	0.0212
2.04	0.4793	0.0207
2.05	0.4798	0.0202
2.06	0.4803	0.0197
2.07	0.4808	0.0192
2.08	0.4812	0.0188
2.09	0.4817	0.0183
2.10	0.4821	0.0179
2.11	0.4826	0.0174
2.12	0.4830	0.0170
2.13	0.4834	0.0166
2.14	0.4838	0.0162
2.15	0.4842	0.0158
2.16	0.4846	0.0154
2.17	0.4850	0.0150
2.18	0.4854	0.0146
2.19	0.4857	0.0143
2.20	0.4861	0.0139
2.21	0.4864	0.0136
2.22	0.4868	0.0132
2.23	0.4871	0.0129
2.24	0.4875	0.0125
2.25	0.4878	0.0122
2.26	0.4881	0.0119
2.27	0.4884	0.0116
2.28	0.4887	0.0113
2.29	0.4890	0.0110
2.30	0.4893	0.0107
2.31	0.4896	0.0104
2.32	0.4898	0.0102
2.33	0.4901	0.0099
2.34	0.4904	0.0096
2.35	0.4906	0.0094
2.36	0.4909	0.0091
2.37	0.4911	0.0089
2.38	0.4913	0.0087
2.39	0.4916	0.0084
2.40	0.4918	0.0082
2.41	0.4920	0.0080
2.42	0.4922	0.0078
2.43	0.4925	0.0075
2.44	0.4927	0.0073
2.45	0.4929	0.0071
2.46	0.4931	0.0069
2.47	0.4932	0.0068
2.48	0.4934	0.0066
2.49	0.4936	0.0064
2.50	0.4938	0.0062
2.51	0.4940	0.0060
2.52	0.4941	0.0059
2.53	0.4943	0.0057
2.54	0.4945	0.0055
2.55	0.4946	0.0054
2.56	0.4948	0.0052
2.57	0.4949	0.0051
2.58	0.4951	0.0049
2.59	0.4952	0.0048
2.60	0.4953	0.0047
2.61	0.4955	0.0045
2.62	0.4956	0.0044
2.63	0.4957	0.0043
2.64	0.4959	0.0041
2.65	0.4960	0.0040

(a) Z	(b) Area Between Mean and Z	(c) Area Beyond Z
2.66	0.4961	0.0039
2.67	0.4962	0.0038
2.68	0.4963	0.0037
2.69	0.4964	0.0036
2.70	0.4965	0.0035
2.71	0.4966	0.0034
2.72	0.4967	0.0033
2.73	0.4968	0.0032
2.74	0.4969	0.0031
2.75	0.4970	0.0030
2.76	0.4971	0.0029
2.77	0.4972	0.0028
2.78	0.4973	0.0027
2.79	0.4974	0.0026
2.80	0.4974	0.0026
2.81	0.4975	0.0025
2.82	0.4976	0.0024
2.83	0.4977	0.0023
2.84	0.4977	0.0023
2.85	0.4978	0.0022
2.86	0.4979	0.0021
2.87	0.4979	0.0021
2.88	0.4980	0.0020
2.89	0.4981	0.0019
2.90	0.4981	0.0019
2.91	0.4982	0.0018
2.92	0.4982	0.0018
2.93	0.4983	0.0017
2.94	0.4984	0.0016
2.95	0.4984	0.0016
2.96	0.4985	0.0015
2.97	0.4985	0.0015
2.98	0.4986	0.0014
2.99	0.4986	0.0014
3.00	0.4986	0.0014
3.01	0.4987	0.0013
3.02	0.4987	0.0013
3.03	0.4988	0.0012
3.04	0.4988	0.0012
3.05	0.4989	0.0011
3.06	0.4989	0.0011
3.07	0.4989	0.0011
3.08	0.4990	0.0010
3.09	0.4990	0.0010
3.10	0.4990	0.0010
3.11	0.4991	0.0009
3.12	0.4991	0.0009
3.13	0.4991	0.0009
3.14	0.4992	0.0008
3.15	0.4992	0.0008
3.16	0.4992	0.0008
3.17	0.4992	0.0008
3.18	0.4993	0.0007
3.19	0.4993	0.0007
3.20	0.4993	0.0007
3.21	0.4993	0.0007
3.22	0.4994	0.0006
3.23	0.4994	0.0006
3.24	0.4994	0.0006
3.25	0.4994	0.0006
3.26	0.4994	0.0006
3.27	0.4995	0.0005
3.28	0.4995	0.0005
3.29	0.4995	0.0005
3.30	0.4995	0.0005
3.31	0.4995	0.0005
3.32	0.4995	0.0005
3.33	0.4996	0.0004
3.34	0.4996	0.0004
3.35	0.4996	0.0004
3.36	0.4996	0.0004
3.37	0.4996	0.0004
3.38	0.4996	0.0004
3.39	0.4997	0.0003
3.40	0.4997	0.0003
3.41	0.4997	0.0003
3.42	0.4997	0.0003
3.43	0.4997	0.0003
3.44	0.4997	0.0003
3.45	0.4997	0.0003
3.46	0.4997	0.0003
3.47	0.4997	0.0003
3.48	0.4997	0.0003
3.49	0.4998	0.0002
3.50	0.4998	0.0002
3.60	0.4998	0.0002
3.70	0.4999	0.0001
3.80	0.4999	0.0001
3.90	0.4999	<0.0001
4.00	0.4999	<0.0001

Appendix B

Distribution of t

Use this table to find the critical region (step 3 of the five-step model) for tests of significance with sample means when the sample size (N) is small. First, choose an alpha level and a one- or two-tailed test. Then, find degrees of freedom to find the t score that marks the beginning of the critical region.

Degrees of Freedom (df)	Level of Significance for One-Tailed Test					
	.10	.05	.025	.01	.005	.0005
	Level of Significance for Two-Tailed Test					
	.20	.10	.05	.02	.01	.001
1	3.078	6.314	12.706	31.821	63.657	636.619
2	1.886	2.920	4.303	6.965	9.925	31.598
3	1.638	2.353	3.182	4.541	5.841	12.941
4	1.533	2.132	2.776	3.747	4.604	8.610
5	1.476	2.015	2.571	3.365	4.032	6.859
6	1.440	1.943	2.447	3.143	3.707	5.959
7	1.415	1.895	2.365	2.998	3.499	5.405
8	1.397	1.860	2.306	2.896	3.355	5.041
9	1.383	1.833	2.262	2.821	3.250	4.781
10	1.372	1.812	2.228	2.764	3.169	4.587
11	1.363	1.796	2.201	2.718	3.106	4.437
12	1.356	1.782	2.179	2.681	3.055	4.318
13	1.350	1.771	2.160	2.650	3.012	4.221
14	1.345	1.761	2.145	2.624	2.977	4.140
15	1.341	1.753	2.131	2.602	2.947	4.073
16	1.337	1.746	2.120	2.583	2.921	4.015
17	1.333	1.740	2.110	2.567	2.898	3.965
18	1.330	1.734	2.101	2.552	2.878	3.922
19	1.328	1.729	2.093	2.539	2.861	3.883
20	1.325	1.725	2.086	2.528	2.845	3.850
21	1.323	1.721	2.080	2.518	2.831	3.819
22	1.321	1.717	2.074	2.508	2.819	3.792
23	1.319	1.714	2.069	2.500	2.807	3.767
24	1.318	1.711	2.064	2.492	2.797	3.745
25	1.316	1.708	2.060	2.485	2.787	3.725
26	1.315	1.706	2.056	2.479	2.779	3.707
27	1.314	1.703	2.052	2.473	2.771	3.690
28	1.313	1.701	2.048	2.467	2.763	3.674
29	1.311	1.699	2.045	2.462	2.756	3.659
30	1.310	1.697	2.042	2.457	2.750	3.646
40	1.303	1.684	2.021	2.423	2.704	3.551
60	1.296	1.671	2.000	2.390	2.660	3.460
120	1.289	1.658	1.980	2.358	2.617	3.373
∞	1.282	1.645	1.960	2.326	2.576	3.291

Source: Table III of Fisher and Yates. *Statistical Tables for Biological, Agricultural and Medical Research,* published by Longman Group Ltd., London (1974), 6th edition (previously published by Oliver & Boyd Ltd., Edinburgh). Reprinted by permission of Addison Wesley Longman Ltd.

Appendix C Distribution of Chi Square

Use this table to find the critical region (step 3 of the five-step model) for tests of significance with chi square. Choose an alpha level and then find the degrees of freedom to find the chi square score that marks the beginning of the critical region.

df	.99	.98	.95	.90	.80	.70	.50	.30	.20	.10	.05	.02	.01	.001
1	$.0^{3}157$	$.0^{3}628$	.00393	.0158	.0642	.148	.455	1.074	1.642	2.706	3.841	5.412	6.635	10.827
2	.0201	.0404	.103	.211	.446	.713	1.386	2.408	3.219	4.605	5.991	7.824	9.210	13.815
3	.115	.185	.352	.584	1.005	1.424	2.366	3.665	4.642	6.251	7.815	9.837	11.341	16.268
4	.297	.429	.711	1.064	1.649	2.195	3.357	4.878	5.989	7.779	9.488	11.668	13.277	18.465
5	.554	.752	1.145	1.610	2.343	3.000	4.351	6.064	7.289	9.236	11.070	13.388	15.086	20.517
6	.872	1.134	1.635	2.204	3.070	3.828	5.348	7.231	8.558	10.645	12.592	15.033	16.812	22.457
7	1.239	1.564	2.167	2.833	3.822	4.671	6.346	8.383	9.803	12.017	14.067	16.622	18.475	24.322
8	1.646	2.032	2.733	3.490	4.594	5.527	7.344	9.524	11.030	13.362	15.507	18.168	20.090	26.125
9	2.088	2.532	3.325	4.168	5.380	6.393	8.343	10.656	12.242	14.684	16.919	19.679	21.666	27.877
10	2.558	3.059	3.940	4.865	6.179	7.267	9.342	11.781	13.442	15.987	18.307	21.161	23.209	29.588
11	3.053	3.609	4.575	5.578	6.989	8.148	10.341	12.899	14.631	17.275	19.675	22.618	24.725	31.264
12	3.571	4.178	5.226	6.304	7.807	9.034	11.340	14.011	15.812	18.549	21.026	24.054	26.217	32.909
13	4.107	4.765	5.892	7.042	8.634	9.926	12.340	15.119	16.985	19.812	22.362	25.472	27.688	34.528
14	4.660	5.368	6.571	7.790	9.467	10.821	13.339	16.222	18.151	21.064	23.685	26.873	29.141	36.123
15	5.229	5.985	7.261	8.547	10.307	11.721	14.339	17.322	19.311	22.307	24.996	28.259	30.578	37.697
16	5.812	6.614	7.962	9.312	11.152	12.624	15.338	18.418	20.465	23.542	26.296	29.633	32.000	39.252
17	6.408	7.255	8.672	10.085	12.002	13.531	16.338	19.511	21.615	24.769	27.587	30.995	33.409	40.790
18	7.015	7.906	9.390	10.865	12.857	14.440	17.338	20.601	22.760	25.989	28.869	32.346	34.805	42.312
19	7.633	8.567	10.117	11.651	13.716	15.352	18.338	21.689	23.900	27.204	30.144	33.687	36.191	43.820
20	8.260	9.237	10.851	12.443	14.578	16.266	19.337	22.775	25.038	28.412	31.410	35.020	37.566	45.315
21	8.897	9.915	11.591	13.240	15.445	17.182	20.337	23.858	26.171	29.615	32.671	36.343	38.932	46.797
22	9.542	10.600	12.338	14.041	16.314	18.101	21.337	24.939	27.301	30.813	33.924	37.659	40.289	48.268
23	10.196	11.293	13.091	14.848	17.187	19.021	22.337	26.018	28.429	32.007	35.172	38.968	41.638	49.728
24	10.856	11.992	13.848	15.659	18.062	19.943	23.337	27.096	29.553	33.196	36.415	40.270	42.980	51.179
25	11.524	12.697	14.611	16.473	18.940	20.867	24.337	28.172	30.675	34.382	37.652	41.566	44.314	52.620
26	12.198	13.409	15.379	17.292	19.820	21.792	25.336	29.246	31.795	35.563	38.885	42.856	45.642	54.052
27	12.879	14.125	16.151	18.114	20.703	22.719	26.336	30.319	32.912	36.741	40.113	44.140	46.963	55.476
28	13.565	14.847	16.928	18.939	21.588	23.647	27.336	31.391	34.027	37.916	41.337	45.419	48.278	56.893
29	14.256	15.574	17.708	19.768	22.475	24.577	28.336	32.461	35.139	39.087	42.557	46.693	49.588	58.302
30	14.953	16.306	18.493	20.599	23.364	25.508	29.336	33.530	36.250	40.256	43.773	47.962	50.892	59.703

Source: Table IV of Fisher and Yates. *Statistical Tables for Biological, Agricultural and Medical Research,* published by Longman Group Ltd., London (1974), 6th edition (previously published by Oliver & Boyd Ltd., Edinburgh). Reprinted by permission of Pearson Education Ltd.

Appendix D Distribution of *F*

Use this table to find the critical region (step 3 of the five-step model) for analysis of variance tests. Choose an alpha level of either 0.05 or 0.01 and then find the degrees of freedom to find the *F* ratio that marks the beginning of the critical region. Values of n_1 and n_2 represent the degrees of freedom associated with the between and within estimates of variance, respectively.

$p = .05$

n_1 / n_2	1	2	3	4	5	6	8	12	24	∞
1	161.4	199.5	215.7	224.6	230.2	234.0	238.9	243.9	249.0	254.3
2	18.51	19.00	19.16	19.25	19.30	19.33	19.37	19.41	19.45	19.50
3	10.13	9.55	9.28	9.12	9.01	8.94	8.84	8.74	8.64	8.53
4	7.71	6.94	6.59	6.39	6.26	6.16	6.04	5.91	5.77	5.63
5	6.61	5.79	5.41	5.19	5.05	4.95	4.82	4.68	4.53	4.36
6	5.99	5.14	4.76	4.53	4.39	4.28	4.15	4.00	3.84	3.67
7	5.59	4.74	4.35	4.12	3.97	3.87	3.73	3.57	3.41	3.23
8	5.32	4.46	4.07	3.84	3.69	3.58	3.44	3.28	3.12	2.93
9	5.12	4.26	3.86	3.63	3.48	3.37	3.23	3.07	2.90	2.71
10	4.96	4.10	3.71	3.48	3.33	3.22	3.07	2.91	2.74	2.54
11	4.84	3.98	3.59	3.36	3.20	3.09	2.95	2.79	2.61	2.40
12	4.75	3.88	3.49	3.26	3.11	3.00	2.85	2.69	2.50	2.30
13	4.67	3.80	3.41	3.18	3.02	2.92	2.77	2.60	2.42	2.21
14	4.60	3.74	3.34	3.11	2.96	2.85	2.70	2.53	2.35	2.13
15	4.54	3.68	3.29	3.06	2.90	2.79	2.64	2.48	2.29	2.07
16	4.49	3.63	3.24	3.01	2.85	2.74	2.59	2.42	2.24	2.01
17	4.45	3.59	3.20	2.96	2.81	2.70	2.55	2.38	2.19	1.96
18	4.41	3.55	3.16	2.93	2.77	2.66	2.51	2.34	2.15	1.92
19	4.38	3.52	3.13	2.90	2.74	2.63	2.48	2.31	2.11	1.88
20	4.35	3.49	3.10	2.87	2.71	2.60	2.45	2.28	2.08	1.84
21	4.32	3.47	3.07	2.84	2.68	2.57	2.42	2.25	2.05	1.81
22	4.30	3.44	3.05	2.82	2.66	2.55	2.40	2.23	2.03	1.78
23	4.28	3.42	3.03	2.80	2.64	2.53	2.38	2.20	2.00	1.76
24	4.26	3.40	3.01	2.78	2.62	2.51	2.36	2.18	1.98	1.73
25	4.24	3.38	2.99	2.76	2.60	2.49	2.34	2.16	1.96	1.71
26	4.22	3.37	2.98	2.74	2.59	2.47	2.32	2.15	1.95	1.69
27	4.21	3.35	2.96	2.73	2.57	2.46	2.30	2.13	1.93	1.67
28	4.20	3.34	2.95	2.71	2.56	2.44	2.29	2.12	1.91	1.65
29	4.18	3.33	2.93	2.70	2.54	2.43	2.28	2.10	1.90	1.64
30	4.17	3.32	2.92	2.69	2.53	2.42	2.27	2.09	1.89	1.62
40	4.08	3.23	2.84	2.61	2.45	2.34	2.18	2.00	1.79	1.51
60	4.00	3.15	2.76	2.52	2.37	2.25	2.10	1.92	1.70	1.39
120	3.92	3.07	2.68	2.45	2.29	2.17	2.02	1.83	1.61	1.25
∞	3.84	2.99	2.60	2.37	2.21	2.09	1.94	1.75	1.52	1.00

$p = .01$

n_1 / n_2	1	2	3	4	5	6	8	12	24	∞
1	4052	4999	5403	5625	5764	5859	5981	6106	6234	6366
2	98.49	99.01	99.17	99.25	99.30	99.33	99.36	99.42	99.46	99.50
3	34.12	30.81	29.46	28.71	28.24	27.91	27.49	27.05	26.60	26.12
4	21.20	18.00	16.69	15.98	15.52	15.21	14.80	14.37	13.93	13.46
5	16.26	13.27	12.06	11.39	10.97	10.67	10.27	9.89	9.47	9.02
6	13.74	10.92	9.78	9.15	8.75	8.47	8.10	7.72	7.31	6.88
7	12.25	9.55	8.45	7.85	7.46	7.19	6.84	6.47	6.07	5.65
8	11.26	8.65	7.59	7.01	6.63	6.37	6.03	5.67	5.28	4.86
9	10.56	8.02	6.99	6.42	6.06	5.80	5.47	5.11	4.73	4.31
10	10.04	7.56	6.55	5.99	5.64	5.39	5.06	4.71	4.33	3.91
11	9.65	7.20	6.22	5.67	5.32	5.07	4.74	4.40	4.02	3.60
12	9.33	6.93	5.95	5.41	5.06	4.82	4.50	4.16	3.78	3.36
13	9.07	6.70	5.74	5.20	4.86	4.62	4.30	3.96	3.59	3.16
14	8.86	6.51	5.56	5.03	4.69	4.46	4.14	3.80	3.43	3.00
15	8.68	6.36	5.42	4.89	4.56	4.32	4.00	3.67	3.29	2.87
16	8.53	6.23	5.29	4.77	4.44	4.20	3.89	3.55	3.18	2.75
17	8.40	6.11	5.18	4.67	4.34	4.10	3.79	3.45	3.08	2.65
18	8.28	6.01	5.09	4.58	4.25	4.01	3.71	3.37	3.00	2.57
19	8.18	5.93	5.01	4.50	4.17	3.94	3.63	3.30	2.92	2.49
20	8.10	5.85	4.94	4.43	4.10	3.87	3.56	3.23	2.86	2.42
21	8.02	5.78	4.87	4.37	4.04	3.81	3.51	3.17	2.80	2.36
22	7.94	5.72	4.82	4.31	3.99	3.76	3.45	3.12	2.75	2.31
23	7.88	5.66	4.76	4.26	3.94	3.71	3.41	3.07	2.70	2.26
24	7.82	5.61	4.72	4.22	3.90	3.67	3.36	3.03	2.66	2.21
25	7.77	5.57	4.68	4.18	3.86	3.63	3.32	2.99	2.62	2.17
26	7.72	5.53	4.64	4.14	3.82	3.59	3.29	2.96	2.58	2.13
27	7.68	5.49	4.60	4.11	3.78	3.56	3.26	2.93	2.55	2.10
28	7.64	5.45	4.57	4.07	3.75	3.53	3.23	2.90	2.52	2.06
29	7.60	5.42	4.54	4.04	3.73	3.50	3.20	2.87	2.49	2.03
30	7.56	5.39	4.51	4.02	3.70	3.47	3.17	2.84	2.47	2.01
40	7.31	5.18	4.31	3.83	3.51	3.29	2.99	2.66	2.29	1.80
60	7.08	4.98	4.13	3.65	3.34	3.12	2.82	2.50	2.12	1.60
120	6.85	4.79	3.95	3.48	3.17	2.96	2.66	2.34	1.95	1.38
∞	6.64	4.60	3.78	3.32	3.02	2.80	2.51	2.18	1.79	1.00

Source: Table V of Fisher and Yates. *Statistical Tables for Biological, Agricultural and Medical Research,* published by Longman Group Ltd., London (1974), 6th edition (previously published by Oliver and Boyd Ltd., Edinburgh). Reprinted by permission of Pearson Education Ltd.

Appendix E

Using Statistics: Ideas for Research Projects

This appendix presents outlines for four research projects, each of which requires you to use SPSS to analyze the 2010 General Social Survey, the database that has been used throughout this text. The research projects should be completed at various intervals during the course, and each project permits a great deal of choice on the part of the student. The first project stresses description and should be done after completing Chapters 2–4. The second involves estimation and should be completed in conjunction with Chapter 6. The third project uses inferential statistics and should be done after completing Part II, and the fourth project combines inferential statistics with measures of association (with an option for multivariate analysis) and should be done after Part III (or IV).

Project 1—Descriptive Statistics

1. Select five variables from the 2010 General Social Survey other than those you have used for end-of-chapter exercises. *(Note: Your instructor may specify a different number of variables.)* Use the **Frequencies** command to get frequency distributions and summary statistics for each variable. Click the **Statistics** button on the **Frequencies** command window and then request the mean, median, mode, standard deviation, and range. Make a note of all relevant information when it appears on screen or make a hard copy. See Appendix G for a list of variables available in the 2010 GSS.

2. For each variable, get bar or line charts to summarize the overall shape of the distribution of the variable.

3. Inspect the frequency distributions and graphs and then choose appropriate measures of central tendency and, for ordinal- and interval-ratio-level variables, dispersion. Also, for interval-ratio and ordinal variables with many scores, check for skew by using the line chart and by comparing the mean and median. Write a sentence or two of description for each variable, being careful to include a description of the overall shape of the distribution (see Chapter 2), the central tendency (Chapter 3), and the dispersion (Chapter 4). For nominal- and ordinal-level variables, be sure to explain any arbitrary numerical codes. For example, on the variable *class* in the 2010 GSS (see Appendix G), a 1 is coded as "lower class," a 2 indicates "working class," and so forth. This is an ordinal-level variable, so you might choose to report the median as a measure of central tendency. For example, the median score on *class* were 2.45, you might place that value in context by reporting that "the median is 2.45, about halfway between 'working class' and 'middle class.'"

4. Here are examples of *minimal* summary sentences, using fictitious data: *For a nominal-level variable* (e.g., gender), report the mode and some detail about the overall distribution. For example: "Most respondents were married (57.5%), but divorced (17.4%) and single (21.3%) individuals were also common." *For an ordinal-level variable* (e.g., occupational prestige), use the median (and perhaps the mean and mode) and the range. For example: "The median prestige score was 44.3, and the range extended from 34 to 87. The most common score was 42, and the average score was 40.8."

   For an interval-ratio level variable (e.g., age), use the mean (and perhaps the median or mode) and the standard deviation (and perhaps the range). For example: "The average age for this sample was 42.3. Respondents ranged from 18 to 94 years old, with a standard deviation of 15.37."

Project 2—Estimation

In this exercise, you will use the 2010 GSS sample to estimate the characteristics of the U.S. population. You will use SPSS to generate the sample statistics and then use either Formula 6.2 or Formula 6.3 to find the confidence interval and state each interval in words.

A. Estimating Means

1. There are relatively few interval-ratio variables in the 2010 GSS, and for this part of the project, you may use ordinal variables that have *at least* three categories or scores. Choose a total of three variables that fit this description *other than* the variables you used in Chapter 6. *(NOTE: Your instructor may specify a different number of variables.)*
2. Use the **Descriptives** command to get means, standard deviations, and sample size (*N*), and use this information to construct 95% confidence intervals for each of your variables. Make a note of the mean, standard deviation, and sample size or keep a hard copy. Use Formula 6.2 to compute the confidence intervals. Repeat this procedure for the remaining variables.
3. For each variable, write a summary sentence reporting the variable, the interval itself, the confidence level, and the sample size. Write in plain language, as if you were reporting results in a newspaper. Most importantly, you should make it clear you are estimating characteristics of the population of the entire United States. For example, a summary sentence might look like this: "Based on a random sample of 1,231, I estimate at the 95% level that U.S. drivers average between 64.46 and 68.22 miles per hour when driving on interstate highways."

B. Estimating Proportions

1. Choose three variables that are nominal or ordinal *other than* the variables you used in Chapter 6. *(Note: Your instructor may specify a different number of variables.)*

2. Use the **Frequencies** command to get the percentage of the sample in the various categories of each variable. Change the percentages (remember to use the "valid percents" column) to proportions and then construct confidence intervals for one category of each variable (e.g., the % female for *sex*) by using Formula 6.3.
3. For each variable, write a summary sentence reporting the variable, the interval, the confidence level, and the sample size. Write in plain English, as if you were reporting results in a newspaper. Remember to make it clear you are estimating a characteristic of the U.S. population.
4. For any one of the intervals you constructed in either Part A or Part B, identify each of the following concepts and terms, and then briefly explain their role in estimation: sample, population, statistic, parameter, EPSEM, representative, confidence level.

Project 3—Significance Testing

In this exercise, you will use the 2010 GSS sample to conduct tests of significance. You will use the five-step model to conduct the tests.

A. Two-Sample *t* Test (Chapter 8)

1. Choose two different dependent variables from the interval-ratio or ordinal variables that have three or more scores. Choose independent variables that might logically be a cause of your dependent variables. Remember that for a *t* test, independent variables can have *only* two categories, but you can still use independent variables with more than two categories by (a) using the **Grouping Variable** box to specify the exact categories (e.g., select scores of 1 and 5 on *marital* to compare married with never-married respondents) or (b) collapsing the scores of variables with more than two categories by using the **Recode** command. Independent variables can be any level of measurement, and you may use the same independent variable for both tests.
2. Click **Analyze**, **Compare Means**, and then **Independent-Samples T Test**. Name your dependent variable(s) in the Test Variable window and your independent variable in the **Grouping Variable** window. You will also need to specify the scores used to define the groups on the independent variable. See Chapter 8 for examples. Make a note of the test results (group means, obtained *t* score, significance, sample size) or keep a hard copy. Repeat the procedure for the second dependent variable.
3. Write up the results of the test. At a *minimum*, your report should clearly identify the independent and dependent variables, the sample statistics, the value of the test statistic (step 4), the results of the test (step 5), and the alpha level you used.

B. Analysis of Variance (Chapter 9)

1. Choose two different dependent variables from the interval-ratio or ordinal variables that have three or more scores. Choose independent variables that might logically be a cause of your dependent variables and that have between

three and five categories. You may use the same independent variables for both tests. Do not use the same variables you used in Part A of this project.

2. Click **Analyze**, **Compare Means**, and then **One-way Anova**. The **One-way Anova** window will appear. Find your dependent variable in the variable list on the left and then click the arrow to move the variable name into the **Dependent List** box. Note that you can request more than one dependent variable at a time. Next, find the name of your independent variable and then move it to the **Factor** box. Click **Options** and then click the box next to **Descriptive** in the Statistics box to request means and standard deviations. Click **Continue** and **OK**. Make a note of the test results or keep a hard copy. Repeat if necessary for your second dependent variable.

3. Write up the results of the test. At a minimum, your report should clearly identify the independent and dependent variables, the sample statistics (category means), the value of the test statistic (step 4), the results of the test (step 5), the degrees of freedom, and the alpha level you used.

C. Chi Square (Chapter 10)

1. Choose two dependent variables of any level of measurement that have five or fewer (preferably two to three) scores. For each dependent variable, choose an independent variable that might logically be a cause. Independent variables can be any level of measurement as long as they have five or fewer (preferably two to three) categories. Output will be easier to analyze if you use variables with few categories. You may use the same independent variable for both tests.

2. Click **Analyze**, **Descriptive Statistics**, and then **Crosstabs**. The **Crosstabs** dialog box will appear. Highlight your first dependent variable and then move it into the **Rows** box. Next, highlight your independent variable and then move it into the Columns box. Click the **Statistics** button at the bottom of the window and then click the box next to chi square. Click **Continue** and **OK**. Make a note of the results or get a hard copy. Repeat for your second dependent variable.

3. Write up the results of the test. At a *minimum*, your report should clearly identify the independent and dependent variables, the value of the test statistic (step 4), the results of the test (step 5), the degrees of freedom, and the alpha level you used. It is also almost always desirable to report the column percentages.

Project 4—Analyzing the Strength and Significance of Relationships

A. Using Bivariate Tables

1. From the 2010 GSS data set, select either:

   a. One dependent variable and three independent variables (possible causes)

   b. One independent variable and three possible dependent variables (possible effects)

Variables can be from any level of measurement but must have only a few (two to five) categories or scores. Develop research questions or hypotheses about the relationships between variables. Make sure the causal links you suggest are sensible and logical.

2. Use the **Crosstabs** procedure to generate bivariate tables. See Chapters 10, 11, and 12 for examples. Click **Analyze**, **Descriptive Statistics**, and **Crosstabs** and then place your dependent variable(s) in the rows and independent variable(s) in the columns. On the **Crosstabs** dialog box, click the **Statistics** button and then choose chi square, phi or Cramer's *V*, and gamma for every table you request. On the **Crosstabs** dialog box, click the **Cells** button to get column percentages for every table you request. Make a note of results as they appear on the screen or get hard copies.

3. Write a report that presents and analyzes these relationships. Be clear about which variables are dependent and which are independent. For each combination of variables, report the test of significance and an appropriate measure of association. In addition, for each relationship, report and discuss column percentages, pattern or direction of the relationship, and strength of the relationship.

B. Using Interval-Ratio Variables

1. From the 2010 GSS, select either:

   a. One dependent variable and three independent variables (possible causes)
   b. One independent variable and three dependent variables (possible effects)

   Variables should be interval-ratio in level of measurement, but you may use ordinal-level variables as long as they have more than three scores. Develop research questions or hypotheses about the relationships between variables. Make sure the causal links you suggest are sensible and logical.

2. Use the **Regression** and **Scatterplot** (click **Graphs**, **Legacy Dialogs**, and then **Scatter**) procedures to analyze the bivariate relationships. Make a note of results (including r, r^2, slope, beta-weights, and a) as they appear on the screen or get hard copies.

3. Write a report that presents and analyzes these relationships. Be clear about which variables are dependent and which are independent. For each combination of variables, report the significance of the relationship and the strength and direction of the relationship. Include r, r^2, and the beta-weights in your report.

4. *Optional Multivariate Analysis*: Pick one of the bivariate relationships you produced in step 2 and then find another logical independent variable. Run **Regression** again with both independent variables and then analyze the results. How much improvement is there in the explained variance after the second independent variable is included? Write up the results of this analysis and then include them in your summary paper for this project.

Appendix F An Introduction to *SPSS for Windows*

Computers have affected virtually every aspect of human society, and as you would expect, their impact on the conduct of social research has been profound. Researchers routinely use computers to organize data and compute statistics—activities that humans often find dull, tedious, and difficult but that computers accomplish with accuracy and ease. This division of labor allows social scientists to spend more time on analysis and interpretation—activities that humans typically enjoy but that are beyond the power of computers (so far, at least).

These days, the skills needed to use computers successfully are quite accessible, even for people with little or no experience. This appendix will prepare you to use a statistics program called *SPSS for Windows*. If you have used a mouse to point and click and run a computer program, you are ready to learn how to use this program. Even if you are completely unfamiliar with computers, you will find this program accessible. After you finish this appendix, you will be ready to do the exercises found at the end of most chapters of this text.

A word of caution before we begin: This appendix is intended only as an *introduction* to SPSS. It will give you an overview of the program and enough information so you can complete the assignments in the text. However, it is unlikely that this appendix will answer all your questions or provide solutions to all the problems you might encounter. Thus, this is a good place to tell you that SPSS has an extensive and easy-to-use Help facility that will provide assistance as you request it. You should familiarize yourself with this feature and use it as needed. To get help, simply click the **Help** command on the toolbar across the top of the screen.

SPSS is a **statistical package** (or **statpak**)—that is, a set of computer programs that work with data and compute statistics as requested by the user (you). Once you have entered the data for a particular group of observations, you can easily and quickly produce an abundance of statistical information without doing any computations or writing any computer programs yourself.

Why bother to learn this technology? The truth is that the labor-saving capacity of computers is sometimes exaggerated, and there are research situations in which they are unnecessary. If you are working with a small number of observations or need only a few, uncomplicated statistics, then statistical packages may not be helpful. However, as the number of cases increases and as your requirements for statistics become more sophisticated, computers and statpaks will become more and more useful.

An example should make this point clearer. Suppose you have gathered a sample of 150 respondents and the *only* thing you want to know about these people is their average age. To compute an average, as you know, you add the scores and divide by the number of cases. How long do you think it would take you to add 150 two-digit numbers (ages) with a hand calculator? If you entered the scores at the rate of one per second— 60 scores a minute—it would take about 3 or 4 minutes to enter the ages and get the average. Even if you worked slowly and carefully and did the addition a second and third time to check your math, you could probably complete all calculations in less than 15 or 20 minutes. If this were all the information you needed, computers and statpaks would not save you any time.

However, such a simple research project is not very realistic. Typically, researchers deal with not one but scores or even hundreds of variables, and samples have hundreds or thousands of cases. While you could add 150 numbers in perhaps 3 or 4 minutes, how long would it take to add the scores for 1,500 cases? What are the chances of adding 1,500 numbers without making significant errors of arithmetic? The more complex the research situation, the more valuable and useful statpaks become. SPSS can produce statistical information in a few keystrokes or clicks of the mouse that might take you minutes, hours, or even days to produce with a hand calculator.

Clearly, this is technology worth mastering by any social researcher. With SPSS, you can avoid the drudgery of mere computation, spend more time on analysis and interpretation, and conduct research projects with very large data sets. Mastery of this technology might be very handy indeed in your senior-level courses, in a wide variety of jobs, or in graduate school.

F.1 Getting Started—Databases and Computer Files

Before statistics can be calculated, SPSS must have some data to process. A **database** is an organized collection of related information, such as the responses to a survey. For purposes of computer analysis, a database is organized into a **file**—a collection of information that is stored under the same name in the memory of the computer, on a disk or flash drive, or on some other medium. Words as well as numbers can be saved in files. If you have ever used a word processing program to type a letter or term paper, you probably saved your work in a file so you could update or make corrections at a later time. Data can be stored in files indefinitely. Because it can take months to conduct a thorough data analysis, the ability to save a database is another advantage of using computers.

For the SPSS exercises in this text, we will use a database that contains some of the results of the General Social Survey (GSS) for 2010. This database contains the responses of a sample of adult Americans to questions about a wide variety of social issues. The GSS has been conducted regularly since 1972 and has been the basis for hundreds of research projects by professional social researchers. It is a rich source of information about public opinion in the United States and includes data on everything from attitudes about abortion to opinions on assisted suicide.

The GSS is especially valuable because the respondents are chosen so the sample as a whole is representative of the entire U.S. population. A representative sample reproduces in miniature form the characteristics of the population from which it was taken (see Chapter 6). Thus, when you analyze the 2010 General Social Survey database, you are in effect analyzing U.S. society as of 2010. The data are real, and the relationships you will analyze reflect some of the most important and sensitive issues in American life.

The complete General Social Survey for 2010 includes hundreds of items of information (age, sex, opinion about social issues, such as capital punishment, and so forth) for about 2,000 respondents. Some of you will be using a student version of *SPSS for Windows*, which is limited in the number of cases and variables it can process. To accommodate those limits, I have reduced the database to 50 items of information and fewer than 1,500 respondents.

The GSS data file is summarized in Appendix G. Please turn to this appendix and familiarize yourself with it. Note that the variables are listed alphabetically by their variable names. In SPSS, the names of variables must be no more than eight characters long. In many cases, the resultant need for brevity is not a problem and variable names (e.g., *age*) are easy to figure out. In other cases, the eight-character limit necessitates extreme abbreviation and some variable names (like *abany* or *fefam*) are not so obvious. Appendix G also shows the wording of the item that generated the variable. For example, the *abany* variable consists of responses to a question about legal abortion: Should it be possible for a woman to have an abortion for "any reason"? Note that the variable name is formed from the question: Should an *ab*ortion be possible for *any* reason? The *fefam* variable consists of responses to the statement "It is much better for everyone involved if the man is the achiever outside the home and the woman takes care of the home and family." Appendix G is an example of a code book for the database because it lists all the codes (or scores) for the survey items along with their meanings.

Notice that some of the possible responses to *abany* and *fefam* and other variables in Appendix G are labeled IAP ("Not applicable"), NA, or DK. The first of these responses means that the item in question was not given to a respondent. The full GSS is very long, and to keep the time frame for completing the survey reasonable, not all respondents are asked every question. NA stands for "No Answer" and means that the respondent was asked the question but refused to answer. DK stands for "Don't Know," which means the respondent did not have the requested information. All three of these scores are *missing values*, and as non information, they should be eliminated from statistical analysis. Missing values are common on surveys, and as long as they are not too numerous, they are not a particular problem.

It is important that you understand the difference between a statpak (SPSS) and a database (the GSS) and what we are ultimately after here. A database consists of information. A statpak processes the information in the database and produces statistics. Our goal is to apply the statpak to the database to produce output (for example, statistics and graphs) that we can analyze and use to answer questions. The process might be diagrammed as in Figure F.1.

Statpaks like SPSS are general research tools that can be used to analyze databases of all sorts; they are not limited to the 2010 GSS. In the same way,

FIGURE F.1 The Data Analysis Process

Database	→	**Statpak**	→	**Output**	→	**Analysis**
(raw information)		(computer programs)		(statistics and graphs)		(interpretation)

the 2010 GSS could be analyzed with statpaks other than those used in this text. Other widely used statpaks include Microcase, SAS, and Stata— each of which may be available on your campus.

F.2 Starting *SPSS for Windows* and Loading the 2010 GSS

If you are using the complete, professional version of *SPSS for Windows*, you will probably be working in a computer lab, and you can begin running the program immediately. If you are using the student version of the program on your personal computer, the first thing you need to do is install the software. Follow the instructions that came with the program and then return to this appendix when installation is complete.

To start *SPSS for Windows*, find the icon (or picture) on the screen of your monitor that has a "SPSS" label attached to it. Use the computer mouse to move the arrow on the monitor screen over this icon and then double-click the left button on the mouse. This will start up the SPSS program.

After a few seconds, the *SPSS for Windows* screen will appear and ask, at the top of the screen, "What would you like to do?" Of the choices listed, the button next to "Open an existing file" will be checked (or preselected), and there will probably be a number of data sets listed in the window. Find the 2010 General Social Survey data set, probably labeled *GSS2010.sav* or something similar. If the data set is on a disk or flash drive, you will need to specify the correct drive. Check with your instructor to make sure you know where to find the 2010 GSS.

Once you have located the data set, click on the name of the file with the left-hand button on the mouse, and SPSS will load the data. The next screen you will see is the SPSS Data Editor screen.

Note that there is a list of commands across the very top of the screen. These commands begin with **File** at the far left and end with **Help** at the far right. This is the main menu bar for SPSS. When you click any of these words, a **menu** of commands and choices will drop down. Basically, you tell SPSS what to do by clicking on your desired choices from these menus. Sometimes, submenus will appear, and you will need to specify your choices further.

SPSS provides the user with a variety of options for displaying information about the data file and output on the screen. I recommend you tell the program to display lists of variables by name (e.g., *age, abany*) rather than by labels (e.g., AGE OF RESPONDENT, ABORTION IF WOMAN WANTS FOR ANY REASON). Lists displayed this way will be easier to read and compare to Appendix G. To do this, click **Edit** on the main menu bar and then click **Options** from the drop-down submenu. A dialog box labeled "Options" will appear with a series of "tabs" along the top. The "General" options should be displayed; if not, click on this tab. On the "General" screen, find the box labeled "Variable Lists" and, if they are not already selected, click "Display names" and "Alphabetical" and then click **OK**.

In this section, you learned how to start up *SPSS for Windows,* load a data file, and set some of the display options for this program. These procedures are summarized in Table F.1.

TABLE F.1 Summary of Commands

To start *SPSS for Windows*	Click the SPSS icon on the screen of the computer monitor.
To open a data file	Double-click the data file name.
To set display options for lists of variables	Click **Edit** from the main menu bar and then click **Options**. On the "General" tab make sure "Display names" and "Alphabetical" are selected and then click **OK**.

F.3 Working With Databases

Note that in the SPSS Data Editor window that the data are organized into a two-dimensional grid, with columns running up and down (vertically) and rows running across (horizontally). Each column is a variable or item of information from the survey. The names of the variables are listed at the tops of the columns. Remember that you can find the meaning of these variable names in the GSS 2010 code book in Appendix G.

Another way to decipher the meaning of variable names is to click **Utilities** on the menu bar and then click **Variables**. The **Variables** window opens. This window has two parts. On the left is a list of all variables in the database, arranged in alphabetical order and with the first variable highlighted. On the right is the **Variable Information** window, with information about the highlighted variable. The first variable is listed as *abany.* The **Variable Information** window displays a fragment of the question that was actually asked during the survey ("ABORTION IF WOMAN WANTS FOR ANY REASON") and shows the possible scores on this variable (a score of 1 = yes and a score of 2 = no), along with some other information.

The same information can be displayed for any variable in the data set. For example, find the variable *marital* in the list. You can do this by using the arrow keys on your keyboard or the slider bar on the right of the variable list window. You can also move through the list by typing the first letter of the variable name you are interested in. For example, type "m" and you will be moved to the first variable name in the list that begins with that letter. Now you can see that the variable measures marital status, that a score of "1" indicates that the respondent was married, and so forth. What do *prestg80* and *marhomo* measure? Close this window by clicking the **Close** button at the bottom of the window.

Examine the window displaying the 2010 GSS a little more. Each row of the window (reading across, or from left to right) contains the scores of a particular respondent on all the variables in the database. Note that the upper left-hand cell is highlighted or is a different color than the other cells. This cell contains the score of respondent 1 on the first variable. The second row contains the scores of respondent 2 and so forth. You can move around in this window with the arrow keys on your keyboard. The highlight moves in the direction of the arrow—one cell at a time.

In this section, you learned to read information in the data display window and to decipher the meaning of variable names and scores. These commands are summarized in Table F.2. We are now prepared to perform some statistical operations with the 2010 GSS database.

TABLE F.2 Summary of Commands

To move around in the Data Editor window	1. Click the cell you want to highlight or 2. Use the arrow keys on your keyboard or 3. Move the slider buttons or 4. Click the arrows on the right-hand and bottom margins.
To get information about a variable	1. From the menu bar, click Utilities and then click Variables. Scroll through the list of variable names until you highlight the name of the variable in which you are interested. Variable information will appear in the window on the right. 2. See Appendix G.

F.4 Putting SPSS to Work: Producing Statistics

At this point, the database on the screen is just a mass of numbers with little meaning for you. That is okay because you will not actually have to read any information from this screen. Virtually all the statistical operations you will conduct will begin by clicking the **Analyze** command from the menu bar, selecting a procedure and statistics, and then naming the variable or variables you would like to process.

To illustrate, let us have *SPSS for Windows* produce a frequency distribution for the variable *sex*. Frequency distributions are tables that display the number of times each score of a variable occurred in the sample (see Chapter 2). Thus, when we complete this procedure, we will know the number of males and females in the 2010 GSS sample.

With the 2010 GSS loaded, begin by clicking the **Analyze** command on the menu bar. From the menu that drops down, click **Descriptive Statistics** and then **Frequencies**. The Frequencies window appears, with the variables listed in alphabetical order in the box on the left. The first variable (*abany*) will be highlighted. Use the slider button or the arrow keys on the right-hand margin of this box to scroll through the variable list until you highlight the variable *sex* or type "s" to move to the approximate location.

Once the variable you want to process has been highlighted, click the arrow button in the middle of the screen to move the variable name to the box on the right-hand side of the screen. SPSS will produce frequency distributions for all variables listed in this box, but for now, we will confine our attention to *sex*. Click the **OK** button in the upper right-hand corner of the Frequencies window, and in seconds, a frequency distribution will be produced.

SPSS sends all tables and statistics to the Output window or SPSS viewer. This window is now "closest" to you, and the Data Editor window is "behind" the Output window. If you want to return to the Data Editor, click on any part of it if it is visible, and it will move to the "front" and the Output window will be "behind" it. To display the Data Editor window if it is not visible, minimize the Output window by clicking the "–" box in the upper right-hand corner.

Frequencies. The output from SPSS is reproduced as Table F.3. What can we tell from this table? The score labels (male and female) are printed at the left, with the number of cases (frequency) in each category of the variable one column

TABLE F.3 An Example of Output from SPSS

		RESPONDENTS SEX			
		Frequency	Percent	Valid Percent	Cumulative Percent
Valid	MALE	654	43.8	43.8	43.8
	FEMALE	840	56.2	56.2	100.0
	Total	1494	100.0	100.0	

to the right. As you can see, there are 654 males and 840 females in the sample. The next two columns give information about percentages and the last column to the right displays cumulative percentages. We will defer a discussion of this last column until a later exercise.

One of the percentage columns is labeled "Percent" and the other is labeled "Valid Percent." The difference between these two columns lies in the handling of missing values. The "Percent" column is based on all cases, including people who did not respond to the item (NA) and people who said they did not have the requested information (DK). The "Valid Percent" column excludes all missing scores. Because we will almost always want to ignore missing scores, we will pay attention only to the "Valid Percent" column. Note that for sex, there are no missing scores and the two columns are identical.

F.5 Printing and Saving Output

Once you have gone to the trouble of producing statistics, a table, or a graph, you will probably want to keep a permanent record. There are two ways to do this. First, you can print a copy of the contents of the Output Window to take with you. To do this, click on **File** and then click **Print**. Alternatively, find the icon of a printer (third from the left) in the row of icons just below the menu bar and then click on it.

The other way to create a permanent record of SPSS output is to save the Output window to the computer's memory or to a flash drive. To do this, click **Save** from the **File** menu. The Save dialog box opens. Give the output a name (some abbreviation such as "freqsex" might do) and, if necessary, specify the name of the drive in which your flash drive is located. Click **OK**, and the table will be permanently saved.

F.6 Ending Your *SPSS for Windows* Session

Once you have saved or printed your work, you may end your SPSS session. Click on **File** from the menu bar and then click **Exit**. If you have not already done so, you will be asked if you want to save the contents of the Output window. You may save the frequency distribution at this point if you wish. Otherwise, click **NO**. The program will close, and you will be returned to the screen from which you began.

Appendix G

Code Book for the General Social Survey, 2010

The General Social Survey (GSS) is a public opinion poll that has been regularly conducted by the National Opinion Research Council. A version of the 2010 GSS is available at the website for this text and is used for all end-of-chapter exercises. Our version of the 2010 GSS includes 50 variables for a randomly selected subsample of about 1,500 of the original respondents. This code book lists each item in the data set. The variable names are those used in the data files. The questions have been reproduced exactly as they were asked (with a few exceptions to conserve space), and the numbers beside each response are the scores recorded in the data file.

The data set includes variables that measure demographic or background characteristics of the respondents, including sex, age, race, religion, and several indicators of socioeconomic status. Also included are items that measure opinion on such current and controversial topics as abortion, capital punishment, and homosexuality.

Most variables in the data set have codes for "missing data." These codes are italicized in the listings below for easy identification. The codes refer to various situations in which the respondent does not or cannot answer the question and are excluded from all statistical operations. The codes are: IAP ("not applicable"—that is, the respondent was not asked the question), DK ("Don't know"—that is, the respondent didn't have the requested information), and NA ("No answer"—that is, the respondent refused to answer).

Please tell me if you think it should be possible for a woman to get a legal abortion if …

abany — She wants it for any reason.

1. Yes 2. No
0. IAP, 8. DK, 9. NA

abpoor — The family has low income and can't afford any more children
(Same scoring as abany)

age — Age of respondent

18–89 Actual years
99. NA

attend — How often do you attend religious services?

0. Never 1. Less than once per year
2. Once or twice a year 3. Several times per year
4. About once a month 5. 2–3 times a month
6. Nearly every week 7. Every week
8. Several times a week
9. DK or NA

cappun — Do you favor or oppose the death penalty for persons convicted of murder?
- 1. Favor
- 2. Oppose
- *0. IAP, 8. DK, 9. NA*

childs — How many children have you ever had? Please count all that were born alive at any time (including any you had from a previous marriage).
- 0–7 Actual Number
- 8. 8 or more
- *9. DK, NA*

class — Subjective class identification
- 1. Lower class
- 2. Working class
- 3. Middle class
- 4. Upper class
- *0. IAP, 8. DK, 9. NA*

closeblk — In general, how close do you feel to blacks?
- 1. Not at all close
- 2.
- 3.
- 4.
- 5. Neither one feeling or the other
- 6.
- 7.
- 8.
- 9. Very close
- *0. IAP, 98. DK, 99. NA*

degree — Respondent's highest degree
- 0. Less than high school
- 1. High school
- 2. Assoc./Junior college
- 3. Bachelor's
- 4. Graduate
- *7. IAP, 8. DK, 9. NA*

educ — Highest year of school completed
- 0–20. Actual number of years
- *97. IAP, 98. DK, 99. NA*

evolved — Human beings, as we know them today, developed from earlier species of animals.
- 1. True
- 2. False
- *0. IAP, 8. DK, 9. NA*

fear — Is there any area right around here—that is, within a mile—where you would be afraid to walk alone at night?
- 1. Yes
- 2. No
- *0. IAP, 8. DK, 9. NA*

fechld — A working mother can establish just as warm and secure a relationship with her children as a mother who does not work.
- 1. Strongly agree
- 2. Agree
- 3. Disagree
- 4. Strongly disagree
- *0. IAP, 8. DK, 9. NA*

fefam — It is much better for everyone involved if the man is the achiever outside the home and the woman takes care of the home and family.

1. Strongly agree
2. Agree
3. Disagree
4. Strongly disagree

0. IAP, 8. DK, 9. NA

goveqinc — It is the responsibility of the government to reduce the differences in income between people with high incomes and those with low incomes.

1. Strongly agree
2. Agree
3. Neither agree nor disagree
4. Disagree
5. Strongly disagree

0. IAP, 8. DK, 9. NA

grncon — Generally speaking, how concerned are you about environmental issues?

1. Not at all concerned
2.
3.
4.
5. Very concerned

0. IAP, 8. DK, 9. NA

grnexagg — Many of the claims about environmental threats are exaggerated.

1. Strongly agree
2. Agree
3. Neither agree nor disagree
4. Disagree
5. Strongly disagree

0. IAP, 8. DK, 9. NA

hapmar — Taking things all together, how would you describe your marriage?

1. Very happy
2. Pretty happy
3. Not too happy

0. IAP, 8. DK, 9. NA

happy — Taken all together, how would you say things are these days—would you say that you are very happy, pretty happy, or not too happy?

1. Very happy
2. Pretty happy
3. Not too happy

0. IAP, 8. DK, 9. NA

hrs1 — How many hours did you work last week, at all jobs?

1–89. Actual hours

–1. IAP, 98. DK, 99. NA

income06 Respondent's total family income from all sources

1. Less than 1,000
2. 1,000 to 2,999
3. 3,000 to 3,999
4. 4,000 to 4,999
5. 5,000 to 5,999
6. 6,000 to 6,999
7. 7,000 to 7,999
8. 8,000 to 9,999
9. 10,000 to 12,499
10. 12,500 to 14,999
11. 15,000 to 17,499
12. 17,500 to 19,999
13. 20,000 to 22,499
14. 22,500 to 24,999
15. 25,000 to 29,999
16. 30,000 to 34,999
17. 35,000 to 39,999
18. 40,000 to 49,999
19. 50,000 to 59,999
20. 60,000 to 74,999
21. 75,000 to 89,999
22. 90,000 to 109,999
23. 110,000 to 129,999
24. 130,000 to 149,999
25. 150,000 or more

98. DK, 99. NA

letdie1 When a person has a disease that cannot be cured, do you think doctors should be allowed by law to end the patient's life by some painless means if the patient and his family request it?

1. Yes
2. No

0. IAP, 8. DK, 9. NA

letin1 Do you think the number of immigrants to America nowadays should be

1. Increased a lot
2. Increased a little
3. Remain the same as it is
4. Reduced a little
5. Reduced a lot

0. IAP, 8. DK, 9. NA

marhomo Homosexual couples should have the right to marry one another.

1. Strongly agree
2. Agree
3. Neither agree or disagree
4. Disagree
5. Strongly disagree

0. IAP, 8. DK, 9. NA

marital Are you currently married, widowed, divorced, separated, or have you never been married?

1. Married
2. Widowed
3. Divorced
4. Separated
5. Never married

9. NA

news How often do you read the newspaper?

1. Every day
2. A few times a week
3. Once a week
4. Less than once a week
5. Never

0. IAP, 8. DK, 9. NA

obey How important is it for a child to learn obedience to prepare him or her for life?

1. Most important
2. 2nd most important
3. 3rd most important
4. 4th most important
5. Least important

0. IAP, 8. DK, 9. NA

paeduc — Father's highest year of school completed
- 0–20. Actual number of years
- *97. IAP, 98. DK, 99. NA*

papres80 — Prestige of father's occupation
- 17–86. Actual score
- *0. IAP, DK, NA*

partnrs5 — How many sex partners have you had over the past five years?
- 0. No partners
- 1. 1 partner
- 2. 2 partners
- 3. 3 partners
- 4. 4 partners
- 5. 5–10 partners
- 6. 11–20 partners
- 7. 21–100 partners
- 8. More than 100 partners
- *9. 1 or more, don't know the number, 95 Several, 98 DK, 99 NA, –1. IAP*

polviews — I'm going to show you a seven-point scale on which the political views that people might hold are arranged from extremely liberal to extremely conservative. Where would you place yourself on this scale?
- 1. Extremely liberal
- 2. Liberal
- 3. Slightly liberal
- 4. Moderate
- 5. Slightly conservative
- 6. Conservative
- 7. Extremely conservative
- *0. IAP, 8. DK, 9. NA*

premarsx — There's been a lot of discussion about the way morals and attitudes about sex are changing in this country. If a man and a woman have sex relations before marriage, do you think it is always wrong, almost always wrong, wrong only sometimes, or not wrong at all?
- 1. Always wrong
- 2. Almost always wrong
- 3. Wrong only sometimes
- 4. Not wrong at all
- *0. IAP, 8. DK, 9. NA*

pres08 — In 2008, did you vote for Obama (the Democratic candidate) or McCain (the Republican candidate)? (Includes only those who said they voted in this election.)
- 1. Obama
- 2. McCain
- *3. Other, 4. No presidential vote, 0. IAP, 8. DK, 9. NA*

prestg80 — Prestige of respondent's occupation
- 17–86. Actual score
- *0. IAP, DK, NA*

racecen1 — Race or ethnicity of respondent
- 1. White
- 2. Black
- 3. American Indian or Alaska Native
- 4. Asian American or Pacific Islander
- 5. Hispanic
- *0. IAP, 98. DK*

rank — In our society there are groups which tend to be towards the top and those that are towards the bottom. Here we have a scale that runs from top to bottom. Where would you put yourself on this scale?

1. Top
2.
3.
4.
5.
6.
7.
8.
9.
10. Bottom

0. IAP, 98. DK, 99. NA

region — Region of interview

1. New England (ME, VT, NH, MA, CT, RI)
2. Mid-Atlantic (NY, NJ, PA)
3. East North Central (WI, IL, IN, MI, OH)
4. West North Central (MN, IA, MO, ND, SD, NE, KS)
5. South Atlantic (DE, MY, WV, VA, NC, SC, GA, FL, DC)
6. East South Central (KY, TN, AL, MS)
7. West South Central (AL, OK, LA, TX)
8. Mountain (MT, ID, WY, NV, UT, CO, AR, NM)
9. Pacific *(WA, OR, CA, AK, HI)*

relig — What is your religious preference? Is it Protestant, Catholic, Jewish, some other religion, or no religion?

1. Protestant
2. Catholic
3. Jewish
4. None
5. Other

8. DK, 9. NA

satjob — All in all, how satisfied would you say you are with the work you do?

1. Very satisfied
2. Moderately satisfied
3. A little dissatisfied
4. Very dissatisfied

0. IAP, 8. DK, 9. NA

sex — Respondent's gender

1. Male
2. Female

sexfreq — About how many times did you have sex during the last 12 months?

0. Not at all
1. Once or twice
2. About once a month
3. 2 or 3 times a month
4. About once a week
5. 2 or 3 times a week
6. More than 3 times a week

–1. IAP, 8. DK, 9. NA

size — Size of place in 1000s. Add three zeros for actual population.

spanking

Do you strongly agree, agree, disagree, or strongly disagree that it is sometimes necessary to discipline a child with a good, hard spanking?

1. Strongly agree
2. Agree
3. Disagree
4. Strongly disagree

0. IAP, 8. DK, 9. NA

spkmslm

There are always some people whose ideas are considered bad or dangerous by other people. Consider a Muslim clergyman who preached hatred of the United States … Should he be allowed to speak or not?

1. Yes, allowed
2. Not allowed

0. IAP, 8. DK, 9. NA

thnkself

How important is it for a child to learn to think for himself or herself to prepare him or her for life?

1. Most important
2. 2nd most important
3. 3rd most important
4. 4th most important
5. Least important

0. IAP, 8. DK, 9. NA

topprob1

Which of these issues is the most important for America today?

1. Health care
2. Education
3. Crime
4. The environment
5. Immigration
6. The economy
7. Terrorism
8. Poverty
9. None of these

0. IAP, 98. DK, 99. NA

trust5

Generally speaking, would you say that most people can be trusted, or that you can't be too careful in dealing with people?

1. You can't be too careful
2.
3.
4.
5. Most people can be trusted

0. IAP, 8. DK, 9. NA

tvhours

On the average day, about how many hours do you personally watch television?

00–22. Actual hours

–1. IAP, DK, NA

wwwhr

Not counting e-mail, about how many minutes or hours per week do you use the Web?

0–120. Actual hours

0. IAP, 998. DK, 999. NA

xmarsex

What is your opinion about a married person having sexual relations with someone other than the marriage partner?

1. Always wrong
2. Almost always wrong
3. Wrong only sometimes
4. Not wrong at all

0. IAP, 8. DK, 9. NA

Answers to Odd-Numbered End-of-Chapter Problems

In addition to answers, this section suggests problem-solving strategies and provides examples of interpretation. You should try to solve and interpret the problems on your own before consulting this section.

In solving these problems, I let my calculator or computer do most of the work. I worked with whatever level of precision these devices permitted and didn't round off until the end or until I had to record an intermediate sum. I always rounded off to two places of accuracy (that is, to two places beyond the decimal point, or to 100ths). If you follow these same conventions, your answers will match mine. However, you should realize that small, generally trivial discrepancies between your answer and mine might occur. If the difference does not seem trivial, you should double-check to make sure you have not made an error or solve the problem again using a greater degree of precision.

Finally, please allow me a brief disclaimer about mathematical errors in this section. Let me assure you, first of all, that I know how important this section is for students and that I worked hard to be certain that these answers are correct. However, human fallibility being what it is, I know that I cannot make absolute guarantees. Should you find any errors, please let me know so I can make corrections in the future.

Chapter 1

1.5
a. Nominal
b. Ordinal (the categories can be ranked in terms of degree of honesty, with "Returned the wallet with money" being the "most honest")
c. Ordinal
d. Interval-ratio ("years" has equal intervals and a true zero point)
e. Interval-ratio
f. Interval-ratio
g. Nominal (the various patterns are different from each other but cannot be ranked from high to low)
h. Interval-ratio
i. Ordinal
j. Number of accidents: interval-ratio; Severity of accident: ordinal

1.7

	Variable	Level of Measurement	Application
a.	Opinion	Ordinal	Inferential
b.	Grade	Interval-ratio	Descriptive (two variables)
c.	Party	Nominal	Inferential
	Sex	Nominal	
	Opinion	Ordinal	
d.	Homicide rate	Interval-ratio	Descriptive (two variables)
e.	Satisfaction	Ordinal	Descriptive (one variable)

Chapter 2

2.1 **a.** Complex A: (5/20) × 100 = 25.00%
Complex B: (10/20) × 100 = 50.00%
b. Complex A: 4:5 = 0.80
Complex B: 6:10 = 0.60
c. Complex A: (0/20) = 0.00
Complex B: (1/20) = 0.05
d. 6/(4 + 6) = 6/10 = 60.00%
e. Complex A: 8:5 = 1.60
Complex B: 2:10 = 0.20

2.3 Bank robbery rate = (47/211,732) × 100,000 = 22.20
Homicide rate = (13/211,732) × 100,000 = 6.14
Auto theft rate = (23/211,732) × 100,000 = 10.86

2.5 For sex:

Sex	Frequency
Male	9
Female	6
Total	15

For age, set $k = 10$. $R = 77 - 23$, or 54, so we can round off interval size to 5 ($i = 5$). The first interval will be 20–24 (to include the low score of 23) and the highest interval will be 75–79.

Age	Frequency
20–24	1
25–29	2
30–34	3
35–39	2
40–44	1
45–49	3
50–54	1
55–59	1
60–64	0
65–69	0
70–74	0
75–79	1
	15

2.9 Set $k = 10$. $R = 92 - 5$, or 87, so set i at 10.

Score	Frequency
0–9	3
10–19	7
20–29	6
30–39	0
40–49	2
50–59	2
60–69	3
70–79	0
80–89	0
90–99	2
	25

2.11 Answers should note and describe the rise in all crime rates, except burglary, up to a peak in the early 1990s, followed by a dramatic decline and then a leveling off in recent years.

Chapter 3

3.1 "Region of birth" and "religion" are nominal-level variables, "support for legalization" and "opinion of food" are ordinal, and "expenses" and "number of movies" are interval-ratio. The mode—the most common score—is the only measure of central tendency available for nominal-level variables. For the two ordinal-level variables, *don't forget to array the scores from high to low* before locating the median. There are 10 freshmen (N is even), so the median for freshmen will be the score halfway between the scores of the two middle cases. There are 11 seniors (N is odd), so the median for seniors will be the score of the middle case. To find the mean for the interval-ratio variables, add the scores and then divide by the number of cases.

Variable	Freshmen	Seniors
Region of birth	Mode = North	Mode = North
Legalization	Median = 3	Median = 5
Expenses	Mean = 48.50	Mean = 63.00
Movies	Mean = 5.80	Mean = 5.18
Food	Median = 6	Median = 4
Religion	Mode = Protestant	Mode = Protestant and None (four cases each)

3.3

Variable	Level of Measurement	Measure of Central Tendency
Sex	Nominal	Mode = male
Social Class	Ordinal	Median = "medium" (the middle case is in this category)
Number of years in the party	I-R	Mean = 26.15
Education	Ordinal	Median = high school
Marital status	Nominal	Mode = married
Number of children	I-R	Mean = 2.39

3.5

Variable	Level of Measurement	Measure of Central Tendency
Marital status	Nominal	Mode = married
Race	Nominal	Mode = white
Age	I-R	Mean = 27.53
Attitude on abortion	Ordinal	Median = 7

3.7

Variable	Level of Measurement	Measure of Central Tendency
Sex	Nominal	Mode = Male
Support for gun control	Ordinal	Median = 1
Level of education	Ordinal	Median = 1
Age	I-R	Mean = 40.93

3.9 Attitude and opinion scales almost always generate ordinal-level data, so the appropriate measure of central tendency would be the median. The median is 9 for the students and 2 for the neighbors. Incidentally, the means are 7.80 for the students and 4.00 for the neighbors.

3.11 Mean = 1.10, median = 1.07. The lower value for the mean indicates a negative skew or a few low scores. For this small group of nations, the skew is caused by the scores of the United States and of Nigeria, which are much lower than the scores of the other nations.

3.13 To find the median, you must first rank the scores from high to low. Both groups have 25 cases (N = odd), so the median is the score of the 13th case. For freshmen, the median is 35, and for seniors, the median is 30. The mean score for freshmen is 31.72. For seniors, the mean is 28.60.

3.15 **a.** 1994: Mean = 7.07 Median = 6.35
2009: Mean = 4.28 Median = 4.30

b. In 1994, the mean was greater than the median, indicating a positive skew. In 2009, the mean and median were virtually identical, indicating very little skew.

c. Removing the case with the highest score (Louisiana) would cause the mean to decrease. The effect is greater for 1994 because the skew is greater. The median would also decrease because you are removing the case with the highest score and changing N to 49.

d. Removing the case with the low score would lower the value of the mean and median.

Chapter 4

4.1 The high score is 50 and the low score is 10, so the range is 50 − 10, or 40. The standard deviation is 12.28.

4.3

	Expenses		Movies		Food	
	Freshmen	Seniors	Freshmen	Seniors	Freshmen	Seniors
Mean	48.50	63.00	5.80	5.18	5.50	4.55
Std. Dev.	9.21	15.92	5.06	3.91	3.35	2.37

4.5

	2000	2006
Mean	48,161.54	63,223.08
Median	47,300	59,000
Standard deviation	6,753.64	10,845.50
Range	23,400	38,300

In this time period, the mean and median increase. The distributions for both years show a positive skew (the mean is greater than the median). The standard deviation and the range increase, indicating a greater variability in 2006.

4.7

Variable	Statistic	Males	Females
Labor force participation	Mean	77.60	58.40
	Standard deviation	2.73	6.73
% high school graduate	Mean	69.20	70.20
	Standard deviation	5.38	4.98
Mean income	Mean	33,896.60	29,462.40
	Standard deviation	4,443.16	4,597.93

Males and females are very similar in terms of educational level, but females are less involved in the labor force and, on average, earn almost $4,500 less than males per year. The females in these 10 states are much more variable in their labor force participation but are similar to males in dispersion on the other two variables.

4.9 $s = 0.35$

4.11 R = 38-5 or 33, s = 8.98. The score for Los Angeles (38) is higher than the other scores, so removing this score would reduce the variation in the data set. The standard deviation of the scores without Los Angeles would be lower in value.

4.13

Division	Range	Standard Deviation
A	$R = 18$	$s = 5.32$
B	$R = 8$	$s = 2.37$
C	$R = 3$	$s = 1.03$
D	$R = 15$	$s = 7.88$

4.15 1994: R = 19.6, s = 4.07. 2009: R = 11.00, s = 2.25

Chapter 5

5.1

X_i	Z Score	% Area Above	% Area Below
5	−1.67	95.25	4.75
6	−1.33	90.82	9.18
7	−1.00	84.13	15.87
8	−0.67	74.86	25.14
9	−0.33	62.93	37.07
11	0.33	37.07	62.93
12	0.67	25.14	74.86
14	1.33	9.18	90.82
15	1.67	4.75	95.25
16	2.00	2.28	97.72
18	2.67	0.38	99.62

5.3

	Z Score(s)	Area
a.	0.10 & 1.10	32.45%
b.	0.60 & 1.10	13.86%
c.	0.60	27.43%
d.	0.90	18.41%
e.	0.60 & −0.40	38.11%
f.	0.10 & −0.40	19.52%
g.	0.10	53.98%
h.	0.30	61.79%
i.	0.60	72.57%
j.	1.10	86.43%

5.5

X_i	Z Score	Number of Students Above	Number of Students Below
60	−2.00	195	5
57	−2.50	199	1
55	−2.83	199	1
67	−0.83	159	41
70	−0.33	126	74
72	0.00	100	100
78	1.00	32	168
82	1.67	10	190
90	3.00	1	199
95	3.83	1	199

Note: The number of students has been rounded off to the nearest whole number.

5.7

	Z Score	Area
a.	−2.20	1.39%
b.	1.80	96.41%
c.	−0.20 & 1.80	54.34%
d.	0.80 & 2.80	20.93%
e.	−1.20	88.49%
f.	0.80	21.19%

5.9

	Z Score	Area
a.	−1.00 & 1.50	.7745
b.	0.25 & 1.50	.3345
c.	1.50	.0668
d.	0.25 & −2.25	.5865
e.	−1.00 & −2.25	.1465
f.	−1.00	.1587

5.11 Yes. The raw score of 110 translates into a Z score of +2.88. 99.80% of the area lies below this score, so this individual was in the top 1% on this test.

5.13 For the first event, the probability is .0919; for the second, the probability is .0655. The first event is more likely.

5.15 a.

Student	Z Score – Freshman Year	Z Score – Senior Year	Student Did Better in
a	0.57	1.25	Senior year
b	−0.29	0.50	Senior year
c	−1.14	−2.50	Freshman year (−1.14 is closer to the mean than −2.50)
d	2.86	2.25	Freshman year
e	1.29	1.50	Senior year

b. Freshman Year

Scores	Z Score(s)	Probability
52	−0.14	.4443
57	0.57	.7157
40, 50	−1.86, −0.43	.3022
51	−0.28	.6103
62	1.29	.0985

Senior Year

Scores	Z Score(s)	Probability
88	−1.00	.1587
98	1.50	.9332
70, 100	−5.50, 2.00	.9771*
97	1.25	.1056
85	−1.75	.9599

*The Z score of −5.50 is not in Appendix A, so use the closest available score (−4.00).

Chapter 6

6.1

a. 5.2 ± 0.11
b. 100 ± 0.71
c. 20 ± 0.40
d. 1020 ± 5.41
e. 7.3 ± 0.23
f. 33 ± 0.80

6.3

Confidence Level	Alpha	Area Beyond *Z*	*Z* Score
95%	.05	.0250	±1.96
94%	.06	.0300	±1.88 or ±1.89
92%	.08	.0400	±1.75 or ±1.76
97%	.03	.0150	±2.17
98%	.02	.0100	±2.33
99.9%	.001	.0005	±2.58

6.5 **a.** 2.30 ± 0.04 **b.** 2.10 ± 0.01, 0.78 ± .07 **c.** 6.00 ± 0.37

6.7 **a.** 478.23 ± 1.97. The students spent between \$476.26 and \$480.20 on books.
b. 1.5 ± 0.04. The students visited the clinic between 1.46 and 1.54 times on average.
c. 2.8 ± 0.13 **d.** 3.5 ± 0.19

6.9. 0.14 ± .07. The estimate is that between 7% and 21% of the population consists of unmarried couples living together.

6.11 **a.** P_s = 823/1496 = 0.55
Confidence interval: 0.55 ± 0.03
Between 52% and 58% of the population agrees with the statement.
b. P_s = 650/1496 = 0.43
Confidence interval: 0.43 ± 0.03
c. P_s = 375/1496 = 0.25
Confidence interval: 0.25 ± 0.03
d. P_s = 1023/1496 = 0.68
Confidence interval: 0.68 ± 0.03
e. P_s = 800/1496 = 0.53
Confidence interval: 0.53 ± 0.03

6.13

Alpha (α)	Confidence Level	Confidence Interval
0.10	90%	100 ± 0.74
0.05	95%	100 ± 0.88
0.01	99%	100 ± 1.15
0.001	99.9%	100 ± 1.48

6.15 The confidence interval is .51 ± .05. The estimate would be that between 46% and 56% of the population prefer candidate A. The population parameter (P_u) is equally likely to be anywhere in the interval (that is, it is just as likely to be 46% as it is to be 56%), so a winner cannot be predicted.

6.17 The confidence interval is 0.23 ± 0.08. At the 95% confidence level, the estimate would be that between 240 (15%) and 496 (31%) of the 1,600 freshmen would be extremely interested. The estimated numbers are found by multiplying *N* (1,600) by the upper (.31) and lower (.15) limits of the interval.

Chapter 7

7.1 **a.**

Alpha	Form	*Z*(Critical)
.05	One-tailed	−.65 or 1.65
.10	Two-tailed	±1.65
.06	Two-tailed	±1.89
.01	One-tailed	−2.33 or 2.33
.02	Two-tailed	±2.33

b.

Alpha	Form	*N*	*t*(Critical)
.10	Two-tailed	31	±1.697
.02	Two-tailed	24	±2.457
.01	Two-tailed	121	±2.576
.01	One-tailed	31	−2.457 or 2.457
.05	One-tailed	61	−1.671 or 1.671

c.
1. *Z*(obtained) = −3.77
2. *t*(obtained) = −2.21
3. *Z*(obtained) = −5.76
4. *Z*(obtained) = 0.66
5. *Z*(obtained) = −0.77

7.3 The research hypothesis would be stated as $\mu > 3.3$, and the *Z*(critical) at alpha = 0.05 would be +1.65.

7.5 *Z*(obtained) = 29.16

7.7 *Z*(obtained) = 6.04

7.9 **a.** *Z*(obtained) = −13.66
b. *Z*(obtained) = 25.50

7.11 *t*(obtained) = 4.50

7.13 *Z*(obtained) = 3.06

7.15 *Z*(obtained) = −1.48

7.17 **a.** *Z*(obtained) = −1.49
b. *Z*(obtained) = 2.19
c. *Z*(obtained) = −8.55
d. *Z*(obtained) = −18.07
e. *Z*(obtained) = 2.09
f. *Z*(obtained) = −53.41

7.19 *t*(obtained) = −1.14

7.21 **a.** Z(obtained) = −1.28
b. Z(obtained) = 3.57
c. Z(obtained) = 34.96
d. Z(obtained) = −19.09
e. Z(obtained) = 35.00
f. Z(obtained) = −140.00

Chapter 8

8.1 **a.** σ = 1.39, Z(obtained) = −2.53
b. σ = 1.61, Z(obtained) = 2.52

8.3 **a.** σ = 10.57, Z(obtained) = 1.70
b. σ = 11.28, Z(obtained) = −2.48

8.5 **a.** σ = 0.08 Z(obtained) = −11.25
b. σ = 0.12 Z(obtained) = −3.33
σ = 0.14 Z(obtained) = 21.43

8.7 These are small samples (combined Ns of less than 100), so be sure to use Formulas 9.5 and 9.6 in step 4.
a. σ = 0.12, t(obtained) = −1.33
b. σ = 0.13, t(obtained) = 14.85

8.9 **a.** (France) σ = 0.12, Z(obtained) = −0.83
b. (Egypt) σ = 0.10, Z(obtained) = −3.00
c. (China) σ = 0.11, Z(obtained) = 0.91
d. (Mexico) σ = 0.10, Z(obtained) = −1.00
e. (Japan) σ = 0.11, Z(obtained) = −1.82

8.11 P_u = .45, σ_p = .06, Z(obtained) = 0.67

8.13 **a.** P_u = .46, σ_p = 0.06, Z(obtained) = 2.17
b. P_u = .80, σ_p = 0.05, Z(obtained) = −2.00
c. P_u = .72, σ_p = 0.08, Z(obtained) = 0.75

8.15 **a.** Z(obtained) = 1.50
b. Z(obtained) = −2.75
c. Z(obtained) = 4.00
d. σ = 0.43, Z(obtained) = 1.86
e. σ = 0.14, Z(obtained) = −5.71
f. σ = 0.08, Z(obtained) = −5.50

Chapter 9

9.1

Problem	Grand Mean	SST	SSB	SSW	*F* Ratio
a.	12.17	231.67	173.17	58.5	13.32
b.	6.87	455.73	78.53	377.20	1.25
c.	31.65	8,362.55	5,053.35	3,309.2	8.14

9.3

Problem	Grand Mean	SST	SSB	SSW	*F* Ratio
a.	4.39	86.28	45.78	40.50	8.48
b.	16.44	332.44	65.44	267.00	1.84

For problem 9.3a, with alpha = 0.05 and dfw = 15 and dfb = 2, the critical F ratio would be 3.68. We would reject the null hypothesis and conclude that decision making *does* vary significantly by type of relationship. Inspecting the group means shows that the "cohabitational" category accounts for most of the differences.

9.5

Grand Mean	SST	SSB	SSW	*F* Ratio
9.28	213.61	2.11	211.50	0.08

9.7

Grand Mean	SST	SSB	SSW	*F* Ratio
5.40	427.32	124.06	303.26	6.00

9.9

Nation	Grand Mean	SST	SSB	SSW	*F* Ratio
Mexico	3.78	300.98	154.08	146.90	12.59
Canada	6.88	156.38	20.08	136.30	1.77
United States	5.13	286.38	135.28	151.10	10.74

At alpha = 0.05 and df = 3, 36, the critical F ratio is 2.92. There is a significant difference in support for suicide by class in Mexico and the United States but not in Canada. The category means for Mexico suggest that the upper class accounts for most of the differences. For the United States, there is more variation across the category means and the working class seems to account for most of the differences. Going beyond the ANOVA test and comparing the grand means, we see that support is highest in Canada and lowest in Mexico.

Chapter 10

10.1 **a.** 1.11
b. 0.00
c. 1.52
d. 1.46

10.3 A computing table is highly recommended as a way of organizing the computations for chi square:

Computational Table for Problem 10.3

(1)	(2)	(3)	(4)	(5)
f_o	f_e	$(f_o - f_e)$	$(f_o - f_e)^2$	$(f_o - f_e)^2/f_e$
6	5	1	1	.20
7	8	−1	1	.13
4	5	−1	1	.20
9	8	1	1	.13
N = 26	N = 26	0		χ^2(obtained) = 0.65

There is 1 degree of freedom in a 2 × 2 table. With alpha set at .05, the critical value for the chi square would be 3.841. The obtained chi square is 0.65, so we fail to reject the null hypothesis of independence between the variables. There is no statistically significant relationship between race and services received.

10.5 a.

Computational Table for Problem 10.5

(1)	(2)	(3)	(4)	(5)
f_o	f_e	$f_o - f_e$	$(f_o - f_e)^2$	$(f_o - f_e)^2/f_e$
21	17.5	3.5	12.25	.70
29	32.5	−3.5	12.25	.38
14	17.5	−3.5	12.25	.70
36	32.5	3.5	12.25	.38
N = 100	N = 100.0	0		χ^2(obtained) = 2.15

With 1 degree of freedom and alpha set at .05, the critical region will begin at 3.841. The obtained chi square of 2.15 does not fall within this area, so the null hypothesis cannot be rejected. There is no statistically significant relationship between unionization and salary.

b. Column percentages:

	Status	
Salary	Union	Nonunion
High	60.00%	44.6%
Low	40.00%	55.4%
Totals	100.00%	100.0%

Although the relationship is not significant, unionized fire departments tend to have higher salary levels.

10.7 The obtained chi square is 5.12, which is significant (df = 1, alpha = .05). The column percentages show that more affluent communities have higher-quality schools.

10.9 The obtained chi square is 6.67, which is significant (df = 2, alpha = .05). The column percentages show that shorter marriages have higher satisfaction.

10.11 The obtained chi square is 12.58 , which is significant (df = 4, alpha = .05). The column percentages show that proportionally more of the students living "off campus with roommates" are in the high-GPA category.

10.13 The obtained chi square is 19.34, which is significant (df = 3, alpha = .05). Legalization was not favored by a majority of any region, but the column percentages show that the West was most in favor.

10.15

Problem	Chi Square	Significant at $\alpha = 0.05$?	Column Percentages
a.	25.19	Yes	The oldest age group was most opposed.
b.	1.80	No	The great majority (72%–75%) of all three age groups were in favor of capital punishment.
c.	5.23	Yes	The oldest age group was most likely to say yes. Although significant, the differences in column percentages are small.
d.	28.43	Yes	The youngest age group was most likely to support legalization.
e.	14.17	Yes	The oldest age group was least likely to support suicide.

Chapter 11

11.1

	Authoritarianism	
Efficiency	Low	High
Low	37.04%	70.59%
High	62.96%	29.41%
Totals	100.00%	100.00%

The conditional distributions change, so there is a relationship between the variables. The change from column to column is quite large, and the maximum difference is (70.59 − 37.04) = 33.55. Using Table 11.5 as a guideline, we can say that this relationship is strong. From inspection of the percentages, we can see that efficiency decreases as authoritarianism increases; workers with dictatorial bosses are less productive (or, maybe, bosses become more dictatorial when workers are inefficient), so this relationship is negative in direction.

11.3

	2004 Election	
2008 Election	Democrat	Republican
Democrat	87.31%	11.44%
Republican	12.69%	88.56%
Totals	100.00%	100.00%

The maximum difference for this table is (87.31 − 11.44), or 75.87. This is a very strong relationship. People are very consistent in their voting habits.

11.5

	Director Experienced?	
Turnover	No	Yes
Low	14.29%	40.91%
Moderate	32.14%	36.36%
High	53.57%	22.73%
Totals	100.00%	100.00%

The maximum difference is (53.57 − 22.73) or 30.84, which indicates a moderate-to-strong relationship. Cramer's V is 0.36, and lambda is 0.13. Cramer's V is more consistent with the maximum difference—both indicate moderate-to-strong relationships—so it would be the preferred measure of association in this case.

11.7 Maximum difference = 60
Phi = 0.59
Lambda = 0.64

All measures of association indicate that this is a strong relationship.

11.9

Problem	Maximum Diff.	Phi	Lambda
a.	1.43	0.02	0.00
b.	9.61	0.11	0.00
c.	7.61	0.08	0.00
d.	0.00	0.00	0.00
e.	1.16	0.02	0.00

Chapter 12

12.1 **a.** $G = 0.71$ **b.** $G = 0.69$ **c.** $G = -0.88$
These relationships are strong. Facility in English and income increase with length of residence. Use the percentages to help interpret the direction of a relationship. In the first table, 80% of the "newcomers" were "Low" in English facility, while 60% of the "old-timers" were "High." In this relationship, low scores on one variable are associated with low scores on the other, and scores increase together (as one increases, the other increases), so this is a positive relationship. In contrast, contact with the old country decreases with length of residence (−0.88). Most newcomers have higher levels of contact, and most old-timers have lower levels.

12.3 $G = 0.40$. This is a moderate-to-strong positive relationship. As city size increases, crime rate increases.

12.5 Canada: $G = -0.28$
United States: $G = 0.21$
Mexico: $G = -0.13$

These relationships are weak to moderate. For Canada and Mexico, agreement increases with age, but for the United States, the reverse is true. Note that the column percentages show that most respondents in all three nations do not agree with the statement.

12.7 $G = -0.17$. This is a weak negative relationship. The more authoritarian the parents, the fewer the symptoms of depression.

12.9 $G = -0.08$ This is a weak negative relationship. The higher the income, the lower the level of happiness.

12.11 $r_s = 0.77$. This is a strong positive relationship. The higher the original score, the higher the performance evaluation.

12.13 $r_s = 0.62$. This is a moderate-to-strong positive relationship.

Chapter 13

13.1 *(HINT: When finding the slope, remember that "Turnout" is the dependent, or Y, variable.)*

	For Turnout (*Y*) and		
	Unemployment	Education	Negative Campaigning
Slope (*b*)	3.00	12.67	−0.90
Y intercept (*a*)	39.00	−94.73	114.01
Reg. Eq.	$Y = 39 + 3X$	$Y = -94.73 + 12.67X$	$Y = 114.01 + (-0.90)X$
R	0.95	0.98	−0.87
r^2	0.90	0.97	0.76

13.3 *(HINT: When finding the slope, remember that "Number of visitors" is the dependent, or Y, variable.)*

Slope (*b*)	−0.56
Y intercept (*a*)	14.04
R	−0.37
r^2	0.14

13.5

Dependent Variables		Independent Variables		
		Poverty	% HS Graduates	Unemployment
Homicide	*a*	−6.52	41.09	3.63
	b	1.09	−0.43	0.13
	r	0.74	0.62	0.08
	r^2	0.54	0.38	0.01
Robbery	*a*	143.15	401.17	154.65
	b	−0.62	−3.13	−2.02
	r	0.03	0.29	0.08
	r^2	0.00	0.09	0.01
Car theft	*a*	168.10	1137.10	23.60
	b	5.56	−10.77	22.73
	r	0.10	0.40	0.35
	r^2	0.01	0.16	0.12

13.7 $b = 0.05, a = 53.18, r = 0.40, r^2 = 0.16$

13.9

		Prestige	Number of Children	Support for Abortion	Hours of TV per Day
Age	*r*	−0.30	0.67	0.08	0.16
	r^2	0.09	0.45	0.01	0.03
Sex	*r*	−0.28	0.19	0.11	−0.29
	r^2	0.08	0.04	0.01	0.08

Chapter 14

14.1 **a.** For turnout (*Y*) and unemployment (*X*) while controlling for negative advertising (*Z*), $r_{yx.z} = 0.95$. The relationship between *X* and *Y* is not affected by the control variable *Z*.

b. For turnout (*Y*) and negative advertising (*X*) while controlling for unemployment (*Z*), $r_{yx.z} = -0.89$. The bivariate relationship is not affected by the control variable.

c. Turnout $(Y) = 70.25 + (2.09)$ unemployment $(X_1) + (-0.43)$ negative advertising (X_2). For unemployment $(X_1) = 10$ and negative advertising $(X_2) = 75$, turnout $(Y) = 58.09$.

d. For unemployment (X_1): $b_1^* = 0.66$. For negative advertising (X_2): $b_2^* = 0.41$. Unemployment has a stronger effect on turnout than negative advertising. Note that the independent variables' effect on turnout is in opposite directions.

e. $R^2 = 0.98$

14.3 **a.** For strife (*Y*) and unemployment (*X*), controlling for urbanization (*Z*), $r_{yx.z} = 0.79$.

b. For strife (*Y*) and urbanization (*X*), controlling for unemployment (*Z*), $r_{yx.z} = 0.20$.

c. Strife $(Y) = -14.60 + (4.94)$ unemployment $(X_1) + (0.16)$ urbanization (X_2). With unemployment = 10 and urbanization = 90, strife (Y') would be 49.19.

d. For unemployment (X_1): $b_1^* = 0.78$. For urbanization (X_2): $b_2^* = 0.13$.

e. $R^2 = 0.65$

14.5 **a.** Turnout $(Y) = 83.80 + (-1.16)$ Democrat (X_1) + (2.89) minority (X_2).

b. For $X_1 = 0$ and $X_2 = 5$, $Y = 98.25$.

c. $Z_y = -1.27Z_1 + 0.84Z_2$.

d. $R^2 = 0.51$.

14.7 **a.** Z_y = (0.31) HS grads + (−0.42) Rank.

b. $R^2 = 0.47$.

14.9 Z_y = (−0.77) Contra. Use + (−0.07) Educ. R = 0.80 and $R^2 = 0.64$

Glossary

Alpha (α). The probability of error or the probability that a confidence interval does not contain the population value. Alpha levels are usually set at 0.10, 0.05, 0.01, or 0.001.

Alpha level (α). The proportion of area under the sampling distribution that contains unlikely sample outcomes, given that the null hypothesis is true. Also, the probability of a Type I error.

Analysis of variance. A test of significance appropriate for situations in which we are concerned with the differences among more than two sample means.

ANOVA. See **Analysis of variance**.

Association. The relationship between two (or more) variables. Two variables are said to be associated if the distribution of one variable changes for the various categories or scores of the other variable.

Bar chart. A graph. The categories of the variable are represented by bars of equal width, with the height of each corresponding to the number (or percentage) of cases in the category.

Beta-weight (b^*). See **Standardized partial slope.**

Bias. A criterion used to select sample statistics as estimators. A statistic is unbiased if the mean of its sampling distribution is equal to the population value of interest.

Bivariate table. A table that displays the joint frequency distributions of two variables.

Cells. The cross-classification categories of the variables in a bivariate table.

Central Limit Theorem. A theorem that specifies the mean, standard deviation, and shape of the sampling distribution, given that the sample is large.

χ^2(critical). The score on the sampling distribution of all possible sample chi squares that marks the beginning of the critical region.

χ^2(obtained). The test statistic as computed from sample results.

Chi square test. A nonparametric test of hypothesis for variables that have been organized into a bivariate table.

Class intervals. The categories used in the frequency distributions for interval-ratio variables.

Coefficient of determination (r^2). The proportion of all variation in Y that is explained by X. Found by squaring the value of Pearson's r.

Coefficient of multiple determination (R^2). The proportion of the total variation explained in the dependent variable by all independent variables combined.

Column. The vertical dimension of a bivariate table. By convention, each column represents a score on the independent variable.

Column percentages. Percentages computed within each column of a bivariate table.

Conditional distribution of *Y*. The distribution of scores on the dependent variable for a specific score or category of the independent variable when the variables have been organized into table format.

Conditional means of *Y*. The mean of all scores on Y for each value of X.

Confidence interval. An estimate of a population value in which a range of values is specified.

Confidence level. A frequently used alternative way of expressing alpha, with the probability that an interval estimate will not contain the population value. Confidence levels of 90%, 95%, 99%, 99.9%, and 99.99% correspond to alphas of 0.10, 0.05, 0.01, 0.001, and 0.0001, respectively.

Cramer's *V*. A chi square–based measure of association for nominally measured variables that have been organized into a bivariate table with any number of rows and columns.

Critical region (region of rejection). The area under the sampling distribution that in advance of the test itself is defined as including unlikely sample outcomes, given that the null hypothesis is true.

Cumulative frequency. An optional column in a frequency distribution that displays the number of cases within an interval and all preceding intervals.

Cumulative percentage. An optional column in a frequency distribution that displays the percentage of cases within an interval and all preceding intervals.

Data. Information expressed as numbers.

Data reduction. Summarizing many scores with a few statistics.

Dependent variable. In a bivariate relationship, the variable that is taken as the effect. The dependent variable is thought to be caused by the independent variable.

Descriptive statistics. The branch of statistics concerned with (1) summarizing the distribution of a single variable or (2) measuring the relationship between two or more variables.

Deviations. The distances between the scores and the mean.

Dispersion. The amount of variety, or heterogeneity, in a distribution of scores.

Dummy variable. A nominal-level variable dichotomized so it can be used in regression analysis. A dummy variable has two scores: one coded as 0 and the other as 1.

Efficiency. The extent to which the sample outcomes are clustered around the mean of the sampling distribution.

EPSEM. The **E**qual **P**robability of **SE**lection **M**ethod for selecting samples. Every element or case in the population must have an equal probability of selection for the sample.

Expected frequency (f_e). The cell frequencies that would be expected in a bivariate table if the variables were independent.

Explained variation. The proportion of all variation in Y that is attributed to the effect of X. Equal to $\Sigma(Y' - \overline{Y})^2$.

Five-step model. A step-by-step guideline for conducting tests of hypotheses. A framework that organizes decisions and computations for all tests of significance.

***F* ratio.** The test statistic computed in step 4 of the ANOVA test.

Frequency distribution. A table that displays the number of cases in each category of a variable.

Frequency polygon. A graph for interval-ratio variables. Class intervals are represented by dots placed over the midpoints, the height of each corresponding to the number (or percentage) of cases in the interval. All dots are connected by straight lines. Same as a **line chart**.

Gamma (*G*). A measure of association appropriate for variables measured with "collapsed" ordinal scales that have been organized into table format.

Histogram. A graph for interval-ratio variables. Class intervals are represented by contiguous bars of equal width (equal to the class limits), with the height of each corresponding to the number (or percentage) of cases in the interval.

Hypothesis. A specific statement, derived from a theory, about the relationship between variables.

Hypothesis testing. Statistical tests that estimate the probability of sample outcomes if the null hypothesis is true.

Independence. The null hypothesis in the chi square test. Two variables are independent if for all cases the classification of a case on one variable has no effect on the probability that the case will be classified in any particular category of the second variable.

Independent random samples. Random samples gathered in such a way that the selection of a particular case for one sample has no effect on the probability that any other particular case will be selected for the other samples.

Independent variable. In a bivariate relationship, the variable that is taken as the cause. The independent variable is thought to cause the dependent variable.

Inferential statistics. The branch of statistics concerned with making generalizations from samples to populations.

Interquartile range (*Q*). The distance from the third quartile to the first quartile.

Lambda (λ). A proportional reduction in error (PRE) measure for nominally measured variables that have been organized into a bivariate table.

Least Squares Principle. The mean is the point of minimum variation of a set of scores.

Level of measurement. The mathematical characteristic of a variable and the major criterion for selecting statistical techniques. Variables can be measured at any of three levels, each permitting certain mathematical operations and statistical techniques. The characteristics of the three levels are summarized in Table 1.2.

Linear relationship. A relationship between two variables in which the observation points (dots) in the scattergram can be approximated with a straight line.

Line chart See **Frequency polygon**.

Marginals. The row and column subtotals in a bivariate table.

Maximum difference. A way to assess the strength of an association between variables that have been organized into a bivariate table. The maximum difference is the largest difference between column percentages for any row of the table.

Mean. The arithmetic average of the scores. $\overline{X}$ represents the mean of a sample, and μ represents the mean of a population.

Mean square estimate. An estimate of the variance calculated by dividing the sum of squares within (SSW) or the sum of squares between (SSB) by the proper degrees of freedom.

Measures of association. Statistics that summarize the strength and direction of the relationship between variables.

Measures of central tendency. Statistics that summarize a distribution of scores by reporting the most typical or representative value of the distribution.

Measures of dispersion. Statistics that indicate the amount of variety, or heterogeneity, in a distribution of scores.

Median (Md). The point in a distribution of scores above and below which exactly half of the cases fall.

Midpoint. The point exactly halfway between the upper and lower limits of a class interval.

Mode. The most common value in a distribution or the largest category of a variable.

μ. The mean of a population.

μ_p. The mean of a sampling distribution of sample proportions.

$\mu_{\bar{X}}$. The mean of a sampling distribution of sample means.

Multiple correlation. A multivariate technique for examining the combined effects of more than one independent variable on a dependent variable.

Multiple correlation coefficient (*R*). A statistic that indicates the strength of the correlation between a dependent variable and two or more independent variables.

Multiple regression. A multivariate technique that breaks down the separate effects of the independent variables on the dependent variable; used to make predictions of the dependent variable.

N_d. The number of pairs of cases ranked in different order on two variables.

N_s. The number of pairs of cases ranked in the same order on two variables.

Negative association. A bivariate relationship in which the variables vary in opposite directions. As one variable increases, the other decreases, and high scores on one variable are associated with low scores on the other.

Nonparametric. A "distribution free" test. These tests do not assume a normal sampling distribution.

Normal curve. A theoretical distribution of scores that is symmetrical, unimodal, and bell shaped. The standard normal curve always has a mean of 0 and a standard deviation of 1.

Normal curve table. A detailed description of the area between a *Z* score and the mean of any standardized normal distribution. See Appendix A.

Null hypothesis (H_0). A statement of "no difference." In the context of single-sample tests of significance, the population from which the sample was drawn is assumed to have a certain characteristic or value.

Observed frequency (f_o). The cell frequencies actually observed in a bivariate table.

One-tailed test. A type of hypothesis test used when (1) the direction of the difference can be predicted or (2) concern focuses on outcomes in only one tail of the sampling distribution.

One-way analysis of variance. Applications of ANOVA in which the effect of a single independent variable on a dependent variable is observed.

P_s. Any sample proportion.

P_u. Any population proportion.

Parameter. A characteristic of a population.

Partial correlation. A multivariate technique for examining a bivariate relationship while controlling for other variables.

Partial correlation coefficient. A statistic that shows the relationship between two variables while controlling for other variables; $r_{yx.z}$ is the symbol for the partial correlation coefficient when controlling for one variable.

Partial slope. In a multiple regression equation, the slope of the relationship between a particular independent variable and the dependent variable while controlling for all other independent variables in the equation.

Pearson's *r* (*r*). A measure of association for variables that have been measured at the interval-ratio level.

Percentage. The number of cases in a category of a variable divided by the number of cases in all categories of the variable and then multiplied by 100.

Percent change. A statistic that expresses the magnitude of change in a variable from time 1 to time 2.

Phi (Φ). A chi square–based measure of association for nominally measured variables that have been organized into a bivariate table with two rows and two columns (a 2 x 2 table).

Pie chart. A graph used for variables with only a few categories. A circle (the pie) is divided into segments proportional in size to the percentage of cases in each category of the variable.

Pooled estimate. An estimate of the standard deviation of the sampling distribution of the difference in sample means based on the standard deviations of both samples.

Population. The total collection of all cases in which the researcher is interested.

Positive association. A bivariate relationship where the variables vary in the same direction. As one variable increases, the other also increases, and high scores on one variable are associated with high scores on the other.

Proportion. The number of cases in one category of a variable divided by the number of cases in all categories of the variable.

Quantitative research. Research project that collects data or information in the form of numbers.

Range (*R*). The highest score minus the lowest score.

Rate. The number of actual occurrences of some phenomenon or trait divided by the number of possible occurrences per some unit of time.

Ratio. The number of cases in one category divided by the number of cases in some other category.

Regression line. The single, best-fitting straight line that summarizes the relationship between two variables. Regression lines are fitted to the data points by the least-squares criterion, whereby the line touches all conditional means of *Y* or comes as close to doing so as possible.

Representative. The quality a sample is said to have if it reproduces the major characteristics of the population from which it was drawn.

Research. Any process of gathering information systematically and carefully to answer questions or test theories. Statistics are useful for research projects that collect numerical information or data.

Research hypothesis (H_1). A statement that contradicts the null hypothesis. In the context of single-sample tests of significance, the research hypothesis says that the population from which the sample was drawn does not have a certain characteristic or value.

Row. The horizontal dimension of a bivariate table, conventionally representing a score on the dependent variable.

Sample. A carefully chosen subset of a population. In inferential statistics, information is gathered from a sample and then generalized to a population.

Sampling distribution. The distribution of a statistic for all possible sample outcomes of a certain size. Under conditions specified in two theorems, the sampling distribution will be normal in shape with a mean equal to the population value and a standard deviation equal to the population standard deviation divided by the square root of *N*.

Scattergram. A graph that depicts the relationship between two variables.

Σ (uppercase Greek letter sigma). "The summation of."

σ_{p-p}. Symbol for the standard deviation of the sampling distribution of the differences in sample proportions.

$\sigma_{\bar{X}-\bar{X}}$. Symbol for the standard deviation of the sampling distribution of the differences in sample means.

Significance testing. See **Hypothesis testing**.

Simple random sample. A method for choosing cases from a population by which every case and every combination of cases has an equal chance of being included.

Skew. The extent to which a distribution of scores has a few scores that are extremely high (positive skew) or extremely low (negative skew).

Slope (*b*). The amount of change in one variable per unit change in the other; *b* is the symbol for the slope of a regression line.

Spearman's rho (r_s). A measure of association appropriate for ordinally measured variables that have many different scores.

Standard deviation. The square root of the sum of the squared deviations of the scores around the mean, divided by *N*. The most important and useful descriptive measure of dispersion: *s* represents the deviation of a sample, and σ represents the standard deviation of a population.

Standard error of the mean. The standard deviation of a sampling distribution of sample means.

Standardized partial slope (beta-weight). The slope of the relationship between a particular independent variable and the dependent variable when all scores have been normalized.

Stated class limits. The class intervals of a frequency distribution.

Statistics. A set of mathematical techniques for organizing and analyzing data.

Student's *t* distribution. A distribution used to find the critical region for tests of sample means when *s* is unknown and the sample size is small.

Sum of squares between (SSB). The sum of the squared deviations of the sample means from the overall mean, weighted by sample size.

Sum of squares within (SSW). The sum of the squared deviations of scores from the category means.

***t*(critical).** The *t* score that marks the beginning of the critical region of a *t* distribution.

Test statistic. The value computed in step 4 of the five-step model that converts the sample outcome into either a *t* score or a *Z* score.

***t*(obtained).** The test statistic computed in step 4 of the five-step model. The sample outcome expressed as a *t* score.

Theory. A generalized explanation of the relationship between two or more variables.

Total sum of squares (SST). The sum of the squared deviations of the scores from the overall mean.

Total variation. The spread of the *Y* scores around the mean of *Y*. Equal to $\Sigma(Y - \bar{Y})^2$.

Two-tailed test. A type of hypothesis test used when (1) the direction of the difference cannot be predicted or (2) concern focuses on outcomes in both tails of the sampling distribution.

Type I error (alpha error). The probability of rejecting a null hypothesis that is in fact true.

Type II error (beta error). The probability of failing to reject a null hypothesis that is in fact false.

Unexplained variation. The proportion of the total variation in *Y* that is not accounted for by *X*. Equal to $\Sigma(Y - Y')^2$.

Variable. Any trait that can change values from case to case.

Variance. The sum of the squared deviations of the scores around the mean divided by N. A measure of dispersion used primarily in inferential statistics and also in correlation and regression techniques; s^2 represents the variance of a sample, and σ^2 represents the variance of a population.

***X*.** Symbol used for any independent variable.

X_i. Any score in a distribution.

***Y*.** Symbol used for any dependent variable.

***Y'*.** Symbol for a predicted score on Y.

***Y* intercept (a).** The point where the regression line crosses the Y axis.

***Z*(critical).** The Z score that marks the beginnings of the critical region on a Z distribution.

Zero-order correlation. Correlation coefficient for bivariate relationships.

***Z*(obtained).** The test statistic computed in step 4 of the five-step model. The sample outcomes expressed as a Z score.

***Z* scores.** Standard scores; the way scores are expressed after they have been standardized to the theoretical normal curve.

Index